CMOS Test and Evaluation

Manjul Bhushan • Mark B. Ketchen

CMOS Test and Evaluation

A Physical Perspective

Springer

Manjul Bhushan
OctEval
Hopewell Junction, NY, USA

Mark B. Ketchen
OcteVue
Hadley, MA, USA

ISBN 978-1-4939-1348-0 ISBN 978-1-4939-1349-7 (eBook)
DOI 10.1007/978-1-4939-1349-7
Springer New York Heidelberg Dordrecht London

Library of Congress Control Number: 2014952022

Printed on acid-free paper

Springer is part of Springer Science+Business Media (www.springer.com)

Preface

Designing, fabricating, and testing of CMOS chips is a multi-billion dollar industry, spanning a multiplicity of engineering fields. Many of the complex tasks at each stage of design and production are handled with automated tools enabling rapid deployment of semiconductor chips in the marketplace at a low cost. Significant engineering resources are devoted to the development of these tools and in generation of associated software. Often engineers engaged in designing and testing of chips rely on automated tools and have limited exposure to the physical behavior of devices and circuits. Generally well proven and efficient, this approach lacks full utilization of university classroom learning in physics and engineering. Although detailed knowledge of a CMOS product may be daunting, a high-level view of various engineering aspects is extremely desirable when it comes to rapid diagnostics, problem resolution, and optimization of the entire production process.

Working with CMOS design, silicon technology development, and manufacturing teams in IBM's 180 to 32 nm CMOS technology nodes, we began deploying special test structures for DC and high-speed characterization of CMOS circuits. Initially our focus was on developing a methodology for bridging observed circuit behavior at high speeds to constituent component properties. Hardware data collected from the fab were compared to the simulated predictions using design automation tools. If the model-to-hardware mismatch was outside the specified range, silicon processes had to be modified to adjust the hardware to match the model. The alternative approach of updating the compact models to match the latest silicon technology was often not viable, considering the high cost of chip redesign and pressure to meet time-to-market demands. In either approach, it was imperative that the feedback from the test structures be accurate and reliable. Special techniques for design, test, and analysis were developed to quickly assimilate the information and to present it in a clear and concise manner so that experts as well as non-experts could follow the presentations and reports. These test structures have been placed in scribe-lines of CMOS chips and on dedicated test vehicles built at IBM and by IBM's partners in CMOS technology development. This integrated approach to design and test of electrical test structures is covered in our book entitled *Microelectronic Test Structures for CMOS Technology* published by Springer in August, 2011.

Some of these test structures, when embedded in CMOS product chips, proved to be very useful in product performance evaluation and debug. These monitors, primarily ring oscillators, are easier to analyze and model than the complex circuitry comprising the chip design itself. By proper configuration of embedded monitors, chip power and performance can be bridged back to device properties and to the EDA models and tools used for designing the chip circuitry. Changes in circuit characteristics can be monitored throughout the life of the product. Additional applications of such embedded test structures are in the areas of sorting and binning of chips and applying test limits with guardbands to meet warranted performance.

It has been our experience that understanding a complex system can be simplified by either observing the aggregate behavior of its components or by comparing the behavior of its key constituents to the system as a whole. Appropriately designed test structures and monitors can play a very important role in predicting the properties of the system. The knowledge derived is physically intuitive making it easier to detect model, design tool, and other software-related errors. A priori knowledge of physical behavior when applied to statistical data mining can considerably reduce the effort in resolving design and test issues. Cross-checking of data collected from embedded monitors, product chips, and system tests for consistency makes the findings conclusive with a high degree of confidence.

In *CMOS Test and Evaluation*: *A Physical Perspective*, we have attempted to describe the relationship between basic circuit components (resistors, capacitors and diodes, and MOSFETs) and a complex CMOS chip with as many as several billion transistors. Our approach is to provide an overview with examples to link various aspects of CMOS technology, design, and test. Simulated data representative of that acquired during electrical testing in product manufacturing and qualification are used to illustrate concepts and to demonstrate data visualization and presentation. Exercises are included at the end of each chapter. Many of the circuits described and incorporated in the examples and exercises enable observation through simulation of features that are not experimentally assessable, often providing clearer insight into aggregate behavior.

We hope that this book will prove useful in preparing physics and electrical engineering students for building a career in the semiconductor industry as it faces new challenges, as well as serving as a useful reference for practitioners in the field.

We are thankful to our former colleagues in IBM's Server and Technology Group and in IBM's Research Division for close collaborations throughout our tenure in IBM.

How to Use This Book

There are at least two effective ways that this book can be used. The first is by practitioners already confronting real problems on the test floor. As emphasized in the 2013 ITRS Roadmap, the CMOS test arena is in a state of rapid change. While this book cannot possibly address all the known changes or foresee many more to come, the underlying physics of it all has not and will not change. In each chapter we attempt to present a high-level summary of the subject matter followed by a number of exercises, many of which relate to actual problems encountered in the field. While none of these may be identical to the crisis of the moment, the approach to resolution that we advance in the formulation and solution of exercises is based on physical insight, is very general in nature, and will apply to a wide range of new problems as they arise.

The second way this book can be used is in the education and training of science and engineering students preparing to work in the semiconductor test enterprise. It is in no way an attempt to replace or compete with any of the fine existing texts that focus in great depth on particular areas of design, fabrication, and test. It is assumed that students will have already mastered the contents of a number of these. It is our intent to build on them to provide an integrated technical view of CMOS test as a whole and to provide students with a set of exercises to help them develop physical insight. As mentioned above, the field of semiconductor test is changing rapidly and will continue to do so. Those who desire to enter and prosper in this field must be prepared to evolve with it. It is our goal to help them prepare for this journey through examples and exercises based on real problems encountered in CMOS manufacturing test, with an emphasis on gaining underlying physical insight, along with high-level topical summaries.

The scope of this book and a brief description of chapter contents are covered in Sect. 1.8. The introductory material in the beginning of the chapters can be covered quickly with much of the time and effort devoted to circuit simulations and data analysis. The aim is to help develop an intuitive physical understanding of the material covered without becoming bogged down in the details.

It is essential to have access to compact device models for different technology nodes or from different foundries, together with a SPICE simulation environment. By working through the examples and exercises, students can learn to cross-check the results and quickly spot and correct errors. Presenting conclusions and the line of reasoning in a clear and unambiguous manner is extremely important. Examples of this are presented in the text as well as in solutions to exercises published on the web.

Hopewell Junction, NY, USA — Manjul Bhushan
Hadley, MA, USA — Mark B. Ketchen

Contents

1 Introduction 1
1.1 Simplicity in Complexity 2
1.2 CMOS Design and Test Overview 4
1.3 Tests Types and Timelines 5
1.4 Test Economics 8
1.5 Future Test Challenges 9
1.6 Silicon Technology and Models 10
1.7 Data Analysis and Characterization 10
1.8 Scope of the Book 11
1.9 Summary and Exercises 13
References 15

2 CMOS Circuits Basics 17
2.1 Circuit Components and Building Blocks 19
2.1.1 MOSFETs 21
2.1.2 Interconnects 28
2.1.3 Passive *R* and *C* Components 30
2.1.4 Logic Gates 31
2.2 SPICE Simulations 34
2.2.1 PTM (BSIM) 37
2.2.2 MOSFET Characteristics 38
2.2.3 Standard Cell Library Book Characteristics 50
2.2.4 Delay Chains 64
2.2.5 Ring Oscillators 71
2.2.6 Comparison of Logic Gate Characterization Methods 75
2.2.7 Monte Carlo Analysis 77
2.3 Summary and Exercises 79
References 83

3 CMOS Storage Elements and Synchronous Logic 85
3.1 CMOS Chip Overview 86
3.1.1 I/O Circuits 87

3.1.2 Combinational Logic 88
3.1.3 Clock Generation and Distribution 91
3.2 Sequential Logic and Clocked Storage Elements 93
3.2.1 Level-Sensitive Latches 95
3.2.2 Edge-Triggered Flip-Flops 98
3.2.3 Setup and Hold Times 100
3.2.4 Register Files 101
3.3 Memory 102
3.3.1 SRAM 103
3.3.2 DRAM 108
3.4 Circuit Simulations 109
3.4.1 SRAM SNM 109
3.4.2 Logic Data Path 112
3.5 Summary and Exercises 121
References 123

4 IDDQ and Power 125
4.1 Silicon Technology Scaling and Power 127
4.2 IDDQ 128
4.2.1 MOSFET Leakage Currents 129
4.2.2 IDDQ of Logic Gates and Memory Cells 130
4.2.3 IDDQ Estimation in Design and Measurements 135
4.2.4 Defect Generated IDDQ 137
4.3 Power 140
4.3.1 Measuring Power 140
4.3.2 AC Power 141
4.3.3 DC Power 146
4.4 Total Power 148
4.5 Power Management 151
4.5.1 Power Management in Chip Design 152
4.5.2 System Power Management 155
4.6 Summary and Exercises 155
References 157

5 Embedded PVT Monitors 159
5.1 Placement and Integration 160
5.2 Silicon Process Monitors 162
5.2.1 MOSFETs 162
5.2.2 Delay Chains 163
5.2.3 Ring Oscillators 166
5.3 Power Supply Voltage and Noise Monitors 170
5.4 Critical Path Monitors 173
5.5 Temperature Monitors 174
5.6 Circuit Stages for ROs and Delay Chains 177
5.6.1 MOSFET Parameter Extraction 182
5.6.2 SRAM Stage Designs 186

5.6.3 Silicon Process-Sensitive Suite 188
5.6.4 Strengths and Limitations of RO-Based Monitors 192
5.7 Data Collection and Characterization 193
5.8 Summary and Exercises 197
References 199

6 Variability 201
6.1 Sources and Impact of Variations 202
6.1.1 Silicon Process Variations 205
6.1.2 Random Variations 211
6.1.3 Voltage Variations 212
6.1.4 Temperature Variations 214
6.2 Variability Characterization 215
6.2.1 Silicon Manufacturing Tests 216
6.2.2 On-Chip Embedded PVT Monitors 216
6.2.3 Functional Parameters 218
6.2.4 Optical Imaging 218
6.2.5 Thermal Imaging 220
6.3 Minimizing Variations 220
6.3.1 Chip Design and Floorplanning 221
6.3.2 Reticle and Wafer Assembly 222
6.3.3 Silicon Process Improvements 222
6.4 Accommodating Variability in Circuit Design 223
6.4.1 Simulation Corners 225
6.4.2 Impact of Random Variability on Circuits 227
6.5 Summary and Exercises 235
References 239

7 Electrical Tests and Characterization in Manufacturing 241
7.1 Digital CMOS Chip Tests 242
7.1.1 Test Flow 243
7.1.2 Test Equipment 245
7.1.3 DC and AC Parametric Tests 246
7.1.4 Structural Faults and ATPG 246
7.1.5 IDDQ Tests 251
7.1.6 DFT and Diagnostics 254
7.1.7 Scan Design 255
7.1.8 Built-in Self-Test 257
7.1.9 Boundary Scan 258
7.1.10 Measurements of T_{cmin}, V_{min}, and AC Power 259
7.2 Yield 260
7.2.1 Defect Limited Yield 261
7.2.2 Cycle Time Limited Yield 263
7.3 Failure Analysis 265
7.4 Product Chip Characterization 267

7.4.1 Silicon Manufacturing Line Tests 267
7.4.2 Silicon Process-Split Hardware 269
7.4.3 Embedded Process Monitors 270
7.4.4 Aggregate Behavior 277
7.4.5 Silicon Manufacturing Process Window 278
7.5 Adaptive Testing and Binning 278
7.6 Summary and Exercises 281
References 284

8 Reliability 285
8.1 Reliability and End-of-Life 286
8.1.1 Accelerated Stress Tests and Failure Rates 288
8.2 CMOS Circuit Performance Degradation Mechanisms 292
8.2.1 Bias Temperature Instability 292
8.2.2 Hot Carrier Injection 300
8.2.3 Time-Dependent Dielectric Breakdown 301
8.2.4 Electromigration 302
8.2.5 Soft Errors 303
8.3 Managing Reliability 303
8.3.1 Voltage Screening 304
8.3.2 Burn-In 305
8.3.3 Guard-Banding 306
8.4 Summary and Exercises 309
References 310

9 Basic Statistics and Data Visualization 311
9.1 Basic Statistics 312
9.1.1 Probability 314
9.1.2 Statistical Distributions 315
9.1.3 Sample Size Effects 318
9.1.4 Non-normal Distributions 321
9.2 Data Filtering, Correlation, and Regression 323
9.3 Statistical Variations 326
9.3.1 Range of Systematic and Random Variations 327
9.3.2 Sensitivity Analysis of a Function 330
9.4 Bayesian Statistics 333
9.5 Data Visualization 334
9.6 Summary and Exercises 343
References 344

10 CMOS Metrics and Model Evaluation 347
10.1 Measurement Standards 348
10.2 Scaling Trends in CMOS Products 351
10.3 CMOS Performance Metrics 355
10.3.1 MOSFET Performance 355

10.3.2 Interconnect Performance 363
10.3.3 Logic Gate Performance 365
10.4 CMOS Power-Performance-Density Metrics 367
10.4.1 Circuit Density 368
10.4.2 Energy and Power Density 369
10.4.3 V_{DD} Dependencies of Different Metric Parameters 373
10.4.4 Summary of Performance Metrics 374
10.5 Compact Models and EDA Tool Evaluation 374
10.5.1 BSIM Models 376
10.5.2 Layout Parasitic Extraction 383
10.5.3 Timing and Power Tools 385
10.6 PD-SOI vs. Bulk Silicon Technology 386
10.7 Closing Comments on CMOS Technology Evaluation 394
10.8 Summary and Exercises 395
References 398

Appendix A: MOSFET and Logic Gate Parameters (PTM HP Models) 399

Appendix B: BSIM4 PTM Models 407

Glossary 413

Index 421

Introduction 1

Contents

1.1 Simplicity in Complexity 2
1.2 CMOS Design and Test Overview 4
1.3 Tests Types and Timelines 5
1.4 Test Economics 8
1.5 Future Test Challenges 9
1.6 Silicon Technology and Models 10
1.7 Data Analysis and Characterization 10
1.8 Scope of the Book 11
1.9 Summary and Exercises 13
References 15

In traditional testing of digital complementary metal-oxide-semiconductor (CMOS) chips, emphasis is placed on functional verification and fault modeling. Push to higher frequencies has led to optimization of circuit properties and chip operating conditions for power/performance and yield. As silicon technology approaches scaling limits, there is a trend towards reducing circuit design margins and product guard-bands to squeeze maximum benefits from higher circuit densities. Such factors have been continually increasing the burden on manufacturing test. Some of these additional test challenges are addressed by examining the underlying physical behaviors and linking silicon technology, circuit design and electrical tests through models and simulations. In this chapter an overview of CMOS test, in conjunction with circuit design methodology and silicon technology performance, is provided as an introduction to the material covered in this book.

M. Bhushan, M.B. Ketchen, *CMOS Test and Evaluation*,
DOI 10.1007/978-1-4939-1349-7_1

1.1 Simplicity in Complexity

There are many examples in nature where the collective behavior of a large number of units, each interacting with the others while maintaining its unique identity and characteristics at all times, can be expressed in a fairly simple way. This allows us to conduct our lives without having a detailed knowledge of how the world operates. If the collective behavior is cyclic, repeated observations make it ever easier to assimilate and reinforce the information and store in memory for ready retrieval.

One example is Boyle's law for ideal gases stating that in a closed system held at a constant temperature, the pressure exerted by gas molecules on the walls of the vessel is inversely proportional to its volume, or pressure $\times$ volume $=$ constant. The law succinctly describes an observation resulting from the motion of a very large number of molecules and their interactions with the vessel. The molecules themselves may have varying chemical and physical properties and their velocities and relative positions are a function of time, but the aggregate behavior is stated in a simple mathematical expression.

A very large scale integrated (VLSI) digital CMOS chip or die is a complex entity comprising hundreds of thousands to billions of circuit elements. These elements are either physically embedded in or layered on top of a crystalline silicon surface. Instructions received by the chip in the form of voltage signals as strings or "1"s and "0"s are processed within the chip and the results communicated to the external world, also in the form of voltage signals. The connectivity of circuit blocks performing the required functions can be altered with voltage signals generated within the chip. The application conditions such as power supply voltages, temperature and frequency of operation may cover a wide range for a single chip design, and for chips of different designs fabricated at the same silicon technology node.

Rules of simplification can be applied to complex CMOS chips as well. Some of the basic building blocks such as logic gates, memory cells, and other storage elements are replicated millions of times within chip. As illustrated in Fig. 1.1a, periodic "clock" signals control arrival timing of signals at the input of the combinational logic block, and set the capture timing windows for the signal in the following clocked storage element. The minimum clock cycle time of a complex digital CMOS microprocessor chip can be estimated as a multiple of the measured or simulated signal propagation delay through an inverter with fanout of four (FO $=$ 4) shown in Fig. 1.1b, representing the signal propagation delay through a data path like that in Fig. 1.1a.

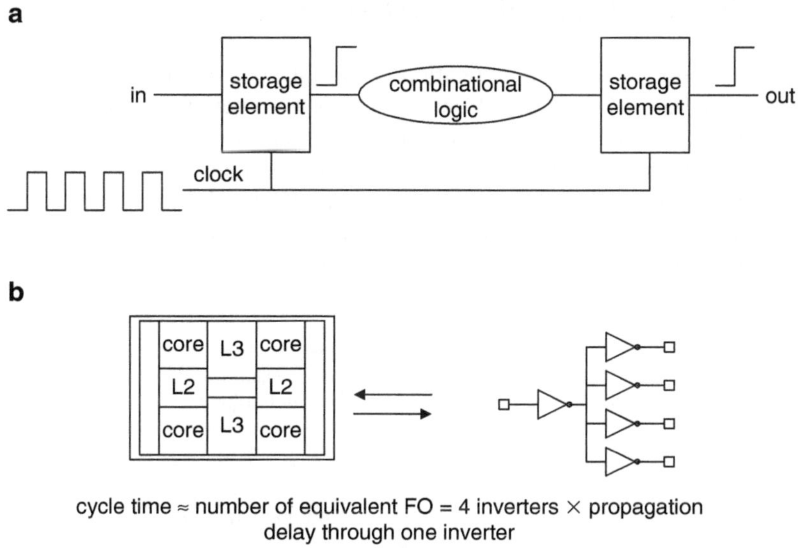

Fig. 1.1 (**a**) Data path with clock signals to control timing, and (**b**) complex microprocessor chip cycle time estimation from logic gate delay

In another example, the aggregate power characteristics of a large number of components, switching at very high frequencies and performing multiple tasks, are expressed with the simple relationship

$$P = V_{\mathrm{DD}} \times I_{\mathrm{avg}}, \tag{1.1}$$

where V_{DD} is the power supply voltage and I_{avg} is the average DC current drawn. In a fashion similar to the first example, an insight into the AC and DC components of power and their dependencies on the properties of circuit components may be developed from modeling a single logic gate.

Knowing and understanding the behaviors of smaller circuit elements proves to be very valuable when moving up the circuit design and test hierarchy. A detailed knowledge of devices, silicon processes, circuit design, logic functions, chip architecture, design for testability features, test and characterization procedures, and staying updated with advances made in all these areas, is daunting even for the most ambitious. However, the learning derived from working with a few smaller representative elements across all the levels mentioned above is extremely useful in developing a methodology for addressing large scale multi-faceted VLSI programs.

1.2 CMOS Design and Test Overview

The major steps in design, fabrication, and test of a digital CMOS chip are illustrated in Fig. 1.2. Starting from the top left corner in the figure, the chip architecture is defined to meet projected product specifications based on market or customer demand. In a digital system, this behavioral description of the design is expressed in a hardware description language (HDL) through abstraction at a register transfer level (RTL). The design is then simulated to verify the logic and converted to a logic gate-level circuit description.

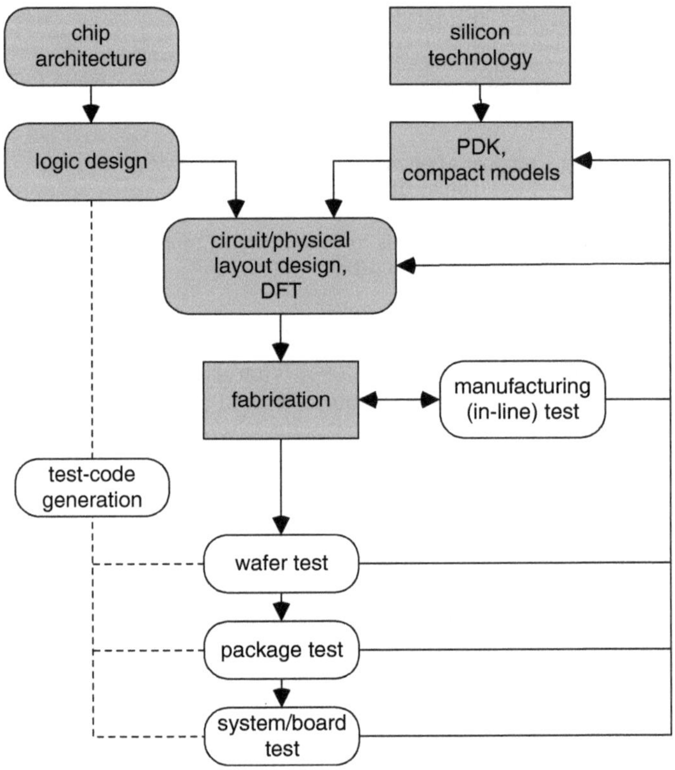

Fig. 1.2 CMOS chip design and test flow illustration

Shown in the top right corner in Fig. 1.2, silicon technology for chip manufacturing is developed by a silicon foundry. Typically, the foundry supplies a process design kit (PDK) containing compact device models for circuit simulations, a parameterized standard cell library and physical design rules for conversion of logic to circuit design, and to enable physical mapping of the circuits onto silicon. Heavy use of electronic design automation (EDA) tools is made throughout the design flow to automate, synthesize, simulate, and validate the design and physical layout. Analog and other sensitive circuits may rely on custom design tools. Design for testability (DFT) features are added to assist in chip test and debug.

Physical design data are used for building photomasks for chip fabrication. Using a photolithographic exposure tool, the pattern on a mask reticle is transferred to a whole wafer in a step-and-repeat manner. Typical reticle exposure field on a 300 mm silicon wafer is ~850 mm^2. The reticle area can accommodate multiple chips. Additional area (scribe-line) between the exposure fields and also between chips is designated for dicing and separating the chips. Test structures are placed in this scribe-line area for process monitoring and quality control in silicon manufacturing.

Wafers are fabricated in a batch process. Chips are probed on the wafer prior to dicing and limited tests are conducted to identify good chips for packaging. Packaged chips go through more elaborate testing which may include environmental tests and burn-in. Based on the test results, chips may be binned for different products or customer applications. This is followed by product assembly and final test prior to shipment.

1.3 Tests Types and Timelines

CMOS chips are tested to guarantee chip functionality within published product specifications throughout the useful lifetime. Rapid debug of problems and determining the right course of action are key elements in developing an optimized test process. With advances in silicon and packaging technologies chip complexity has been increasing, yet the cost of testing must be contained to assure desired profitability.

Electrical tests begin early in the silicon fabrication cycle and continue on at different stages in production prior to shipping to the customer. Key characteristics may even be monitored in the field throughout the chip lifetime. Additional tests are applied to ensure long-term reliability and mechanical robustness of packaged chips. Chips for military applications are subjected to environmental stress such as temperature, humidity, shock, vibration, acceleration, and resistance to chemicals and radiation. In the USA, these specifications are issued by the US Department of Defense and known as MIL-STD, MIL-Spec, or MILSpecs.

Types of tests and test conditions are optimized to eliminate defective chips in the beginning of a long test sequence, and early in the production cycle. Failing chips are removed from further tests to reduce total test time and cost of handling. These chips are either scrapped or utilized for diagnostics, and to provide feedback for future improvements.

In Fig. 1.3 different aspects of tests and their progression with time are shown. Electrical tests are conducted at several stages during manufacturing as shown in Fig. 1.3a. Test structures for monitoring the silicon process and defect densities, and for technology model-to-hardware correlation of devices and small circuit blocks are placed in the scribe-line area between chips. Contact to the scribe-line tests structures is made with cantilever probes landing on metal pads, each ~10^3–10^4 μm^2 in area. Data collected from electrical measurements are used for process quality control and to ensure all circuit component properties are within the range described in the compact models supplied by the silicon foundry for chip design.

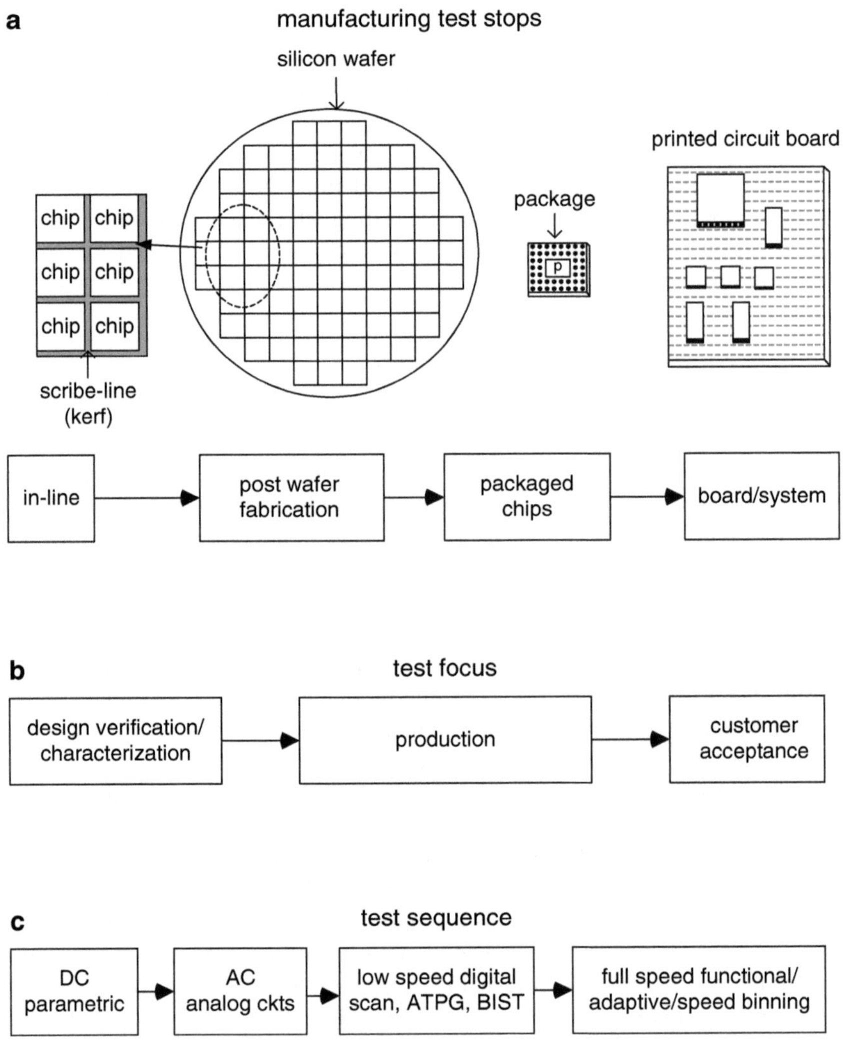

Fig. 1.3 Test timelines for (**a**) manufacturing test stops, (**b**) product development to production and (**c**) test sequence from DC to functional

After completion of the silicon manufacturing process, CMOS product chips are tested on the wafer. Electrical contacts to the chip I/Os may be made via probe pads or solder bumps. Defective chips are isolated and rejected. After dicing, good chips also called known good die (KGD) are individually packaged or placed in multichip modules and tested again. These chips may undergo further tests in a printed circuit board assembly or at system level.

The extent to which tests are performed and the type of tests vary as the production program moves from development to the manufacturing phase. Chips delivered to customers may undergo further tests for acceptance. This timeline is illustrated in Fig. 1.3b. The chip function as defined by the architecture, logic and circuit design is validated in the development phase prior to full-scale production. This is to ensure that the chip is performing all of its intended tasks correctly and that it will meet customer specifications over the full operating window of voltage, temperature, and other environmental variables. Chip yield, which is the ratio of the number of good chips to the total number of chips, must also be within an acceptable limit over the full range of silicon process variations. To verify this, the silicon process may be intentionally skewed to cover the range of expected process variations over time.

The number and types of tests conducted during the chip design-verification phase are typically more extensive than in routine production. In the early stages, more resources may be devoted to characterization, diagnostics, and failure analysis. The findings provide feedback to the design and silicon technology teams for making any modifications if necessary. In production mode, the number of tests is reduced while keeping the essential parametric and functional tests in the flow. If the yield falls below the set target, characterization test mode may be turned on to assist with debug. Customers may retest the chips prior to acceptance.

The sequence of different test types at wafer and package levels is outlined in Fig. 1.3c. DC parametric tests include tests for opens and shorts in the power grid and other circuit blocks, I/O pin leakage and drive currents/voltages, and leakage currents in the quiescent state (IDDQ) for different power supply domains. These tests serve to eliminate chips with gross defects from the production flow. The next set of tests is conducted on circuits that provide vital functions, such as clock generators (phase locked loops), I/O interfaces, and monitors for recording the state of the chip.

Functionality is first validated at lower frequencies for clocked storage elements (scan tests), logic and memory. The tests are generated with automated test pattern generation (ATPG), and built-in self-test (BIST) for logic and memory. For each set of input test vectors (strings of "1"s and "0"s), the output signals are compared with expected values obtained from simulation or from a KGD. Mismatches reveal errors in chip logic and design, silicon process skew or defects causing the chip to malfunction. Chips passing all earlier tests are then subjected to functional workloads at full speed and may cover the extreme ranges of operating conditions in the field. The yield may be maximized by binning chips to be operated at different frequencies, voltages, or power levels for different market offerings.

Properties of some circuit elements degrade over time; the threshold voltage of a MOSFET increases with time due to bias temperature instability (BTI), reducing its current drive, and wire resistances may increase due to electromigration. Ensuring chip operability over lifetime requires accelerated stress tests and burn-in (BI) by subjecting the chip to higher voltages and temperatures. Such tests accelerate silicon process induced defects and serve to screen "weak" chips from being shipped to customers where they may potentially fail in the field.

1.4 Test Economics

Chip fabrication is a batch process, with simultaneous processing of hundreds of chips on a wafer and typically 5–25 wafers in a lot. Chips are tested individually. The cost of testing is therefore a significant part of the total manufacturing cost. For many CMOS products, test cost may exceed silicon manufacturing cost. Major contributors to the cost of testing are shown in Fig. 1.4.

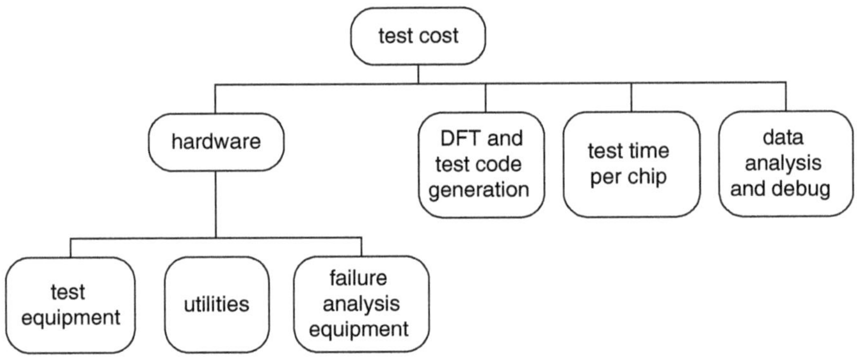

Fig. 1.4 Factors contributing to test cost

Automated test equipment (ATE), probing fixtures, wafer and package handlers, thermal control and the cost of housing and maintaining the equipment are fixed costs which may be shared among different products. Failure analysis facilities feature scanning and transmission electron microscopy, mechanical pico-probing, optical and thermal imaging, and equipment for sample preparation, delayering and materials analysis. This cost may also be shared among product lines or covered by outsourcing to failure analysis service companies.

Costs associated with test-code generation, software support for automated testing and the infrastructure for data handling, storage and analysis are considerable as well. Some of these are shared among products, but cost associated with routine test analysis and custom test-code generation unique to a chip design must be fully absorbed by each product line individually. The cost of using the test facility and running the tests, such as utilities, is proportional to the test time. Additional resources include engineering support for diagnostics and managing throughput.

It is therefore most economical to reject bad chips early in the manufacturing process and with a minimum number of tests. However, tests conducted at full speed and those involving I/O interfaces with other components may only be conducted at package or board level, so some fallout at later test stops is inevitable. A simple rule of thumb is that the cost of rejecting a bad chip increases by a factor $>10\times$ at each subsequent test stop shown in Fig. 1.3a.

Test structures, sensors, and monitors are placed in the scribe-line and embedded on-chip for early and on-going diagnostics. The cost of design, integration, and testing of these elements is small compared to the benefits derived from their use. Parameters averaged over large circuit blocks, such as leakage current (IDDQ) and AC power give a quick readout of silicon technology and circuit design margins. Characterization and customized data analysis techniques for rapid debug may also provide a large return on investment (ROI).

1.5 Future Test Challenges

The international technology roadmap for semiconductors (ITRS) document issued once every 2 years devotes one full section to test and test equipment [1]. Advances in silicon technology and shorter time-to-market demands have generated additional challenges in high volume manufacturing (HVM) test. Some of the new key challenges in test and diagnostics as outlined in the 2013 ITRS roadmap, and related to the material covered in this book, are listed below:

- Test data feedback to tune silicon manufacturing
- Detecting systematic defects from CMOS technology, design model limitations, and changing circuit sensitivities
- Detecting variability induced defects and device degradation over time
- Adaptive testing using in-situ, feed-forward and feedback in test flow
- Incorporating on-chip test structures and sensors in the test flow to set test content and test limits
- Managing large data volumes and data traceability

The recognition and addition of these difficult challenges in the semiconductor test roadmap highlights the upcoming changes in the test arena beyond the traditional go/no-go methodology used in digital testing.

1.6 Silicon Technology and Models

The characteristics of a specific silicon technology are described through compact electrical device models, and through ground rules for mapping the devices and circuits into planar physical layers for photomask generation. In digital circuit designs, to better manage circuit simulation time, a significant level of simplification and abstraction of compact models is introduced in EDA tools for chip timing and power analysis. Design margins are imposed to account for many sources of variations in devices, silicon processes, voltages, temperatures and environmental conditions, along with aging of CMOS chips.

How well do the models and design assumptions represent the production hardware? Such questions are frequently raised when chip power/performance or yield fall below expectations, directly affecting product delivery or profit margins. To answer these questions, it is necessary to know if the models describe correct physical behavior over the entire process and application space, that this accuracy is maintained in abstractions used in EDA tools, that the design margins cover all other sources of variations such as noise and clock jitter, and that the silicon process has remained within the parameter ranges included in the models for all of the hardware.

Some of the issues mentioned above can be resolved proactively. Accuracy of models installed in the circuit design environment can be checked prior to their incorporation in more sophisticated EDA tools, and tools and design assumptions can be scrutinized before design activity begins. Model-to-hardware correlation with appropriately designed test structures may be conducted throughout the production cycle. By identifying or eliminating any existing technology or design issues first, focus can then be placed on possible test issues. With this approach the root cause of yield loss in test can be quickly and correctly ascribed.

In selecting a CMOS technology node for a particular product or in selecting a foundry for manufacturing, customers want to evaluate relative merits of the technologies being offered. Such comparisons are typically based on compact models supplied by the foundries. Hardware-to-hardware comparisons of CMOS chips manufactured in different technology generations or foundries can only be carried out if appropriately designed monitors are embedded on-chip and tested under the same conditions to bridge between technology and chip functional characteristics. A modest investment in building an infrastructure for model evaluation and model-to-hardware correlation on an on-going basis goes a long way towards making the right decisions.

1.7 Data Analysis and Characterization

Analysis, evaluation, and characterization of large volumes of electrical test data are integral parts of the test methodology. First, data collected on individual chips are analyzed. Digital test data may be automatically filtered using pass/fail criteria by matching the output vector patterns with expected signatures. Analog measurements of currents and voltages, maximum frequency of operation of the chip, and data collected from embedded monitors and test structures placed in the scribe-line need to be correlated with design model predictions.

Statistical analysis is carried out on data collected from a large number of chips. The data are charted for visualization. Some of the charting techniques are illustrated in Fig. 1.5. These include parameter distributions, trends in parameter spreads over time, correlating parameter *Y* with *X*, and wafer maps to relate parameter variations with the geographical locations of chips on wafers. Commercial statistical analysis tools can be adapted to routinely generate a set of standard charts. Warning and alarm levels are set to alert the test team of potential or real problems. Early in the product manufacturing cycle, and when unexpected problems arise, a more detailed hands-on analysis becomes necessary. In this arena, characterization, test, chip design, and silicon manufacturing teams work together to find a solution. Cross-disciplinary knowledge is valuable in guiding the collective team to achieve rapid resolution.

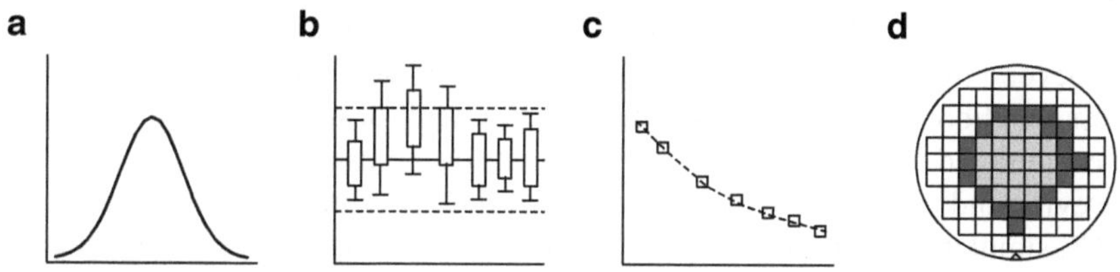

Fig. 1.5 Example charts used in characterization showing (**a**) statistical distribution, (**b**) parameter trend, (**c**) *XY* scatter plot and (**d**) gray-scaled wafer map of a parameter measured on each chip

1.8 Scope of the Book

University courses in CMOS circuit design and fabrication and VLSI testing are included in electrical engineering curricula. There are a number of excellent textbooks on VLSI design and test [2–7]. Advances in silicon technology and further miniaturization have added more complexity to design and test. There is now a greater need to combine silicon process data with chip test data to improve manufacturing efficiency and to assist in debug. This knowledge is spread across many books and publications. These serve well for in-depth views of different aspects of this multi-faceted topic, but make it challenging to get an integrated view.

Our aim in writing this book is to provide a single source for an overview of test and data analysis methodology for CMOS chips, covering circuit sensitivities to MOSFET characteristics, impact of silicon technology process variability, applications of product representative test structures and monitors, product yield, and reliability over the lifetime of the chip.

The organization of this book is similar to our book *Microelectronic Test structures for CMOS Technology* [8]. There are ten chapters, covering a full range of topics, from characteristics of circuit building blocks and impact of variability and reliability, to CMOS chip test methods and statistical data analysis. Examples are provided with circuit simulations using a simulation program with integrated circuit emphasis (SPICE). By including the full range of device parameter variations in circuit simulations, the examples emulate electrical test data typically seen in hardware.

Exercises at the end of the chapters feature practical examples of the material presented. A strong emphasis is placed on the physical behavior of circuits and on statistical methods. Knowledge derived from the physical behaviors of CMOS logic gates and memory elements can be extended to complex circuits ranging in transistor count from a few thousand to several billion.

We have used MOSFET predictive technology models (PTM) released by Arizona State University for 45, 32, and 22 nm technology nodes, and LTspice available from Linear Technology as a circuit simulator. These models and tools are freely available to all. Although the PTM models serve well to exemplify the concepts and methodologies presented for digital circuits, they have limited accuracy beyond the nominal operating range and for analog applications. Readers with access to silicon foundry models and CMOS circuit design and tool infrastructure are encouraged to use their own environment for the exercises.

Device and circuit basics are covered in Chap. 2. A methodology for setting up SPICE simulations for characterizing MOSFETs and logic gates is described. Key device and circuit parameters which can be measured during electrical testing are extracted from circuit simulations. By observing the impact of variations in these MOSFET parameters on logic gate delays, one begins to correlate device and circuit behaviors. This sets the stage for relating electrical test data back to circuit design models and tools.

Chapter 3 gives an overview of CMOS chip building blocks, from I/O and clock signal distribution to clocked storage elements (latches) and static memory arrays, emphasizing the repeated nature in circuit design and operations. Circuit simulation examples include static noise margins of SRAM cells and extraction of minimum clock cycle time and minimum operating voltage of a logic data path. We now begin to connect CMOS chip test data to characteristics of latches, logic gates, and memory elements.

In Chapter 4 MOSFET leakage current components and defect generated contributions to the measured current in the quiescent state (IDDQ) are described. Circuit simulations are carried out to model DC and AC components of power. Strategies for reducing power and on-chip power management schemes are discussed.

Embedded monitors for tracking variations in silicon process and in local power supply voltage and silicon temperature during test and operation are described in Chap. 5. Circuit simulation examples and sensitivity analysis to select an optimum set of silicon process monitors are included. Data collected from these monitors are used in variability analysis described in the next chapter. A description of sources of variations and methods for characterizing, minimizing and accommodating variability in CMOS chips are covered in Chap. 6.

Basics of CMOS test from DC parametric to logic verification tests are covered in Chap. 7. Manufacturing yield and characterization methods to optimize chip power and performance are discussed. Adaptive test methods and binning as a means of improving chip yield are introduced.

Chapter 8 deals with reliability models and degradation mechanisms in MOSFETs and interconnects. Methods of eliminating defective chips from the manufacturing test flow by accelerated tests and burn-in, and thereby improving the CMOS chip reliability, are discussed. Strategies for guard-banding during test to assure product functionality throughout specified lifetime are included.

Basic statistics and methods for effective data visualization are presented in Chap. 9. Although some knowledge of normal statistical distributions is assumed in the examples presented in other chapters, the treatment here includes deviations from normality, small sample sizes, probabilities of multiple events and relative parameter sensitivities. Examples presented are those typically encountered in CMOS circuit simulations and product test.

In Chap. 10, methodologies for setting up device and circuit performance metrics based on models and hardware data are described. This is an essential part of evaluating both technology and product performance and comparing different products, technology nodes and technology enhancements. A methodology for evaluation of BSIM MOSFET models and circuit design tools, based on physical behaviors, highlights the need to ensure correctness of design assumptions. Circuit performance has been the subject of many debates in the industry, and it is important to have the correct measures in place for such evaluations.

Working through the examples and the exercises in each chapter with device models and simulation tools will provide readers a broadened and more detailed view of the material and reinforce a physical approach and methodology to problem resolution. We hope that engineering students as well as professional engineers working in silicon manufacturing, circuit design, and test will find this book a helpful resource in preparing to confront existing and emerging challenges in these fields.

1.9 Summary and Exercises

A brief overview of CMOS design and test is provided along with the various stages in development and production testing of CMOS chips. Economic constraints in the face of silicon process variability and shrinking design margins dictate growing emphasis on accurate device models, model-to-hardware correlation, and rapid feedback from electrical test data analysis as described in the semiconductor test roadmap. Examples provided at the beginning of the chapter relate the importance of obtaining physical insight from low complexity product representative circuit blocks that capture aggregate chip behavior. The scope of the book and contents of each chapter are described.

The following exercises are designed to develop an appreciation for modeling of complex behavior, test flow and economics, and upcoming challenges in the CMOS test arena.

1.1. A microprocessor chip with an area of 4 cm^2 consumes an average power of 100 W when running a full workload. An appreciation of how some aggregate properties of a multi-billion transistor chip relate to the properties of domestic and laboratory equipment can be gained by considering the following:
   (a) Compare the average power density of this chip under the above conditions to that of the surface of a 100 W incandescent light bulb.
   (b) If the power supply voltage V_{DD} is 1.0 V, what is the average current that must be delivered to the chip by the power supply?
   (c) What is the effective resistance of the chip as viewed from the power supply?
   (d) If the total device and interconnect capacitance between V_{DD} and ground is 1.0 μF how much charge Q is stored on the chip with $V_{DD} = 1.0$ V?
   (e) Assuming that that the microprocessor is operating at a frequency of 2 GHz, how does Q compare with the charge provided by the power supply during one machine cycle?

1.2. The fabrication cost per wafer for a product with 1,000 chips per wafer is $10,000. The test cost is $2.0 per chip. The number of chips/wafer increases by 2× per technology node, the cost of fabrication increases by 20 %, and the cost of test increases by 10 %. At this rate, after how many technology generations will the cost of test per chip exceed its cost of fabrication? Illustrate your findings graphically in one chart (assume 100 % yield).

1.3. A new EDA tool is released to measure power consumption of CMOS circuits from SPICE simulations at different voltages, temperatures, and operating frequencies. The tool developer suspects there is a bug in the software. Describe a test suite with resistors and capacitors to validate the tool without requiring complex circuit simulations.

1.4. Manufacturing decisions are often closely tied with the economic model for the product. Such decisions should always be consistent with physics based models and sound engineering judgment.
   (a) Macroeconomics is a branch of economics dealing with the aggregate behavior of the economy at the national and global levels. List fundamental differences between building a national economic model and an aggregate physical model of a CMOS chip. Can the growth in consumption and national wealth be as precisely predictable as power consumption of a CMOS chip?
   (b) A silicon foundry is consistently yielding 98 % of the wafers processed for a specific product. Management would like to increase profits and throughput by eliminating electrical tests at intermediate steps during processing. Is this a good choice? If not, build a case in favor of intermediate test stops.

1.5. The test and test equipment chapter in the ITRS roadmap describes key drivers and challenges in the test industry [1].
   (a) Select two focus areas covered in this book from Sect. 2.2.3 (Detecting Systematic Defects) and Sect. 3.1 (Electrical Test Based Diagnostics) of the 2013 roadmap.
   (b) If later ITRS roadmaps are available, i.e., ITRS 2015 and beyond, note the changes in these two focus areas over time.

References

1. International technology roadmap for semiconductors: ITRS 2013 edition (2013). http://www.itrs.net/Links/2013ITRS/2013Chapters/2013Test.pdf. Accessed 21 Jul 2014
2. Wang LT, Wu C-W, Wen X (2006) VLSI test principles and architectures: design for testability. Morgan Kaufmann, Burlington
3. Abramovici M, Breuer MA, Friedman AD (1994) Digital systems testing and testable designs. Wiley, New York
4. Bushnell M, Agrawal V (2000) Essentials of electronic testing for digital, memory and mixed-signal VLSI circuits. Springer, Berlin
5. Jha NK, Gupta S (2003) Testing of digital systems. Cambridge University Press, Cambridge
6. Weste NH, Harris D (2010) CMOS VLSI design: a circuit and systems perspective, 4th edn. Addison-Wesley, Boston
7. Rabaey JM, Chandrakasan A, Nikolic B (2003) Digital integrated circuits, 2nd edn. Prentice Hall, Upper Saddle River
8. Bhushan M, Ketchen MB (2011) Microelectronic test structures for CMOS technology. Springer, Berlin

CMOS Circuits Basics

2

Contents

2.1 Circuit Components and Building Blocks ... 19
2.1.1 MOSFETs ... 21
2.1.2 Interconnects ... 28
2.1.3 Passive *R* and *C* Components ... 30
2.1.4 Logic Gates ... 31
2.2 SPICE Simulations ... 34
2.2.1 PTM (BSIM) ... 37
2.2.2 MOSFET Characteristics ... 38
2.2.3 Standard Cell Library Book Characteristics ... 50
2.2.4 Delay Chains ... 64
2.2.5 Ring Oscillators ... 71
2.2.6 Comparison of Logic Gate Characterization Methods ... 75
2.2.7 Monte Carlo Analysis ... 77
2.3 Summary and Exercises ... 79
References ... 83

Although a CMOS chip is a complex object comprising logic, memory, analog, and I/O functions, significant insight can be gained from the simulated and measured behaviors of circuit elements and small circuit blocks. The basic components and building blocks of digital logic circuits and their electrical properties are described. Circuit simulations are set up with BSIM models for plotting *I*–*V* and *C*–*V* characteristics of MOSFETs and extracting their key parameters. A methodology to characterize logic gates typically found in a standard cell library is introduced using an inverter as an example. Lookup tables for computing signal delays in combinational logic circuits with different input signal waveforms and load capacitances are generated, highlighting their interdependencies. Delay chains and ring oscillator configurations used for model validation in silicon hardware are described and simulated to extract delay parameters of logic gates. The foundations laid here including Monte Carlo analysis for determining parameter spreads are used throughout the book.

M. Bhushan, M.B. Ketchen, *CMOS Test and Evaluation*,
DOI 10.1007/978-1-4939-1349-7_2

The spreads in propagation delays and power levels of circuit blocks reflect the ranges of cycle time and operating power of a digital CMOS chip as a whole. The behavior of relatively simple circuit blocks can also be easily related to their constituent MOSFETs, interconnects, and parasitics. Small circuit blocks, which may be characterized in detail, therefore provide direct links to both the product chip and the underlying silicon technology elements as indicated in Fig. 2.1. With the physical insight acquired through such an approach, a common platform for communication among silicon technology, circuit design, design tools, test and characterization teams emerge.

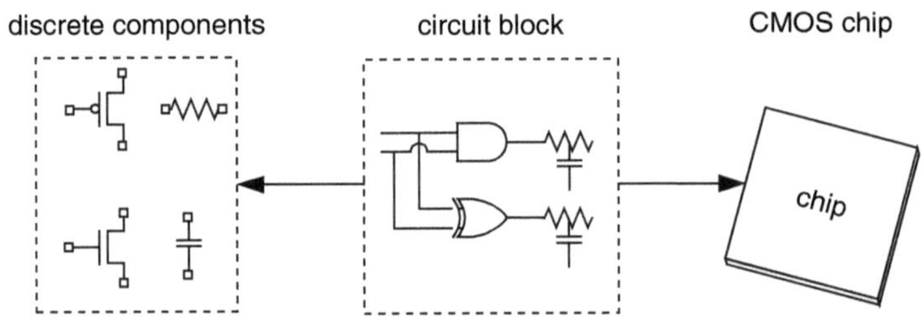

Fig. 2.1 Linking circuit blocks to silicon technology and CMOS product

Computer modeling and simulation play an important role in both design and test of microelectronic products. While this has been true all along, the compute resources now available generally far exceed the requirements of most such tasks. Indeed the compute power present in a personal computer purchased for a few hundred dollars in 2014 far exceeds that on the original Apollo lunar lander in late 1960s! The real challenge facing the design and test community today is the effective use of compute power to get the job done in a systematic, efficient, and accurate manner. Sophisticated quantum mechanical models may be helpful and even essential in understanding some of the basic physical behavior of highly scaled devices as we approach the nano-regime, yet this complexity can be simplified and encapsulated in a compact model, along with parasitic resistances, capacitances, and inductances, for efficient and accurate representation of device behavior over a practical range of use.

Typically CMOS chip designs are carried out on a workstation equipped with custom vendor supported tools. Licensing fees for use of these tools can be significant. Recognizing that not everyone has access to such tools, we have opted to use LTspice IV released by Linear Technology as the simulation tool [1]. LTspice is widely used by circuit designers and can be downloaded for free to run on a personal computer. All the examples and problems described in this book may be adapted to other versions of SPICE simulation tools [2, 3] as well.

SPICE simulators require compact models for MOSFETs, diodes, interconnects, and other parasitic components. Berkeley short-channel IGFET models (BSIM) for MOSFETs are presently the industry standard [4]. We have used BSIM predictive technology models (PTM) released by Arizona State University [5]. These models use a set of simplified equations to describe critical electrostatic behavior and carrier transport rather than the full set used in more complex BSIM models [6, 7]. Published data from early technology development as well as from previous technology generations are used for building more realistic models in advance of full technology development. The PTM models take into account limits of scaling due to manufacturability and fabrication cost and some new features introduced in successive technology nodes. These models do not represent any particular silicon foundry.

The 45-nm technology PTM models for high-performance (HP) and low-power (LP) devices are used in circuit simulation examples. While generally realistic in their representation of 45-nm technology these models do occasionally exhibit unusual behavior. As an example the temperature dependence of MOSFET drive current in the saturation mode is much stronger than normally observed. Most circuit simulation examples and problems are carried out at 25 °C to avoid operating conditions where the models are weak.

In this chapter circuit simulation techniques for MOSFET and logic gate characterization are introduced. The foundations laid here will be used in other chapters. For completeness, a brief overview of circuit components, chip design methodology, and test are included. Circuit components and their basic properties are described in Sect. 2.1. Circuit simulations and SPICE commands for characterization and model-to-hardware correlation of MOSFETs and logic gates are covered in Sect. 2.2.

In-depth treatment of CMOS circuits can be found in many excellent textbooks [8–11]. Other useful text book references cover CMOS devices [12, 13] and silicon fabrication technology [14, 15].

2.1 Circuit Components and Building Blocks

A schematic cross section for a CMOS planar process with four metal interconnecting layers is shown in Fig. 2.2. MOSFETs and diodes, the two active elements in CMOS circuits, are delineated in the silicon substrate. Metal and dielectric isolation layers are deposited on top of the active devices and patterned for making wire interconnections. Connections to the chip package are made through solder balls at the top of the metal wire stack by flip-chip bonding, or by wire bonding to I/O pads in the top metal layer.

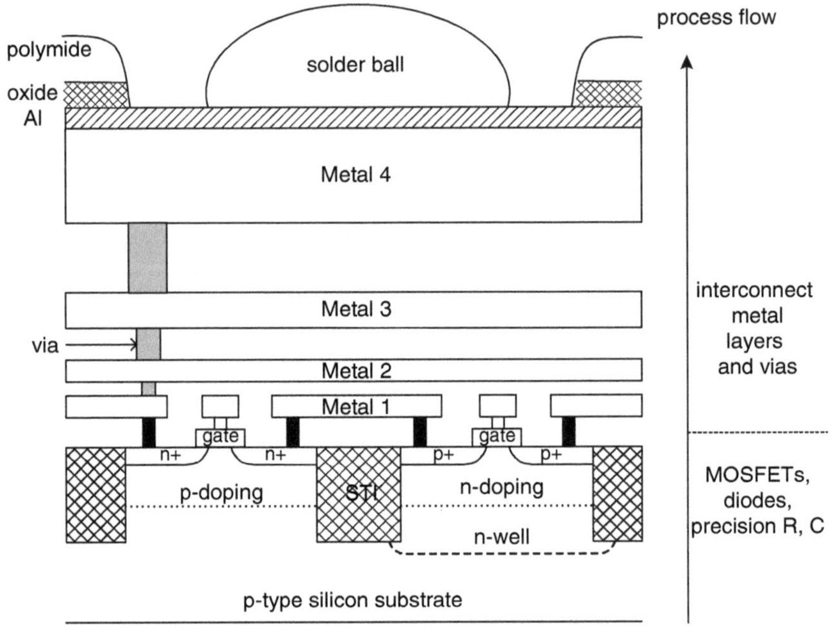

Fig. 2.2 Schematic cross section of part of a CMOS circuit showing an n-FET, a p-FET, four metal layers with inter-level vias and a solder ball for connecting to the chip package

An integrated CMOS fabrication process is very complex with many flavors of active devices differing in their electrical properties, and up to 15 or more interconnecting metal layers. The translation from circuit schematic to data input required for fabrication is through physical mapping to layers comprising two-dimensional geometric shapes and alignment of each layer with respect to other layers. Each layer is assigned a key and a color or shading to distinguish it from other layers in the drawing, and is defined with opaque and transparent areas on a photomask. This photomask is in turn used to transfer the layer shapes to a silicon wafer coated with a photosensitive material (photoresist). The wafer is exposed to ultraviolet light through the photomask and the exposed photoresist chemically processed to develop the pattern. Subsequent processing such as dopant implantation, material deposition, reactive-ion etching to remove material from unwanted areas, chemical mechanical polishing to obtain a planar surface for forming metal interconnects, and thermal and other treatments accomplish the three-dimensional physical realization of each layer with its desired properties.

Electrical and other material properties of the layers, circuit components and devices defined by the layers, and parasitic elements associated with the circuit components are described in compact models released by the silicon foundry. The models include the nominal values of the parameters, and the range of expected variations in key parameters of circuit elements introduced during manufacturing. CMOS chips are generally designed to operate with circuit properties varying within their published range.

Physical layer dimensions and compact models are two of the key inputs to circuit simulation tools. As an example, the circuit symbol of a resistor shown in Fig. 2.3a is an electrical representation of a metal wire. The physical layout of the metal wire, drawn as a rectangle with a layer key designated to a specific metal layer (e.g., metal layer M3) is shown in Fig. 2.3b. It has a drawn or design length l, width w, and is placed at a lateral distance s from a neighboring wire on the same layer. Properties of the metal wire, its actual dimensions when printed on silicon (l, w_{eff}), its temperature dependence defined by a temperature coefficient of resistance (TCR), and parameter tolerances ($\pm 3\sigma$) are included in the compact models. These parameters are used in model equations of the type shown in Fig. 2.3c.

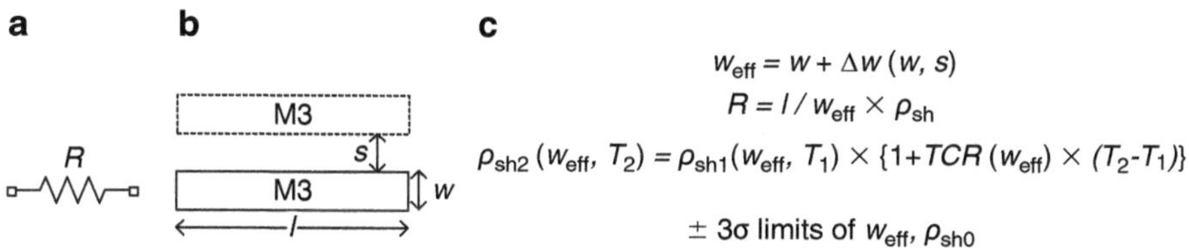

Fig. 2.3 (**a**) Circuit symbol of a resistor and (**b**) top view of the physical layout of a metal wire resistor of length l, width w, and nearest neighbor spacing s drawn in metal layer M3. (**c**) Model equations describing effective width w_{eff}, wire resistance R, sheet resistance ρ_{sh}, and parameter tolerances

In circuit simulations and in analyzing CMOS electrical test data, it is important to understand the relationships among the physical dimensions of circuit components and their electrical properties. At the circuit simulation stage, the details of the silicon fabrication process need not be considered. However, some knowledge of silicon processing is needed when analyzing process-induced variability and its impact on chip functionality and yield. A basic description of silicon process steps and sources of variations in the manufacturing process is covered in Chap. 6. Detailed description of CMOS processing can be found in any one of several books on this topic [14, 15].

2.1.1 MOSFETs

A cross section schematic view of a MOSFET with its source (S), drain (D), gate (G), and body (B) terminals is shown in Fig. 2.4a. The gate electrode is separated from the body by a thin insulating layer of silicon oxide or an alternative dielectric material of thickness t_{ox}. When a voltage of appropriate polarity with respect to source is applied to the G terminal (positive for n-FET and negative for p-FET), minority carriers in the body are pulled towards the surface and a conducting channel is formed. With the source and drain regions contacting the channel, carrier transport occurs across the channel in the presence of an electric field.

a

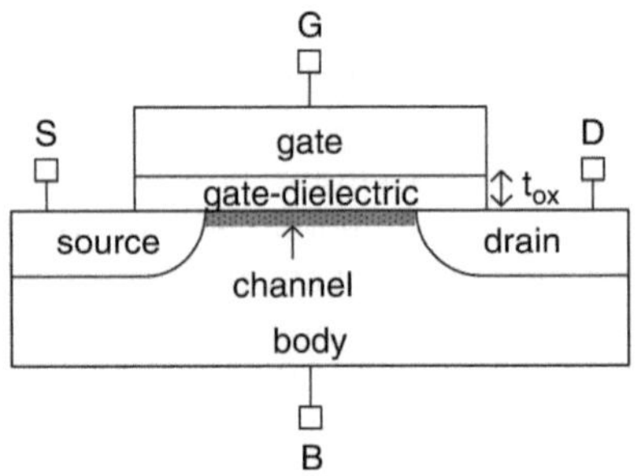

b

	n-FET	p-FET
body	p	n
source	n+	p+
drain	n+	p+
gate	p+ polySi / metal	n+ polySi / metal

Fig. 2.4 (**a**) Schematic cross section of a MOSFET and (**b**) table listing the doping types of the body, source, drain, and gate in n-FETs and p-FETs

In complementary MOS (CMOS) technology, an n-type MOSFET (n-FET) is formed in a p-type body and a p-type MOSFET (p-FET) is formed in an n-type body. Conduction in the channel is primarily by electrons in an n-FET and by holes in a p-FET. The source and drain regions are heavily doped (n+ for n-FET and p+ for p-FET) and make low-resistance contact to the channel. The gate material is doped polysilicon, p+ for n-FET and n+ for p-FET. A highly conductive silicide film covers the gate, source, and drain regions to reduce the parasitic resistances. In advanced technologies with a high-K dielectric gate insulator, the gate electrode is a metal stack with tailored work functions for n-FETs and p-FETs. The doping types of different MOSFET regions are included in the table in Fig. 2.4b.

A schematic cross section with an n-FET and a p-FET in a single p-type silicon substrate is shown in Fig. 2.5. An n-type doped region (n-well) is created for the body of the p-FET and the two MOSFET types are isolated by shallow trench oxide (STI) regions. A polysilicon or metal layer (PS) forms the gate electrode. Connection to the substrate or body (B) is made through a heavily doped silicon layer (p+ for n-FET and n+ for p-FET). All of the MOSFET terminals are contacted by the first metal layer (M1) through H0 vias in a dielectric isolation layer (not shown).

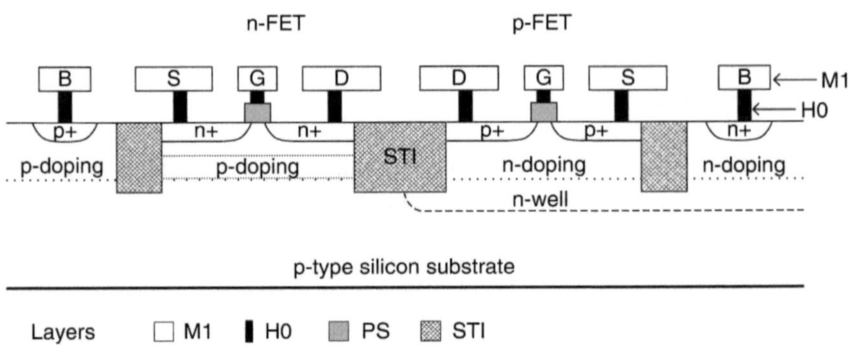

Fig. 2.5 Schematic cross sections of an n-FET and a p-FET with body contacts

Circuit symbols for the n-FET and p-FET with all four terminals are shown in Fig. 2.6a. Bias voltages are measured with respect to the S terminal: drain-to-source voltage V_{ds}, gate-to-source voltage V_{gs} and body-to-source voltage V_{bs}. The S terminal is typically held at GND for the n-FET and at V_{DD} for the p-FET. The B terminal is generally tied to the source terminal ($V_{bs} = 0$). The MOSFET schematic can then be drawn with three terminals (D, G, S). In twin well-bulk silicon CMOS technology, the B terminal may be biased independently. The nominal voltages in the conducting state of the MOSFETs are listed in Fig. 2.6b.

a

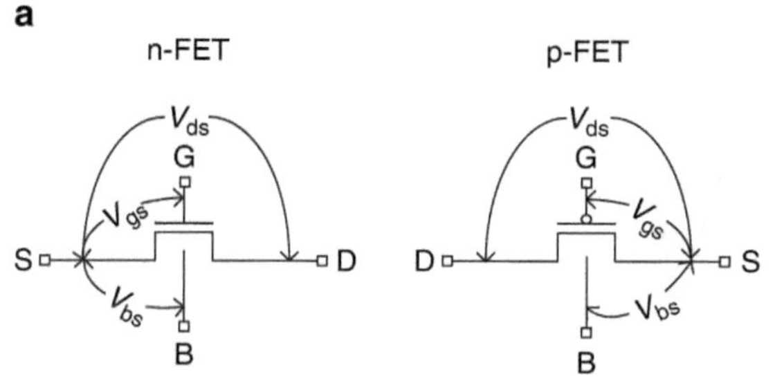

b

	n-FET	p-FET
source	GND	V_{DD}
drain	V_{DD}	GND
gate*	V_{DD}	GND

* in full saturation

Fig. 2.6 (**a**) Circuit symbols for n-FET and p-FET indicating voltage bias parameters and (**b**) nominal voltage bias values in conducting state

Physical layer mappings of an n-FET and a p-FET are shown in Fig. 2.7. Layer DF defines the source and drain regions and the gate region is defined by the PS layer adjacent to DF. Metal M1 wires connect to PS and DF layers through H0 vias in the dielectric isolation layer. Key parameters are drawn channel length L_p widths W_n (n-FET), and W_p (p-FET), and length of source and drain regions L_{ds}.

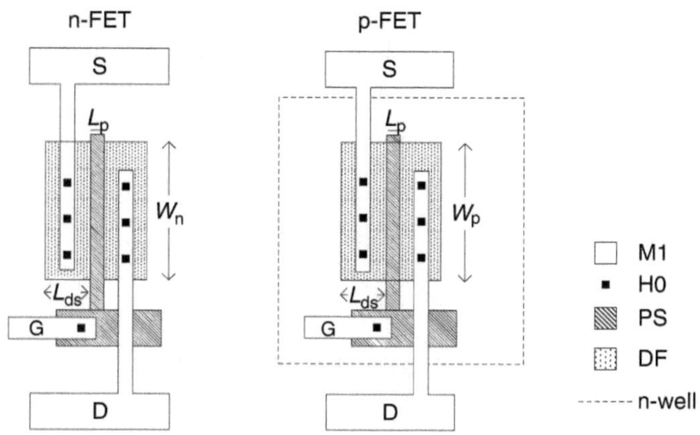

Fig. 2.7 Physical layout of an n-FET and a p-FET. Source and drain diffusion areas are implemented in the DF layer and PS is the gate layer

MOSFET (n-FET) drain-to-source current I_{ds} is measured by holding V_{gs} constant and sweeping V_{ds} from GND to the power supply voltage V_{DD}, or by holding V_{ds} constant and sweeping V_{gs}. Normalized I_{ds}–V_{ds} and I_{ds}–V_{gs} curves obtained in this manner are shown in Fig. 2.8a and b respectively. The I_{ds}–V_{ds} characteristics are separated into linear and saturation regions. In the linear region, I_{ds} varies linearly with V_{gs} and V_{ds}. In the saturation region, in an ideal MOSFET, I_{ds} is nearly independent of V_{ds} but increases linearly with (V_{gs}–V_t), where V_t is the threshold voltage at which the channel begins to conduct.

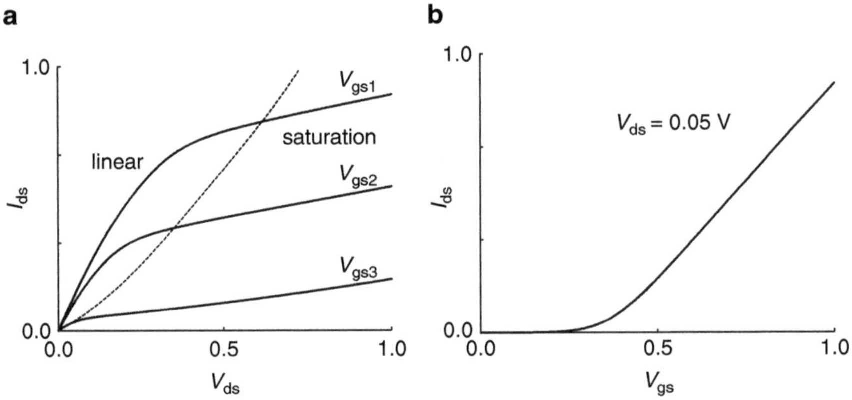

Fig. 2.8 An n-FET (**a**) I_{ds}–V_{ds} characteristics and (**b**) I_{ds}–V_{gs} characteristics

The range over which I_{ds} varies as V_{gs} is increased spans several decades. This can be viewed by plotting I_{ds}–V_{gs} characteristics on a log-linear scale as shown in Fig. 2.9. Ideally a MOSFET draws current only in the "on" state but there is also a small current component in the off-state ($V_{gs} = 0$). The current in the off-state I_{off} is measured at $V_{gs} = 0$. As V_{ds} is increased, I_{off} increases and V_t is lowered. At a low V_{ds} (typically 0.05 V), a MOSFET is in the linear region of its operating characteristics and $V_t = V_{tlin}$. When V_{ds} is set equal to V_{DD}, the MOSFET is in the saturation region and $V_t = V_{tsat}$.

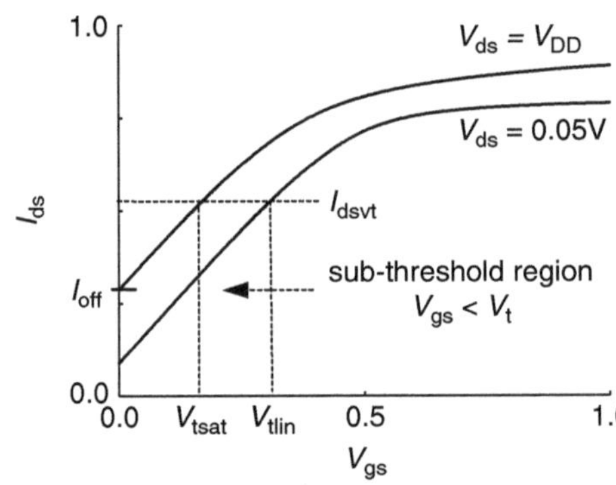

Fig. 2.9 Log(I_{ds})–V_{gs} characteristics of an n-FET with $V_{ds} = V_{DD}$ (saturation region) and $V_{ds} = 0.05$ V (linear region)

A convenient way to measure the threshold voltage V_t in both linear and saturation regions is illustrated in Fig. 2.9. Here V_t is defined as the value of V_{gs} required to reach a fixed I_{ds} ($=I_{dsvt}$). The value of I_{dsvt} is selected to be in the range of ~$100 \times I_{off}$. In another method, V_{tlin} is obtained by extrapolating the linear section of the I_{ds}–V_{gs} characteristic to $I_{ds} = 0$, and V_{tsat} is obtained by extrapolating the linear section of the $\sqrt{I_{ds}}$–V_{gs} characteristic to $I_{ds} = 0$.

Note that the definition of V_t is not precise and depends on the value of V_{ds} and the method of extraction. It does serve as a very useful single parameter for comparing MOSFET properties in subthreshold, linear and saturation regions even though its absolute value may vary with the choice of extraction.

MOSFET *I*–*V* characteristics are highly nonlinear. Long-channel MOSFET behavior in the linear and saturation regions is described by Eqs. 2.1 and 2.2 below:

Linear (non-saturation) region:

$$I_{ds} = \beta\left\{\left(V_{gs} - V_t\right)V_{ds} - \frac{V_{ds}^2}{2}\right\}, \quad 0 < V_{ds} < \left(V_{gs} - V_t\right). \tag{2.1}$$

Saturation region:

$$I_{ds} = \frac{\beta}{2}\left\{\left(V_{gs} - V_t\right)^2\right\}, \quad 0 < \left(V_{gs} - V_t\right) < V_{ds}. \tag{2.2}$$

The gain factor β in Eqs. 2.1 and 2.2 is given by

$$\beta = \frac{\mu_{eff}\varepsilon\varepsilon_0}{t_{ox}}\left\{\frac{W}{L_p}\right\}, \tag{2.3}$$

where μ_{eff} is the effective carrier mobility, t_{ox} is the thickness of the gate-dielectric, ε is the gate-dielectric constant, and ε_0 ($=8.854 \times 10^{-12}$ F/m) is the permittivity of free space.

In the subthreshold region shown in Fig. 2.9, the slope of V_{gs} vs. log(I_{ds}) is the subthreshold slope (SS) defined as

$$\mathrm{SS} = \frac{dV_{gs}}{d(\log_{10} I_{ds})}, \tag{2.4}$$

and expressed in mV/decade. It indicates the increase in V_{gs} in mV corresponding to a 10× increase in I_{ds}. Although I_{off} is a strong function of V_{ds}, SS is nearly independent of V_{ds}.

As MOSFETs are scaled to smaller dimensions, phenomena such as short-channel effect (SCE), drain-induced barrier lowering (DIBL), gate-induced drain leakage (GIDL), and strain-enhanced mobility effects become increasingly significant. The basic expressions in Eqs. 2.1–2.4 are modified and many additional terms added to incorporate these effects.

In any CMOS technology node, several different types of MOSFET pairs are offered. These MOSFET types differ in nominal values of L_p, V_t, t_{ox}, and mobility μ_{eff} to meet circuit and reliability requirements of different circuit topologies and applications. Additional physical layers are added to the layouts shown in Fig. 2.7 for silicon processing. In advanced technologies, a nominal value of L_p is specified for logic circuits and a wider range of L_p or a discrete set of L_p values is allowed for analog circuit applications. MOSFET width dimensions W_n and W_p vary in the designs and typically a minimum allowed width is specified.

A MOS capacitor is formed by the gate and the body of the MOSFET. It has an intrinsic capacitance arising from the inversion and depletion layers in the channel region. In addition there are parasitic capacitances associated with overlap regions between the gate and source/drain, and with the p/n junctions at the source/drain boundaries as shown in Fig. 2.10a.

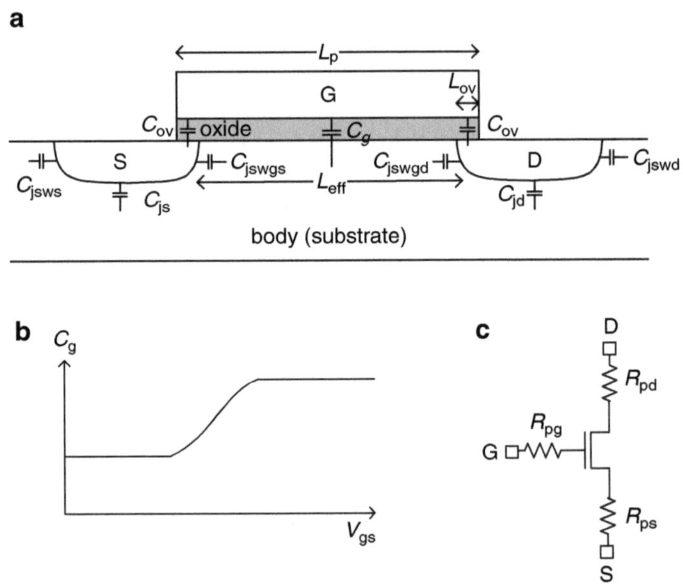

Fig. 2.10 An n-FET: (**a**) gate, source, and drain capacitances, (**b**) gate capacitance C_g as a function of V_{gs} and (**c**) parasitic series resistances

The intrinsic MOSFET gate capacitance is a function of V_{gs}, increasing with the formation of the inversion layer. The gate-dielectric capacitance C_{ox} per unit area varies inversely with the oxide (gate-dielectric) thickness t_{ox}. In the subthreshold region of operation, the capacitances associated with the gate-dielectric and the depletion layers in the channel are in series and the gate capacitance per unit width C_g is given by

$$C_{\mathrm{g}} = L_{\mathrm{p}} \left(\frac{1}{C_{\mathrm{ox}}} + \frac{1}{C_{\mathrm{d}}} \right)^{-1}, \tag{2.5}$$

where C_d is the depletion layer capacitance per unit area. In the linear region, the inversion layer screens the body of the MOSFET from the gate electrode. The MOS capacitance is then just the gate-dielectric capacitance C_{ox}. In the saturation region with a significant V_{ds}, the inversion charge layer density decreases in the drain region and C_g is reduced [12].

The parasitic capacitances associated with the gate-to-source and gate-to-drain overlap regions C_{ov}, have a direct overlap component and fringing components at the outer and inner boundaries. These capacitances are in parallel with the gate-to-channel capacitance and add to the intrinsic gate capacitance. The overlap capacitance on the drain side is a weak function of gate-to-drain voltage. The contribution of C_{ov} to the total gate capacitance C_g increases as the effective channel length L_{eff}, is decreased. Its contribution is also affected during switching of a CMOS circuit when the gate and source/drain voltages vary in opposite sense with time (Miller effect). The variation in C_g with V_{gs} is shown in Fig. 2.10b.

The p/n junction capacitance of the source and drain diffusion regions have an area component, and perimeter components for the STI and gate bounded perimeters. The total capacitance of source diffusion region, C_{ds} is given by

$$C_{\mathrm{ds}} = L_{\mathrm{ds}} W C_{\mathrm{js}} + (2L_{\mathrm{ds}} + W) C_{\mathrm{jsws}} + W C_{\mathrm{jswgs}}, \tag{2.6}$$

where C_{js} is the capacitance per unit area and C_{jsws} and C_{jswgs} are the capacitances per unit perimeter length on the STI and gate sides respectively. A similar expression holds for the capacitance of the drain diffusion region C_{dd}, with corresponding parameters C_{jd}, C_{jswd}, and C_{jswgd}.

In addition, there are parasitic series resistances associated with the source, drain, and gate contacts which include H0 vias and M1 and higher metal layers and spreading resistance of the diffusion regions (Fig. 2.10c). These resistances are extracted from the physical layout using parasitic component models provided by the silicon foundry and included in the circuit netlist. Of these, the resistance in series with the S node, R_{ps}, is the most significant as it reduces V_{gs} by the *IR* drop across it. Since the onset of conduction is brought about by an exponential increase in I_{ds} with V_{gs}, a small change in V_{gs} has a significant impact.

Equations describing MOSFET *I–V* and *C–V* characteristics contain many terms to incorporate device physical dependencies on various process and operating parameters. BSIM compact models used in circuit simulations are generated by fitting the measured *I–V* characteristics of representative MOSFETs in a technology to the MOSFET equations. There are typically hundreds of fitting parameters required to obtain a suitable fit over the desired range of MOSFET dimensions, voltages, and temperatures. However, the basic shape of the curves can be summarized in terms of a few point values as described in Sect. 2.2.2. These parameters are correlated with circuit switching speeds and power drawn in the AC and standby states.

2.1.2 Interconnects

Metal wires are used for (1) interconnecting terminals of MOSFETs and other devices, (2) interconnecting circuit blocks, (3) distributing clock signals across chip, (4) providing a power supply grid and (5) connecting to chip I/Os. Metal wire resistances and capacitances are dependent on the physical properties of metal and dielectric layer composites, layer dimensions, and near neighbor interactions.

The resistance R of a wire of length l, width w, and thickness d shown in Fig. 2.11a is given by

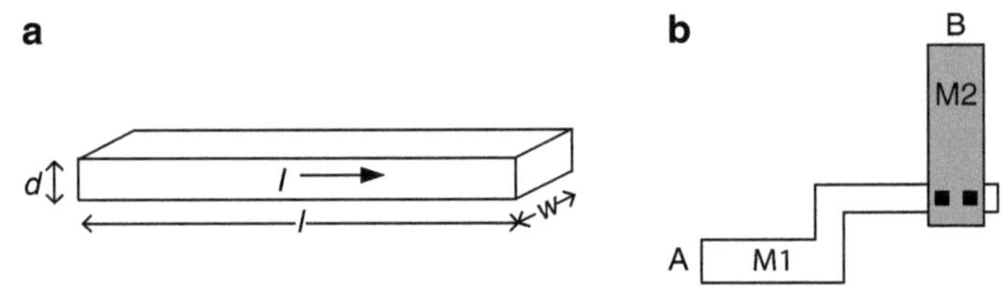

Fig. 2.11 (**a**) An isolated metal wire of length l, width w, and thickness d, and (**b**) top view of an interconnect path from point A to B through metals M1 and M2 and two inter-level vias

$$R = \rho\frac{l}{wd} = \rho_{\text{sh}}\frac{l}{w} = \rho_{\text{sh}}n_{\text{sq}}\,, \tag{2.7}$$

where ρ is the resistivity of the metal, ρ_{sh} is the sheet resistance, and n_{sq} is the number of squares. The quantity n_{sq} represents the number of squares ($l = w$) that fill up the resistor geometry irrespective of the magnitude of l and w. Thus a section of wire with $l = w = 10$ μm has the same resistance as a section with $l = w = 1$ μm. Resistances of vias connecting metal wires in one layer to those in metal layers above and/or below are added to get the resistance of a wire path in multiple layers. The values of ρ_{sh} of metal layers and resistances of vias of fixed dimensions are available from the silicon foundry for estimating interconnect resistances.

Although it is straightforward to compute the resistance of a wire of rectangular geometry, many wires have complex shapes with width variations and bends as illustrated in Fig. 2.11b. Wire models are provided by the silicon foundry and used in conjunction with physical extraction tools to determine interconnect resistances.

Wire resistance is a function of temperature. For metal wires the resistance varies linearly with temperature over practical temperature ranges

$$R_2 = R_1\{1 + \text{TCR}(T_2 - T_1)\}, \tag{2.8}$$

where R_1 and R_2 are resistances at temperatures T_1 and T_2, and TCR is the temperature coefficient of resistance.

In addition to parasitic resistances, metal interconnects also add parasitic capacitances to circuits. A capacitor formed by two parallel plates of length l and width w and separated by a dielectric of thickness h is shown in Fig. 2.12a.

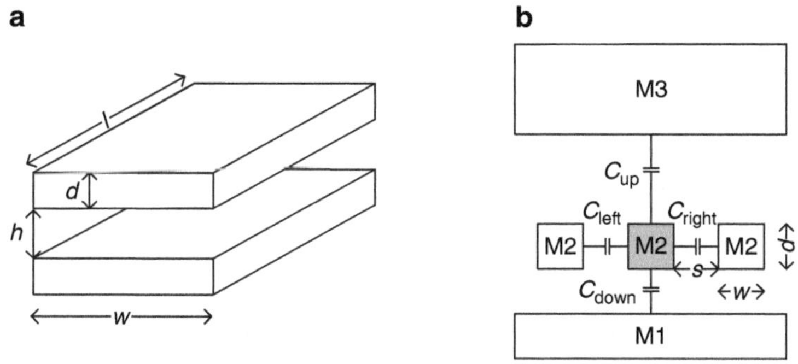

Fig. 2.12 (**a**) A parallel plate capacitor with plates of length l, width w, and separation h, and (**b**) M2 signal wire (shown in *gray tone*) and its capacitance components C_{left}, C_{right}, C_{up}, and C_{down} with its nearest neighbors

In the simplest case of a parallel plate capacitor with plate dimensions larger than the separation h, the capacitance C, is given by

$$C = \frac{\varepsilon\varepsilon_0 lw}{h} = \frac{\varepsilon\varepsilon_0 A}{h} \quad \text{for} \quad h \ll l, w. \tag{2.9}$$

In general, the wire capacitance has components to the nearest neighbors on the same layer and to the wires in layers above and below. These capacitance components are depicted as C_{left}, C_{right}, C_{up}, and C_{down} in Fig. 2.12b. The total switching capacitance of a wire during a transition is determined by the relative node voltages of the surrounding wires. If all neighboring nodes are at a fixed potential, and the signal switches between V_{DD} and GND, the total switching capacitance per unit length is given by

$$C_{\text{w}} = C_{\text{up}} + C_{\text{down}} + C_{\text{left}} + C_{\text{right}}. \tag{2.10}$$

The capacitance between wires is doubled when switching simultaneously in opposite sense and is zero when switching in the same sense.

Interconnect resistances and capacitances add to signal propagation delay in a circuit. A lumped element equivalent circuit shown in Fig. 2.13a may be used in circuit simulations. Here R_{w} and C_{w} are the resistance and capacitance per unit length of the wire. For very long wires, inductance L comes into play as well and in that case a distributed transmission line *RLC* model shown in Fig. 2.13b is used. The inductance per unit length is given by

$$L_{\text{w}} = g\mu_0, \tag{2.11}$$

where $\mu_0 = 1.26 \times 10^{-6}$ H/m (=1.26 pH/μm) is the permeability of free space and g is a geometric factor with a value in the range of 0.5–1.5.

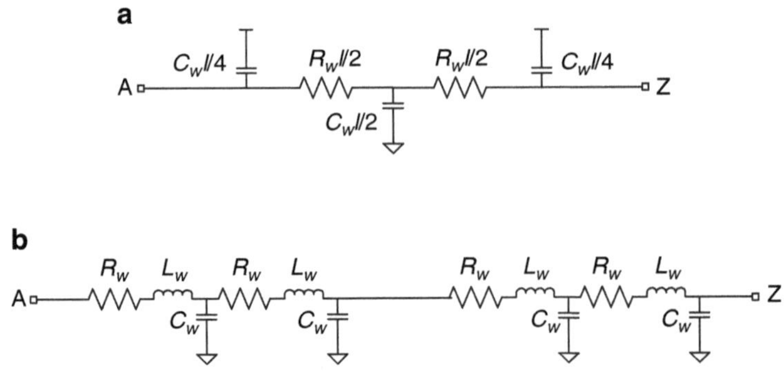

Fig. 2.13 (**a**) Lumped *RC* model of a wire of length *l*, and (**b**) distributed transmission line *RLC* model of a long wire showing four sections with length $l = 1$ in each section

2.1.3 Passive *R* and *C* Components

Silicon DF or polysilicon PS layers with higher sheet resistances than metal interconnect layers are utilized for defining precision resistors. Such resistors are used in analog and I/O circuits. The temperature dependence of these resistors is nonlinear and quadratic and higher order terms may be required in an equation analogous to Eq. 2.8.

On-chip decoupling capacitors (DECAPs) are used to help maintain a stable power supply voltage during switching of AC circuits. These are placed in unused silicon area on the chip in close proximity to regions of high switching activity. The schematic cross section of a metal-dielectric-silicon DECAP is shown in Fig. 2.14a. This configuration gives a high capacitance per unit area (~10 fF/μm^2). Resistance of the diffusion and PS layers appears in series with the capacitor. The equivalent circuit in Fig. 2.14b is used to compute the *RC* time constant and the maximum frequency for effective operation of the DECAP.

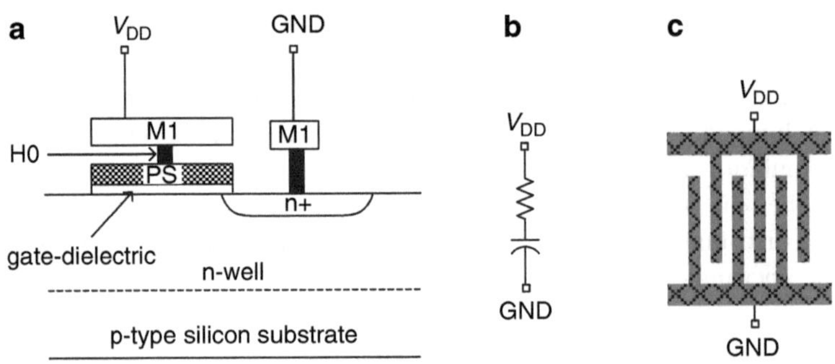

Fig. 2.14 (**a**) Schematic cross section of a metal-dielectric-silicon DECAP and (**b**) its equivalent circuit, and (**c**) interdigitated metal DECAP

Interdigitated metal capacitors are also offered for use as DECAPs in many CMOS technologies. The metal fingers in several metal layers are stacked to increase the capacitance per unit area. This type of capacitor may be used when the metal wiring is less dense and large open areas in the metal layers are available. Typical capacitance per unit area is a factor of 10 lower than for MOS capacitors.

2.1.4 Logic Gates

Static CMOS circuits comprise combinational logic gates with one or more inputs and one output. The simplest logic gate is an inverter with a single input. The circuit schematic of an inverter and its symbol are shown in Fig. 2.15a and b respectively. An inverter performs the function of inverting the input signal voltage at its output, i.e., if the input voltage is at V_{DD} ("1" or high), its output voltage is at GND ("0" or low), and conversely if the input voltage is at "0" its output voltage is at "1". The logic truth table describing this function is shown in Fig. 2.15c.

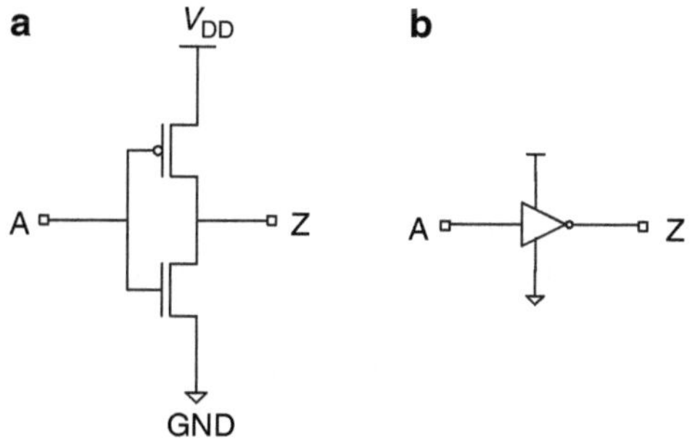

c

A	Z
0	1
1	0

Fig. 2.15 An inverter: (**a**) circuit schematic, (**b**) symbol and (**c**) logic truth table

Input and output voltage waveforms of an inverter driving a fixed capacitive load C_L are shown in Fig. 2.16. The input signal is rising and the output voltage level is pulled down from a "1" to a "0" as the signal passes through the inverter. This is called a pull-down (PD) transition. Conversely, with input signal falling and output rising, the transition is a pull-up (PU) transition. The input signal has a rise time for the signal to increase from $0.1 \times V_{DD}$ to $0.9 \times V_{DD}$, of τ_r. The signal propagation delay τ through the inverter is measured between the $V_{DD}/2$ points of the input and output waveforms.

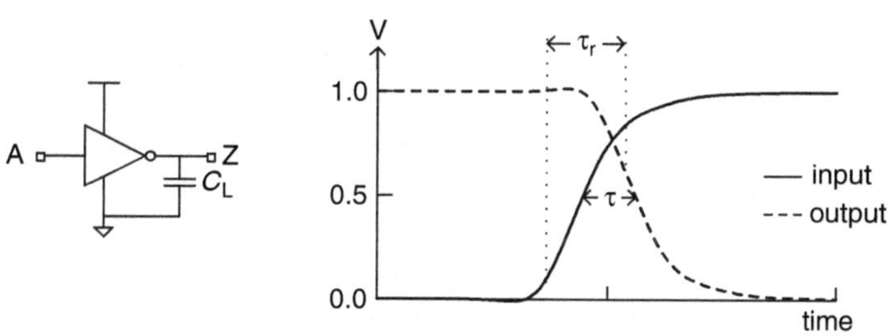

Fig. 2.16 Voltage waveforms at the input and output of an inverter driving a fixed load capacitance C_L

Even though the function of an inverter is a simple logical inversion, this structure's characteristics, such as delay and power consumption, are representative of other static CMOS logic gates described in this section and relate directly to CMOS chip cycle time and power.

Two other basic logic functions are AND and OR. Circuit symbol, logical function, and truth tables for 2-input AND and OR gates are shown in Fig. 2.17. The output Z of an AND gate is a logical "1" only if both inputs are at "1" and expressed as $Z = A \cdot B$, where A and B represent the input levels. The output of an OR gate is a logical "1" if either or both inputs are at "1" and expressed as $Z = A + B$.

a

$Z = A \bullet B$

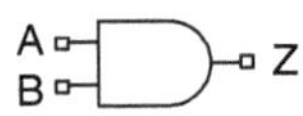

A	B	Z
0	0	0
0	1	0
1	0	0
1	1	1

b

$Z = A + B$

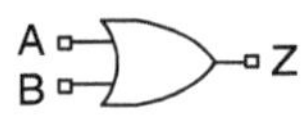

A	B	Z
0	0	0
0	1	1
1	0	1
1	1	1

Fig. 2.17 Circuit symbol and truth table of (**a**) AND gate and (**b**) OR gate

Static CMOS circuits use inversion logic in which the outputs of the logic gates are inverted. Hence the basic functions become NAND (AND + inverter) and NOR (NOR + inverter).

NAND and NOR logic gates may have two, three or more inputs, and a single output. Circuit schematics of a 3-input NAND gate (NAND3) and a 3-input NOR gate (NOR3) are shown in Fig. 2.18. The output signal voltage is a function of the states of all the inputs as listed in the truth tables. A gray ellipse highlights the inverter section of the schematic. A NAND gate can be described as an inverter with one or more n-FETs in series with the inverter n-FET and p-FETs in parallel with the inverter p-FET. Similarly a NOR gate can be described as an inverter with series p-FETs and parallel n-FETs.

Other commonly used logic gates are XOR (exclusive OR), XNOR (exclusive NOR), AOI (AND-OR-Invert), and OAI (OR-AND-Invert). Their symbols and logical functions are shown in Fig. 2.19. The output of an XOR2 or exclusive OR gate is a "0" when both of its inputs are at the same logic level, either "1" or "0". Its output is a "1" when only one of its two inputs is at "1". XNOR2 inverts the output of XOR2.

Another set of circuits that is useful for evaluating silicon technology performance comprises inverters driving a single ended n-FET (n-passgate or NPG) or p-FET (p-passgate or PPG) or a transmission gate (TG). These circuit configurations are shown in Fig. 2.20. The passgates act as switches to pass or to block signals.

A significant fraction of the signal propagation delay through a logic gate can be attributed to the internal resistances and capacitances of the MOSFETs and the gate capacitance of the following logic gates. Interconnecting wires add to the *RC* delay, their contribution increasing with interconnect length. Short interconnects are

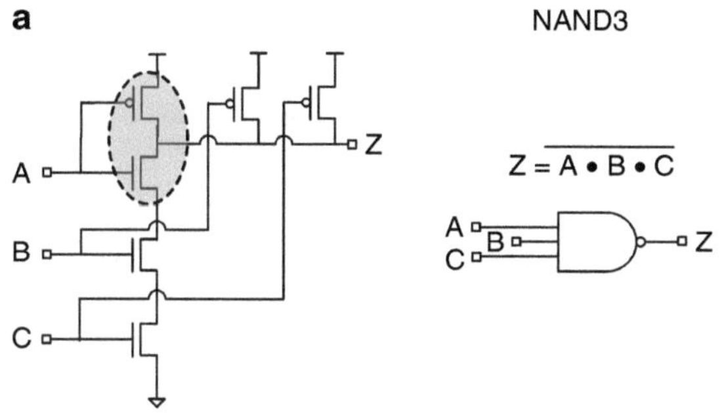

A	B	C	Z
0	0	0	1
0	0	1	1
0	1	0	1
0	1	1	1
1	0	0	1
1	0	1	1
1	1	0	1
1	1	1	0

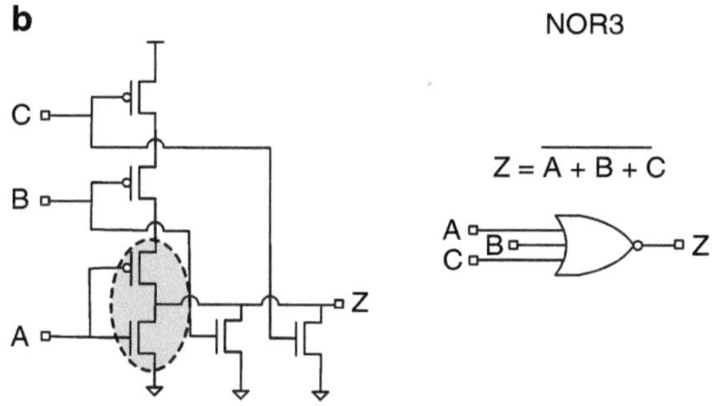

A	B	C	Z
0	0	0	1
0	0	1	0
0	1	0	0
0	1	1	0
1	0	0	0
1	0	1	0
1	1	0	0
1	1	1	0

Fig. 2.18 Circuit schematic, symbol, and truth table of (**a**) 3-input NAND gate (NAND3) and (**b**) 3-input NOR gate (NOR3). *Gray ellipses* indicate inverter schematic

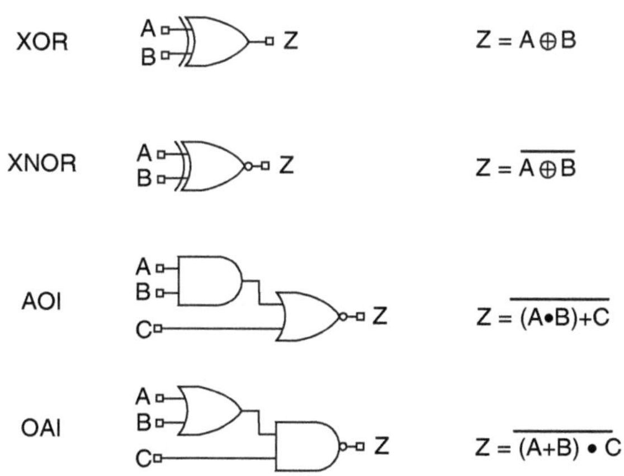

Fig. 2.19 Circuit symbols and logic functions of XOR2, XNOR2, and three input AOI and OAI gates

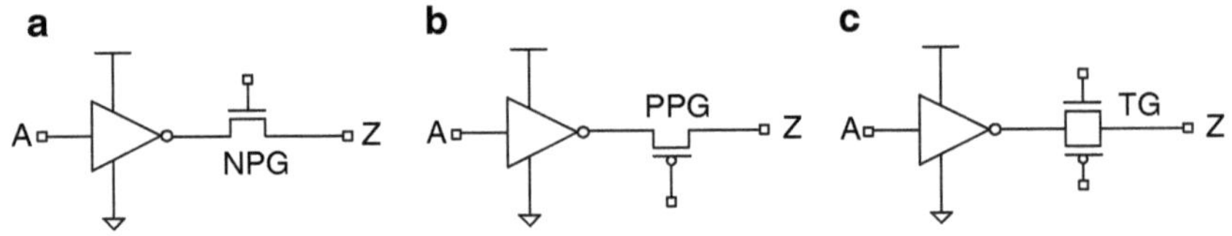

Fig. 2.20 Circuit schematic of an inverter driving (**a**) an n-FET passgate (NPG), (**b**) a p-FET passgate (PPG) and (**c**) a transmission gate (TG)

primarily capacitive whereas long interconnects are modeled as distributed *RC* networks as shown in Fig. 2.13a.

Interconnect delay responds differently to changes in voltage and temperature than do MOSFET delays. It is instructive to simulate the behavior of a logic gate driving both interconnect and MOSFET gate loads. Two circuit configurations for studying logic gate response to interconnect loads are shown in Fig. 2.21. The ratio of interconnect to gate load may be varied to characterize their relative contributions to propagation delay.

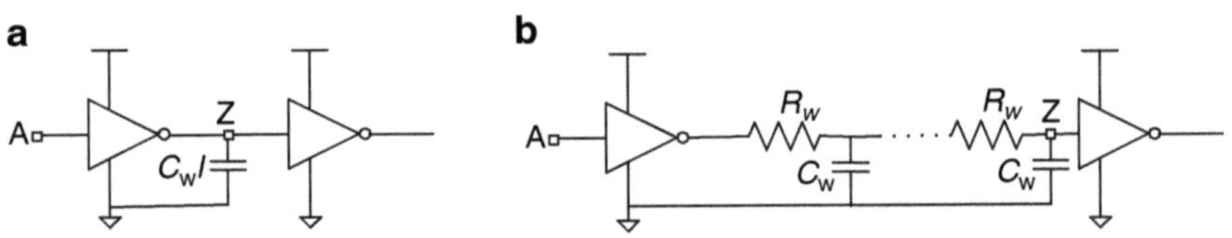

Fig. 2.21 Inverter driving (**a**) interconnect capacitive load and (**b**) interconnect distributed *RC* load

2.2 SPICE Simulations

In this section, methodologies for setting up circuit simulations of MOSFETs and logic gates are described. The simulations are geared towards evaluating technology models, circuit performance, and power. The small circuit blocks investigated here can be simulated on a personal computer using simulation tools available in the public domain or with licensed software and tools for designing large CMOS chips on work stations. Emphasis is placed on the role of the basic underlying physics at the device level and on illustrating how this propagates to a physical understanding of circuit behavior.

The essential elements of a circuit simulation setup are shown in Fig. 2.22. A schematic editor may be used for a pictorial representation of a circuit. A netlist describes the circuit elements and their interconnections as a text description of a circuit schematic. Although not essential, the netlist can be generated automatically with a netlist generator coupled to a schematic editor. This automates the tedious task of generating a netlist manually and facilitates hierarchical views of complex circuits which are then easier to debug.

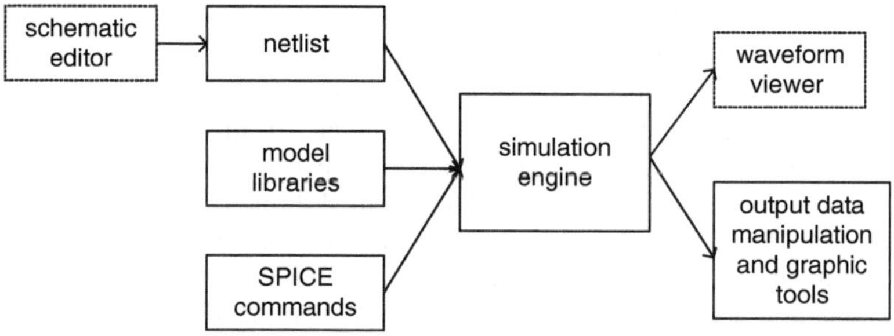

Fig. 2.22 Block diagram showing elements of a circuit simulation setup

Model libraries contain parameters and equations describing the behavior of each circuit element. SPICE commands describe the settings under which circuits are simulated and assert output requests. The simulation engine processes these inputs and generates the requested outputs. A waveform viewer is a convenient tool for viewing the voltages and currents at different nodes. Raw data output may be exported to a data analysis tool or a spreadsheet for further manipulation and for generating customized graphical displays.

The sources of MOSFET models and tools for simulation and analysis used in this book are listed in Table 2.1.

Table 2.1 Models and tools used for circuit simulations and analysis

Components and tools	Source	Model/version
MOSFET models	BSIM PTM (Arizona State University)	45 nm (HP & LP), 32 nm (HP & LP), 22 nm (HP & LP)
Circuit component library	LTspice	IV
Schematic editor	LTspice	IV
Netlister	LTspice	IV
Simulation engine	LTspice	IV
Waveform viewer	LTspice	IV
Data export and graphics	Microsoft Office Excel	2003 or later

BSIM models for MOSFETs issued by U.C. Berkley are charted by a consortium of semiconductor companies and simulation tool vendors. Silicon foundries release BSIM models for each technology offering. These models are used for chip design and for manufacturing process control. There may be several model releases by a foundry for the same technology node reflecting technology modifications that impact circuit design. The models are proprietary and typically available only to customers of the foundry.

The predictive technology models (PTM) are distributed by Arizona State University. The model cards can be downloaded from their website [5]. These models are based on CMOS scaling rules in advance of full technology maturity for early circuit design analysis. They do not represent any particular silicon foundry offering but are for the most part a reasonable representation of a technology node. Hence, the circuit simulation data presented here are for illustration purposes only.

The schematic editor and waveform viewing capabilities of LTspice IV are very useful for generating circuit netlists and for debug. LTspice can be run on Microsoft Windows or OS X operating systems and contains model libraries for standard circuit elements such as resistors, capacitors, and voltage sources. The examples given here are generic although some SPICE commands may be specific to LTspice. With minor adaptation, the commands can be used with other versions of SPICE such as HSPICE and Ngspice [2, 3].

Each circuit component in a schematic is identified with a unique symbol. A circuit component may have multiple instances. Each instance in a circuit is given a unique instance name. Some properties of an instance may be assigned in the schematic as attributes. The attributes may be values or variables. The circuit element names and their equivalent LTspice names, symbols, and key attributes are listed in Fig. 2.23, with further elaboration in Fig. 2.24 and Table 2.3. Note that in the three terminal symbols for nmos and pmos the body is tied to the source terminal.

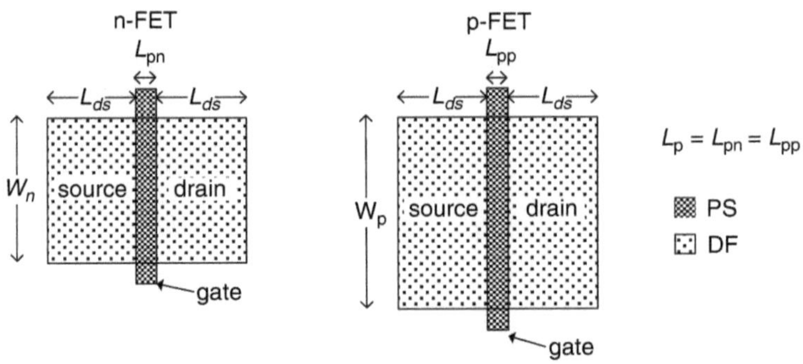

Fig. 2.24 MOSFET physical layouts showing the dimensions of gate, source, and drain regions

We begin with simulating MOSFET characteristics and follow this with circuit simulations of small circuit blocks. Three different circuit configurations are used for characterizing such circuit blocks. These are:

- Individual circuit blocks with variable input waveforms and output load
- Delay chains comprising series-connected circuit blocks
- Ring oscillators comprising series-connected circuit blocks configured in a closed loop

Individual circuit blocks are simulated in a manner similar to that used for characterization of logic gates in a standard cell library. Delay chain and ring

circuit element	LTspice component name	symbol	attributes
n-FET	nmos	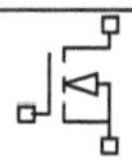	*w l lds ad as pd ps nf delvto*
p-FET	pmos	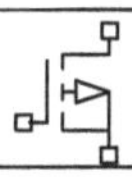	*w l lds ad as pd ps nf delvto*
resistor	res		*R*
capacitor	cap		*C*
voltage source	bv, voltage		*V*
current source	bi, current		*I*

Fig. 2.23 Symbols and attributes used in LTspice simulations for different circuit elements

oscillator configurations are commonly used in test structures placed on silicon to monitor the CMOS process and for model-to-hardware correlation. Such applications of delay chains and ring oscillators are covered in Chaps. 4–10.

2.2.1 PTM (BSIM)

The PTM releases are available from the 180 nm technology node through the 7 nm node. Models in version V2.1 have been released in 16, 22, 32, and 45 nm for high performance (HP) and low power (LP) applications. Models for future technologies (7, 10, 14, 16, and 20 nm nodes) at the time of publication of this book (2014) use multi-gate transistors. Model releases for these advanced technologies are for high performance (HP) and low standby power (LSTP).

The 45 nm PTM for high performance (HP) and low power (LP) MOSFETs were released in September of 2008 and November of 2008 respectively [5]. The BSIM4 model cards for these MOSFETs are included in Appendix B. The model cards have been modified to run in LTspice. The model level has been changed from 54 to 14. A prefix "m" has been added to the MOSFET model names: nmos to **m**nmos and pmos to **m**pmos to meet LTspice requirements.

The nominal operating voltage V_{DD} and channel length L_p for the 45 nm PTM models are listed in Table 2.2. It is assumed that these parameters are normally distributed with a mean corresponding to the nominal value and a standard deviation σ specified in the BSIM models (Sect. 9.1.2). Systematic process variations in L_p and V_t are expressed as standard deviations of their distributions, σL_p and σV_t respectively. Typically circuit properties are examined over $\pm 3\sigma L_p$ and $\pm 3\sigma V_t$ ranges. Other sources of variations are covered in Chap. 6.

Table 2.2 Nominal parameters for PTM models for the 45 nm technology node

MOSFET	PTM	V_{DD} (V)	L_p (μm)	σL_p (μm)	σV_t (V)
n-FET	HP	1.0	0.045	0.015	0.020
	LP	1.1	0.045	0.015	0.020
p-FET	HP	1.0	0.045	0.015	0.020
	LP	1.1	0.045	0.015	0.020

Parasitic resistances and capacitances which are normally extracted from physical layouts are generally not included in the examples in this book. This simplifies the netlists, and the results are independent of physical layout styles. Full parasitic extraction is essential for precise evaluation of circuit behavior and for model-to-hardware correlation.

2.2.2 MOSFET Characteristics

MOSFET parameters related to a physical layout are specified as attributes to the instance symbol. Representative layout of an n-FET and a p-FET along with their key dimensions are shown in Fig. 2.24. Channel length, L_{pn} for n-FET and L_{pp} for p-FET, is the gate dimension in the direction of current flow. Generally a common value of gate length L_p ($=L_{pn}=L_{pp}$) is used for both n-FET and p-FET. The channel widths and corresponding widths of source and drain regions are W_n for n-FET and W_p for p-FET. The n-FET and p-FET widths are allowed to vary independently. In this symmetric layout, the lengths of source and drain regions are both equal to L_{ds}. The source and drain area (*as* and *ad*) and perimeter (*ps* and *pd*) are defined in terms of W_n, W_p, and L_{ds}. These relationships are listed in Table 2.3. The LTspice parameters are in normal italic font (lower case) whereas the parameters in the text, figures, and equations use the normal convention of subscripts, superscripts, and upper and lower cases.

Table 2.3 MOSFET parameters in LTspice, and corresponding symbols used in the text

Parameter	Description	n-FET	p-FET
w	Channel width	W_n	W_p
l	Channel length	L_{pn}	L_{pp}
lds	Diffusion area length	L_{ds}	L_{ds}
as	Source diffusion area	$L_{ds} \times W_n$	$L_{ds} \times W_p$
ad	Drain diffusion area	$L_{ds} \times W_n$	$L_{ds} \times W_p$
ps	Source diffusion perimeter	$2 \times (L_{ds} + W_n)$	$2 \times (L_{ds} + W_p)$
pd	Drain diffusion perimeter	$2 \times (L_{ds} + W_n)$	$2 \times (L_{ds} + W_p)$

It is sometimes convenient to assign variables to the MOSFET attributes. Values of the variables are then assigned in LTspice commands. In a hierarchical schematic view, values of the variables can be assigned at the top level of the hierarchy as described in Sect. 2.2.4. The variables assigned to MOSFET attributes are listed in Table 2.4. The variables for n-FET have a prefix 'n' and for p-FET have a prefix 'p'.

Table 2.4 Variables used in n-FET (prefix 'n') and p-FET (prefix 'p') attributes

Parameter	Description	n-FET	p-FET
w	Channel width	*nw*	*pw*
l	Channel length	*nl*	*pl*
lds	Diffusion area length	*lds*	*lds*
nf	Number of fingers	*nnf*	*pnf*
delvto	ΔV_t	*ndelvto*	*pdelvto*

The circuit schematic of an n-FET for LTspice simulation of I_{ds}–V_{ds} and I_{ds}–V_{gs} characteristics is shown in Fig. 2.25a, and for a p-FET in Fig. 2.25b. Note that the MOSFET node voltages are measured with respect to source, which is at GND for n-FET and at V_{DD} for p-FET. The current and voltages, I_{ds}, V_{ds}, and V_{gs}, are positive for n-FET and negative for p-FET in their on-states.

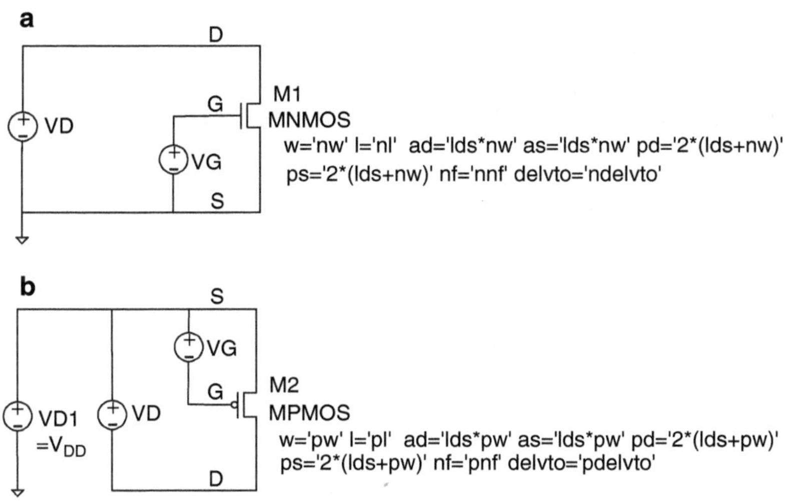

Fig. 2.25 Circuit schematic for simulating I_{ds}–V_{ds} characteristics in LTspice for (**a**) an n-FET and (**b**) a p-FET

We start with a complete LTspice input file for simulating the I_{ds}–V_{ds} characteristics of an n-FET. The text is separated into sections each beginning with a comment line in bold and preceded by *. First the LTspice, BSIM library paths, and model cards are specified. Simulator options are set to aid convergence and improve error tolerance. These values override the default values in the simulator. The netlist gives connectivity of the two power supplies (VD and VG) with drain, gate, source, and body terminals of the NMOS instance M1, in that order.

The variables used in the attributes of n-FET instance M1 are assigned values characteristic of the 45 nm technology node. The nominal value of the channel length $L_p = nl$ is 0.045 μm. The n-FET width $W_n = nw$ is set to 1.0 μm, convenient for obtaining the current drive per unit width. The temperature is set at room temperature, 25 °C, overriding the default temperature of 27 °C.

The simulation is run by sweeping the power supply voltage VD from 0.0 V to V_{DD} in steps of 0.01 V. Five such sweeps are made with VG power supply voltages at fractional values of V_{DD}, i.e., V_{gs}/V_{DD} of 0.0 V, 0.25, 0.50, 0.75, and 1.0. Alternatively five instances of identical n-FETs may be included in the netlist, each with a unique value of VG. For plotting I_{ds}–V_{gs} characteristics of an n-FET in the linear and saturation regions, the power supply VG in Fig. 2.25a is swept from 0.0 V to V_{DD} and VD is stepped from 0.05 V to V_{DD}.

An LTspice deck for simulating n-FET I_{ds}–V_{ds} characteristics is listed below:

```
* n-FET Ids-Vds characteristics using 45 nm PTM HP models
* measure n-FET Ion in µA/µm at 1.0V, 25°C
* library paths and models cards
.lib C:\Program Files\LTC\LTspiceIV\lib\cmp\standard.mos
.inc C:\Users\model_library
.model NMOS NMOS
*simulator options
.option gmin=1e-14 abstol=0.1pA reltol=1e-5 noopiter
* netlist: voltage sources VG, VD and MOSFET M1 (drain. gate, source, body)
VG g 0
VD d 0
M1 d g 0 0 MNMOS w='nw' l='nl' ad='lds*nw' as='lds*nw' pd='2*(lds+nw)' ps='2*(lds+nw)' nf='nnf' delvto='ndelvto'
* variables
.param nw=1u nl=0.045u lds=0.12u nnf=1 ndelvto=0.0
.temp 25
* DC sweep
.dc VD 0 1.0 0.01
.step VG 0 1.0 0.25
.end
```

LTspice features output currents flowing through the MOSFET instance. These are labeled Id(m1), Ig(m1), Is(m1), and Ib(m1) as the drain, gate, source, and body currents in instance M1. The currents as a function of the voltage being swept (VD or VG) may be viewed in the waveform viewer. For customized graphics, the data points may be exported to a spreadsheet (e.g., Microsoft Office Excel) or another software tool with similar capabilities.

In the case of a p-FET, D is at GND potential and all voltages are measured with respect to the S terminal which is at a higher potential. The current flow is via holes from source to drain and the sign of the conventional current is negative, opposite that in an n-FET. It is convenient to multiply the p-FET I_{ds} and V_{ds} by -1 so that the n-FET and p-FET characteristics may be overlaid for comparison.

The I_{ds}–V_{ds} characteristics of an n-FET at V_{gs}/V_{DD} values of 0.5, 0.75, and 1.0 are shown in Fig. 2.26a. The V_{ds} range is from 0 to V_{DD}. In Fig. 2.26b, the I_{ds}–V_{gs} characteristics are plotted at V_{ds} values of 0.05 and 1.0 V.

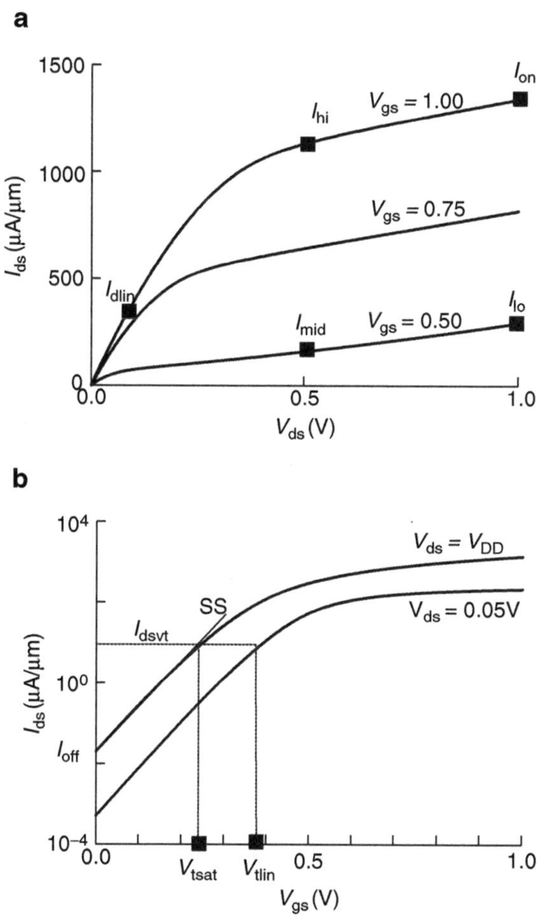

Fig. 2.26 n-FET: (**a**) I_{ds}–V_{ds} characteristics and (**b**) I_{ds}–V_{gs} characteristics. 45 nm PTM HP models at 25 °C

The I_{ds}–V_{ds} and I_{ds}–V_{gs} characteristics are summarized in terms of a few locations on the curves. The location of the parameters on I_{ds}–V_{ds} and I_{ds}–V_{gs} plots are shown in Fig. 2.26. The definitions of measurement locations for I_{on}, I_{hi}, I_{lo}, I_{mid}, and I_{off} are given in Table 2.5. In thin gate-dielectric MOSFETs, the leakage current through the gate-dielectric I_{gl} can become significant. It is dependent on both V_{gs} and V_{ds} and is maximum when $V_{gs} = 1.0$ V and $V_{ds} = 0.0$ V.

Table 2.5 Source, drain, and gate voltages for measuring n-FET parameters with the body tied to source

MOSFET parameter	Source	Drain	Gate
I_{off}	GND	V_{DD}	0
I_{dlin}	GND	0.05–0.10 V	V_{DD}
I_{mid}	GND	$V_{DD}/2$	$V_{DD}/2$
I_{lo}	GND	V_{DD}	$V_{DD}/2$
I_{hi}	GND	$V_{DD}/2$	V_{DD}
I_{on}	GND	V_{DD}	V_{DD}
I_{gl}	GND	GND	V_{DD}
V_{tlin}	GND	0.05 V	
V_{tsat}	GND	V_{DD}	

LTspice commands for measuring I_{on} in μA/μm are shown below. Normalized values of other I_{ds} parameters (I_{hi}, I_{lo}, I_{mid}, I_{off}, and I_{gl}) with appropriate V_{ds} and V_{gs} bias voltages are obtained in a similar fashion.

```
* measure Ion
.measure dc Ion find id(M1) at=1.0          *set VG=1.0
.measure Ion_uA_um param = 'Ion/nw'
```

The threshold voltage V_t is obtained from I_{ds}–V_{gs} characteristics. Here V_t is defined at a fixed current, I_{dsvt} (positive for n-FET, negative for p-FET). For the 45-nm high-performance (HP) technology node the I_{dsvt} values are selected to be

$$I_{dsvt} = 300\ \text{nA} \times \frac{W_n}{L_p};\quad \text{n-FET,} \tag{2.12}$$

and

$$I_{dsvt} = 100\ \text{nA} \times \frac{W_p}{L_p};\quad \text{p-FET.} \tag{2.13}$$

The subthreshold slope (SS) is calculated as the slope of $\log_{10}(I_{ds})$ vs. V_{gs} line between the values of $V_{gs} = 0$ and $V_{gs} = V_t$.

The netlist for power supplies and LTspice commands to measure I_{off} in nA/μm, V_t and SS for a fixed value of VD are listed below:

```
* measure n-FET Ioff, Vt and SS
.dc VG 0 1.0 0.01
.step VD 0.05 1.0 0.95
.measure dc Ioff_n find id(M1)*1000000000 at=0
.measure dc Vtn find V(G) when id(M1)=(300e-9*(nw/nl))
.measure SS param=Vtn/(log10(300*(nw/nl))-log10(Ioff_n))*1000

* measure p-FET Vt
.measure dc Vt_g find V(G) when id(M2)=(-1*100e-9*(pw/pl))
.measure Vtp param = 'V(Vdd)-Vt_g'
```

Key derived parameters for MOSFETs are listed in Table 2.6. The effective current, I_{eff} is a measure of the current drive of the MOSFET during switching. It correlates well with the signal propagation delay of logic gates as discussed in Sect. 2.2.3. In short-channel MOSFETs, V_t is lowered with increase in V_{ds}. This effect is known as drain-induced barrier lowering (DIBL), the difference between V_{tlin} and V_{tsat} is a measure of its magnitude.

Table 2.6 Definitions of calculated electrical parameters of a MOSFET

Parameter	Definition	Comments
I_{eff}	$(I_{hi}+I_{lo})/2$	Correlates to circuit delay
SS	$dV_{gs}/d(\log_{10} I_{ds})$	mV/decade, $0<V_{gs}<V_t$
DIBL	$(V_{tlin}-V_{sat})$	Varies with L_p (SCE)
g_m	dI_{ds}/dV_{gs} at constant V_{ds}	Saturation region
g_{ds}	dI_{ds}/dV_{ds} at constant V_{gs}	Saturation region
γ	$1+dV_t/dV_{bs}$	$V_{bs}\approx 0$

The output conductance g_{ds} is the change in output current I_{ds} in the saturation region with output voltage V_{ds}. The transconductance g_m is the change in I_{ds} in response to a change in input voltage V_{gs} and is a measure of the gain of a MOSFET. Both g_{ds} and g_m are important parameters in the design of analog circuits.

The subthreshold characteristics of a MOSFET can be modulated by independently controlling its body voltage, V_{bs}. With the body at a negative voltage with respect to the source in an n-FET, there is an increase in V_t and hence a decrease in I_{off} as $|V_{bs}|$ is increased. The body-effect coefficient γ gives a measure of the sensitivity of V_t to V_{bs} near $V_{bs}=0$. This effect can be utilized in reducing off-state power as discussed in Chap. 4.

The I_{ds}–V_{ds} and I_{ds}–V_{gs} plots for an n-FET and p-FET are shown in Fig. 2.27 (all p-FET currents and voltages are shown as positive). The n-FET and p-FET characteristics are overlaid for ease of visual comparison. From the I_{ds}–V_{ds} plots it is apparent that the n-FET current is higher than the p-FET, and this difference varies within the I_{ds}–V_{ds} space.

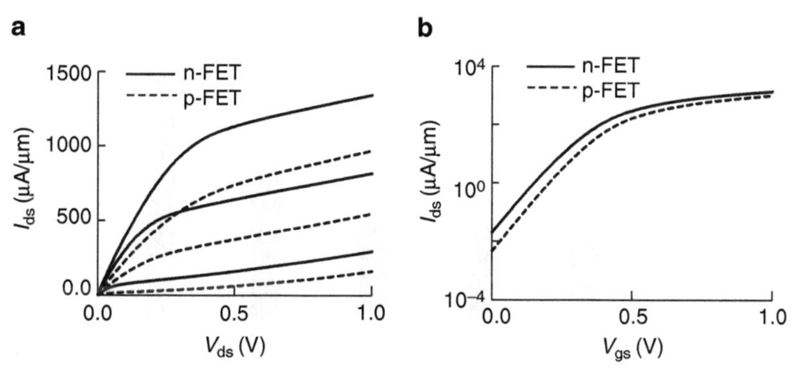

Fig. 2.27 n-FET and p-FET characteristics: (**a**) I_{ds}–V_{ds} for $V_{gs}=0.5$, 0.75 and 1.0 V and (**b**) I_{ds}–V_{gs} plots for $V_{ds}=1.0$ V. 45 nm PTM HP models at 25 °C

Visual inspection of the I_{ds}–V_{ds} curves gives a qualitative assessment of the relative strengths of the n-FET and the p-FET. For a quantitative assessment, the I_{ds} values specified in Tables 2.5 and 2.6 are recorded. These values for HP and LP models are shown in Table 2.7. The HP model has >3× higher current in the saturation region than LP resulting in faster switching speeds in CMOS circuits. This comes at a cost of higher I_{off} and higher power in the off-state by a factor of >200.

Table 2.7 Properties of nominal p-FETs and n-FETs in 45 nm PTM models for HP @ 1.0 V and LP @ 1.1 V, 25 °C

	Model	V_{DD} (V)	I_{on} (μA/μm)	I_{eff} (μA/μm)	I_{dlin} (nA/μm)	I_{mid} (μA/μm)	I_{off} (nA/μm)
n-FET	HP	1.0	1,339	711	394	159	20
p-FET	HP	1.0	968	450	229	60	4.5
n-FET	LP	1.1	525	239	186	7.6	0.024
p-FET	LP	1.1	310	136	92	4.9	0.020

The relative strengths or current drive capabilities of n-FET and p-FET in a technology are critical in sizing the MOSFET widths in logic gates. Typically the I_{on} ratio is used as a measure of n/p strength. However, it is apparent from Fig. 2.27a that relative current drives of n-FET and p-FET are functions of V_{ds} and V_{gs}. By taking the ratios of the n-FET and p-FET I_{ds} parameters listed in Table 2.7, the relative strengths of the two MOSFET types I_{ds}(n/p) are evaluated in different regions of the I_{ds}–V_{ds}–V_{gs} space. These ratios are listed in Table 2.8. The same data are presented in a bar chart in Fig. 2.28.

Table 2.8 Ratios of n-FET and p-FET key I_{ds} values in 45 nm PTM models @ 1.0 V, 25 °C

Model	V_{DD} (V)	I_{on} (n/p)	I_{eff} (n/p)	I_{dlin}(n/p)	I_{mid} (n/p)	I_{off} (n/p)
HP	1.0	1.38	1.58	1.72	2.65	4.44
LP	1.1	1.69	1.75	2.02	1.55	1.20

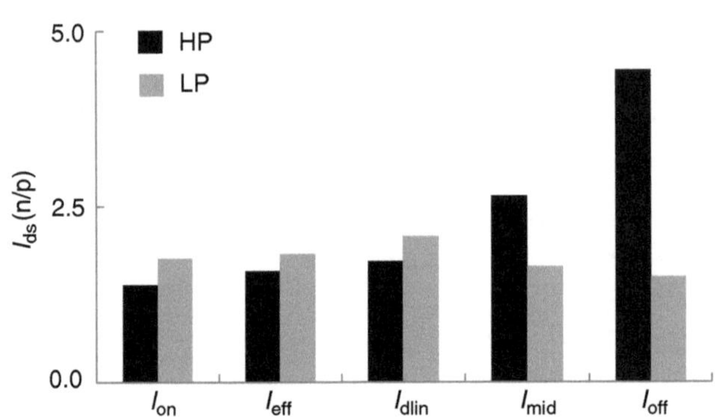

Fig. 2.28 Bar chart showing n/p ratios in Table 2.8

Let us consider the variation in two of the most significant MOSFET parameters, L_p and V_t. The parameter values may be changed in LTspice simulations by using an adder to the nominal value. An adder parameter for V_t is provided in the BSIM model. This parameter, *delvto*, has a nominal value of 0. For varying L_p, adder parameters *ndl* for n-FET and *pdl* for p-FET are introduced. The LTspice commands for these variables are listed below:

*** specifying parameter values for ΔL_p and ΔV_t**
.param nl = '(0.045u+ndl)'
.param ndl = 0.015u ** increase n-FET Lp by 0.015 μm*
.param ndelvto = 0.020 ** increase n-FET Vt by 20 mV*
.param pdelvto = - 0.020 ** increase p-FET Vt by 20 mV*

Note that to increase $|V_t|$, *ndelvto* is positive and *pdelvto* is negative.

The variations of I_{on} and I_{off} for an n-FET as a function of L_p are shown in Fig. 2.29. The range of L_p shown corresponds to a variation of $\pm 3\sigma L_p$ as listed in Table 2.2 and includes 99.7 % of the population. Over this range, the I_{on} swing with respect to nominal is from +22 % (L_p at its -3σ value) to -14 % (L_p at its $+3\sigma$ value). At these same extreme values of L_p, I_{off} varies from 27× to 0.2× of its nominal value.

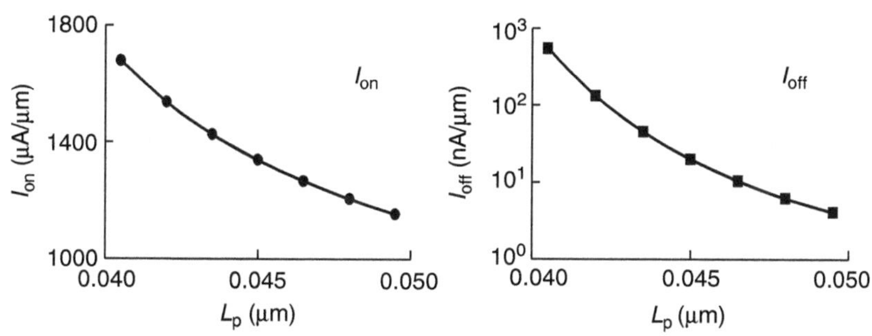

Fig. 2.29 n-FET current vs. channel length, L_p: (**a**) I_{on} and (**b**) I_{off}. 45 nm PTM HP models @ 1.0 V, 25 °C

The variations of I_{on} and I_{off} for an n-FET as a function of a shift in V_{tn} from nominal (*ndelvto* $= \Delta V_{tn}$) over the $\pm 3\sigma V_{tn}$ range are shown in Fig. 2.30. I_{on} varies linearly with V_{tn}, with a slope of 2.41 μA/μm per mV shift in V_{tn}. I_{off} vs. $\Delta|V_{tn}|$ is plotted on a log-linear scale in Fig. 2.30b. The value of SS obtained from the slope of the line fit to $\log_{10} I_{off}$ vs. $\Delta|V_{tn}|$ is 82.5 mV/decade.

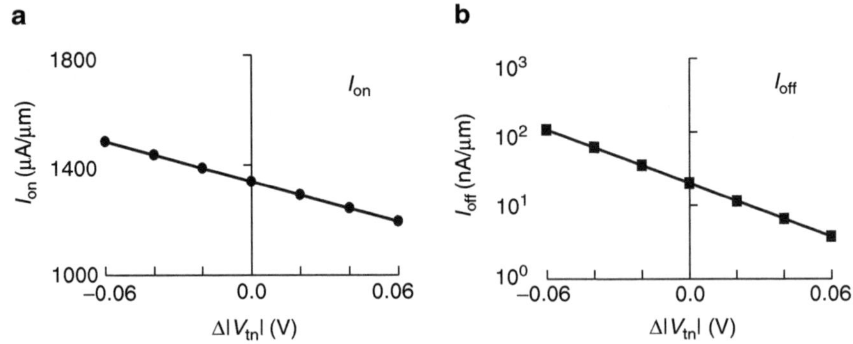

Fig. 2.30 n-FET current vs. $\Delta|V_{tn}|$: (**a**) I_{on} and (**b**) I_{off}. 45 nm PTM HP models @ 1.0 V, 25 °C

MOSFET parameters influencing the DC characteristics such as L_p and V_t vary nearly independently within their specified distributions. There is some interplay between these parameters such as V_t roll-off at short channels ($L_p <$ nominal). To obtain the full range of variations, Monte Carlo simulations are carried out by independently varying all of the MOSFET parameters for which distributions are available. This is described in Sect. 2.2.7.

Another important subject of MOSFET characterization and simulation is capacitance. The C_g–V_{gs} characteristics of an n-FET may be simulated using the circuit schematic shown in Fig. 2.31a. The S and D terminals of the n-FET are connected together and a DC voltage bias VGB is applied to the G terminal. A small signal voltage source VM connected in series with VGB swings the voltage at terminal G from (VGB − VM/2) to (VGB + VM/2) as shown in Fig. 2.31b. The current flowing through VM is integrated to give the charge transferred in raising the voltage across the capacitor by VM. The capacitance, C_g as a function of VGB = V_{gs} is calculated using the equation

$$C_g = \frac{Q_c}{\mathrm{VM}} = \frac{\int I(\mathrm{VM})dt}{\mathrm{VM}}. \tag{2.14}$$

Here Q_c is the total charge transferred during the VM transition. The pulse rise time is selected to be sufficiently short so that the charging current is at least a few orders of magnitude larger than the gate-to-body leakage current.

The netlist and LTspice commands for simulating the circuit shown in Fig. 2.31a are listed below.

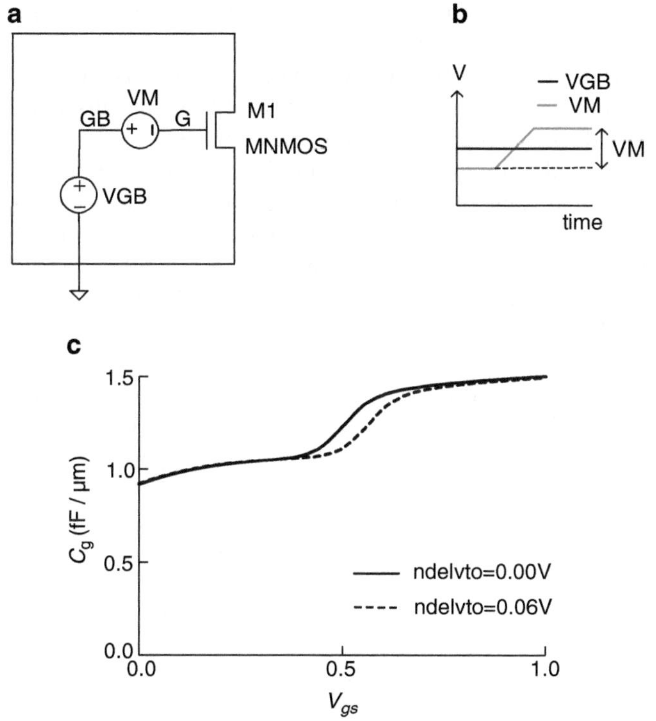

Fig. 2.31 (**a**) Circuit schematic for simulating C_g–V_{gs} characteristics of an n-FET, (**b**) voltage levels for VGB and VM, and (**c**) C_g vs. V_{gs} for an n-FET for nominal V_{tn} and with $\Delta|V_{tn}| = 0.06$ V. 45 nm PTM HP models at 25 °C

```
* netlist and commands for measuring n-FET Cg in fF/µm at Vgs = pvdd
.VGB GB 0 =pvdd
.VM  GB G pulse(-0.005 0.005 0 100e-12 100e-12 5e-9 10e-9 1)
.step param pvdd 0 1.0 0.05
.measure tran i_charge_rise integ i(VM) from=0 to 100E-12
.measure cg param= 'i_charge_rise/0.01/nw*1e15'
.tran 1E-9
```

Simulation results for a 45 nm HP n-FET with $W_n = 1.0$ μm, and $L_p = 0.045$ μm are shown in Fig. 2.31c. The plots compare C_g vs. V_{gs} for a nominal V_t n-FET with an n-FET whose V_t is raised by 0.06 V. The C_g values in the off- and on-states for the two cases are the same but the transition region for the higher V_t n-FET is shifted to the right as expected.

The circuit configuration described in Fig. 2.31a is similar to the arrangement used for measuring C_g of MOSFETs in silicon. These measurements are used for model building and for monitoring the process in the manufacturing line.

A circuit for directly obtaining a plot of C_g vs. V_{gs} in an LTspice waveform viewer is shown in Fig. 2.32. The voltage of the VG power supply is increased linearly from 0 to V_{DD} with VD = 0. The charging current is measured through the zero-voltage source VM. As C_g is charged to 1.0 V, the instantaneous current through VM is directly proportional to the instantaneous value of C_g, while V_{gs} is linearly proportional to time. By integrating the charging current through VM and dividing by the final value of VGB, the average switching capacitance is obtained.

In logic gates, when an n-FET is in the on-state, its V_{ds} = "0" and when it is in the off-state, its V_{ds} = "1". The circuit schematic in Fig. 2.32a can also be used for measuring n-FET capacitance under different switching conditions. The rising waveform for V_{gs} and the simultaneously falling waveform for V_{ds} are shown in Fig. 2.32b. This imitates the case of an n-FET in an inverter as its output switches from a "1" to a "0" (PD transition). In Fig. 2.32c, C_g–V_{gs} characteristics during an inverter PD transition and with V_{ds} = 0 are compared. Clearly the capacitance during switching, with V_{gs} and V_{ds} transitioning in opposite sense, is higher due to the Miller effect.

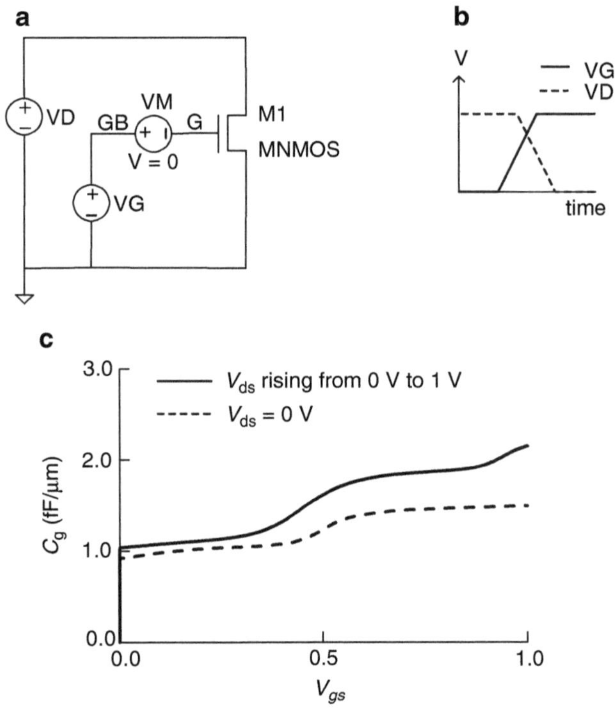

Fig. 2.32 (**a**) Circuit schematic for simulating C_g–V_{gs} characteristics of an n-FET with V_{gs} rising and V_{ds} falling, (**b**) voltage levels for VG and VD, and (**c**) C_g vs. V_{gs} for an n-FET during a PD transition and with V_{ds} = 0. 45 nm PTM HP models at 25 °C

The C_g values of an n-FET and a p-FET from different simulation conditions are summarized in Table 2.9. The average capacitance during a switching transition of an inverter is nearly equal to the on-state capacitance if V_{gs} and V_{ds} are changing in the opposite sense and ~20 % lower if $V_{ds} = 0$. Average gate capacitances for different V_{gs}, V_{ds}, and V_{bs} waveforms may be extracted in a similar fashion.

Table 2.9 C_g for n-FET and p-FET at static points and averaged over different transitions. 45 nm PTM HP models @ 25 °C

Method	$\lvert V_{gs}\rvert$ (V)	$\lvert V_{ds}\rvert$ (V)	n-FET C_{gn}(fF/μm)	p-FET C_{gp} (fF/μm)
Static	0.0	0.0	0.99	0.91
Static	1.0	0.0	1.49	1.53
Average	0.0–1.0	0.0	1.24	1.21
Average	0.0–1.0	1.0–0.0	1.53	1.51

A circuit schematic for obtaining g_m and g_{ds} of an n-FET from simulations is shown in Fig. 2.33. For measuring g_m, I_{ds} is measured with a DC bias applied to the gate and modulated by VMG at a fixed V_{ds} (=VDB). For measuring g_{ds}, I_{ds} is measured with a DC bias applied to the drain and modulated by VMD at a fixed V_{gs} (=VGB). LTspice commands for determining g_m are listed below:

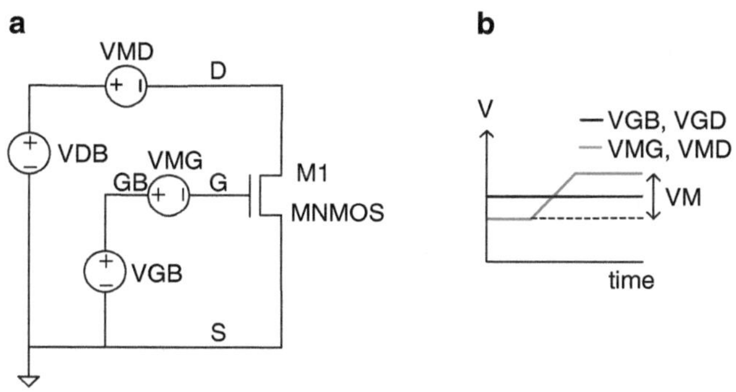

Fig. 2.33 (**a**) Circuit schematic, and (**b**) voltage waveforms for simulating n-FET transconductance g_m, and output conductance, g_{ds}

```
* voltage sources and gm measurement
.VMG GB G PULSE(-0.01 0.01 100e-12 20e-12 20e-12 5e-9 10e-9 1)
.measure tran ids1 avg id(M1) from= 0E-12 to =50E-12
.measure tran ids2 avg id(M1) from =150E-12 to =200E-12
.measure del_ids param=(ids2 -ids1)
.measure gm param = del_ids/0.02*1e6
```

MOSFETs for analog circuits may be engineered with longer channel lengths than high-performance MOSFETs and are described by different BSIM models.

2.2.3 Standard Cell Library Book Characteristics

In CMOS circuit design, individual logic gates are characterized by measuring the propagation delays through the gate and the output waveform rise and fall times while varying the input signal waveform shape and output capacitance load C_L. This type of characterization can be easily carried out in simulation and is convenient for building a set of lookup tables for gate delays which are then used to compute cycle time for logic paths comprising multiple gates. However, the data obtained from simulation at the gate level cannot be easily validated in hardware. The difficulty of accurately measuring the input and output waveform shapes and propagation delays of the order of a few picoseconds on silicon hardware prohibits direct experimental verification of simulation results on individual gates.

A standard cell library comprises physical layouts, circuit schematics and symbols for commonly used logic gate families and small circuit blocks. There are multiple books of each logic gate type with different strengths (MOSFET widths, width ratios, and V_t values). Netlists of logic books with all of the parasitic resistances and capacitances extracted from physical layouts are included.

Electrical characteristics of each logic book are determined from circuit simulations as shown in Fig. 2.34. The library book X1, represented by a box in Fig. 2.34a, may be a logic gate LG, or a circuit block, and may have an inverted or non-inverted output signal. The book drives a fixed capacitive load of value C_L. A zero-voltage source VI is connected in series with the input node A to measure the current drawn for charging the input capacitance of the logic book. The value of the input capacitance C_{in}, is determined by integrating the current through the voltage source VI $\{\int I(\mathrm{VI})dt\}$ during a transition.

The voltage sources for the power supply and for launching a pulse at input node A are shown in Fig. 2.34b. The voltage waveform of the input pulse is depicted in Fig. 2.34c. After time T_d, the signal rises to V_{DD} or "1" with a transition time from GND to V_{DD} of T_r. After time T_{on}, the signal falls to GND or "0" with a transition time of T_f. The waveform is generated using a PULSE function in SPICE and the signal rises and falls linearly with time. Alternatively other waveform shapes described by exponential or PWL functions may be used. The standard definition of signal rise and fall times τ_r and τ_f is the time for the voltage level to rise or fall between $0.1 \times V_{DD}$ and $0.9 \times V_{DD}$. In case of the linear ramp shown here, $\tau_{ri} = 0.8 \times T_r$ and $\tau_{fi} = 0.8 \times T_f$.

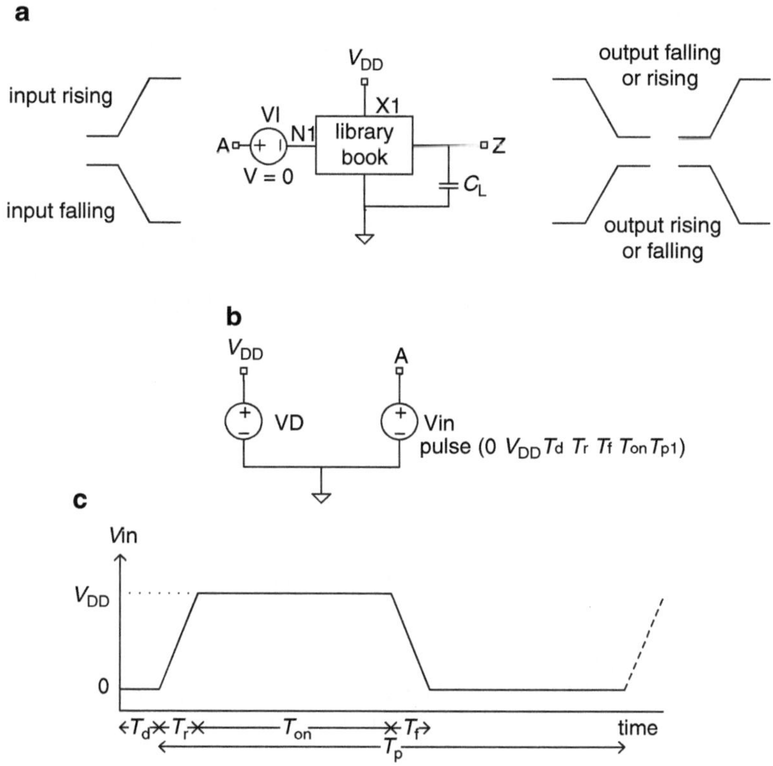

Fig. 2.34 Circuit simulation setup for library book characterization: (**a**) circuit schematic, (**b**) voltage sources and (**c**) input signal waveform for one cycle with T_d = time delay, T_r = rise transition time, T_{on} = on time, T_f = fall transition time, and T_p = pulse period

Circuit simulations are carried out by varying the input rise time τ_{ri}, fall time τ_{fi}, and output load C_L. The propagation delays of the signal through the logic book for pull-down (PD) and pull-up (PU) transitions, τ_{pd} and τ_{pu}, and the corresponding output signal fall time τ_{fo}, and rise time τ_{ro} are measured. The parameter names and corresponding LTspice names are listed in Table 2.10.

An inverter logic book circuit schematic is shown in Fig. 2.35a. The attributes of the n-FET and the p-FET of the inverter are displayed in the schematic. As in the case of MOSFET characterization in Sect. 2.2.2, variables are assigned to the MOSFET properties. The physical layout of an inverter with one PS finger ($nnf = pnf = 1$) is shown in Fig. 2.35b. This design with $W_n = 0.4$ μm and $W_p = 0.6$ μm is used as a "standard" inverter design for characterization using 45 nm PTM HP models. The sum of these widths, $(W_n + W_p) = 1.0$ μm is selected for the convenience of directly expressing capacitance in units of fF/μm. The ratio, $W_p/W_n = 1.5$ is selected to render nearly equal delays for PD and PU transitions. The length of the source and drain regions L_{ds} is 0.12 μm corresponding to a contacted PS pitch of 0.165 μm. The source and drain capacitance values are

Table 2.10 Circuit and LTspice parameter names and descriptions

	Parameter	LTspice parameter	Description
Inputs	τ_{fi}	*tfi*	Input signal fall time
	τ_{ri}	*tri*	Input signal rise time
	C_L	*cl*	Load capacitance
Outputs	τ_{pd}	*tpd*	PD delay
	τ_{pu}	*tpu*	PU delay
	τ_p	*delay_stage*	Average delay = $(\tau_{pu} + \tau_{pu})/2$
	τ_{fo}	*tfo*	Output signal fall time
	τ_{ro}	*tro*	Output signal rise time
	C_{in}	integ *i(vi)/pvdd*	Input capacitance
	IDDQ	*i(vd)* at time $< T_d$	Current in the off-state

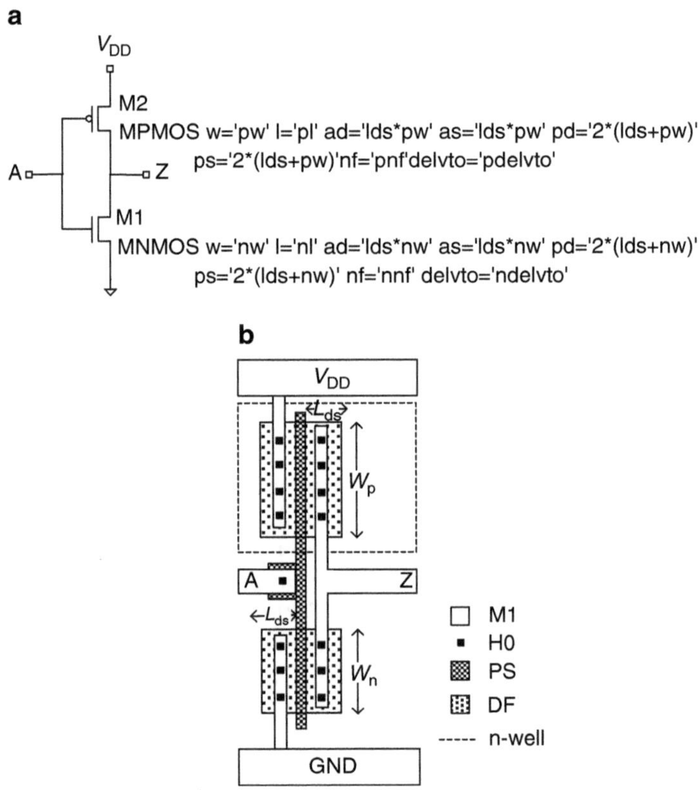

Fig. 2.35 (**a**) Circuit schematic of an inverter with MOSFET attributes, (**b**) physical layout of an inverter with one PS finger

computed during simulations from the BSIM model equations for junction areas and perimeters listed in Table 2.3. Other parasitic resistances and capacitances associated with interconnect wiring are not included in the simulation results described here.

The inverter is defined as a subcircuit (subckt) "inv" and instantiated as X1 in the circuit schematic in Fig. 2.34a. The netlist of "inv" subckt in LTspice is listed below:

```
* netlist for inverter subcircuit
.subckt inv A VDD 0 Z
M1 Z A 0 0 MNMOS w='nw' l='nl' ad='lds*nw' as='lds*nw' pd='2*(lds+nw)'
ps='2*(lds+nw)' nf='nnf' delvto='ndelvto'
M2 Z A VDD VDD MPMOS w='pw' l='pl' ad='lds*pw' as='lds*pw' pd='2*(lds+pw)'
ps='2*(lds+pw)' nf='pnf' delvto='pdelvto'
```

A full netlist with representative parameter values and commands follows:

```
*inverter characterization
*netlist
XX1 N1 VDD 0 Z inv
C1  Z 0 =cl
VI A N1 0
VD VDD 0 =pvdd
Vin A 0 PULSE(0 =pvdd 100e-12 =tr =tf 400e-12 2e-9 1)
*parameter values
.param pvdd=1.0
.temp 25
.param nw=0.4u pw=0.6u lds=0.12u nnf=1 pnf=1
.param nl='(0.045u+ndl)' pl='(0.045u+pdl)' ndl=0.0u pdl=0.0u
.param ndelvto=0 pdelvto= 0
.param tr=20E-12 tf=20E-12
.tran 1E-9
*CL parameter sweeps
.step param cl 0f 15f 3f
*measure τpd, τri, τfo
.measure tran tpd + trig v(A) val=0.5*pvdd rise=1 + targ v(Z) val=0.5*pvdd fall=1
.measure tran tri + trig v(A) val=0.1*pvdd rise=1+ targ v(A) val=0.9*pvdd rise=1
.measure tran tfo + trig v(Z) val=0.1*pvdd fall=1+ targ v(Z) val=0.9*pvdd fall=1
*compute input capacitance Cin in fF for signal rising (PD transition)
.measure tran in_charge_rise integ i(VI) from=100e-12 to=150e-12
.measure Cin_rise param = 'in_charge_rise/pvdd*1e15'
*measure leakage current (IDDQ) in off-state
.measure tran IDDQ avg i(VD)*-1 from=10e-12 to=50E-12
```

Characteristics of the standard inverter PD transitions with input rise times τ_{ri} of 6, 16, and 32 ps, and C_L values of 3 and 15 fF are obtained from circuit simulations using 45 nm PTM HP models at 1.0 V and 25 °C. The findings are recorded in Table 2.11. From a closer inspection of the data in Table 2.11 it is apparent that τ_{pd} increases with τ_{ri} and C_L, while C_{in} remains nearly constant. IDDQ is measured with input A at "0", prior to any transition. Lookup tables of this type for PD and PU transitions over a range of τ_{ri} and C_L are generated using design automation software tools for a wide variety of gates.

Table 2.11 Standard inverter parameters for PD transitions with variable inputs τ_{ri} and C_L. 45 nm PTM HP models @ 1.0 V, 25 °C

τ_{ri} (ps)	C_L (fF)	τ_{pd} (ps)	τ_{fo} (ps)	C_{in} (fF)	IDDQ, A at "0" (nA)
8	3.0	10.3	15.9	1.50	9.35
8	15.0	19.5	34.4	1.51	9.35
32	3.0	15.3	21.5	1.50	9.35
32	15.0	24.9	37.6	1.51	9.35
56	3.0	18.1	28.2	1.50	9.35
56	15.0	30.2	43.4	1.50	9.35

The simulation sets described above are repeated at different values of V_{DD}, temperature, and MOSFET parameters L_p and V_t. In order to limit the number of simulation runs and the size of lookup tables, a few fixed simulation corners are defined as nominal (NOM), worst-case (WC), and best-case (BC) for delays.

Typical scenarios for defining NOM, WC, and BC simulation corners are shown in Table 2.12. Here it is assumed that delay decreases with V_{DD} and increases with temperature, L_p and V_t. Although generally true, the delay trend with temperature may reverse in some technologies. The V_{DD} and temperature values in these corners are based on product specifications and environmental operating conditions. MOSFET parameter values are based on the parameter distributions in the models as illustrated in Table 2.2. Circuit design methodology varies from product to product and simulation corners may be defined at $\pm 2\sigma$, $\pm 3\sigma$ or other suitable parameter values. Additional corners may be defined as needed.

Table 2.12 Three commonly used simulation corners to cover product application conditions and process variations

Simulation corner	V_{DD}	Temperature	L_p	V_{tn}, V_{tp}
Nominal (NOM)	Nominal	Nominal	Nominal	Nominal
Worst-case (WC)	Minimum	Maximum	Maximum	High
Best-case (BC)	Maximum	Minimum	Minimum	Low

The total delay of a logic path comprising different logic gate library books is calculated using the delay of each gate in the lookup tables. C_L is replaced by C_{in} of the following gate, and the output rise/fall time of the preceding gate becomes the input rise/fall time of the following gate. A logic path may be represented as a black box characterized by an input capacitance and output rise/fall times, similar in many respects to Fig. 2.34a. The path may, however, have multiple inputs and outputs. This methodology continues up the circuit hierarchy.

Although lookup tables are extremely useful in timing tools for circuit design, it is difficult to obtain physical insight into the behavior of logic gates and circuits from lengthy tables. In addition, a software bug in generating such tables can introduce timing errors which may require costly fixes if detected later in hardware test. It is therefore important to understand the dependencies of circuit delays and capacitances on the properties of MOSFETs and parasitic elements under different environmental conditions. This knowledge can then be applied to validate timing and power EDA tools with test cases designed to catch nonphysical circuit behavior.

An equivalent *RC* model of a logic gate is helpful in gaining physical insight and in relating gate delay and power characteristics to the constituent MOSFET and parasitic elements. We will demonstrate this with an inverter. This methodology is extended to other logic gates in Chap. 5 for more extensive model-to-hardware correlation of CMOS chip circuitry.

The circuit schematic and an equivalent *RC* circuit for an inverter are shown in Fig. 2.36a and b respectively. The inverter changes state as its internal switch is toggled between two possible positions in response to changes in voltage levels at input A. With input A at "0", output Z is at "1", connected to V_{DD} through R_{swp}.

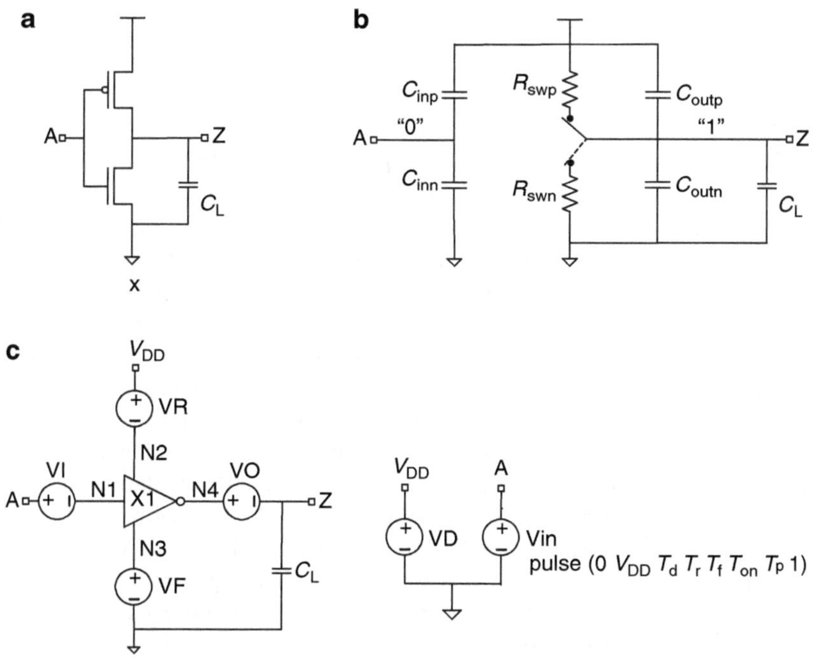

Fig. 2.36 (**a**) Circuit schematic of an inverter driving a capacitive load C_L, (**b**) equivalent *RC* model, and (**c**) circuit simulation setup for inverter characterization

When input A transitions from a "0" to a "1", the switch toggles to its lower position, and output Z is pulled down to a "0" with an *RC* time delay determined by R_{swn} and the total capacitance being discharged. When input A subsequently transitions from a "1" to a "0", the switch closes through R_{swp} connecting Z to V_{DD} and the capacitance is charged. Input capacitances C_{inn} and C_{inp} are charged and discharged through the power supply Vin.

The capacitances C_{in} and C_{out} are the sum of their n-FET and p-FET components:

$$C_{\text{in}} = C_{\text{inn}} + C_{\text{inp}}, \tag{2.15}$$

$$C_{\text{out}} = C_{\text{outn}} + C_{\text{outp}}. \tag{2.16}$$

The PD and PU delays are given by

$$\tau_{\text{pd}} = R_{\text{swn}}(C_{\text{out}} + C_{\text{L}}), \tag{2.17}$$

$$\tau_{\text{pu}} = R_{\text{swp}}(C_{\text{out}} + C_{\text{L}}). \tag{2.18}$$

The average delay τ_{p} is given by

$$\tau_{\text{p}} = \frac{\tau_{\text{pu}} + \tau_{\text{pd}}}{2} = \frac{1}{2}(R_{\text{swn}} + R_{\text{swp}})(C_{\text{out}} + C_{\text{L}}), \tag{2.19}$$

or

$$\tau_{\text{p}} = R_{\text{sw}}(C_{\text{out}} + C_{\text{L}}), \tag{2.20}$$

where R_{sw} is the average switching resistance for PD and PU transitions.

The effective resistances of the n-FET and the p-FET during switching are R_{swn} and R_{swp} and are related to their current drive strengths. Input capacitance C_{in} of the inverter is the sum of gate, overlap and other parasitic capacitances of the n-FET and p-FET denoted by C_{inn} and C_{inp} respectively. On the output side, internal capacitances C_{outn} and C_{outp} of the inverter add to the load. The main contributors to C_{outn} and C_{outp} are the junction capacitances on the drain side of the MOSFETs with additional contributions from gate and overlap capacitances [12].

Keeping the *RC* model in mind, simulations of an inverter are carried out to determine key parameters that characterize its properties and performance. A circuit schematic to determine C_{in}, C_{out}, R_{swn}, R_{swp}, and their relationships to gate delay and to the properties of the constituent MOSFETs is shown in Fig. 2.36c. Zero-voltage sources VR, VF, and VO are included in addition to VI introduced in the schematic in Fig. 2.34a. The charging and discharging currents through these sources are integrated over time to obtain the corresponding capacitances. The charging current for total output capacitance $(C_{\text{L}} + C_{\text{out}})$ flows through VR during a PU transition and the discharging current for $(C_{\text{L}} + C_{\text{out}})$ during a PD transition flows through VF. A small additional current flowing between V_{DD} and GND when both n-FET and p-FET are momentarily "on" during switching adds to the extracted capacitance. This is discussed in more detail in Sect. 4.3.2.

C_{in} is determined by integrating the charging current through zero-voltage source VI. Similarly, C_L can also be measured from the charging current through VO. As C_L is an input parameter in the simulation runs, its measured value may be compared with the input value to validate the simulation setup and SPICE commands. R_{swn} and R_{swp} are calculated from measured τ_{pd}, τ_{pu}, and capacitances using Eqs. 2.17 and 2.18. The average delay τ_p and average switching resistance R_{sw} are also computed from the measured data.

Circuit simulations are carried out while varying input fall and rise times, τ_{fi} and τ_{ri}, capacitive load C_L, total n-FET and p-FET width ($W_n + W_p$), and width ratio W_p/W_n. All simulation runs are made at the nominal V_{DD} of 1.0 V for 45 nm PTM HP models. Instead of viewing the simulation results in a table, the data are displayed graphically in Figs. 2.37, 2.38, 2.39, 2.40, 2.41, 2.42, and 2.43. Linear equations describing a least square fit to the data and the regression coefficient R^2 (Sect. 9.2) indicating the goodness of fit are displayed on the plots wherever applicable.

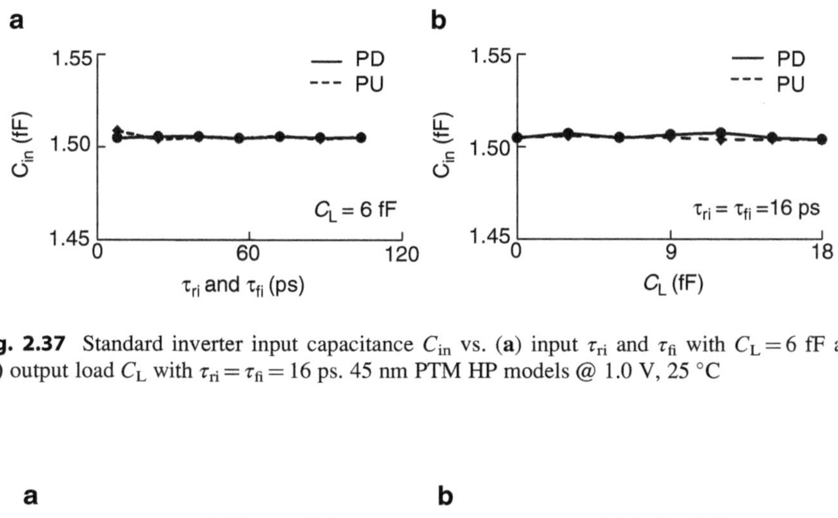

Fig. 2.37 Standard inverter input capacitance C_{in} vs. (**a**) input τ_{ri} and τ_{fi} with $C_L = 6$ fF and (**b**) output load C_L with $\tau_{ri} = \tau_{fi} = 16$ ps. 45 nm PTM HP models @ 1.0 V, 25 °C

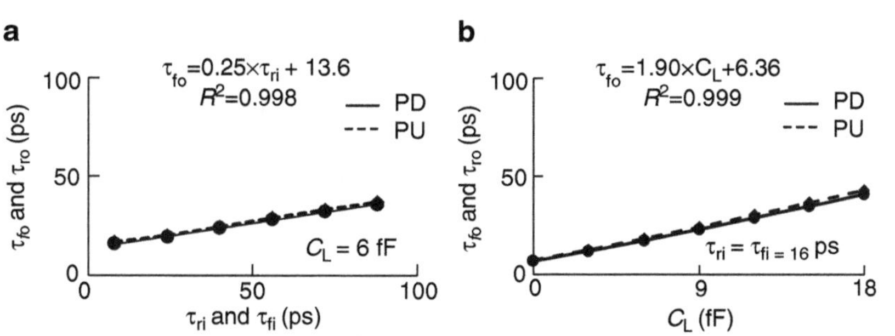

Fig. 2.38 Standard inverter output τ_{fo} and τ_{ro} vs. (**a**) input τ_{ri} and τ_{fi} with $C_L = 6$ fF and (**b**) output load C_L with $\tau_{ri} = \tau_{fi} = 16$ ps. 45 nm PTM HP models @ 1.0 V, 25 °C

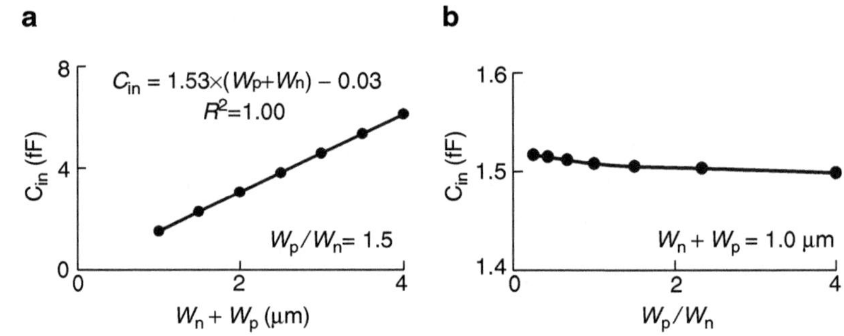

Fig. 2.39 Inverter C_{in} as a function of: (**a**) $(W_n + W_p)$ with $W_p/W_n = 1.5$ and (**b**) W_p/W_n with $(W_n + W_p) = 1.0$ µm. 45 nm PTM HP models @ 1.0 V, 25 °C

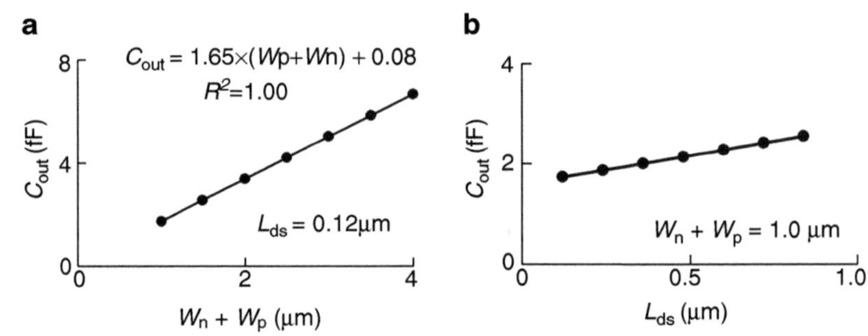

Fig. 2.40 Inverter C_{out} as a function of (**a**) $(W_n + W_p)$, with $W_p/W_n = 1.5$ and $L_{ds} = 0.12$ µm and (**b**) L_{ds} with $(W_n + W_p) = 1.0$ µm. 45 nm PTM HP models @ 1.0 V, 25 °C

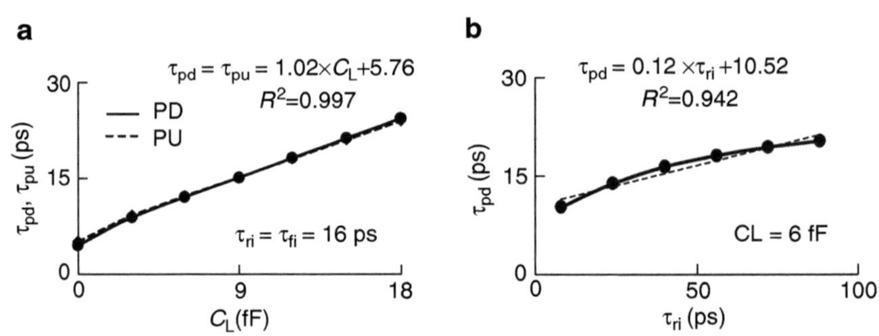

Fig. 2.41 Standard inverter: (**a**) signal propagation delays, τ_{pd} and τ_{pu} vs. C_L and (**b**) τ_{pd} vs. τ_{ri} for $C_L = 6$ fF. 45 nm PTM HP models @ 1.0 V, 25 °C

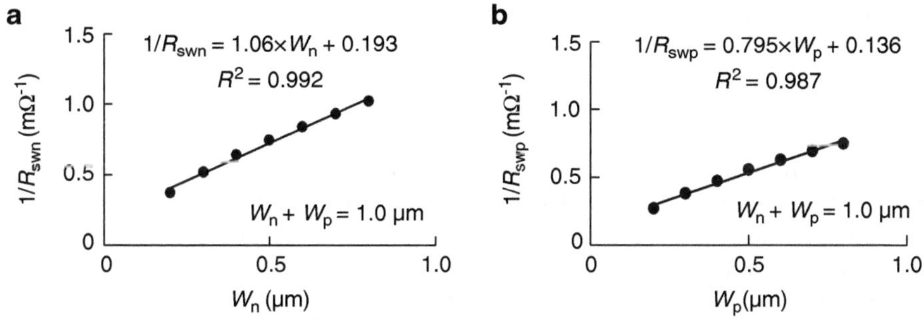

Fig. 2.42 Inverter (FO = 4): (**a**) $1/R_{swn}$ as a function of W_n and (**b**) $1/R_{swp}$ as a function of W_p. 45 nm PTM HP models @ 1.0 V, 25 °C

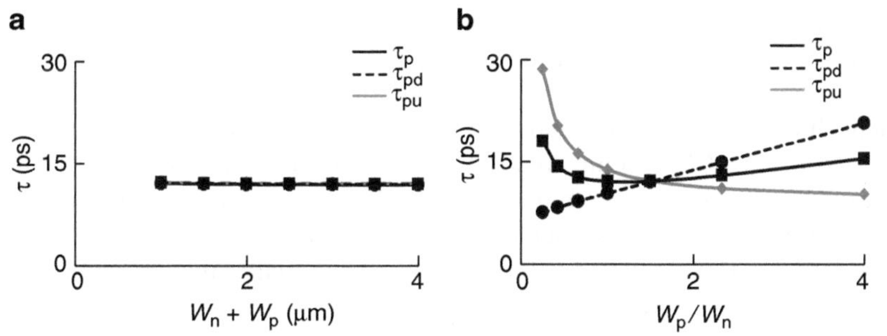

Fig. 2.43 τ_p, τ_{pd}, and τ_{pu} of inverter (FO = 4) with $\tau_{fi} = \tau_{ri} = 16$ ps: (**a**) vs. $(W_n + W_p)$ and (**b**) vs. W_p/W_n. 45 nm PTM HP models @ 1.0 V, 25 °C

In Fig. 2.37a, C_{in} is plotted as a function of τ_{fi} and τ_{ri} with C_L fixed at 6 fF. In Fig. 2.37b, C_{in} is plotted as a function of C_L with τ_{fi} and τ_{ri} fixed at 16 ps. From the plots we can see that C_{in} is nearly constant (~1.51 fF) over a wide range of τ_{fi} and τ_{ri}, and C_L. Since C_{in} is the total gate capacitance of the inverter MOSFETs, an approximate value of C_{in} can be obtained from the equation below:

$$C_{in} = C_{gn} \times W_n + C_{gp} \times W_p. \tag{2.21}$$

With $W_n = 0.4$ μm and $W_p = 0.6$ μm in this inverter design and using n-FET and p-FET C_g values listed in Table 2.9 for the inverter PD case (V_{gs} rising, V_{ds} falling), $C_{in} = (1.53 \times 0.4 + 1.51 \times 0.6) = 1.52$ fF, close to the value of 1.51 fF in Fig. 2.37a. Further simplification is made by assuming $C_{gp} = C_{gn}$ as a first approximation, and C_{in} expressed as

$$C_{in} = c_{in}(W_n + W_p), \tag{2.22}$$

where c_{in} is the normalized input capacitance per unit width. A normalized output capacitance per unit width c_{out} is defined in a similar fashion.

Knowing C_{in}, C_{L} is calibrated in units of fanout (FO), where

$$\text{FO} = \frac{C_{\text{L}}}{C_{\text{in}}}. \tag{2.23}$$

Approximating C_{in} as 1.5 fF, a load capacitance $C_{\text{L}} = 6$ fF is equivalent to FO = 4. In CMOS technology evaluation and in circuit design an inverter FO = 3 or FO = 4 is used as a reference. Following this general practice, in many of the examples C_{L} is set at 6 fF for a standard inverter (FO = 4) design.

In, Fig. 2.38a output parameters τ_{fo} and τ_{ro} are plotted as a function of τ_{ri} and τ_{fi} with C_{L} fixed at 6 fF. In Fig. 2.38b, τ_{fo} and τ_{ro} are plotted as a function of C_{L} with τ_{ri} and τ_{fi} fixed at 16 ps. The range of C_{L} covers FO = 0 (unloaded) to FO = 12. Simulated data points are fit to linear equations to estimate the output parameters from known input parameters.

Linear superposition of these equations allows approximate estimations of τ_{fo} and τ_{ro} at any combination of τ_{fi}, τ_{ri}, and C_{L}. As an example, with $\tau_{\text{ri}} = 30$ ps and $C_{\text{L}} = 12$ fF, a τ_{fo} estimation of 32.5 ps $\{= 0.25 \times 30 + 13.6 + 1.9 \times (12 - 6)\}$ matches the simulated value of 32 ps within the accuracy required in circuit designs. Typically, for noise and delay considerations, maximum allowed values of τ_{fo} and τ_{ro} on a CMOS chip are limited to one-third of the clock cycle time. For a clock cycle time of 250 ps (frequency = 4 GHz), τ_{fo} and τ_{ro} are limited to 83 ps. Hence, with $\tau_{\text{ri}} = 83$ ps, this standard inverter can drive a load of ~33 fF (FO = 21) for $\tau_{\text{fo}} = 83$ ps. The simulated value of τ_{fo} is 77 ps, smaller than estimated because of small deviations from linearity.

Next, the inverter *RC* parameters are related to inverter design dimensions. All the simulations are done with $\tau_{\text{fi}} = \tau_{\text{ri}} = 16$ ps and $C_{\text{L}} = 6$ fF (FO = 4) except when these input parameters are the variables being studied.

In Fig. 2.39a, C_{in} is plotted as a function of $(W_{\text{n}} + W_{\text{p}})$ while maintaining $W_{\text{p}}/W_{\text{n}} = 1.5$ and still using a one finger inverter design, ignoring any width bias in the model. C_{in} increases with $(W_{\text{n}} + W_{\text{p}})$ as expected from Eq. 2.22. A linear fit of the simulated data gives $C_{\text{in}} \sim 1.53 \times (W_{\text{n}} + W_{\text{p}})$ fF, and $c_{\text{in}} = 1.53$ fF/μm.

In Fig. 2.39b, C_{in} is plotted as a function of $W_{\text{p}}/W_{\text{n}}$ while keeping $(W_{\text{n}} + W_{\text{p}}) =$ 1.0 μm. There is a small decrease in C_{in} as $W_{\text{p}}/W_{\text{n}}$ increases. From Table 2.9, C_{gn} (=1.53 fF/μm) is slightly larger than C_{gp} (=1.51 fF/μm). Hence, from Eq. 2.21, C_{in} is expected to decrease with increase in $W_{\text{p}}/W_{\text{n}}$.

The inverter output capacitance C_{out} is determined from the difference between the charging currents through voltage sources VF (output falling) or VR (output rising), and VO. Its dependence on source and drain diffusion region areas is seen by plotting C_{out} as a function of $(W_{\text{n}} + W_{\text{p}})$ and L_{ds} in Fig. 2.40. Linear fits of the data provide expressions to compute C_{out} for any value of $(W_{\text{n}} + W_{\text{p}})$ or L_{ds}. As in the case of C_{in}, for a fixed L_{ds}, C_{out} is also proportional to $(W_{\text{n}} + W_{\text{p}})$.

In Fig. 2.41a, PD and PU delays τ_{pd} and τ_{pu} are plotted as a function of C_{L}. The delays increase linearly with C_{L} at the rate of 1.02 ps/fF which corresponds to ~1.54 ps/FO for $\tau_{\text{ri}} = \tau_{\text{fi}} = 16$ ps. The slope of this plot in delay/FO defines the logical effort of a gate. The concept of logical effort for determining gate delays has been described in engineering textbooks [8, 11]. Here we have extended the delay model to explore the dependencies of gate delays on MOSFET properties and input signal waveform shapes.

In Fig. 2.41b, τ_{pd} is plotted as a function of input rise time τ_{ri}. The delay increases with increase in τ_{ri}, and the behavior can be approximated as a linear dependence. From the linear fit over the range simulated, the estimated inverter delay τ_p (FO = 4, C_L = 6 fF), with $\tau_{ri} = \tau_{fi} = 16$ ps, is 11.88 ps. The results obtained from simulations for these precise input values are $\tau_p = 12.14$ ps and $\tau_{fo} = 17.4$ ps.

Effective switching resistances R_{swn} and R_{swp} are calculated from the measured delays and capacitances using Eqs. 2.17 and 2.18. R_{swn} and R_{swp} decrease with increase in the corresponding current drives and in turn MOSFET widths. In Fig. 2.42a, $1/R_{swn}$ is plotted as a function of W_n, and in Fig. 2.42b $1/R_{swp}$ is plotted as a function of W_p. The dependence of $1/R_{swn}$ and $1/R_{swp}$ on corresponding MOSFET widths can also be described by linear equations displayed on the plots. Thus $R_{swn} \times W_n$ and $R_{swp} \times W_p$ are nearly constant for practical MOSFET finger widths.

Rewriting Eq. 2.19 and expressing C_L in terms of FO gives

$$\tau_p = \frac{1}{2}\left(R_{swn} + R_{swp}\right)\left(\text{FO} \times C_{in} + C_{out}\right). \tag{2.24}$$

It is convenient to introduce normalized parameters r_{swn} and r_{swp},

$$r_{swn} = R_{swn} \times W_n, \tag{2.25}$$

$$r_{swp} = R_{swp} \times W_p. \tag{2.26}$$

When MOSFET widths are selected to give $\tau_{pd} \approx \tau_{pu}$, $R_{swn} = R_{swp}$, and

$$\tau_p \approx r_{sw} \times (\text{FO} \times c_{in} + c_{out}), \tag{2.27}$$

where r_{sw} $\{=R_{sw} \times (W_n + W_p)\}$ in Ω-μm is the average switching resistance normalized to total width. Following Eqs. 2.22 and 2.27, the average delay τ_p is nearly independent of $(W_n + W_p)$.

In Fig. 2.43a, τ_{pd}, τ_{pu}, and τ_p are plotted as functions of $(W_n + W_p)$, with $W_p/W_n = 1.5$, and the load C_L adjusted for FO = 4 (= $4 \times c_{in}[W_n + W_p]$). The delays are nearly independent of $(W_n + W_p)$ as predicted by Eq. 2.27. In Fig. 2.43b, τ_{pd}, τ_{pu}, and τ_p are plotted as a function of W_p/W_n, with $(W_n + W_p) = 1.0$ μm. In this case the r_{sw} value in Eq. 2.27 varies with W_n and W_p excursions from that needed to maintain $\tau_{pd} \approx \tau_{pu}$. The PD delay increases with W_p/W_n (W_n decreasing, R_{swn} increasing), and PU delay increases with decreasing W_p/W_n (W_p decreasing, R_{swp} increasing). The average delay τ_p, increases as W_p/W_n is allowed to move away from its ideal value of 1.5. The variation of τ_p with W_p/W_n around this value is fairly small.

The R_{sw} of a logic gate is the average of its PD and PU resistances during switching and inversely proportional to its average current drive capability. In a PD transition R_{swn} is related to the I_{on} or I_{eff} of the n-FET and in a PU transition

R_{swp} is related to the I_{on} or I_{eff} of the p-FET. In Fig. 2.44, $1/R_{swn}$ vs. I_{effn} and $1/R_{swp}$ vs. I_{effp} for an inverter (FO = 4) are plotted. Variations in R_{swn} and R_{swp} are obtained by varying W_p/W_n at constant $(W_n + W_p)$. I_{effn} and I_{effp} for the corresponding widths are obtained from I_{ds}–V_{ds} characterization of MOSFETs as described in Sect. 2.2.2.

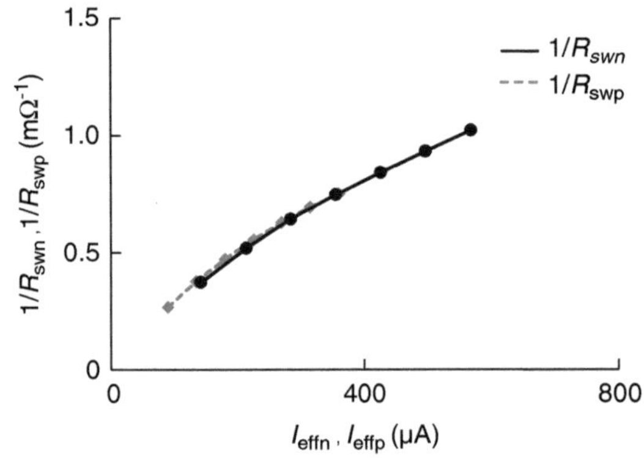

Fig. 2.44 Inverter (FO = 4) $1/R_{swn}$ vs. I_{effn}, and $1/R_{swp}$ vs. I_{effp} obtained by varying W_p/W_n with $(W_n + W_p) = 1.0$ μm. 45 nm PTM HP models @ 1.0 V, 25 °C

It is instructive to track I_{ds} of MOSFETs during a switching transition. Let us consider an n-FET in an inverter undergoing a PD transition. Initially, the inverter input voltage or V_{gs} of the n-FET is "0", and the inverter output voltage or V_{ds} of the n-FET is "1". The n-FET is in the off-state, with $I_{ds} = I_{off}$. As the input voltage transitions from "0" to "1", n-FET V_{gs} and I_{ds} increase while V_{ds} is decreasing. As the n-FET passes through the saturation region, its I_{ds} begins to decrease with V_{ds} ultimately going to zero as the transition is completed. The instantaneous resistance of the n-FET, at time t, $V_{ds}(t)/I_{ds}(t)$, averaged over the switching time is equivalent to the R_{swn} of the inverter.

By overlaying the $I_{ds}(t)$–$V_{ds}(t)$ trajectory on the DC I_{ds}–V_{ds} characteristics of a MOSFET, $I_{ds}(t)$ values can be compared to the MOSFET parameters listed in Table 2.7. Such plots for an n-FET and a p-FET during PD and PU transitions are shown in Fig. 2.45. I_{ds} values in the plots are normalized to the n-FET and p-FET widths for direct visual comparison. The $I_{ds}(t)$ trajectory passes near the I_{lo} and I_{hi} locations for $V_{ds} > V_{DD}/2$.

From Fig. 2.45 it is apparent that the average current during switching is better approximated as $I_{eff} = (I_{hi} + I_{lo})/2$ than as I_{on}. As I_{eff} is inversely proportional to R_{sw}, by selecting $W_p/W_n = I_{effn}/I_{effp}$, $R_{swp} \approx R_{swn}$, and $\tau_{pu} \approx \tau_{pd}$. As an example, the I_{eff} (n/p) ratio from Table 2.8 is 1.58, slightly larger than the W_p/W_n value of 1.50 in our standard inverter.

The increase in τ_{pd} with τ_{ri} is explained by comparing I_{ds} trajectories of the n-FET in the inverter for $\tau_{ri} = 16$ and 80 ps in Fig. 2.46a. As τ_{ri} increases, V_{gs} increases more gradually with time and the current drive of the n-FET is lowered. As a result both τ_{pd} and τ_{fo} increase. Similar plots for n-FET trajectories for

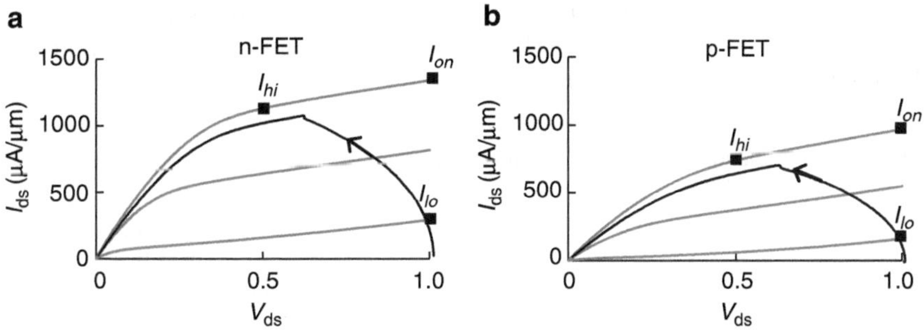

Fig. 2.45 Normalized I_{ds}–V_{ds} characteristics and inverter (FO = 4) trajectory during switching for (**a**) an n-FET, and (**b**) a p-FET. 45 nm PTM HP models @ 1.0 V, 25 °C

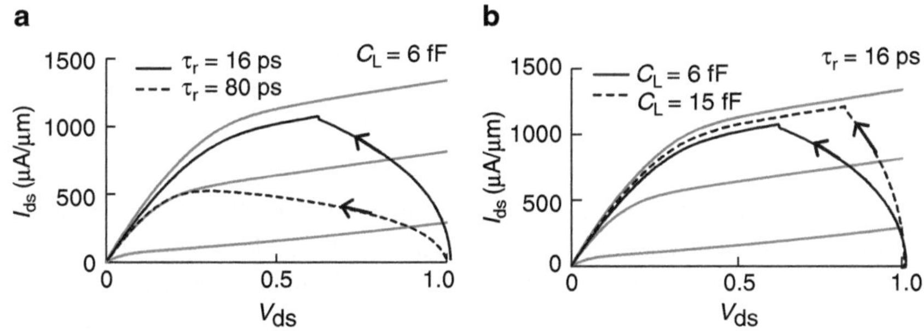

Fig. 2.46 (**a**) I_{ds}–V_{ds} characteristics and the inverter (FO = 4) trajectory during switching for an n-FET (**a**) with τ_{ri} = 16 and 80 ps, and (**b**) with C_L = 6 and 15 fF. 45 nm PTM HP models @ 1.0 V, 25 °C

C_L = 6 fF (FO = 4) and 15 fF (FO = 10) with $I_{ds}(t)$ increasing with C_L are shown in Fig. 2.46b.

With the circuit simulation results described above we can see that the highly nonlinear behavior of MOSFETs is transformed into a set of circuit parameters that are either independent of other parameters or have a near linear dependence on them. These observations are stated below:

- C_{in} is independent of input τ_{ri}, τ_{fi} and output load C_L
- C_{in} is proportional to $(W_n + W_p)$
- C_{out} is proportional to $(W_n + W_p)$ and L_{ds}
- τ_p, τ_{pu}, and τ_{pd} vary linearly with C_L and FO
- τ_{pu} varies linearly with τ_{fi}, and τ_{pd} varies linearly with τ_{ri}
- τ_{pu} and τ_{pd} are nearly equal when $W_p/W_n = I_{effn}/I_{effp}$
- $1R_{swn}$ in a PD transition varies linearly with W_n and I_{effn}
- $1/R_{swp}$ in a PU transition varies linearly with W_p and I_{effp}

Delay parameters of our standard inverter, with $W_n = 0.4$ μm and $W_p = 0.6$ μm and driving a fixed load C_L of 6 fF (FO = 4), obtained from circuit simulations using 45 nm PTM HP models are shown in Table 2.13.

Table 2.13 Simulation results for the standard inverter book. 45 nm PTM HP models @ 1.0 V, 25 °C

Parameter	Nominal value	Variation with τ_{ri}	Variation with C_L	Variation with $(W_n + W_p)$, $W_p/W_n = 1.5$
c_{in}	1.50 fF/μm	None	None	Very weak
c_{out}	1.73 fF/μm	None	None	Very weak
r_{sw}	1,574 Ω μm	Weak	Weak	Weak
τ_{pd}	12.1 ps	~Linear	~Linear	Weak
τ_{pu}	12.2 ps	~Linear	~Linear	Weak
τ_{fo}	17.4 ps	~Linear	~Linear	Weak
τ_{ro}	18.2 ps	~Linear	~Linear	Weak

Nominal values are for standard inverter $W_n = 0.4$ μm and $W_p = 0.6$ μm with $C_L = 6$ fF (FO = 4) and $\tau_{ri} = \tau_{fi} = 16$ ps. Dependencies of delay parameters with τ_{ri}, C_L, and $(W_n + W_p)$ are indicated

The current drawn by the inverter in its quiescent state, IDDQ, is dependent on MOSFET leakage currents and the input node voltage. Measurements of standby and active power are covered in Sects. 2.2.4 and 2.2.5, and in more detail in Chap. 4.

2.2.4 Delay Chains

The signal propagation delay through a circuit can be increased to a few hundred picoseconds or more by connecting a number of logic gates in series. Delays in this range can be measured on silicon for model-to-hardware correlation with an accuracy of a few %. If all the logic gates in the chain are nominally identical, the average delay τ_p through a single gate is obtained by dividing the total delay across the chain by the number of gates.

The circuit schematic in Fig. 2.47 shows four inverters with different values of $(W_n + W_p)$ and W_p/W_n connected in series. An input signal with a rise time of τ_{ri1} at A1 is propagated through inverter X1 with a load capacitance equal to input capacitance C_{in2} of inverter X2. Output signal fall time τ_{fo1} and the propagation delay through X1, τ_{pd1}, are determined by the characteristics of inverter X1, τ_{ri1} and C_{in2}. The falling signal at A2 propagates through the second inverter X2. The output rise time τ_{ro3} and delay τ_{pu2} through X2 are determined by the characteristics of inverter X2, τ_{fi2} $(=\tau_{fo1})$ and C_{in3}. The process of alternate stages going through PD and PU transitions continues until the signal reaches the end of the chain. The delay across the chain τ, is given by

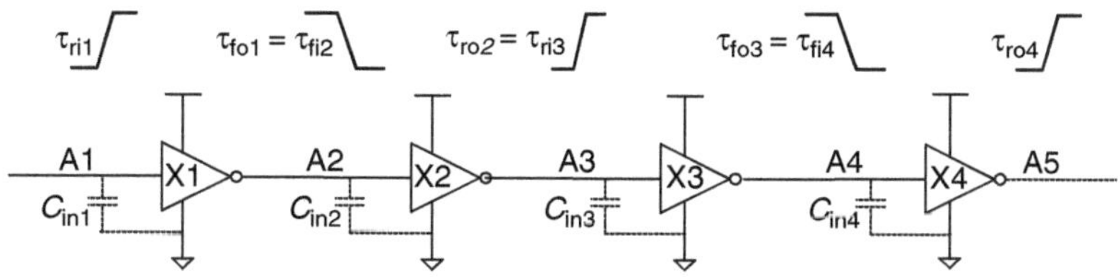

Fig. 2.47 Four inverters of different designs in a delay chain. Input and output waveforms at each node and input capacitances are indicated

$$\tau = \tau_{pd1} + \tau_{pu2} + \tau_{pd3} + \tau_{pu4} + \cdots. \tag{2.28}$$

The value of τ can be calculated from lookup tables similar to Table 2.11 for each inverter as explained in the previous section. Alternatively, a circuit simulation can be carried out treating the entire chain as a single unit. However, measurement of τ in the hardware does not provide sufficient information to characterize each inverter in the chain individually.

If all the inverters in a long delay chain are identical, the capacitive loads on each inverter are also identical. The input signal waveform shape and MOSFET widths of the inverter may be carefully adjusted such that the rise times and fall times at the inputs of alternating stages are nearly the same ($\tau_{ri1} = \tau_{fi2} = \tau_{ri3} = \cdots$). In such an arrangement $\tau_{pd1} = \tau_{pu2} = \tau_{pd3} = \cdots$. The average delay τ_p ($= \tau_{pd} = \tau_{pu}$) across one inverter is obtained by dividing τ by the number of inverter stages in the chain. This approach may be used for any logic gate with inverting or non-inverting characteristics.

The load on a logic gate stage (LG), in addition to the C_{in} of the subsequent stage in a chain, may be realized in several different ways. A fixed capacitive load C_L at the output of each stage can represent a desired FO following the relationship in Eq. 2.23. For a correct representation of a logic gate driving other logic gates, the fixed capacitor is replaced with additional logic gates operating as capacitive loads.

The circuit schematic corresponding to true logic gate FO is compacted with the use of a current multiplier circuit shown in Fig. 2.48. The charging current during switching through zero-voltage source VX, I(VX), is measured. The value of a current source connected between the output of the logic gate and GND terminal is set as (XL − 1) × I(VX). This circuit behaves as if the total gate load being driven is XL times the capacitive load of the single logic gate at the output node. The parameter XL (=FO) is a user-defined variable.

The three schemes described above for adding a capacitive load at the output of a logic gate are shown in Fig. 2.49. For a pure C_L load in Fig. 2.49a, the value of C_{in} for an LG has to be determined for each gate type and each simulation condition (V_{DD}, temperature, L_p, V_t, etc.) to get equivalent FO. Schematics of one LG driving two additional identical LGs for FO = 3 is shown in Fig. 2.49b. In this representation, a schematic is created for a specific FO value and C_L of equivalent FO load added to the load LGs. The schematic with an XL load in Fig. 2.49c is compact and its FO value is set as a variable so that the same schematic may be used for simulating LGs with different FOs.

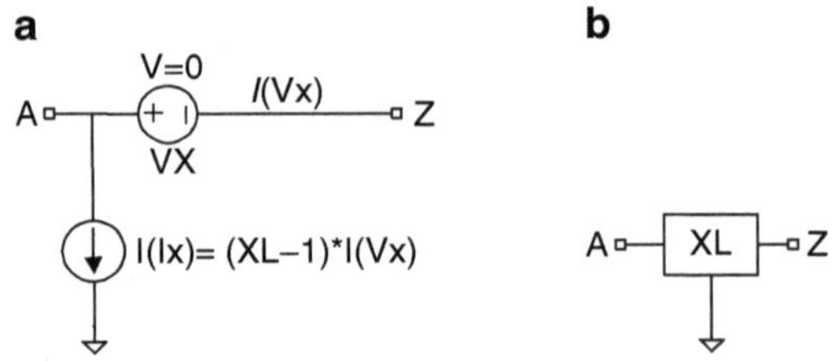

Fig. 2.48 A capacitance load multiplier (**a**) circuit schematic and (**b**) symbol

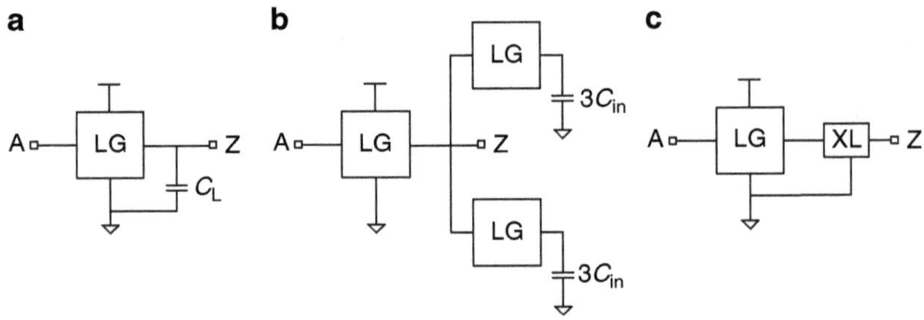

Fig. 2.49 Circuit schematics of logic gates (LGs) driving an additional capacitive load: (**a**) fixed capacitive load C_L, (**b**) logic gate loads for FO = 3 and (**c**) variable XL load

The schematic of a delay chain may be further compacted by creating a hierarchy. The schematic of one stage in the delay chain comprising a logic gate LG, load current multiplier XL, and a fixed capacitive load C_L to represent interconnect capacitance is shown in Fig. 2.50a. This subcircuit (subckt) is represented by a symbol LGXL in Fig. 2.50b. The sources to provide the power supply voltage V_{DD} and the voltage signal pulse at node A are shown in Fig. 2.50c. A delay chain with ten identical LGXL stages is shown in Fig. 2.50d. This hierarchical schematic is less cluttered and provides the option of defining MOSFET parameters, XL and C_L in LTspice commands at the top level. The schematic may be used as a template in which different types of logic gates are instantiated.

If the logic gate LG has more than one input, the additional inputs are tied to a "1" or "0" voltage level such that the output of the LG has the desired level (inverting or non-inverting). The power supply nodes in each stage are vdd and vss and may be connected to V_{DD} and GND respectively or biased at different voltage levels.

The circuit schematic for stage 7 in Fig. 2.50d is expanded to show the wiring of this stage. A zero-voltage source V7 is placed at the output of the LG in stage 7 to measure the charging and discharging current during a transition for computing the load capacitance. Another zero-voltage source V7S is placed in the GND connection of this stage to measure the total discharging current during a PD transition. The charging current during a PU transition can be measured by inserting a third zero-voltage source (not shown) in the vdd connection to stage 7.

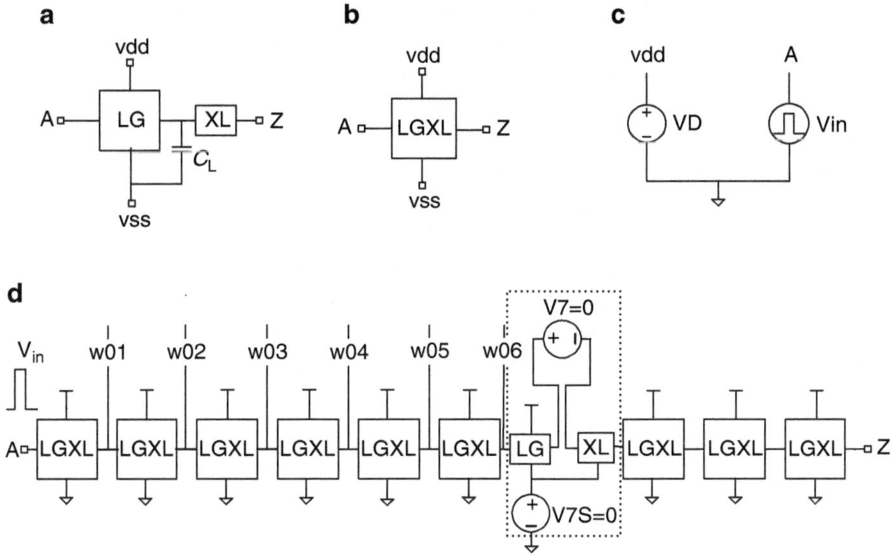

Fig. 2.50 (**a**) Circuit schematic of a logic gate with load multiplier XL and interconnect capacitance C_L, (**b**) symbol of the stage LGXL in (**a**), (**c**) DC power supply and signal source to enable signal propagation through the delay chain, and (**d**) a chain of 10 LGXL stages with zero-voltage sources added in stage 7

A single pulse is launched at the input A of the delay chain comprising identical inverters. The rise and fall times may be different as the signal traverses the delay chain. A typical scenario with standard inverter stages is shown in Fig. 2.51 with $\tau_{ri} = 80$ ps at input A. At the output of the first stage, node w01, $\tau_{fo1} = 36$ ps, at the output of the second stage, node w02, $\tau_{ro2} = 24$ ps, at the output of the third stage, node w03, $\tau_{fo3} = 21$ ps, and at w04, $\tau_{ro4} = 20$ ps. The signal waveform shape stabilizes after the first three stages and $\tau_r \approx \tau_f$. If there is no load on the last stage in the chain at node Z, $\tau_{roz} \approx \tau_{foz} \approx 8.5$ ps. This has a small impact on the delay of the previous stage in the chain.

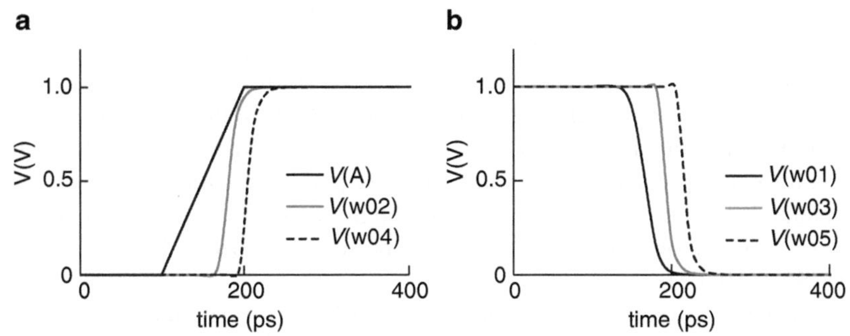

Fig. 2.51 Signal waveforms in an inverter FO = 4 delay chain at nodes shown in Fig. 2.50: (**a**) rising edges at A, w02 and w04 and (**b**) falling edges at w01, w03 and w05. 45 nm PTM HP models @ 1.0 V, 25 °C

In the middle of the chain, the output waveforms for all PD transitions are identical. Similarly, output waveforms of all PU transitions are identical. This holds for any value of W_p/W_n in the inverter, an inverter design with $\tau_{pd} = \tau_{pu}$ being a special case. The characteristics of a logic gate in the middle of a chain are equivalent to those of the single logic gate described in Sect. 2.2.3 for the special case of τ_{ri}, τ_{fi}, and C_L (=XL × C_{in}) being matched.

Circuit simulations are carried out to measure signal propagation delays through stages 5 and 6 to avoid any end-effects. The waveforms at nodes w04, w05, and w06 are shown in Fig. 2.52. The delay across a stage is measured at signal levels of $V_{DD}/2$ at the input and output of a stage. The signal rise and fall times are measured between signal levels of 10 and 90 % of V_{DD}. PD and PU delays are measured across stage 5 (nodes w04 and w05). The delay between nodes w04 and w06 gives

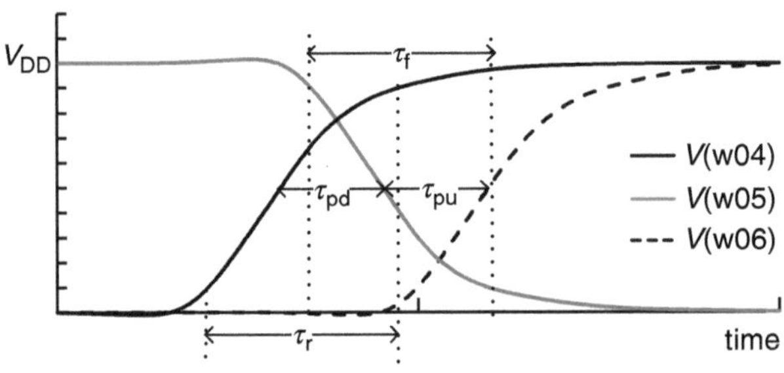

Fig. 2.52 Voltage waveforms at nodes w04, w05, and w06 and measured parameters τ_{pd}, τ_{pu}, τ_r, and τ_f. 45 nm PTM HP models @ 1.0 V, 25 °C

$$2\tau_p = \left(\tau_{pu} + \tau_{pd}\right). \tag{2.29}$$

Using Eqs. 2.20 and 2.24, average delay τ_p, with any FO (=XL) and additional capacitive load C_L is expressed as

$$\tau_p = R_{sw}(\text{FO} \times C_{in} + C_{out} + C_L). \tag{2.30}$$

The parameter FO × C_{in} is determined by integrating the charging current through zero-voltage source V7 in stage 7. The total switching capacitance, C_{sw} (= FO × C_{in} + C_{out} + C_L), is measured by integrating the discharging current for a PD transition through the zero-voltage source V7S. With $C_L = 0$, C_{in}, C_{out}, and R_{sw} can be determined for the PD transition. Similarly, by measuring the charging current through V_{DD} connection to stage 7, these parameters for a PU transition may be obtained. Interconnect wire load of length l is modeled as a capacitance ($C_L = C_w l$) for short wires and as an RC network shown in Fig. 2.13 for long wires (Sect. 10.3.2).

The leakage current in the quiescent state, IDDQ, is measured prior to the arrival of the first pulse edge. An average IDDQ per stage (IDDQ/stage) is obtained by dividing the measured IDDQ by the number of stages (=10 in this example). This is an average current of inverters with inputs a "0" and at "1". Subthreshold leakage current contributions to IDDQ are from an n-FET when the inverter input is a "0" and a p-FET when its input is a "1". Gate-dielectric leakage contributions may become significant in some CMOS technologies and add to IDDQ (Sect. 4.2.1)

A netlist of the delay chain circuit shown in Fig. 2.50 can be generated from the schematic. LTspice commands for measuring τ_p, C_{in}, and IDDQ are listed below:

***delay chain voltage sources and measurements**

```
.Vin A 0 PULSE(0 =pvdd 100e-12 20e-12 20e-12 400e-12 1000e-12 1)
.param XL =4
.measure tran tpd+trig v(w04) val=0.5*pvdd rise=1+ targ v(w05) val=0.5*pvdd fall=1
.measure tran tpu+trig v(w04) val=0.5*pvdd fall=1+ targ v(w05) val=0.5*pvdd rise=1
.measure tran tprr+trig v(w04) val=0.5*pvdd rise=1+targ v(w06) val=0.5*pvdd rise=1
.measure tran i_charge_fall integ I(v7) from=100e-12 to=300e-12
.measure tran i_charge_rise integ i(V7) from=600e-12 to 800e-12
.measure cin param ='(i_charge_rise - i_charge_fall)/(2*pvdd*xl)*1e15'
.measure tp param='tprr/2*1e12'
.measure tran IDDQ avg I(VD)*-1 from=10e-12 to=50E-12
.measure IDDQ_stage param=' IDDQ/10 '
```

Circuit simulations are carried out for the standard inverter with $W_n = 0.4$ μm, $W_p = 0.6$ μm, $L_p = 0.045$ μm, $L_{ds} = 0.12$ μm at nominal V_t values using 45 nm PTM HP models. The MOSFET widths are selected to give equal PD and PU delays at $V_{DD} = 1.0$ V. With $C_L = 0$, parameter XL is used to define FO.

The simulated delay per stage τ_p and the load capacitance as a function of FO are shown in Fig. 2.53. Note that the load increases linearly with FO (=XL) and the increase in delay with FO is 2.13 ps/FO, a value higher than 1.54 ps/FO for the inverter book obtained with $\tau_{ri} = \tau_{fi}$ held constant at 16 ps The load capacitance measured using the zero-voltage source V7, also varies linearly with FO, giving 1.50 fF/FO. Its value is consistent with the simulation results on a single inverter in Sect. 2.2.3. This calibration can be used to compute propagation delay and C_{in} of an inverter for any FO. The value of C_{out} is independent of FO.

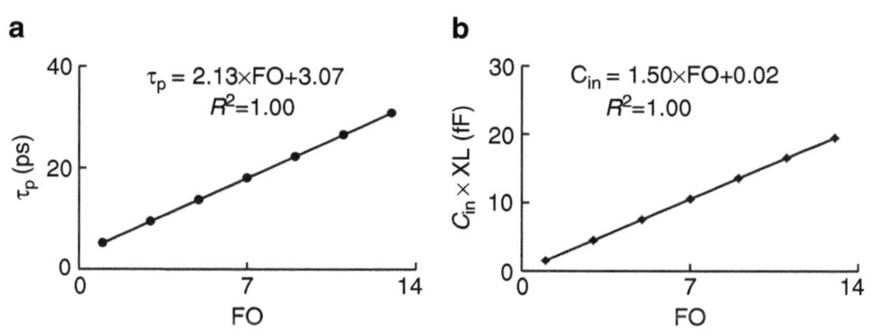

Fig. 2.53 Simulated values of a standard inverter as a function of FO (=XL): (**a**) τ_p and (**b**) load capacitance (=C_{in} × XL). 45 nm PTM HP models @ 1.0 V, 25 °C

From the known value of C_{out} (=1.83 fF for our standard inverter), R_{sw} is determined using Eq. 2.30. R_{sw} is plotted vs. FO in Fig. 2.54a. As FO increases, R_{sw} becomes nearly constant. The normalized I_{ds}–V_{ds} trajectories of the n-FET during a PD transition in the inverter with FO = 1 and FO = 13, superimposed on n-FET I_{ds}–V_{ds} DC characteristics are shown in Fig. 2.54b. These trajectories are slightly different than those shown in Fig. 2.46b because of differences in input and output waveforms for the transitions and the C–V characteristics of the load capacitance.

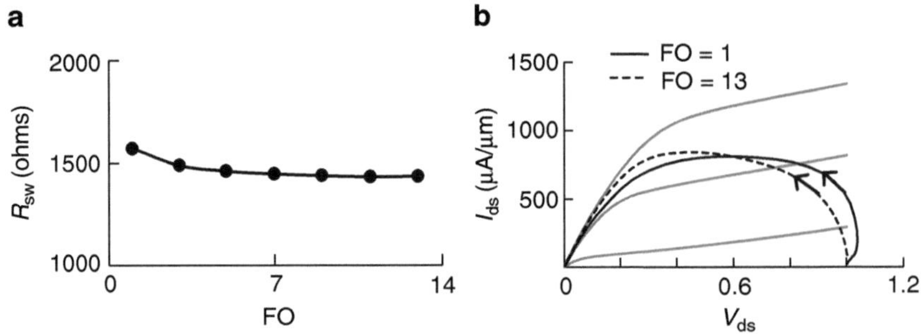

Fig. 2.54 (**a**) R_{sw} vs. FO, and (**b**) inverter n-FET I_{ds}–V_{ds} trajectory during switching. 45 nm HP models @ 1.0 V, 25 °C

Variations in MOSFET parameters impact signal waveforms, propagation delay, and power during switching and in the quiescent state. The two parameters with the largest impact are channel length L_p and threshold voltage V_t. Other parameters such as overlap capacitances of the MOSFETs, gate-dielectric thickness, diffusion area capacitance, and source-drain series resistance also introduce variability in circuit delays.

In Fig. 2.55a, average inverter (FO = 4) delay τ_p is plotted as a function of L_p for the $\pm 3\sigma L_p$ range of 0.0405 to 0.0495 μm. In Fig. 2.55b, τ_p is plotted as a function of the shift in V_t from the nominal $\Delta|V_t|$ of both n-FET and p-FET over the $\pm 3\sigma V_t$

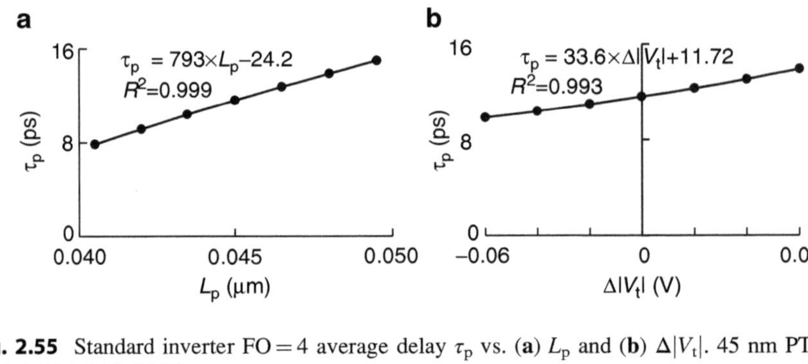

Fig. 2.55 Standard inverter FO = 4 average delay τ_p vs. (**a**) L_p and (**b**) $\Delta|V_t|$. 45 nm PTM HP models @ 1.0 V, 25 °C

range. Data are fit to get a linear relationship between τ_p and L_p and τ_p and $\Delta|V_t|$. The equations displayed in Fig. 2.55a, b are used for hand calculations of τ_p for any value of L_p or $\Delta|V_t|$ under the simulation conditions of $V_{DD} = 1.0$ V, 25 °C.

The τ_p value increases from 7.8 ps for minimum L_p to 15.0 ps for maximum L_p, a 1.92× increase in delay over the full range L_p. This is comparable to 1.88× increase in ($W_n \times I_{effn} + W_p \times I_{effp}$) over the same L_p range. Similarly, τ_p increases from 9.8 to 13.9 ps, a 1.41× increase over the full $\Delta|V_t|$ range compared with 1.49× increase in ($W_n \times I_{effn} + W_p \times I_{effp}$). Hence, the spread in ($W_n \times I_{effn} + W_p \times I_{effp}$) gives a rough estimate of the spread in τ_p.

The variations in C_{in} and IDDQ/stage as a function of L_p are shown in Fig. 2.56a, b respectively. Over the $\pm 3\sigma$ L_p range, IDDQ varies by >40× whereas C_{in} remains within ±4 %. Change in IDDQ affects the off-state leakage power, an important parameter in low-power electronic applications and in high-performance applications where the off-state power is a significant fraction of total power.

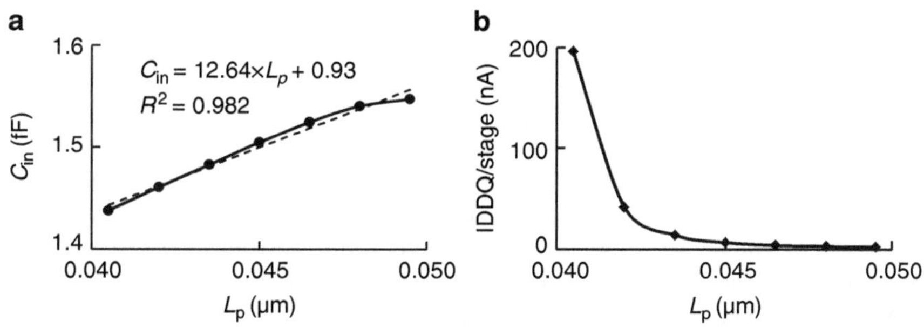

Fig. 2.56 Standard inverter (FO = 4): (**a**) C_{in} vs. L_p and (**b**) IDDQ/stage vs. L_p. 45 nm PTM HP models @ 1.0 V, 25 °C

2.2.5 Ring Oscillators

Although a delay chain is very useful for obtaining model-based logic gate delays, it is a nontrivial task to measure the signal propagation delays in a chain embedded in a CMOS chip with a high degree of accuracy. As described in Chap. 5, delay chain measurements require clock signals, latches, and control circuitry to capture voltage signals and facilitate relative timing measurements.

An alternate method to measure logic gate delays in silicon is to use ring oscillators. A ring oscillator (RO) is formed by connecting the output of a delay chain to its input, thus forming a closed loop. The signal edge launched at any node of the RO travels around the loop continuously. As a result the voltage at any node of the ring oscillates with a period proportional to the signal propagation delay through the circuit stages.

The circuit schematic of a ring oscillator with 51 inverting stages is shown in Fig. 2.57a. It comprises five instances of the 10-stage delay chain block, LGXL10, shown in Fig. 2.50d. A 2-input NAND gate is included in the loop to enable and disable the oscillations with the aid of an external input signal at node EBL.

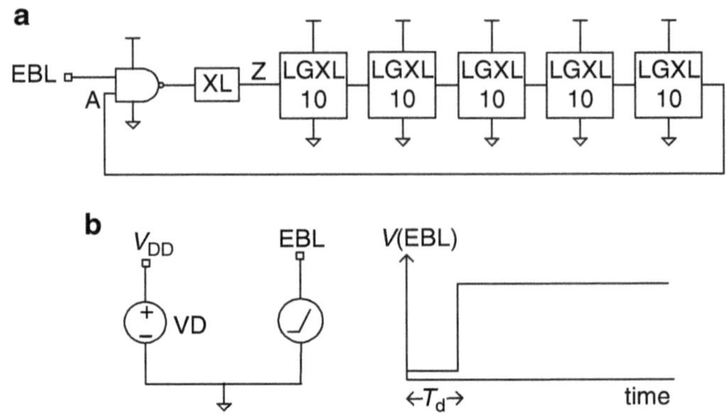

Fig. 2.57 (**a**) Circuit schematic of a ring oscillator with 51 inverting stages: each LGXL10 is a chain of 10 inverting logic gate stages with FO = XL, (**b**) voltage sources to power and enable oscillations

The signal input at node EBL is shown in Fig. 2.57b. At time $<T_d$, the input level at node EBL is a "0" and the output node of the NAND2, Z is a "1" as is node A. If the gates in the chain follow inverting logic, the voltage level at the output of each gate alternates between "0" and "1". At time $=T_d$, the voltage at EBL rises to a "1" and node Z switches to a "0". This signal edge falling from a "1" to a "0" at Z travels through the 50 stages and arrives at node A. With the EBL input of the NAND2 at "1", node Z switches to a "1". Again the second signal edge rising from a "0" to a '1" at Z travels through the loop and node Z falls back to "0" again. This cycle is repeated as long as EBL is held at "1". The voltage at any node oscillates as shown in Fig. 2.58a. For sustainable oscillations, the number of inverting stages must be odd.

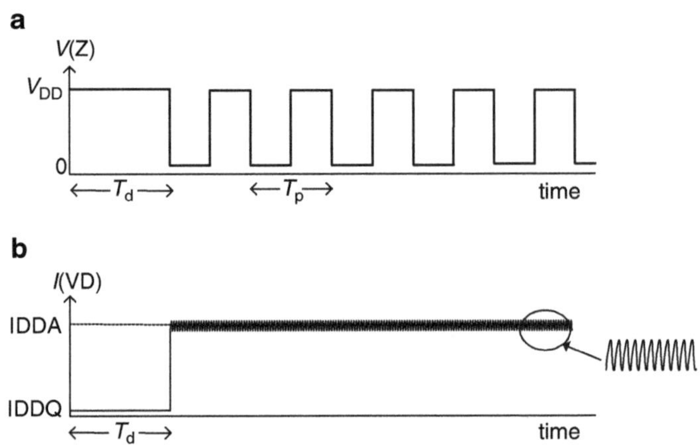

Fig. 2.58 Ring oscillator (**a**) node voltage oscillation, and (**b**) current drawn by the power supply VD in quiescent and active modes. The IDDA ripple frequency corresponds to the stage delay

If the RO comprises $(2\alpha+1)$ inverting stages, the period of oscillation T_p is given by

$$T_p = 2\tau_p(2\alpha+1) = \frac{1}{f}, \tag{2.31}$$

where τ_p is the average of PU and PD delays of a stage, f is the frequency of oscillation and α is an integer. Measurements of f in the MHz range for an RO embedded in a CMOS chip can be easily made with an off-the-shelf frequency counter. Typically RO frequencies for $\alpha < 50$ are in the GHz range. A frequency divider circuit at the output of the RO is used to lower the frequency for a direct measurement. Alternatively, a counter circuit using the on-chip clock as a reference is employed for measuring RO frequencies, and the data streamed out for analysis.

The time variation of current drawn by the RO power supply VD, is shown in Fig. 2.58b. At time $<T_d$, with the RO in its quiescent state, this current is the sum of the leakage currents of all the stages in the RO. Since the input nodes of successive stages are alternating between "0" and "1", the leakage current contributions of successive stages alternate in value. The measured quiescent current IDDQ may be divided by $(2\alpha+1)$ to obtain the average IDDQ/stage. In the oscillating state, with EBL held at "1", current is drawn by each stage as it switches. The magnitude of the current varies on the time scale of the switching transitions (~10 to 100 ps). The current measured externally with a longer integration time is essentially constant. Measurement of currents drawn by an RO in the oscillating state, IDDA, and in the quiescent state, IDDQ, along with frequency of oscillation are used to determine the power drawn during switching and the switching capacitances and resistances of the stages in the RO.

As the signal edge traverses the loop, there is only one stage switching at any instant and all other stages are in the quiescent state. The total current drawn is the sum of switching current I_{sw} of one stage and the IDDQ of the remaining stages. For an RO with $(2\alpha+1)$ stages

$$\text{IDDA} = I_{sw} + \frac{\text{IDDQ}}{\text{stage}} \times 2\alpha. \tag{2.32}$$

The power drawn in the quiescent state is given by

$$P_{off} = \frac{\text{IDDQ}}{\text{stage}} \times (2\alpha+1) \times V_{DD}. \tag{2.33}$$

The AC component of power (ignoring short-circuit power discussed in Section 4.3.2) is

$$P_{\mathrm{sw}} = I_{\mathrm{sw}} \times V_{\mathrm{DD}} = \left(\mathrm{IDDA} - \frac{\mathrm{IDDQ}}{\mathrm{stage}} \times 2\alpha\right) \times V_{\mathrm{DD}}\,. \tag{2.34}$$

From Eq. 2.34 it is apparent that if IDDQ $\ll I_{\mathrm{sw}}$, P_{sw} is essentially independent of the number of stages whereas P_{off} increases with the number of stages. Power is an important consideration when designing stand-alone ROs for model-to-hardware correlation (Sect. 5.2.3) and for technology performance evaluation (Sect. 10.4).

The power consumed during switching can also be expressed in terms of the switching capacitance, C_{sw}

$$P_{\mathrm{sw}} = \frac{1}{2} C_{\mathrm{sw}} V_{\mathrm{DD}}^2 \{2(2\alpha + 1)\} f = \frac{1}{2\tau_{\mathrm{p}}} C_{\mathrm{sw}} V_{\mathrm{DD}}^2\,, \tag{2.35}$$

where C_{sw} is the switching capacitance in each transition and there are $2(2\alpha + 1)f$ transitions per second.

For $\alpha \gg 1$, the factor $2\alpha/(2\alpha + 1) \approx 1$, and using Eqs. 2.31, 2.34, and 2.35, C_{sw} is expressed in terms of measured IDDA and IDDQ as

$$C_{\mathrm{sw}} = \frac{(\mathrm{IDDA} - \mathrm{IDDQ})}{\{2 \times (2\alpha + 1) V_{\mathrm{DD}} f\}} = \frac{(\mathrm{IDDA} - \mathrm{IDDQ})}{V_{\mathrm{DD}}} \times 2\tau_{\mathrm{p}}\,. \tag{2.36}$$

The average switching resistance R_{sw} can thus be estimated from the measured parameters IDDA, IDDQ, and τ_{p} as

$$R_{\mathrm{sw}} = \frac{V_{\mathrm{DD}}}{2 \times (\mathrm{IDDA} - \mathrm{IDDQ})}\,. \tag{2.37}$$

The RO circuit parameters and their relationships to electrical measurements in silicon hardware are summarized in Table 2.14.

Table 2.14 RO circuit parameter names, units, and description

Parameter	Unit	Expression	Description
T_{p}	ps		Period of oscillation
IDDQ	mA	Avg I(VD) at time $< T_{\mathrm{d}}$	Current in quiescent state
IDDA	mA	Avg I(VD) at time $> T_{\mathrm{d}}$	Current in active state
f	MHz	$1/T_{\mathrm{p}}$	Frequency of oscillation
τ_{p}	ps	$T_{\mathrm{p}}/\{2(2\alpha + 1)\}$	Average delay/stage
P_{off}	mW	IDDQ $\times V_{\mathrm{DD}}$	Off-state power
P	mW	IDDA $\times V_{\mathrm{DD}}$	Active power
P_{sw}	mW	(IDDA – IDDQ) $\times V_{\mathrm{DD}}$	Switching power
C_{sw}	fF	$2\tau_{\mathrm{p}} \times$ (IDDA – IDDQ)/V_{DD}	Switching capacitance/stage
R_{sw}	Ω	$V_{\mathrm{DD}}/\{2 \times$ (IDDA – IDDQ)$\}$	Switching resistance of a stage

LTSpice commands for measuring period T_p and signal rise and fall times at any node are listed below. The period of oscillation of the RO is measured in the tenth cycle to allow sufficient time for the oscillations to stabilize. In PD-SOI technology, MOSFET body voltages change with time during switching, and sufficient time must be allowed for the body potentials to equilibrate. Techniques for achieving steady-state conditions in SOI are described elsewhere [17].

```
*commands for RO measurements
.measure tran tp + trig v(Z) val=0.5*pvdd rise=10+ targ v(Z) val=0.5 rise=11
.measure tran tr + trig v(Z)=0.1*pvdd rise=10+ targ v(Z)=0.9*pvdd rise=10
.measure tran tf + trig v(Z)=0.1*pvdd fall=10+ targ v(Z)=0.9*pvdd fall=10
```

In estimating τ_p, C_{sw} and IDDQ per stage, it is assumed that the insertion of a NAND2, or another scheme to enable the oscillations, is equivalent to adding one more identical stage in the loop. Generally, the delay through the NAND2 is different than the delay of a standard stage, and this introduces an error in computing average delay parameters per stage.

There are four ways to reduce the error in computing average delay parameters per stage:

1. Increase the number of stages in the RO to reduce the effect of the NAND2 imbalance
2. Match the NAND2 delay with the LGXL block delay by adjusting XL at the output of the NAND2
3. Replace the NAND2 with a different enable scheme that matches LGXL delay
4. Use circuit simulations to estimate the error correction for the NAND2 for each RO design

Increasing the number of stages in the RO has the advantage that the same circuit schematic template may be used for different LGXL topologies and simulation conditions. However, the simulation time increases with the number of stages.

Measurement of IDDQ and IDDA in silicon requires ROs with independent power supplies. This can be done for RO test structures placed in the scribe-line. In the case of ROs embedded in a CMOS chip and tied to the chip power grid, differential measurement techniques are used to characterize the delay parameters C_{sw} and R_{sw} per stage [16]. These and other considerations in the design and measurement of RO-based test structures for model-to-hardware correlation and for monitoring CMOS technology are covered in Chaps. 5 and 10.

2.2.6 Comparison of Logic Gate Characterization Methods

Each of the three circuit templates for characterization of logic circuit blocks as described in Sect. 2.2.3 (single block), Sect. 2.2.4 (delay chain), and Sect. 2.2.5 (ring oscillator) have unique features. A single circuit block has complete flexibility in its input and output settings in circuit simulations, but the characterization results cannot be directly validated in silicon. A delay chain is useful for both simulations and measurements in silicon. Ring oscillator characteristics are relatively easy to measure in silicon compared with delay chains. Circuit simulations for ROs, however, take considerably longer than for open-ended delay chains. The simulator may take a long time to find a DC operating point for a closed loop, and timing tools generally cannot handle this circuit configuration.

The simulation results obtained from these three approaches may differ slightly. A comparison of simulation results for a single inverter book, an inverter delay chain, and a ring oscillator comprising 50 inverter stages and a NAND2 is shown in Table 2.15. The inverter schematic and design dimensions used for all three methods are identical. The NAND2 gate in the ring oscillator is counted as equivalent to one inverter stage, thereby resulting in small differences in τ_p from the delay chain and RO. Related differences are seen in derived parameters C_{sw} and R_{sw}.

Table 2.15 Simulated parameters for inverter FO = 4, $W_p = 0.6$ μm, $W_n = 0.4$ μm. 45 nm PTM HP models @ 1.0 V, 25 °C

Parameter	Single gate[a]	Delay chain	Ring oscillator
τ_p (ps)	12.13	11.59	11.68
C_{in} (fF)	1.50	1.50	1.51
IDDQ (nA) (FO = 1)	5.95[b]	6.15	6.18
C_{sw} (fF)	7.70	7.87	7.88
R_{sw} (Ω)	1,574	1,471	1,481
$\tau_i \approx \tau_f$ (ps)	~16 to 18	~20	~20

[a]$C_L = 6$ fF (FO = 4 equivalent)
[b]Average IDDQ with inputs at "1" and at "0"

In a single inverter book, the input signal waveform is user defined as a linear ramp (pulse function in LTspice) whereas the input waveform shapes in the delay chain and RO are as shown in Fig. 2.52. Also, the inverter book has a fixed load capacitance whereas in the delay chain and RO the load capacitance is a function of time-varying voltage on the gate terminal during switching (MOSFET gate load). IDDQ/stage values for the delay chain and RO include gate-dielectric leakage current whereas the single inverter does not.

The simulator is able to provide outputs with a high degree of accuracy. However, the method used for simulating circuit characteristics must be clearly specified when comparing different BSIM models. Measurements made on silicon with different test structure designs and test environments add another level of complexity to accurate model-to-hardware correlation.

2.2.7 Monte Carlo Analysis

In Figs. 2.55 and 2.56 the variations of circuit parameters with L_p or $\Delta|V_t|$ are demonstrated by assigning identical values of L_p or $\Delta|V_t|$ to all the n-FETs and p-FETs in the inverter chain. In silicon processing, across chip, across wafer and lot-to-lot variations described in Chap. 6 cause L_p and V_t of n-FETs and p-FETs to vary independently over their $\pm 3\sigma$ ranges. In addition, random dopant fluctuations result in V_t variations in identical MOSFETs on the chip even when placed in close proximity. This random variation in V_t is given by σV_{tr} where

$$\sigma V_{tr} = \frac{A_{vt}}{\sqrt{WL_p}}. \tag{2.38}$$

Here A_{vt} is a technology-dependent constant that is assigned a value of 0.004 V-µm when using 45 nm PTM HP models. The value of σV_{tr} increases as the MOSFET area is reduced.

Circuit simulations for taking into account, systematic process variations and random V_t variations are carried out with Monte Carlo analysis. The simulations are run repeatedly, and in each run randomly picked values are assigned to the variables. The results contain all possible outcomes over the variable ranges, similar to the data for silicon hardware manufactured over a period of time.

Monte Carlo simulations are run by assigning a nominal value to a parameter of interest and its range of variation. If the parameter is normally distributed (Gaussian distribution) then a mean (nominal) and a σ value are assigned. The probability that a specific parameter value is picked follows the probability in the normal distribution. The probability of a variable having a value within $\pm 1\sigma$ is 68.3 % whereas the probability of having a value beyond $\pm 3\sigma$ is 0.3 % (Sect. 9.1.2).

LTspice commands for running Monte Carlo simulations for n-FET V_t adder variable *ndelvto* are listed below.

```
*commands for Monte Carlo simulations with variables ΔVtn (ndelvto)
.param ndelvto={normal(ndelvtonom, ndelvtosigma)}
.function normal(nom, sigma) if (run==1, nom, nom+(gauss(sigma)))
.param ndelvtonom=0
.param ndelvtosigma=0.02
.step param run 1 500 1                    * number of cases=500
```

Here *ndelvto* is a Gaussian (normal) function with a mean of *ndelvtonom* and a standard deviation of *ndelvtosigma*. A value of 0.0 V is assigned to *ndelvtonom* and *ndelvtosigma* = 0.02 V. The 'run' parameter specifies the number of cases to simulate. In the first simulation run, *ndelvto* is assigned the nominal value of 0.0 V. The results of the first run may be compared with a single standard nominal run to validate the simulation setup. In subsequent runs, the value of *ndelvtosigma* is randomly picked from its normal distribution with $\sigma = 0.02$ V.

In LTspice the assignment of variables in a Monte Carlo simulation is dependent on the circuit topology and its hierarchy. Each instance in the schematic is treated as one unit. Constituent circuit components of a unit with the same variable names are assigned identical parameter values. Parameter values assigned to components within each unit are independent of the values assigned within any other unit. The outcome of this feature is illustrated with two different hierarchical arrangements in Fig. 2.59.

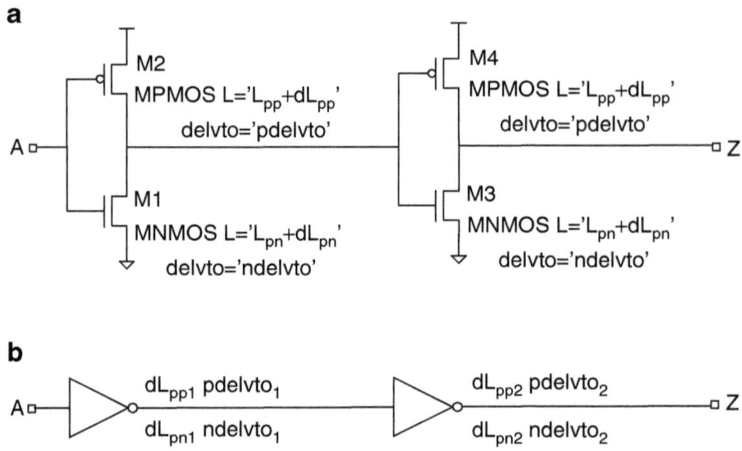

Fig. 2.59 Circuit schematics with different hierarchical arrangements for running Monte Carlo simulations in LTspice: (**a**) all n-FETs (and p-FETs) with identical parameter values in each run, and (**b**) different randomly assigned parameter values for n-FET (and p-FET) in each inverter in each run

In the circuit schematic in Fig. 2.59a, there are two inverters with two n-FETs and two p-FETs. When running Monte Carlo simulations, the randomly picked values for the n-FET V_t adder *ndelvto*, as an example, are identical for both n-FETs. Similarly, both p-FETs are assigned identical V_t values using *pdelvto*. This type of circuit schematic is used for simulating systematic process variations (Sect. 6.4.1). The nominal value *ndelvtonom* is set $=0$, and *ndelvtosigma* $=\sigma V_{ts}$ defines the standard deviation for systematic V_t variations in both n-FETs.

The circuit schematic in Fig. 2.59b shows two inverter symbols at the top level schematic. In Monte Carlo simulations for *ndelvto*, the values of *ndelvto* (*ndelvto*1 and *ndelvto*2) are picked at random for each n-FET in each inverter. This type of schematic is used for simulating a systematic offset in the nominal value (e.g., ΔV_{ts}) along with *ndelvtosigma* $=\sigma V_{tr}$. In this case, all n-FETs in the circuit schematic will have V_t shifted by ΔV_{ts}. In addition, the V_t of each n-FET will be modulated by the random σV_{tr} assignment in a simulation run. For random variation alone, *ndelvtonom* is set $=0$.

The methodology described above applies only to LTspice simulators. Other simulators have different ways of handling random variable assignments when running Monte Carlo simulations. It is prudent to output the assigned values of variables for the circuit components of interest with 100 or more Monte Carlo runs and verify the desired parameter distributions.

In Fig. 2.60 the probability distribution of τ_p for 500 cases for the circuit in Fig. 2.50d, with systematic L_p ($L_{pp} = L_{pn}$), V_{tn} and V_{tp} variations in their respective $\pm 3\sigma$ ranges, is shown. The mean of the distribution is 11.49 ps and $\sigma\tau_p = 1.32$ ps. The maximum and minimum values of τ_p are 15.32 and 7.57 ps respectively. The distribution may get wider when variations in other parameters such as C_{ov} and drain-to-source resistance R_{ds} are included.

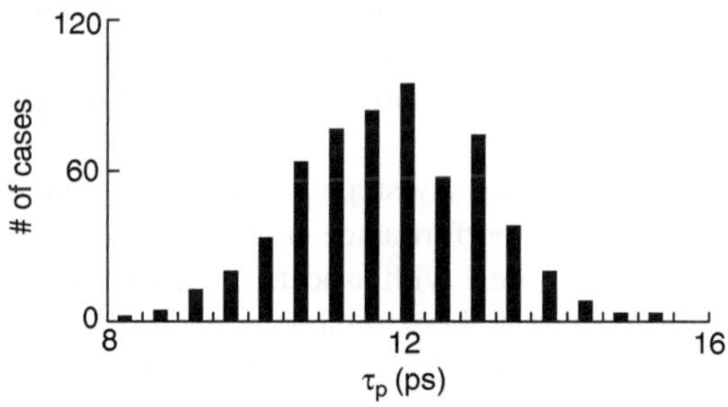

Fig. 2.60 Histograms of average delay τ_p of a standard inverter (FO = 4) for systematic L_p, V_{tn}, and V_{tp} variations. Monte Carlo simulations with 500 cases. 45 nm PTM HP models @ 1.0 V, 25 °C

The number of simulation runs may be reduced dramatically by performing a worst-case analysis. In this case, only values at the extremes of the parameter distributions, (nominal +3σ) and (nominal -3σ), are assigned. The LTspice command for the worst-case function is

.function wc(nom,tol) if (run == 1, nom, if(flat(1)z> 0,nom + tol, nom − tol))

The SPICE function *flat*(1) returns a random number between −1 and +1 and *tol* may be set to the 3σ value of the parameter. The parameter value is (*nom* + *tol*) if *flat*(1) is between −1 and 0, and (*nom-tol*) if *flat*(1) is between 0 and +1.

Monte Carlo simulations serve well to get an insight into the effect of process and other environmental variations on circuit behavior. If the range and shape of parameter distributions are realistic, the simulation results provide a fair representation of measured electrical test data in the hardware.

2.3 Summary and Exercises

The basic properties of CMOS circuit components and logic gates are reviewed. Circuit simulation setups using LTspice and PTM models, as described in this chapter, are used in examples and exercises throughout the book. Key MOSFET I_{ds}

and capacitance parameters are defined and extracted from simulated I–V and C–V characteristics. Parameter values for n-FETs and p-FETs are compared to assess their relative strengths.

Logic gate characterization is exemplified with a standard inverter design for generating a lookup table of signal propagation delays by varying capacitive load C_L and signal input signal rise and fall times (τ_{ri}, τ_{fi}). The dependencies of signal PD and PU delays, input capacitance of the inverter, and output signal rise and fall times on τ_{ri}, τ_{fi}, and C_L are graphically illustrated and analyzed. An equivalent RC model of the inverter is described in terms of average switching resistances and capacitances.

Special features and circuit simulation setups of commonly used test structures comprising delay chains and ring oscillators for measuring logic gate delays and power in silicon are described. Examples are provided to show the variation of average delay with MOSFET parameter variations. A methodology for running Monte Carlo simulations with systematic and random silicon process variations is introduced.

Exercises are designed for step-by-step evaluation of simulation setups, device models, and circuit power and performance, and for cross-checking and validating the results. Exercises 2.1–2.3 deal with aspects of MOSFET characterization, and Exercises 2.4–2.6 with library book characterization. Exercises 2.7–2.10 are designed to set up delay chain and RO circuit simulations as well as Monte Carlo simulations. These circuit simulation configurations are used in many of the examples and exercises in subsequent chapters.

Note: It would be helpful to create a reference table similar to Table A.4 in Appendix A for the available models at the nominal corner

2.1. (a) Simulate n-FET and p-FET DC I_{ds}–V_{ds} characteristics, and generate a table similar to Table 2.7. Do the I_{ds} parameter values match the published values if available? If not, are these within a reasonable range for the corresponding technology node?

(b) Simulated n-FET I_{on} is found to be lower by 1 %, and I_{off} higher by 20 %, than published values. What would you look for in the simulation setup to correct these errors?

(c) If physical layout/parasitic extraction models are available, simulate n-FET and p-FET I_{ds}–V_{ds} characteristics with a netlist including parasitics and calculate the % change due to parasitics in each of the I_{ds} parameters. Which parameters are affected most and why? If parasitic extraction models are not available, add resistances ($=0.02 * V_{DD}/I_{on}$) in series with S, D, and G terminals for this exercise.

2.2. (a) Simulate n-FET C–V characteristics and generate a table similar to Table 2.9.

(b) Simulate C–V characteristics and determine average C_g with (1) D held at V_{DD}, S at GND, and (2) D and S held at V_{DD}. How do you explain the difference in C_g values?

(c) Measure capacitance using the method described in Fig. 2.32 with VD = 0 and VG ramped from GND to V_{DD} and then back to GND. Compare this with the C_g values in the table and explain the difference?

2.3. The mean I_{on} values for isolated n-FETs provided by a silicon foundry are ~5 % higher than the measured values on a test chip designed by the product team. The physical layouts of individual n-FETs in the test chip are identical to the foundry designs except that the test structure is configured to measure 10 n-FETs in parallel on the test chip.

(a) Propose at least four possible causes of this discrepancy?

(b) Recommend additional information/measurements to assist in debug.

2.4. Using the methodology described for library book characterization, the τ_{pd} and τ_{pu} values for a standard inverter for different $\tau_{ri} = \tau_{fi}$ and $C_L = 0.1$ and 1.5 fF are listed in the table below:

(a) Note that τ_{pd} decreases at long rise times whereas τ_{pu} continues to increase. Explain why.

(b) What design or simulation parameters can be changed so that τ_{pd} will continue to increase with τ_{ri}. How does this affect τ_{pu} behavior with τ_{ri}?

(c) Using the data in the table, estimate approximate delay across three inverter books connected in series, each with $C_L = 0.1$ fF, and $\tau_{ri} = \tau_{fi} = 8$ ps?

$\tau_{ri} = \tau_{fi}$ (ps)	τ_{pd} (ps) $C_L = 0.1$ fF	τ_{pu} (ps) $C_L = 0.1$ fF	τ_{pd} (ps) $C_L = 1.5$ fF	τ_{pu} (ps) $C_L = 1.5$ fF
8	4.1	4.3	7.2	7.2
48	5.7	8.3	12.1	14.3
88	5.5	10.5	13.6	18.2
128	4.9	12.4	14.1	21.2
168	4.1	14.0	14.2	23.7
208	3.1	15.5	14.0	25.9
248	2.1	16.9	13.6	28.0

2.5. (a) Design and characterize a NAND3B gate with $(W_n + W_p) = 1.0$ μm and $\tau_{pd} \approx \tau_{pu}$. Obtain plots similar to Figs. 2.37, 2.38, 2.39, 2.40, and 2.41 in the nominal corner.

(b) Set SPICE commands for the following input signal waveforms: (1) linear ramp (pulse function), (2) sine-squared function, (3) output signal of an inverter (FO = 3).

(c) Adjust the waveform parameters to get the same measured τ_r and τ_f.

(d) Measure τ_{pd} and τ_{pu} across the NAND3B using the three different waveform shapes and compare the results.

(e) Using the fitting parameters from the plots from (a), calculate the signal propagation delay from A to Z for the circuit shown below. Compare the calculated delay with simulated delay. The $(W_n + W_p)$ values for each gate in the chain are as indicated, with constant W_p/W_n as sized in (a).

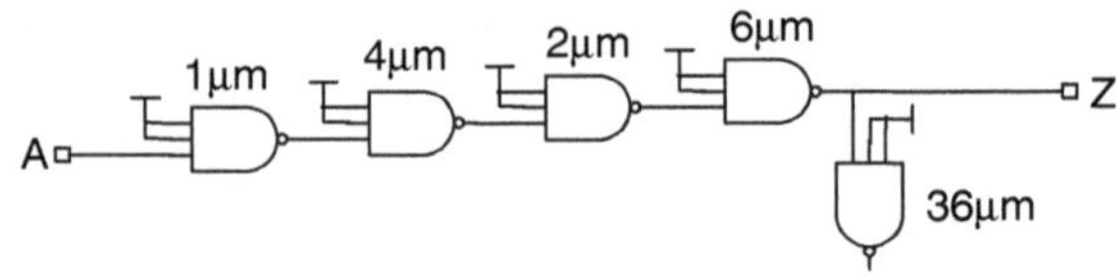

2.6. (a) Estimate C_L/FO for an inverter, NAND2T, NAND3T, and NOR2T. Set $(W_n + W_p) = 1.0$ μm, and W_p/W_n sized to get $\tau_{pd} \approx \tau_{pu}$ for each gate with $\tau_{ri} = \tau_{fi} = 16$ ps.

(b) Set W_p and W_n of non-switching MOSFETs to 0.5× of the value determined in (a). What is the new C_L/FO for these gates? Explain the results.

2.7. (a) Using a delay chain for simulations, design a standard inverter for the model library such that $(W_n + W_p) = 1.0$ μm and $\tau_{pd} \approx \tau_{pu}$ at nominal V_{DD} and 25 °C (nominal corner). How does W_p/W_n compare with I_{onn}/I_{onp} and I_{effn}/I_{effp}?

(b) Measure τ_{pd} and τ_{pu} for this standard inverter at (1) $0.8 \times V_{DD}$, 25 °C, (2) $1.2 \times V_{DD}$, 25 °C, (3) $1.0 \times V_{DD}$, −25 °C and (4) $1.0 \times V_{DD}$, 75 °C. Compare τ_{pd}, τ_{pu}, and τ_{pd}/τ_{pu} in these four simulation corners. Plot τ_{pd} and τ_{pu} as a function of V_{DD} and temperature including the data at $1.0 \times V_{DD}$, 25 °C. Can you use the plots to predict τ_{pd} and τ_{pu} at other V_{DD} and temperature values?

(c) Set *ndelvto* = +0.06 V and *pdelvto* = −0.06 V. Is τ_{pd}/τ_{pu} expected to increase or decrease at nominal V_{DD}? Simulate and compare τ_{pd}/τ_{pu} with the nominal V_t inverter.

2.8. (a) Set up a template for an RO circuit with 50 LGXL stages and a NAND2 enable gate.

(b) Determine τ_p of the NAND3B logic gate used in Exercise 2.5(a) in the RO configuration and compare with previous results.

(c) An RO circuit fails to oscillate in simulation. What are the possible sources of error in (1) RO circuit, and (2) simulation setup?

2.9. (a) Set up ROs with 4, 10, 24, 50, and 100 standard inverter (FO = 3) stages and a NAND2 gate (FO = 1, $W_n + W_p = 1$ μm) enable gate. Determine τ_p and P_{ac} in the nominal corner for each design and compare.

(b) Compare τ_p values of the ROs with the τ_p value from the delay chain simulation in Exercise 2.7.

(c) It is desirable to minimize RO area on silicon. How many stages would you recommend for the RO to reduce error in τ_p to <0.5 %?

2.10. (a) Run Monte Carlo simulations for 500 cases to measure τ_{pd} and τ_{pu} with ±3σ systematic variations (Gaussian distribution) in L_p ($L_{pn} = L_{pp}$), V_{tn}, and V_{tp} for a NAND2T logic gate book designed to have $\tau_{pd} \approx \tau_{pu}$ in the nominal corner.

(b) Plot histograms of L_p, V_{tn}, and V_{tp} measure the range. How do these compare with their corresponding 6σ ranges?

(c) Plot histograms of τ_{pd} and τ_{pu} compare their ranges.

References

1. Linear technology. Design simulation and device models. http://www.linear.com/designtools/software/. Accessed 21 Jul 2014
2. Synopsys HSPICE. http://www.synopsys.com/Tools/Verification/AMSVerification/CircuitSimulation/HSPICE/Pages/default.aspx. Accessed 21 Jul 2014
3. NGSPICE Mixed mode-mixed signal simulator. http://ngspice.sourceforge.net. Accessed 21 Jul 2014
4. BSIM group. http://www-device.eecs.berkeley.edu/bsim/. Accessed 21 Jul 2014
5. Predictive technology models website. http://ptm.asu.edu/latest.html. Accessed 21 Jul 2014
6. Cao Y, Sato T, Sylvester D, Orshansky M, Hu C (2000) New paradigm of predictive MOSFET and interconnect modeling for early circuit design. In: Proceedings of the IEEE 2000 custom integrated circuits conference, pp 201–204
7. Zhao W, Cao Y (2006) New generation of predictive technology model for sub-45 nm early design exploration. IEEE Trans ED-53:2816–2823
8. Weste NH, Harris D (2010) CMOS VLSI design: a circuit and systems perspective, 4th edn. Addison-Wesley, Boston
9. Baker RJ (2010) CMOS circuit design, layout and simulation, 3rd edn. Wiley, Hoboken
10. Rabaey JM, Chandrakasan A, Nikolic B (2003) Digital integrated circuits, 2nd edn. Prentice Hall, Upper Saddle River
11. Sutherland I, Sproull B, Harris D (1999) Logical effort: designing fast CMOS circuits. Morgan Kaufmann, New York
12. Taur Y, Ning TH (2009) Fundamentals of modern VLSI devices, 2nd edn. Cambridge University Press, New York
13. Sze SM (2006) Semiconductor devices: physics and technology, 3rd edn. Wiley, Hoboken
14. Campbell SA (2001) The science and engineering of microelectronic fabrication, 2nd edn. Oxford University Press, Oxford
15. Jaeger RC (2001) Introduction to microelectronic fabrication, vol 5, 2nd edn, Modular series on solid state devices. Prentice Hall, Upper Saddle River
16. Bhushan M, Ketchen MB (2011) Microelectronic test structures for CMOS technology. Springer, Berlin
17. Joshi RV, Kroell K, Chuang CT (2004) A novel technique for steady state analysis for VLSI circuits in partially depleted SOI. In: Proceedings of the 17th international conference on VLSI design, pp 832–836

CMOS Storage Elements and Synchronous Logic

3

Contents

3.1	CMOS Chip Overview	86
	3.1.1 I/O Circuits	87
	3.1.2 Combinational Logic	88
	3.1.3 Clock Generation and Distribution	91
3.2	Sequential Logic and Clocked Storage Elements	93
	3.2.1 Level-Sensitive Latches	95
	3.2.2 Edge-Triggered Flip-Flops	98
	3.2.3 Setup and Hold Times	100
	3.2.4 Register Files	101
3.3	Memory	102
	3.3.1 SRAM	103
	3.3.2 DRAM	108
3.4	Circuit Simulations	109
	3.4.1 SRAM SNM	109
	3.4.2 Logic Data Path	112
3.5	Summary and Exercises	121
References		123

The number of MOSFETs in a single CMOS microprocessor chip exceeded one billion in the year 2010 and this trend has continued. Tracking the behavior of individual MOSFETs on CMOS chips in electrical testing is a daunting task. This task is simplified by following the hierarchical nature of chip architecture. Repetitive patterns in data transactions and in writing and reading data in memory arrays are implemented with a small subset of building blocks. At the next level down in the hierarchy, logic gates, storage elements, and memory cells can be independently characterized and their behaviors related to the properties of their constituent MOSFETs, interconnects, and parasitic resistances and capacitances.

M. Bhushan, M.B. Ketchen, *CMOS Test and Evaluation*,
DOI 10.1007/978-1-4939-1349-7_3

These concepts are illustrated with circuit simulation techniques for directly extracting minimum cycle time of a data path, and noise margins of an SRAM cell.

The electrical properties of MOSFETs, interconnects, and basic logic gates are described in Chap. 2. These elements are used in building more complex functional blocks on CMOS chips. Adders and multipliers perform arithmetic operations and decoders and multiplexers are used for steering signals. Clock signals are used for timing and synchronization of data transactions. Latches and flip-flops form a family of clocked storage elements for temporarily storing data and instruction sets. Arrays of memory cells which can be accessed at random by specifying a cell address are used for storing larger data volumes. Input/output (I/O) circuits connect chip circuitry with the external world. Additional circuitry is added for failure diagnostics of various functions and circuit blocks. Control circuits to tune and optimize chip performance and power are useful in compensating for weaknesses in circuit designs, signal propagation timing errors, variability in silicon processes, and for managing defect induced failures in large memory blocks.

An overview of chip functions, I/O circuits, combinational logic functional blocks, and clock distribution is given in Sect. 3.1. Clocked storage elements including latches and register files are described in Sect. 3.2. SRAM and DRAM memory cells and arrays are covered in Sect. 3.3. Circuit simulation examples are presented in Sect. 3.4. These include determination of noise margins in SRAM cells, and determination of maximum frequency of operation $f_{\max}$ and minimum operating voltage $V_{\min}$ from timing analysis of data paths.

Textbooks on microprocessor chip architecture, CMOS circuits, and memory are excellent resources for overview of chip design and functional blocks [1–4]. A book edited by Xanthopoulos covers clocking strategies and storage elements in modern VLSI systems [5]. Citations for other relevant representative technical literature are included.

3.1 CMOS Chip Overview

CMOS circuitry is in use for a range of applications, from high performance microprocessors to a variety of application specific integrated circuits (ASICs). The majority of the functions implemented can be grouped into a few units. As an illustration, a block diagram of primary functional units of a microprocessor is shown in Fig. 3.1 [1]. In this high level architectural view, instructions are received through I/Os as a string of "1"s and "0"s after powering up the chip. Instruction sets for performing arithmetic logic, addressing memory, and for conditional branching are loaded in the instruction memory and executed on data fetched from registers.

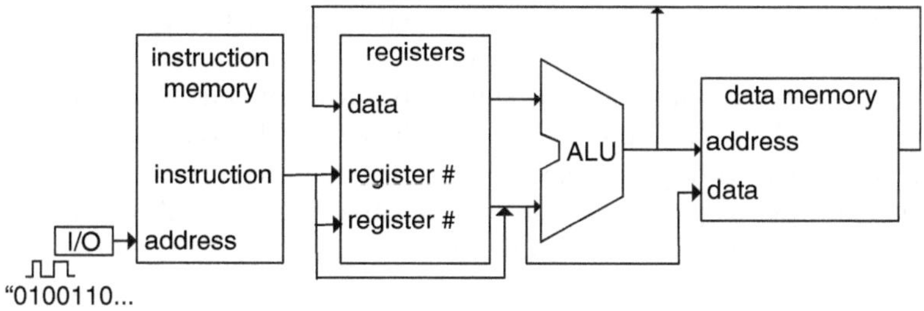

Fig. 3.1 A schematic showing major functional units for receiving and executing instructions and for data storage in a microprocessor unit

There are many common features in the implementation of different instruction sets. An instruction address register or program counter holds the address of the instruction to be fetched from registers. The arithmetic logic unit (ALU) is used for memory address calculations, for arithmetic calculations, for logical operations, and for comparing data from different branches. Data from the memory block are retrieved and loaded in registers as directed. The program counter is then advanced to carry out the next instruction.

From the high level view of Fig. 3.1, it becomes apparent that the functions on the chip can be partitioned into a few classes of operations. Logic circuits are used to carry out arithmetic and branching functions for addressing register files and memory cells and for maintaining integrity of signals travelling across the chip. Data are stored in registers and memory units which comprise regular arrays of unit cells along with control circuitry to address each cell and to read and write data. Clock signals, latches, and flip-flops are used for timing and synchronization of signals on the chip.

There is additional circuitry for I/O drivers and receivers, clock generation and distribution, and other peripheral functions. Analog circuits used in I/Os and phase-locked loops (PLLs) for clock generation are essential for chip operation and need to be operated at different power supply voltage levels. In advanced CMOS technologies several different MOSFET pairs are available for digital logic, memory, and analog applications.

3.1.1 I/O Circuits

Connections from the chip I/Os to the package pins are made through wire bonds to metal pads or through solder balls in a flip-chip configuration. In flip-chip technology, an array of controlled collapse chip connections (C4s) is distributed across the top surface of the chip. The chip is flipped over on the package substrate with metallized areas matching the C4 pattern. The C4s are subsequently bonded to the package by applying heat and pressure.

I/Os for power supply and GND connections to the chip power grid are distributed across the chip to meet current handling and *IR* drop constraints. Some I/Os are used for DC voltage and current measurements. Signals are received and transmitted through I/O circuitry. All signal I/O circuits are designed to provide protection against electrostatic discharge (ESD).

A circuit for receiving external signals is shown in Fig. 3.2a. A dual-diode structure is used for ESD protection. The diodes are forward biased and act as current sinks and voltage clamps. The p^+/n diode suppresses positive voltage transients and the n^+/p diode suppresses negative transients. A resistor in series with the I/O pad serves as a current limiter.

A bidirectional I/O circuit is shown in Fig. 3.2b. With the en_out signal at "0", n-FET N1 and p-FET P1 are in the off-states and offer a high impedance to the I/O pad which can then be used as an input pad by exercising enable signal en_in. With the en_out signal at "1", signal DAT is propagated to the I/O pad.

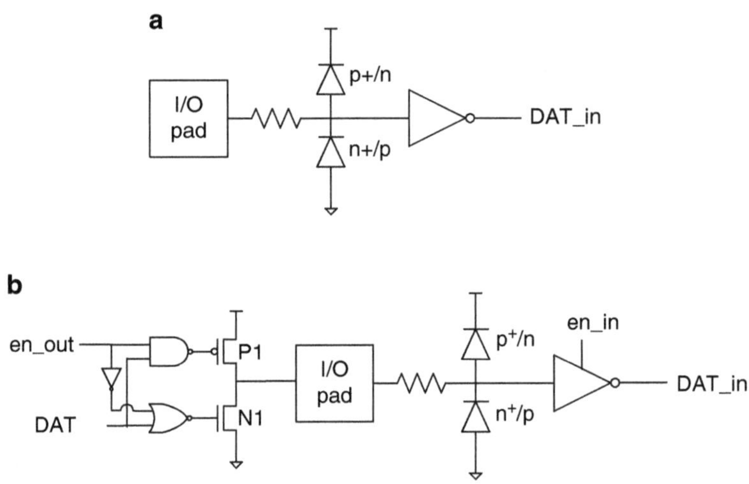

Fig. 3.2 I/O circuits (**a**) for receiving off-chip signals and (**b**) for bi-directional operation

3.1.2 Combinational Logic

Static CMOS logic gates are used for carrying out more complex functions than simple combinations of AND, OR, and INVERT logic described in Chap. 2. Such functions include multiplexing, decoding and adding. Example implementations of these functions are described below to get a flavor of the logic operations and circuit configurations. These functions may be implemented with other circuit topologies and logic families as well.

A multiplexer selects and passes one and only one of its input signals to the output. A 4-input multiplexer circuit constructed with *n*-passgates is shown in Fig. 3.3a. It has two control inputs A0 and A1 and four data inputs E1, E2, E3, and E4. A symbol for this multiplexer is shown in Fig. 3.4b along with a logic truth table in Fig. 3.4c. With A0 and A1 at "0", only the path from E1 to Z has both n-FETs turned on, and the signal from E1 propagates to output Z. With A0 and A1 at "1", only the path from E4 to Z has both n-FETs turned on, and the signal arriving at E4 propagates to Z. The multiplexer may also be used as a de-multiplexer where with control signals A0 and A1, a single input Z may pass to only one of the outputs E1, E2, E3, or E4. With N control signals (A0, A1, . . ., A_N), any one of 2^N signal inputs or outputs may be propagated.

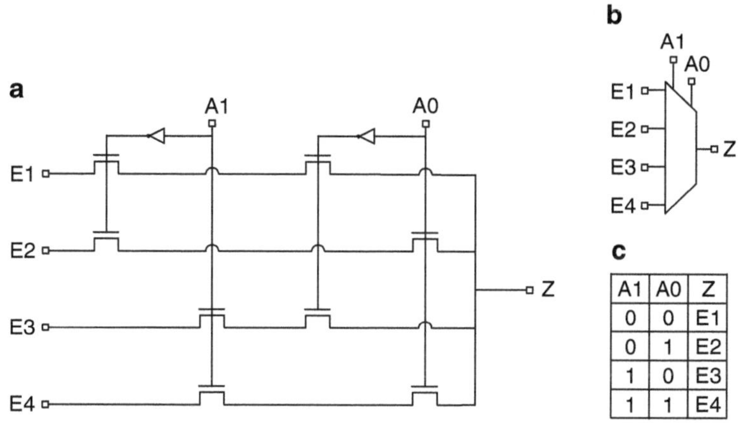

A1	A0	Z
0	0	E1
0	1	E2
1	0	E3
1	1	E4

Fig. 3.3 A 4-input multiplexer/de-multiplexer: (**a**) circuit schematic, (**b**) symbol, and (**c**) logic truth table

A decoder sets one and only one of its outputs at "1" (or "0"), and all the other outputs are set at a complementary level of "0" (or "1"). A circuit schematic, symbol and logic truth table for a 2-bit decoder are shown in Fig. 3.4a–c. Inputs A0 and A1 are used for selecting the output. As in the case for a multiplexer, in an N-bit decoder, N control signals (A0, A1, . . ., A_N) are required for 2^N outputs. Decoders with outputs travelling in orthogonal directions are used for selecting a specific row (x) and a specific column (y) of a memory array to address a cell with coordinates (x, y). The use of row and column decoders in memory arrays is illustrated in Sect. 3.3.1.

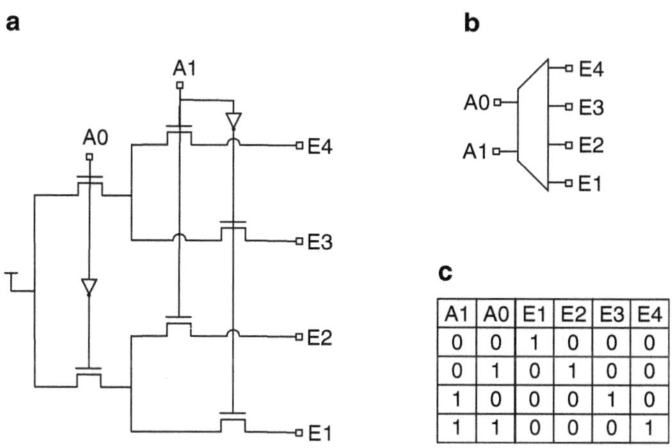

A1	A0	E1	E2	E3	E4
0	0	1	0	0	0
0	1	0	1	0	0
1	0	0	0	1	0
1	1	0	0	0	1

Fig. 3.4 A 2-bit decoder: (**a**) circuit schematic, (**b**) symbol, and (**c**) logic truth table

An arithmetic logic unit (ALU) performs arithmetic as well as logical operations [1, 3]. A binary adder is an essential part of an ALU. The block diagram of a single bit adder is shown in Fig. 3.5a. Addition of two bits A and B results in a sum (S) bit and a carry bit. The bit carry-in (C_n) from a previous addition step is the third input. The outputs are C_{n+1} (carry-out) and S (sum). The logic truth table for this adder function is shown in Fig. 3.5b. The logic equations for sum and carry-out and example circuits for logic gate implementation are shown in Fig. 3.5c. Subtraction is performed by adding the complement value of the bit to be subtracted. A 32-bit adder may be generated by serially linking 32 single bit adders. Different schemes are used for speeding up the adding function such as carry look-ahead and ripple carry.

a

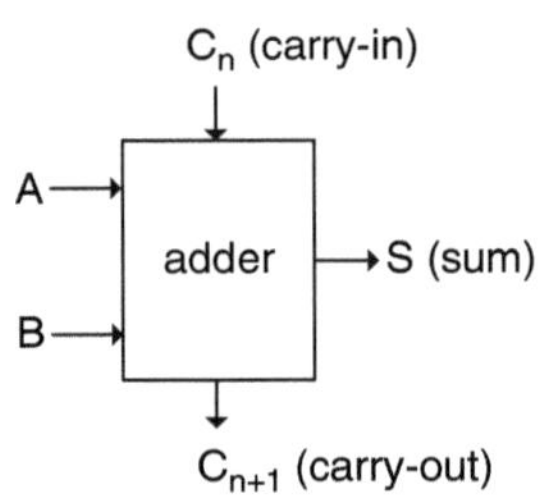

b

A	B	C_n	C_{n+1}	S
0	0	0	0	0
0	0	1	0	1
0	1	0	0	1
0	1	1	1	0
1	0	0	0	1
1	0	1	1	0
1	1	0	1	0
1	1	1	1	1

c

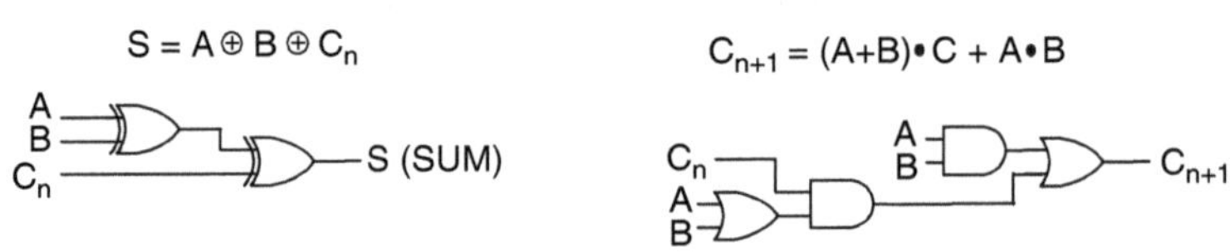

Fig. 3.5 (**a**) Block diagram of a single bit adder, (**b**) truth table, and (**c**) circuit implementation of sum and carry-out functions

In order for these circuit blocks to function correctly in silicon, their specifications as described in truth tables must be met at all times. In circuit design and test, these logic functions are validated by setting the input bits and comparing the output bits to the expected results. The time delay of a signal travelling across a circuit block is the sum of delays through the logic gates in its path. The total power of the circuit block in the off-state and during operation is a function of MOSFET widths and circuit topologies as described in Chap. 4.

The relative timing of arrival of input signals in a circuit block must be synchronized for correct operation. Timing of data signals is controlled and managed by clock signals and clocked storage elements.

3.1.3 Clock Generation and Distribution

The clock is the heartbeat of a CMOS chip as it regulates the flow of data and sets the relative timing of events. A clock signal is a free running periodic signal similar to the output of a ring oscillator. An ideal clock signal is shown in Fig. 3.6a. It has a cycle time of T_c and a duty cycle of 50 % with its voltage at "1" for half the time during each cycle and at "0" for the other half of the cycle. The rising (positive) and falling (negative) edges of clock waveforms are used for triggering events related to storing or launching data signals.

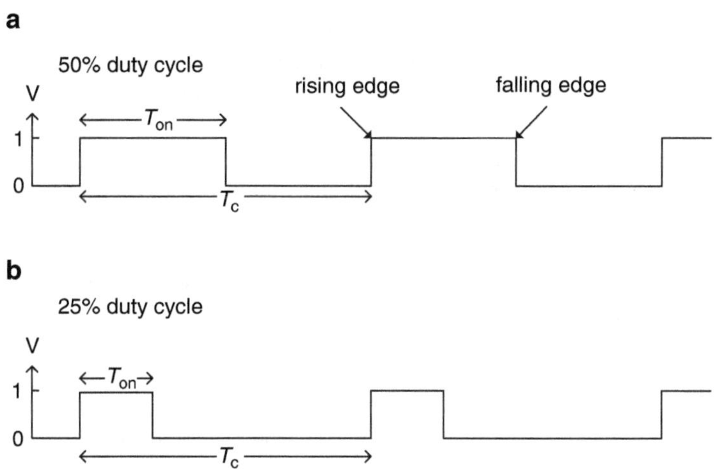

Fig. 3.6 Clock signal waveforms showing rising and falling edges: (**a**) 50 % duty cycle and (**b**) 25 % duty cycle

The duty cycle may deviate from 50 % because of imbalance in path delays. It may also be intentionally set to a different value as in the case of pulsed clock signals. A clock signal with 25 % duty cycle shown in Fig. 3.6b.

An external reference clock signal is provided by a clock chip using a crystal oscillator or LC circuit. During electrical testing, the reference clock is supplied by the tester. A phase-locked loop (PLL) circuit on the chip generates a stable high frequency clock signal.

The block diagram of a PLL circuit is shown in Fig. 3.7. The PLL comprises a voltage controlled oscillator (VCO), a phase detector, and a filter to maintain a constant phase between the PLL output and the external reference clock signal. A frequency divider circuit sets the output frequency $= N \times f_{ref}$, where f_{ref} is the reference signal frequency and N is an integer. Multiple values of N may be selected to generate different frequency signals. For a tester clock reference frequency of 200 MHz and $N = 20$, an on-chip clock frequency of 4 GHz is obtained.

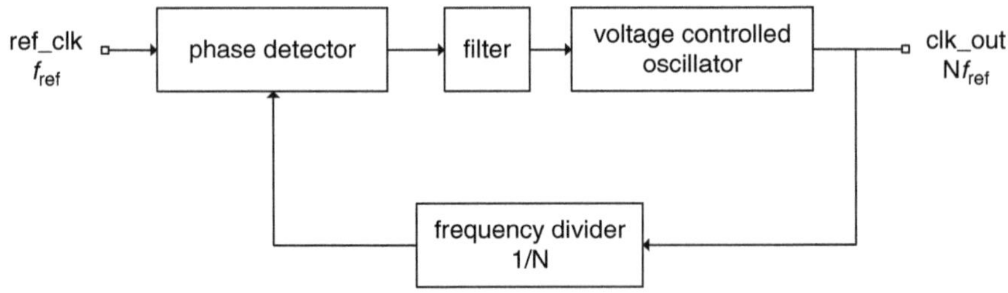

Fig. 3.7 Block diagram of a phase-locked loop (PLL) circuit

Clock signals from the PLL are distributed across the chip or within circuit blocks sharing a common PLL. Buffers are inserted in the interconnecting distribution network to maintain signal integrity. The wire length assignment and buffer sizing are done carefully to balance the propagation delays such that the clock signals are synchronized across the chip.

Two commonly used schemes for clock distribution in a synchronous design are shown in Fig. 3.8a, b. In the H-tree distribution network in Fig. 3.8a, delays through the buffer and interconnects are balanced in each branch in both vertical and horizontal directions. The binary tree in Fig. 3.8b has balanced clock only in the horizontal direction.

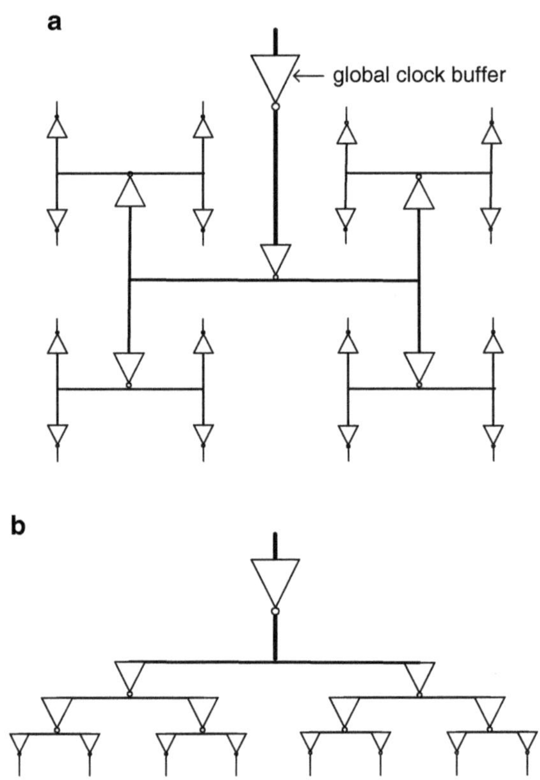

Fig. 3.8 (**a**) H-tree clock network and (**b**) binary tree clock network

The arrival times of clock edges at different sections of a chip or circuit block may be offset due to delay differences in the clock paths, silicon process induced variability in circuit components and jitter. In Fig. 3.9, clock waveforms CLK1 and CLK2 are on the average shifted in time by T_{skew} due to imbalance in the clock tree branches. In addition, each individual edge may have a relative displacement in time due to noise or power supply fluctuations. This jitter or dynamic shift in the clock edges is statistical in nature as indicated by shaded areas centered at the clock edges. It is quantified by a parameter T_{jitter}, which represents the maximum displacement in the clock edge over time [5].

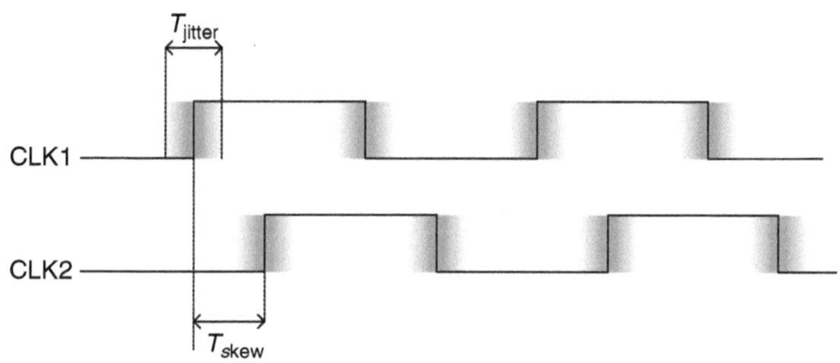

Fig. 3.9 Clock waveforms CLK1 and CLK2 with clock skew T_{skew} and jitter T_{jitter}

Global clock buffers (GCBs) regenerate the clock signal in the clock distribution network. GCBs may be designed with device width controls to tune current-drive strengths. This allows compensation for silicon process variations to reduce clock skew. Local clock buffers (LCBs) regenerate the clock and deliver it to functional blocks. LCBs may be designed with controls to move relative clock edges to compensate for duty cycle imbalances [6]. The LCB control signals may be utilized for addressing timing errors during electrical testing as well as in establishing the final operating configuration for the chip. LCBs may also be designed with control signals for clock gating to reduce active power consumption (Sect. 4.5).

3.2 Sequential Logic and Clocked Storage Elements

In combinational logic, the output is a function of the inputs at all times and hence the outcome is deterministic. As an example, the output of a NAND gate is always a "0" when all its inputs are at "1". If instead the observed output is a "1" on a chip, either the circuit wiring is incorrect or there is a defect in silicon. Such defects are detected by fault modeling and testing as described in Sect. 7.1.4.

In sequential logic, the state of an element is dependent not only on its inputs but also the past states of its inputs. Elements for storing data such as clocked storage elements (CSEs) fall in this classification. CSEs have at least two inputs and one output, and can be in one of two states, storing either a "0" or a "1", or in some cases transparent. Clock signal levels or clock signal edges determine when input data are stored or written and propagated to the output.

A schematic for signal propagation in a data path within a circuit block is shown in Fig. 3.10. Here CSE1 and CSE2 are triggered by the rising or positive edge of the clock signals CLK1 and CLK2 respectively. Input data passes through CSE1, at the first rising edge of CLK1. After a delay through the combinational logic path, the data arrives at CSE2 and waits for the second rising edge of CLK2 before being propagated through CSE2. If the signal arrives at CSE2 after the second rising edge of CLK2, it will only be propagated through CSE2 on the following or third rising edge (not shown in Fig. 3.10), and a timing error occurs.

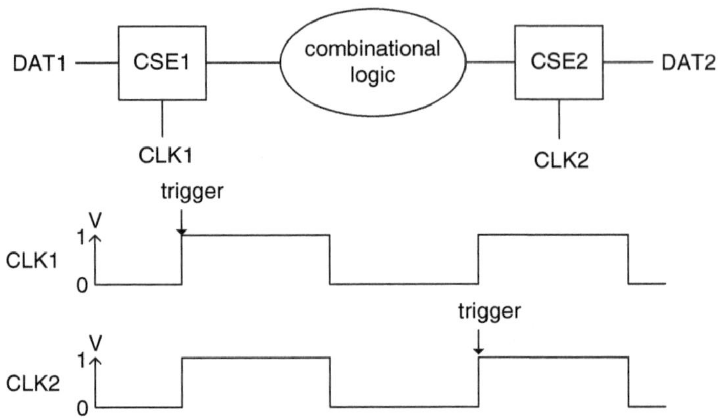

Fig. 3.10 Configuration for data propagation from DAT1 to DAT2 through the combinational logic path with rising clock signals at CSE1 and CSE2

A timing error due to late arrival of the data may be corrected in design by tuning the circuits in the combinational logic to reduce the path delay. This approach to circuit design allows each logic path to be tuned to meet timing requirements based on clock cycle time. It partitions a complex design into smaller manageable units. All such units are tuned in a similar manner to meet timing requirements, while minimizing power consumption.

If timing errors are detected during test, the clock cycle time may be increased to delay the clock edge arrival at CSE2 with respect to the data signal. Alternatively, V_{DD} may be raised to reduce data path delay. Both of these approaches impose a penalty of either reduced operating frequency or higher power.

The two commonly used CSEs are level-sensitive latches and edge-triggered flip-flops. In these types of CSEs, the output level is configured to be the same as that of the data stored in the element. The clock is said to be "asserted" for enabling a change in the state of a CSE and its output. The basic difference between a latch and a flip-flop is that in a latch the output is updated as long as the clock signal is high whereas in a flip-flop the output is updated only at either the rising or the falling edge of the clock signal. These designations are loosely followed and edge-triggered flip-flops may be referred to as latches.

3.2.1 Level-Sensitive Latches

The circuit schematic and symbol of a single stage level-sensitive latch are shown in Fig. 3.11a, b. The latch has two inputs, clock (C) and data (D), and a single output (O). It comprises two cross-coupled inverters, one of which is enabled by the clock signal, a transmission gate switch in the data path, and an inverter to get a complimentary clock signal. When the clock is high, with C at "1", data is propagated to the output node O through the transmission gate TG. When the clock goes low, with C falling from a "1" to a "0", the clocked inverter in the latch is enabled and a data value is stored in the latch. At the same time the TG switch is turned off, and any further changes in the data signal are not propagated to O until the next clock cycle.

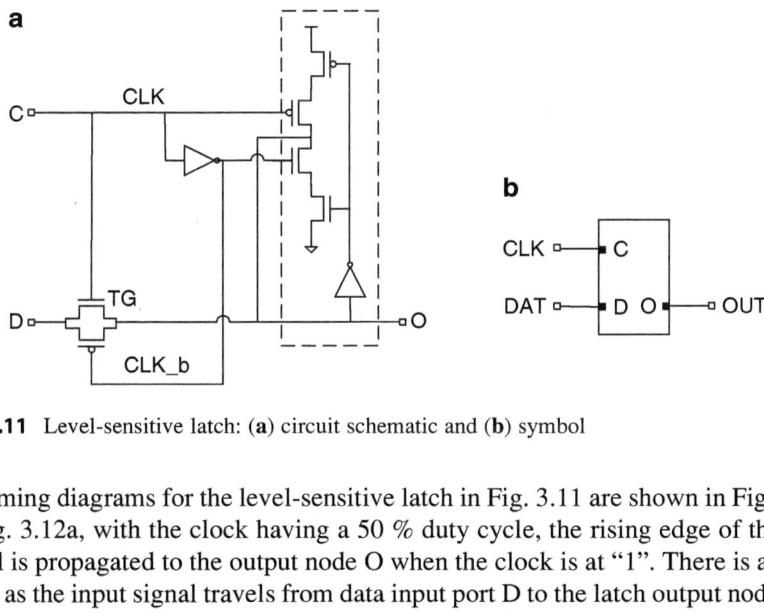

Fig. 3.11 Level-sensitive latch: (**a**) circuit schematic and (**b**) symbol

Timing diagrams for the level-sensitive latch in Fig. 3.11 are shown in Fig. 3.12. In Fig. 3.12a, with the clock having a 50 % duty cycle, the rising edge of the data signal is propagated to the output node O when the clock is at "1". There is a small delay as the input signal travels from data input port D to the latch output node. The falling edge of the data signal arrives at D when the clock is at "0" and must wait for the clock to rise to a "1" again to be propagated to the output. In Fig. 3.12b, after a small delay, the clock is held at "1" at all times. In this case the latch is transparent, and the data signal is propagated to the output.

In a level-sensitive latch, the clock must be high while the data signal is travelling through the latch. Hence, data must arrive at the input to the latch before the falling edge of the clock. The minimum time between data arrival and falling clock edge to capture the data at the output is called setup time. Key properties of the latch are setup time and signal delay through the latch. Physical dimensions and power consumption are also important considerations as latches are used extensively on the chip and add to silicon area and chip power. A more detailed discussion of setup time is included in Sect. 3.2.3 and of power consumption in Sect. 4.2.2.

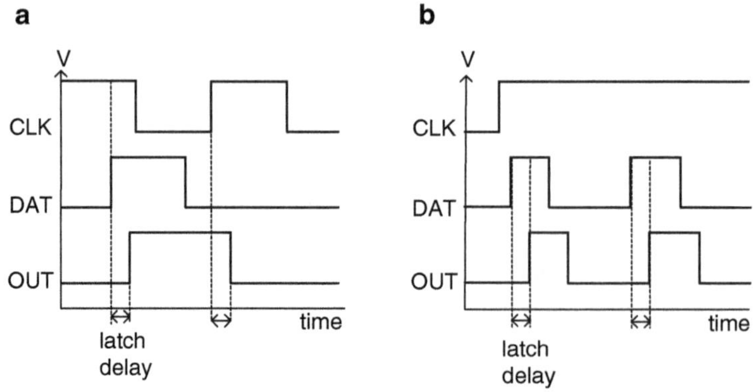

Fig. 3.12 Timing diagram for level-sensitive latch: (**a**) with CLK duty cycle of 50 % and (**b**) with CLK held at "1"

Data arrival and capture times are synchronized with the clock for a master–slave latch (MSL) shown in Fig. 3.13. An MSL comprises two level-sensitive latches connected in series and having complementary clocks. The input clock to the first or master latch is labeled dCLK. The input clock to the second or slave latch is labeled lCLK. The clock signals dCLK and lCLK are typically 180° out of phase, although in some designs there is a slight overlap between the clocks to capture late

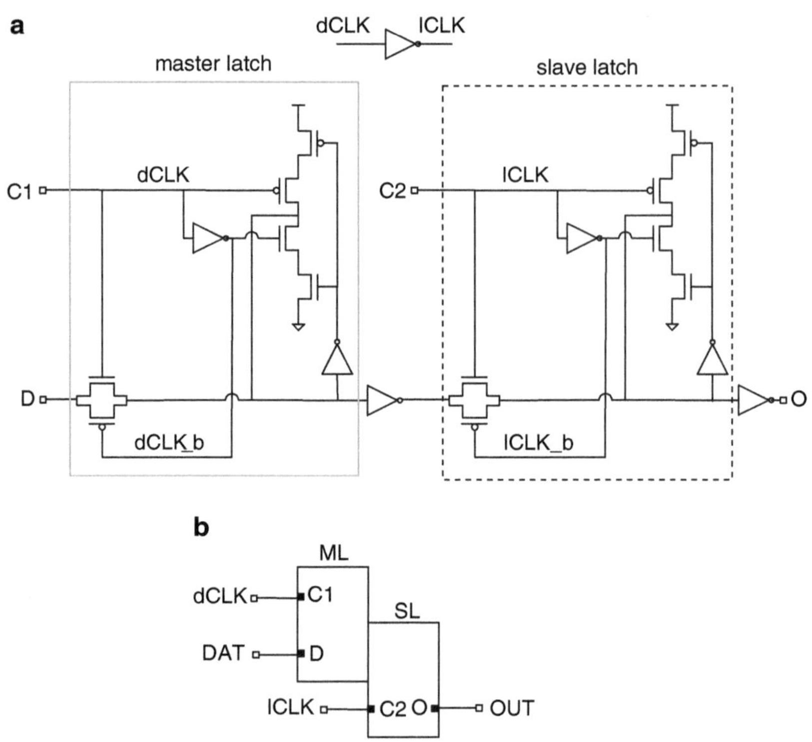

Fig. 3.13 Level-sensitive master–slave latch (MSL): (**a**) circuit schematic and (**b**) symbol

arriving data. In pulsed operation, dCLK is held at "1" at all times, and lCLK is pulsed to propagate the data signal.

In a scannable MSL (SMSL), an additional data port is added to enable resetting the latch or to store a predefined value independent of the incoming data flow from the preceding combinational logic. A corresponding additional clock port S for the scan clock, sCLK, similar in phase to dCLK, gives independent control of the scan port. The circuit schematic and symbol of a SMSL are shown in Fig. 3.14a, b. In electrical testing and during chip initialization, SMSLs are connected in series to test MSL integrity and to load specific data patterns in the latches in the scan mode. This method of testing latches and setting initial states is discussed in Sect. 7.1.7.

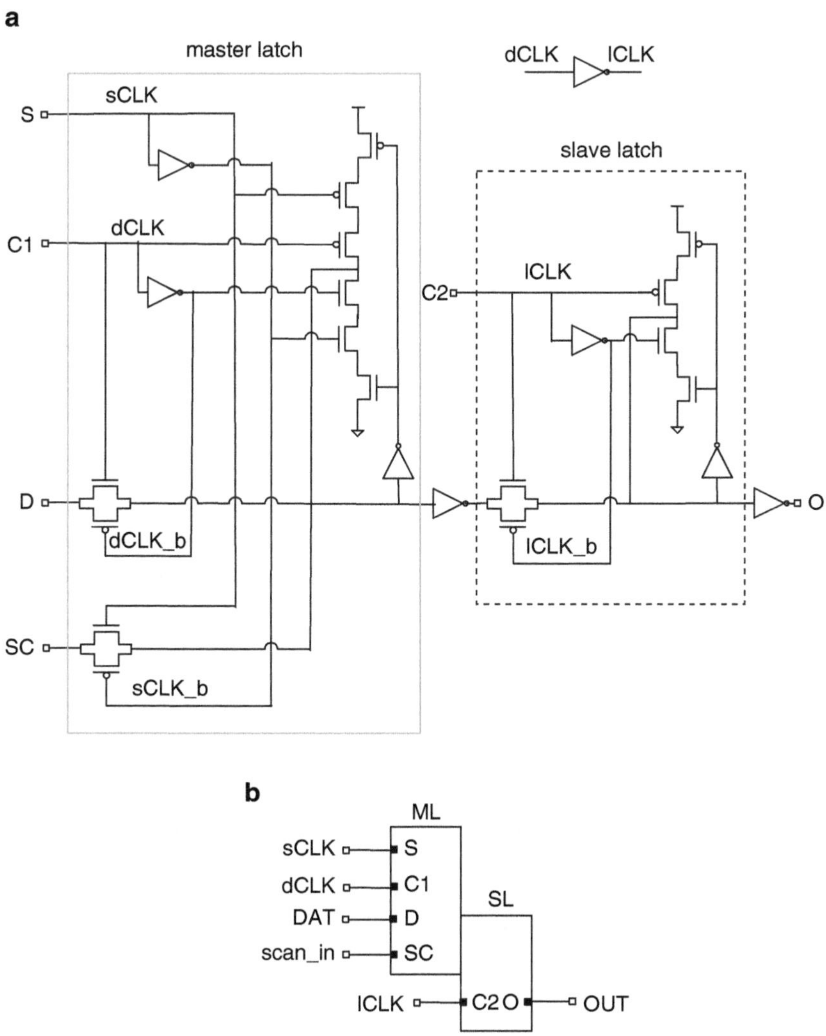

Fig. 3.14 Scannable master–slave latch: (**a**) circuit schematic and (**b**) symbol

A timing diagram for an MSL is shown in Fig. 3.15a. The dCLK and the lCLK are 180° out of phase. A rising signal edge arrives at DAT when dCLK is high. It is propagated to OUT when lCLK transitions to a "1" while dCLK transitions to a "0" to prevent propagation of any further changes in the DAT signal.

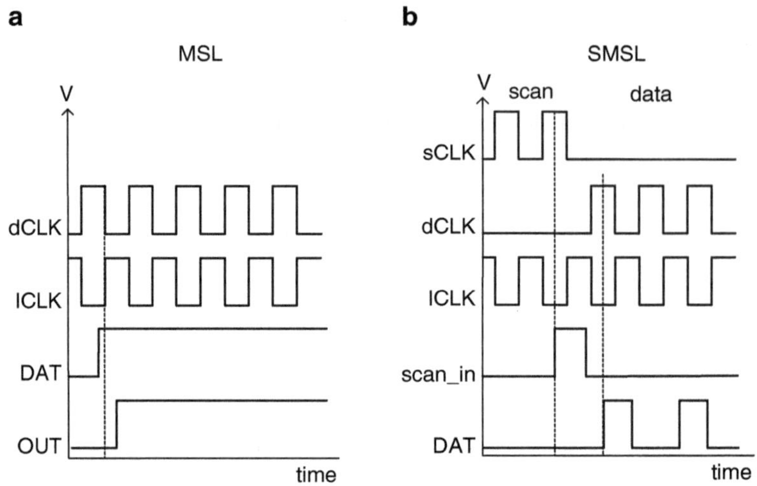

Fig. 3.15 Timing diagrams for (**a**) MSL clocks, DAT, and OUT signals, and (**b**) scannable MSL clocks, scan_in, and DAT signals

In Fig. 3.15b, the clock and data waveforms for an SMSL are shown. The two clocks, sCLK and dCLK are never asserted (at "1") simultaneously. With sCLK at "1" and dCLK at "0", the signal at the SC (scan_in) port is propagated. Conversely with sCLK at "0" and dCLK at "1", the signal at the D port is propagated. In both cases, the output O is updated only when lCLK goes high.

3.2.2 Edge-Triggered Flip-Flops

Flip-flops are generally constructed to be clock-edge triggered. Edge-triggered flip-flops may be either positive or negative-edge-triggered. Positive-edge-triggered flip-flops are activated by the rising edge of the clock and negative-edge triggered by the falling edge of the clock. The circuit schematic of a negative-edge-triggered MSL flip-flop constructed with inverters and transmission gate logic is shown in Fig. 3.16. This circuit has data and clock inputs and complementary outputs Q and Q_b.

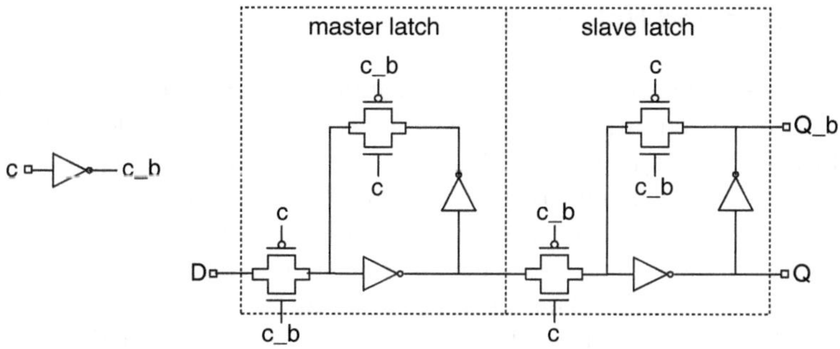

Fig. 3.16 Circuit schematic of a negative-edge-triggered flip-flop

The timing diagrams of positive and negative-edge-triggered flip-flops are shown in Fig. 3.17a, b. In the case of a positive-edge-triggered flip-flop, the data signal is propagated to the output at the arrival of a rising clock edge. The output level is held until the next rising edge of the clock at which time the output is refreshed with the current state of the data input signal. In a negative-edge-triggered flip-flop in Fig. 3.17b, the data signal is propagated to the output at the falling edge of the clock and held until the arrival of the next falling edge. The symbol of the flip-flop indicates the flip-flop type, with an invert symbol at the clock input for the negative-edge-triggered flip-flop.

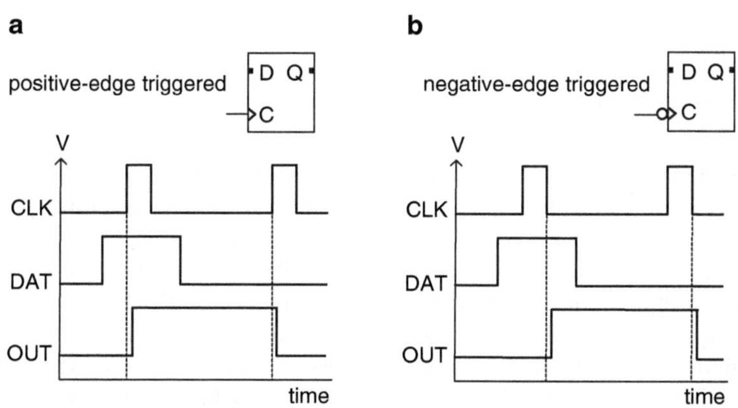

Fig. 3.17 Timing diagrams of (**a**) a positive-edge-triggered flip-flop and (**b**) a negative-edge-triggered flip-flop

3.2.3 Setup and Hold Times

The time delay between the arrival of data at the latch and capture at the output is called latency. In a transparent latch, latency is the propagation delay through the latch. In a clocked latch or a MSL, latency is a function of data arrival time before the clock edge. The latch delay as a function of data arrival time with respect to the clock edge for a level-sensitive latch is shown in Fig. 3.18a. As the data edge moves closer to the falling clock edge, there is a rapid increase in latch delay. In order to maintain timing, data must arrive before a prescribed setup time with respect to the clock edge. The setup time of a latch, T_S, is defined as the minimum DAT-to-CLK delay for a specified latch delay. From Fig. 3.18a it is apparent that the exact definition of T_S is dependent on design methodology and timing margin allocations. In a robust circuit design, the setup time criteria are met at all application corners.

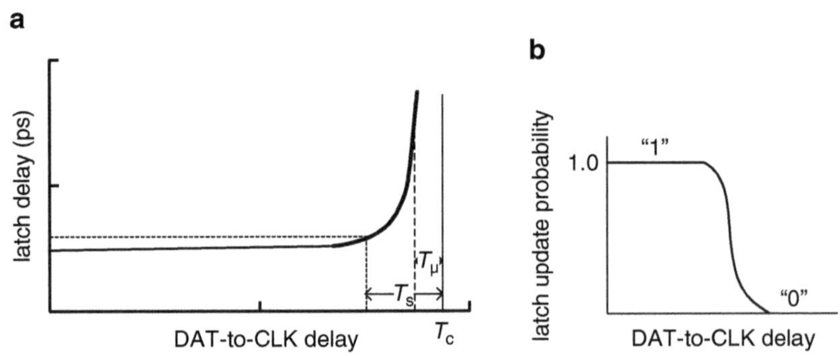

Fig. 3.18 (a) Propagation delay through a latch as a function of DAT-to-CLK delay and (b) latch update probability at the metastable point in the presence of jitter

As the DAT-to-CLK delay is reduced further ($<T_S$) the latch delay exponentially increases, theoretically diverging to infinity at the so-called metastability point, with a critical DAT-to-CLK delay of T_μ. If the DAT-to-CLK delay is $<T_\mu$, the latch does not update. The area in the vicinity of T_μ is called the latch metastability region. In the absence of any noise or jitter, for a time difference of T_μ, the latch has a 50 % probability of updating after an infinite amount of time. From a practical perspective, in simulation or in test, it is difficult to observe a situation where the latch output does change and the latch delay is greater than $\sim 3\times$ that for a transparent transition. For such delays the latch is already <1 ps from the metastability point [7].

In the presence of jitter there is no longer a sharp cut-off at T_μ. Instead, for a nominal delay of T_μ the latch update probability is as shown in Fig. 3.18b, where the derivative of the curve shown directly reflects jitter. Noise effects including cross-talk and thermal noise will further broaden and reshape this curve.

For edge-triggered flip-flops, updating data must be held for a certain time after the arrival of the clock edge to ensure data integrity at the output of the flip-flop.

This is defined as the hold time T_h for the flip-flop. Timing diagrams indicating setup and hold times are shown Fig. 3.19a, b.

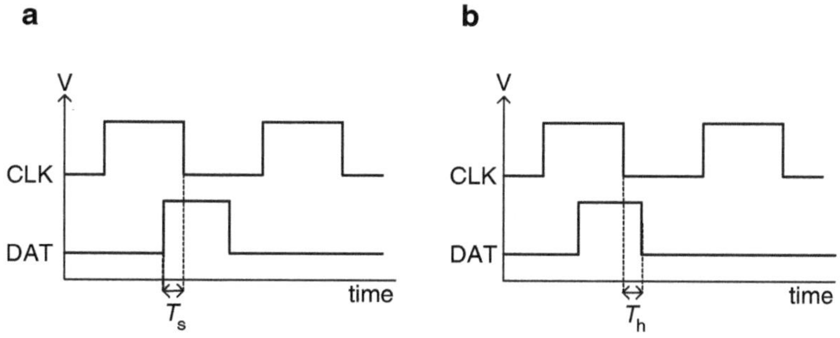

Fig. 3.19 (**a**) Setup time T_s in a level-sensitive latch or an edge-triggered flip-flop and (**b**) hold time T_h in a negative-edge-triggered flip-flop

3.2.4 Register Files

A CMOS chip has small memory units for fast access of data and instructions. These are known as register files. Registers comprise flip-flops activated by a common clock signal and serve as memory elements.

The basic element of a flip-flop (FF) is a set–reset (SR) latch. One implementation of an SR flip-flop (SRFF) element is shown in Fig. 3.20a. It comprises a pair of cross-coupled NAND2 gates with inputs S and R and complementary outputs Q and Q_b. The truth table for the SRFF is shown in Fig. 3.20b. With complementary inputs S and R, the outputs follow S and are complementary. With both inputs high, the previous states of output Q and Q_b are maintained. With both inputs low, the FF is in an indeterminate and undesirable state.

a

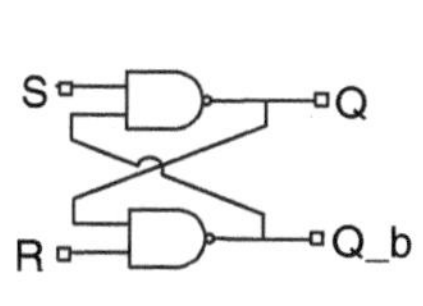

b

S	R	Q	Q_b	
0	0	X	X	not allowed
1	0	1	0	
0	1	0	1	
1	1	Q	Q_b	hold

Fig. 3.20 (**a**) An SR (Set–Reset) flip-flop and (**b**) its truth table

Register files are used in microprocessors for storing instructions and data. A register file contains a set of registers (flip-flops), each with a unique address. The registers are addressed with a decoder and the outputs from the registers are multiplexed. The block diagram of a register file is shown in Fig. 3.21. It has address inputs to read two registers, two outputs, write register address and data inputs to write the outcome of logic manipulation of the two readouts, and an enable input. The register file is updated after reading the two registers and performing a logic function when the enable signal is on.

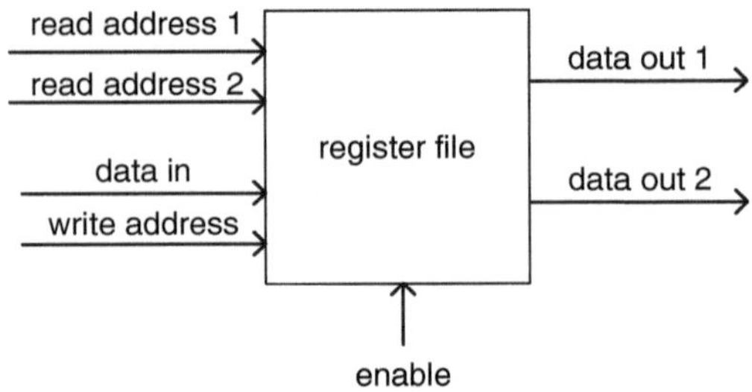

Fig. 3.21 Register file showing ports for addressing and reading two registers, and writing data in a third register when the enable signal is on

Just as the behavior of combinational logic is described as a truth table, the behavior of sequential logic is described as a finite state machine [1]. The finite state machine has a set of internal states, a *next-state* function that defines its output given its current state and its input, and an *output* function. These two functions are implemented with combinational logic while a register file holds its internal state. In a synchronous finite state machine, the internal states are updated once every clock cycle.

3.3 Memory

Storage of data is accomplished in a hierarchical fashion. Using a microprocessor as an example, a representative hierarchy is listed in Table 3.1. Information is rapidly stored and accessed in small volumes in register files of the central processing unit (CPU). Cache memory in the CPU stores frequently or recently used information bits. There may be multiple caches such as instruction cache and data cache. The data cache is hierarchical with L1, L2 and more levels, increasing in size and access times. Register files and cache memory are volatile. Hard disks are nonvolatile, stand-alone memories and used for permanent data storage.

Table 3.1 Memory hierarchy arranged in the order of increasing density and decreasing speed indicated by shading intensity

Memory Type	Usage	Density	Speed	Comments
Flip-flops	Register files			custom
SRAM	L1, L2 cache			standard process technology, no refresh
DRAM	stand-alone or embedded			deep-trench capacitor, higher V_{DD}, refresh cycle
Hard disk drive	non-volatile			bulky, large capacity

On-chip memory units are built with arrays of cells, each cell storing one logic bit. In random access memory (RAM) used for data storage, any cell in a memory array may be accessed at random for storing or retrieving bits. There are two types of commonly used memory cells, static random access memory (SRAM) and dynamic random access memory (DRAM). SRAM cells store data in a latch formed by a pair of cross-coupled inverters. In a DRAM, the data storage element is a capacitor which is accessed through an n-FET. Fabrication technology for SRAM is compatible with standard CMOS logic. DRAM requires additional processing steps for the formation of deep-trench capacitors for charge storage.

The memory cells for on-chip storage are small in area, pushing the limits of silicon technology, and are more prone to failures due to particulate and process induced defects, noise, and soft-errors. With built-in redundancy and error-correction algorithms, defective cells can be identified and replaced.

3.3.1 SRAM

A standard SRAM cell comprises six MOSFETs and is generally referred to as a 6T cell. The circuit schematic of a 6T SRAM cell is shown in Fig. 3.22a. In Fig. 3.22b, the schematic is redrawn to highlight the cross-coupled pair of inverters forming the latch. The p-FET in the inverter is referred to as the cell's PU FET and the n-FET as the cell's PD FET. There are two n-FET passgates, NPL and NPR, to control access to the latch. The passgates are momentarily turned on for reading and writing a cell by raising WD from a low to a high. The cell has twofold mirror symmetry with MOSFETs PL, NL, and NPL forming the left half and PR, NR, and NPR forming the right half. In circuit simulation for SRAM characterization, it is often convenient to use a half-cell as indicated by dotted lines in Fig. 3.22a.

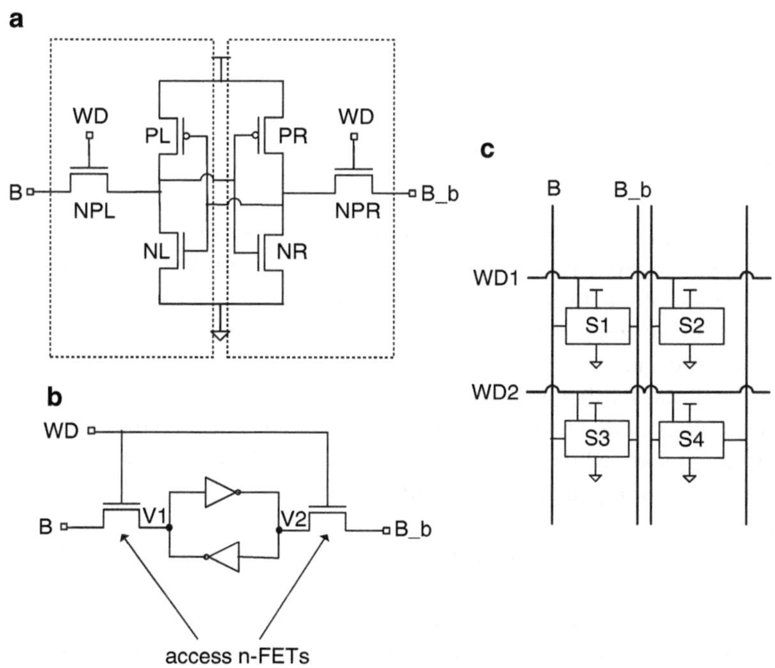

Fig. 3.22 (**a**) Circuit schematic of a 6T SRAM cell, (**b**) SRAM cell schematic showing access n-FETs and latch, and (**c**) symbolic representation of a 2 × 2 SRAM array

With the access n-FET pass-transistors enabled by setting the gate terminals WD at "1", data are written in the latch and read out through input ports B and B_b of the true and complimentary bit lines. A 2 × 2 SRAM memory array block is shown in Fig. 3.22c. All cells in a row share a common WD or "word" line and cells in a column share input/output bit line terminals, B and B_b.

Let us examine the operation of a single SRAM cell. Referring to Fig. 3.22b, if the voltage at node V1 ($V1$) is low, the voltage at node V2 ($V2$) is high forcing $V1$ to stay low. Conversely, if the voltage $V1$ is high, $V2$ is low forcing $V1$ to stay high. The two cross-coupled inverters therefore provide positive feedback to hold $V1$ and $V2$ in stable complementary states.

The state of the latch is changed by flipping $V1$ and $V2$ voltage levels through the two access n-passgates, NPR and NPL. These two n-passgates are also used to access the latch for reading the stored data. The sizing of the MOSFETs in the SRAM is an important consideration in optimizing cell area, performance, and stability.

The design and physical layout of the SRAM cell is optimized by selecting the widths and channel lengths of each MOSFET type. The ratio of current-drive strengths of the PD to passgate (PASS) devices is referred to as the cell's beta ratio. Expressing the current drive strength as W/L_{p}, the beta ratio is given by

$$\text{Beta ratio} = \frac{(W/L_{\mathrm{p}})_{\mathrm{PD}}}{(W/L_{\mathrm{p}})_{\mathrm{PASS}}}. \tag{3.1}$$

A higher beta ratio improves cell stability during read operation but increases the write time. The write performance is improved by increasing the cell ratio expressed as

$$\text{Cell ratio} = \frac{(W/L_{\mathrm{p}})_{\mathrm{PASS}}}{(W/L_{\mathrm{p}})_{\mathrm{PU}}}. \tag{3.2}$$

Typical recommended values are beta ratio >2.5, and cell ratio >1.2.

It is apparent from Eqs. 3.1 and 3.2 that a robust SRAM cell design requires complex trade-offs between cell stability and performance (read and write times). The cell stability and performance are also impacted by the operating V_{DD} and by systematic and random variations in MOSFET properties, which affect both beta ratio and cell ratio. Silicon foundries offer several custom designed SRAM cells at each technology node in which the MOSFETs and physical layouts are tailored for density and yield, while meeting the stability and performance objectives. In advanced technology nodes, each SRAM MOSFET type may be specifically engineered.

A circuit schematic to simulate the write and read operations of an SRAM cell is shown in Fig. 3.23. Six MOSFETs are added to the SRAM cell schematic for pre-charging the bit lines and for writing and reading the cell. The corresponding timing diagrams for writing a "1" and then reading the cell are shown in Fig. 3.24.

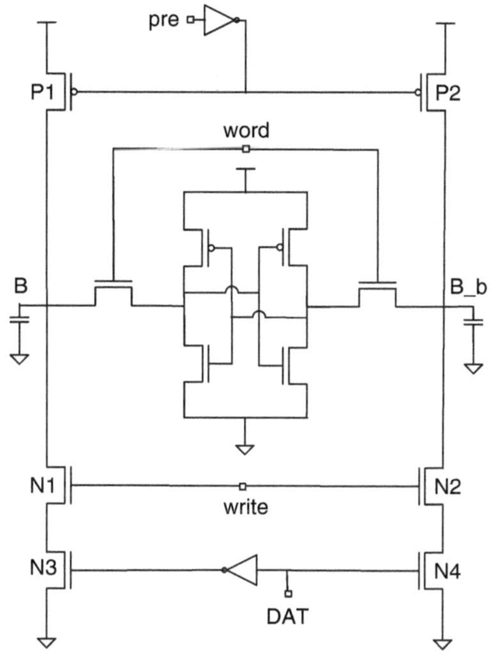

Fig. 3.23 Circuit schematic for SRAM read and write simulations

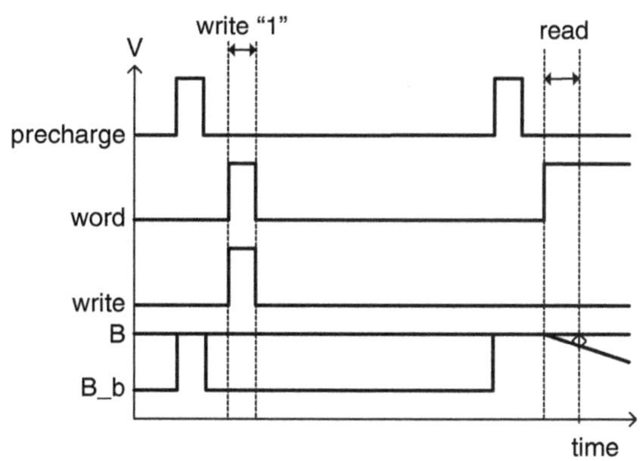

Fig. 3.24 Timing diagrams for simulating SRAM write and read operations

In order to write a "1" in the SRAM cell, the bit lines B and B_b are first pre-charged to a "1" by turning on p-FETs P1 and P2 (node pre at "1"). During the pre-charge state, n-FETs N1 and N2 and the two access n-FETs of the cell are all in their off-states. To write data, pre is set to a "0", and the word line is raised to a "1" to turn on the access n-FETs. The n-FETs N1 and N2 are also turned on while p-FETs P1 and P2 are maintained in their off-states. With DAT at "1", N4 is turned on, pulling B_b low while B remains high, writing a "1" in the left node of the cell.

The read operation is done by setting the word line high after the pre-charge cycle in the second half of the timing diagram. The B_b node starts to discharge slowly through the right access n-FET. When a specified voltage threshold is crossed, this drop in voltage at B_b is resolved with a sensing circuit to determine the state of the cell. The difference voltage is sensed and amplified to get a full V_{DD} rail signal. An example of a differential sense amplifier circuit is shown in Fig. 3.25.

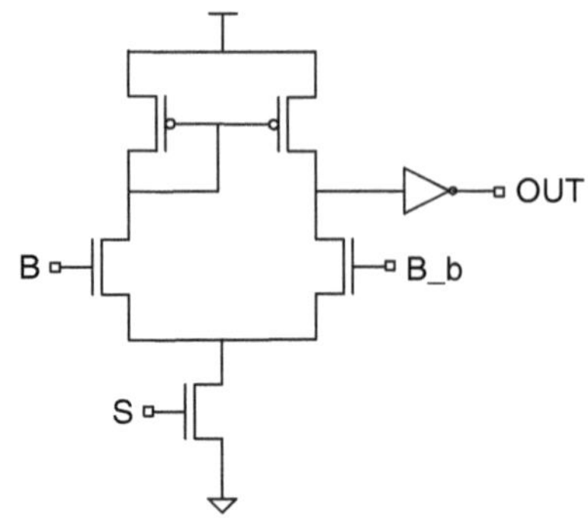

Fig. 3.25 Circuit schematic of a differential sense amplifier

A 4 × 4 SRAM memory block is shown in Fig. 3.26. Row decoders are used to address a selected word line using the input bits a0 and a1. Column decoders are used for reading and writing a selected column.

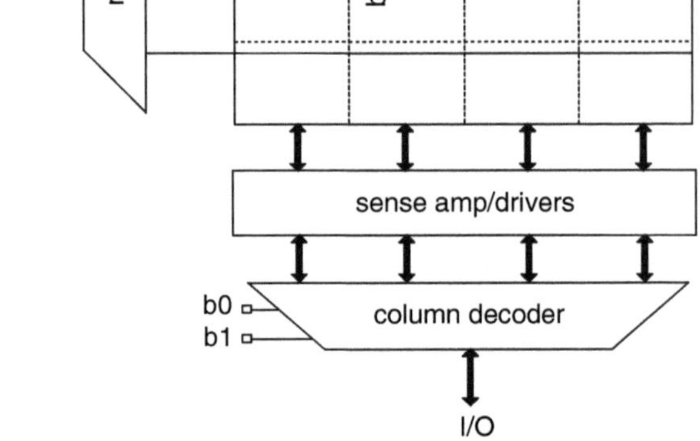

Fig. 3.26 4 × 4 SRAM array with sense amplifiers and row and column decoders to select a cell

Each cell must maintain its state when a read operation occurs as well as in the presence of noise. The immunity of the cell to noise is therefore a very important consideration. The noise margins of the cell are evaluated by simulating the transfer characteristics of one latch inverter and then mirroring its terminals. This plot for the latch alone is shown in Fig. 3.27a. It is called a butterfly curve in reference to its shape. The point on the curves indicated by V_F is the voltage at which the cell can flip its state. Conversely, the cell is in a stable state as long as $V1$ and $V2$ are kept within one or the other wing of the butterfly curve. The static noise margin (SNM) for hold of a cell is defined as the side of the largest square that can fit inside the butterfly wing.

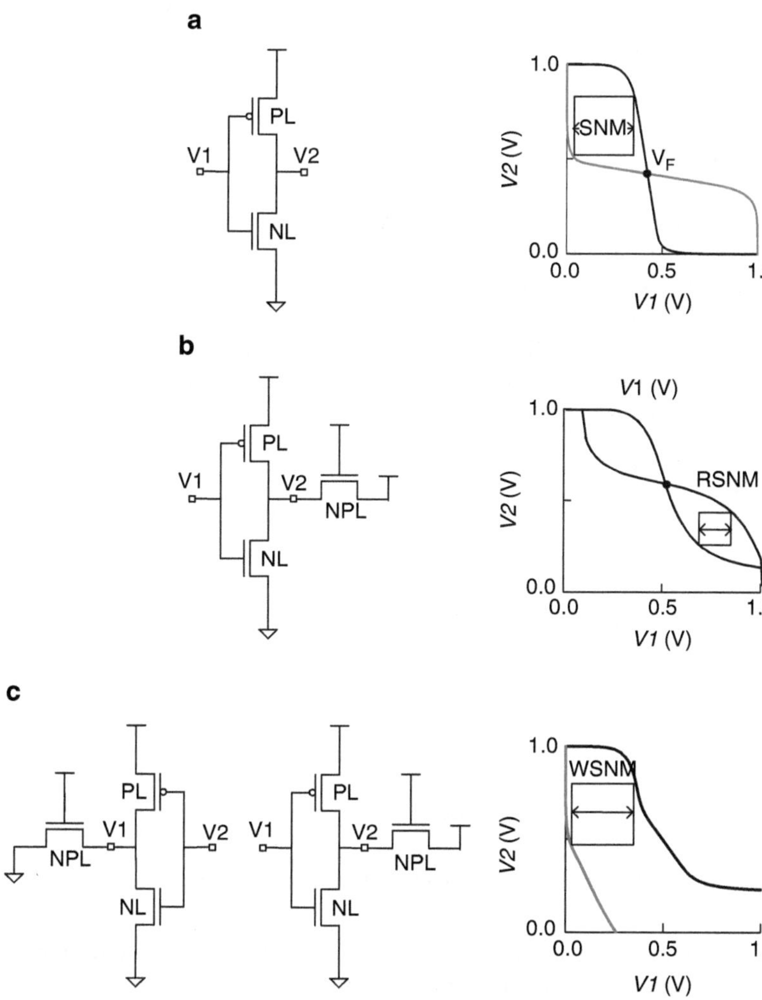

Fig. 3.27 Circuit schematics and corresponding butterfly curves for determining static noise margins: (**a**) SNM for hold, (**b**) read SNM of the cell (RSNM) and (**c**) write SNM of the cell (WSNM)

The read static noise margin (RSNM) is determined by plotting the butterfly curve for the inverter and an access n-FET. The access n-FET is turned on with its source held at "1", as in the case of a read operation. The corresponding circuit diagram and butterfly curve are shown in Fig. 3.27b. The RSNM of the cell in the example shown here is smaller than the SNM of the latch.

The write static noise margin (WSNM) is determined by plotting *V*1 vs. *V*2 for the read configuration and then the "mirror" obtained by setting the source of the access n-FET low. This is equivalent to the write operation where B and B_b are true and compliment signals. The circuit configurations for simulating the write plots are shown in Fig. 3.27c, along with the corresponding *V*2–*V*1 curves. The WSNM in this case is much larger than the RSNM, as the requirements for wide read and write margins are not balanced for this cell.

3.3.2 DRAM

Dynamic random access memory (DRAM) is used in dense stand-alone memory chips. Embedded DRAM, eDRAM, is fabricated as an integral part of microprocessor and ASIC chips. The basic circuit of a DRAM cell comprises a capacitor to store charge and a single n-FET for write and read access to the capacitor node. The circuit schematic of a 1T cell is shown in Fig. 3.28a. The transistor is turned on with the word line, WD at "1" and the bit line is used for reading and writing the data. An array of DRAM cells is shown in Fig. 3.28b.

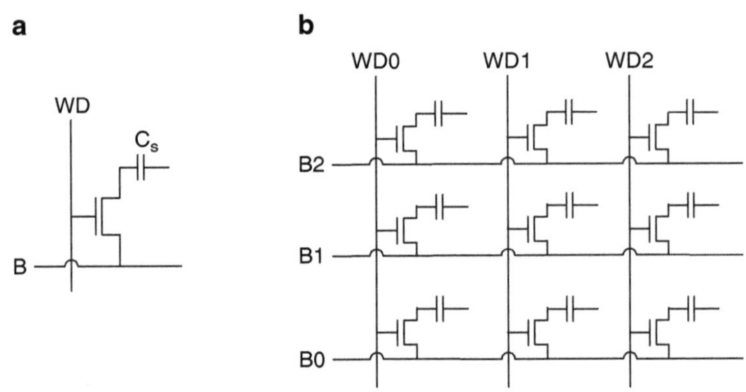

Fig. 3.28 Circuit schematic of (**a**) 1T DRAM cell and (**b**) DRAM array

The DRAM storage capacitor is formed as deep-trench structure in silicon to get a high capacitance while maintaining a small surface area. The capacitor slowly leaks charge and the DRAM cell must be periodically refreshed to hold its state.

3.4 Circuit Simulations

Two examples of circuit simulations are given to provide additional insight into logic and memory circuit operations. First, circuit simulations to determine static noise margins of an SRAM cell are described in Sect. 3.4.1. Next, timing of a data signal launched from a CSE and captured by a second CSE after combinational logic operations is described in Sect. 3.4.2. The outcome is a minimum cycle time T_{cmin} (or maximum clock frequency f_{max}) at a fixed V_{DD} or a minimum operating V_{DD}, V_{min}, for a fixed T_c. In both examples circuit schematics and SPICE commands are configured to obtain the final result without the need for further data manipulation. This approach is favorable for running Monte Carlo simulations to study the impact of silicon process induced variability or variations in operating conditions (V_{DD}, temperature) described in Chap. 6.

3.4.1 SRAM SNM

In simulating the butterfly curves in Fig. 3.27, it is assumed that the SRAM cell is perfectly symmetrical and that the transfer characteristics of the half-cell can be mirrored to represent the full cell. Process variations and random variations in MOSFET characteristics described in Chap. 6 introduce asymmetry in the cell. A technique to simulate the butterfly curves for both halves of the cell simultaneously and to extract the SNM directly from these SPICE simulations is described below [8]. This approach will be illustrated in the context of the 4T latch alone, but is directly applicable for analyzing the read and write operations of the SRAM cell as well. With the SNM being a direct simulation output, no graphical construction is required. This feature enables, for example, an automated approach to obtaining SNM distributions of a large number of cells in the face of statistical variations in threshold voltages of the MOSFETs.

Fig. 3.29a shows the butterfly curve for the 4T latch of a perfectly symmetrical SRAM cell, as previously shown in Fig. 3.27a. Also indicated in Fig. 3.29a is the empirically placed largest square that can fit inside the upper wing of the butterfly curve, the side of that square being the static noise margin (SNM) for hold. The object of this exercise is to perform a simulation that will determine for an asymmetric cell the length of the side of that square as well as that of the complementary square in the other wing, and then deliver as a single output the shorter of the two lengths, which is the SNM of the latch.

The first step in the procedure is to rotate the butterfly curve and its $V1$–$V2$ coordinate system counterclockwise by 45° ($\pi/4$) as shown in Fig. 3.29b. A second coordinate system is then introduced with horizontal and vertical axes of $U1$ and $U2$ respectively. The transformation equations relating ($V1$, $V2$) to ($U1$, $U2$) are as follows.

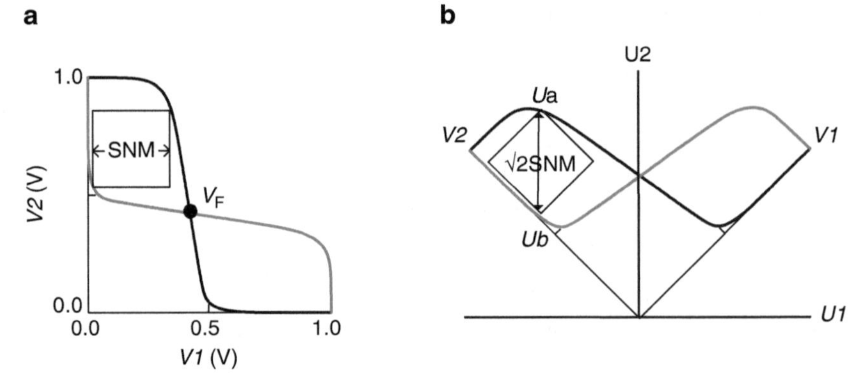

Fig. 3.29 (**a**) SRAM latch butterfly curve, and (**b**) butterfly curve in (**a**) rotated by 45°

$$V1 = \frac{1}{\sqrt{2}}U1 + \frac{1}{\sqrt{2}}U2, \tag{3.3}$$

which can be rewritten as $U2 = -U1 + \sqrt{2} \times V1$, and

$$V2 = -\frac{1}{\sqrt{2}}U1 + \frac{1}{\sqrt{2}}U2, \tag{3.4}$$

which can be rewritten as $U1 = U2 - \sqrt{2} \times V2$, where $1/\sqrt{2} = \cos(\pi/4)$.

The counterclockwise rotated inverter transfer curve for the left side inverter of the SRAM cell can be expressed in the $V1$–$V2$ coordinate system as

$$V2 = FL(V1). \tag{3.5}$$

This same curve in the $U1$–$U2$ coordinate system can be expressed as $U2 = FL1(U1)$. While $FL(V1)$ can be directly simulated in LTspice, obtaining $FL1(U1)$ is not so straightforward. However, by combining $FL(V1)$ with the U–V coordinate system transformation equations, an expression relating $U1$ and $U2$ along the rotated transfer curve of the left side inverter can be written and subsequently simulated as:

$$U2 = U1 + \sqrt{2}FL(V1) = U1 + \sqrt{2}\,FL1\left(\frac{1}{\sqrt{2}}U1 + \frac{1}{\sqrt{2}}U2\right). \tag{3.6}$$

In a similar fashion an expression relating $U1$ and $U2$ along the mirrored ($U1 \rightarrow -U1$) clockwise rotated transfer curve of the right side inverter can be written and subsequently simulated as:

$$U2 = -U1 + \sqrt{2}\,FR1\left(-\frac{1}{\sqrt{2}}U1 + \frac{1}{\sqrt{2}}U2\right). \tag{3.7}$$

The circuit schematic for simulations in LTspice that will generate the rotated transfer curves in the $U1$–$U2$ coordinate system is shown in Fig. 3.30. The left side (L) and right side (R) inverters comprising the full latch may or may not have identical transfer characteristics. With $V_{DD} = 1.0$ V across the inverters, $VU1$ is swept from $-1/\sqrt{2}$ V to $+1/\sqrt{2}$ V. Voltage sources $BV1_L$, $BV1_R$, $BU2_L$, $BU2_R$, and $BVY_{L\text{–}R}$ are arbitrary behavioral voltage sources whose output voltages are expressed to reflect the inverter transfer functions in conjunction with the above equations for axes transformation. The expressions for the voltages are also included in Fig. 3.30.

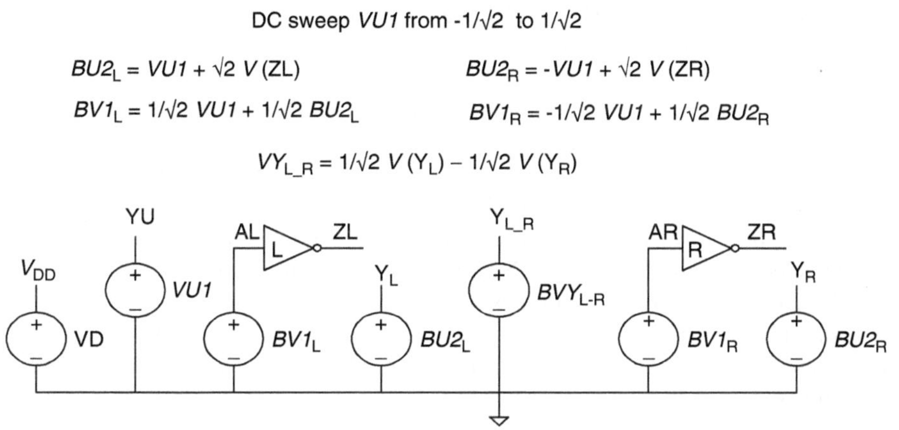

Fig. 3.30 Circuit schematic for simulating SRAM latch butterfly curves and determining SNM

As the voltage source $VU1$ is swept through $\pm V_{DD}/\sqrt{2}$, the voltages for inputs to behavioral sources $BU2_L$ and $BU2_R$ are derived from $VU1$ using Eqs. 3.6 and 3.7 and the output voltages of the corresponding inverters. The input voltages of the inverters supplied by $BV1_L$ and $BV1_R$ are expressed using Eq. 3.3 and its reflected counterpart as indicated in Fig. 3.30. The butterfly plot rotated by 45° is obtained by plotting output voltages $V(Y_L)$ and $V(Y_R)$ as a function of $VU1$. An example plot is shown in Fig. 3.31a.

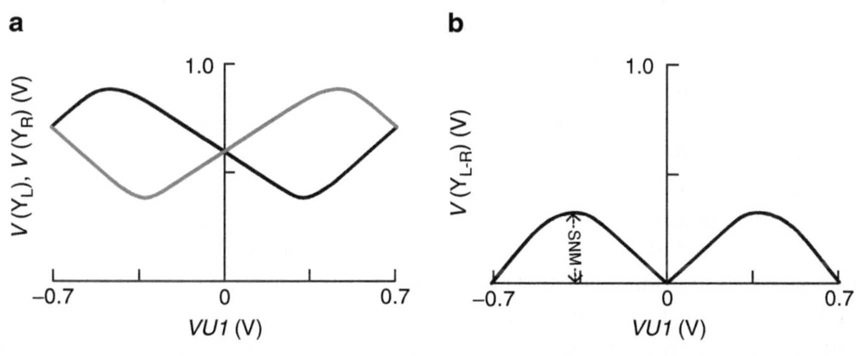

Fig. 3.31 (**a**) Rotated SRAM butterfly curve simulated using the circuit in Fig. 3.33, and (**b**) difference plot from (**a**) to determine SNM

The difference voltage $\{V(Y_L) - V(Y_R)\}$ is measured by including a fifth behavioral source $BVY_{L\text{-}R}$ in the schematic, where the factor of $1/\sqrt{2}$ converts from the SNM square's diagonal to its edge. The output voltage at the node $Y_{L\text{-}R}$ is plotted in Fig. 3.31b. The SNM of the cell's latch is given by the minimum of the maximum values of $Y_{L\text{-}R}$ in the two wings of the butterfly curve. SPICE commands for measuring the SNM are included below:

```
.DC VU1 -0.707 0.707 0.01
.measure dc snml max abs(V(YL_R)) trig V(YU) val=-0.70  targ V(YU) val=0.00
.measure dc snmr max abs(V(YL_R)) trig V(YU) val= 0.00  targ V(YU) val=0.70
.measure snm param = min(snml, snmr)
```

The single output from this simulation is thus the SNM of the SRAM cell's 4T latch corresponding to the device parameters used in the simulation. Specific parameter values can be replaced with distributions, and Monte Carlo simulations performed to obtain a realistic distribution of cell SNMs corresponding to the 4T latch's response to variability. As previously mentioned this same methodology can be extended to the RSNM and WSNM using the schematics shown in Fig. 3.27b, c.

The circuit simulation setup described above is used for studying the impact of V_{DD} and V_t variations (systematic and random) on the cell stability for hold, read, and write operations. This study is included as an exercise at the end of Chapter 6.

3.4.2 Logic Data Path

Functional units are partitioned to operate in multiples of a fundamental cycle time. As an example, in a microprocessor, the cores operate at the fundamental cycle time T_c. Memory circuits and supporting logic may operate at nT_c, where n is an integer. Chip timing is carried out hierarchically starting at the macro level (small circuit block), moving up to a full functional unit and finally to the complete integrated chip design.

Let us examine a signal path traversed in one clock cycle. The circuit schematic of the logic path for this example is shown in Fig. 3.32. It comprises two level-sensitive MSLs, X1 and X2, and a chain of 20 identical inverters (FO = 4) within block "inv20" followed by two additional inverters (FO = 1) to achieve a sharp signal edge at the input of X2. We will first assume that the delay components through the path corresponding to the rising and falling edges of the input data pulse are identical. For a real path this is not always the case as discussed in detail later.

A timing diagram for the circuit in Fig. 3.32 is shown in Fig. 3.33. A signal arriving at DAT before the falling edge of the dCLK in the first clock cycle is propagated to the output of the first MSL, X1, following the rising edge of the lCLK. The signal travels through the chain of inverters and arrives at Z, the D input of the second MSL, X2. If the signal arrives at Z before the falling edge of

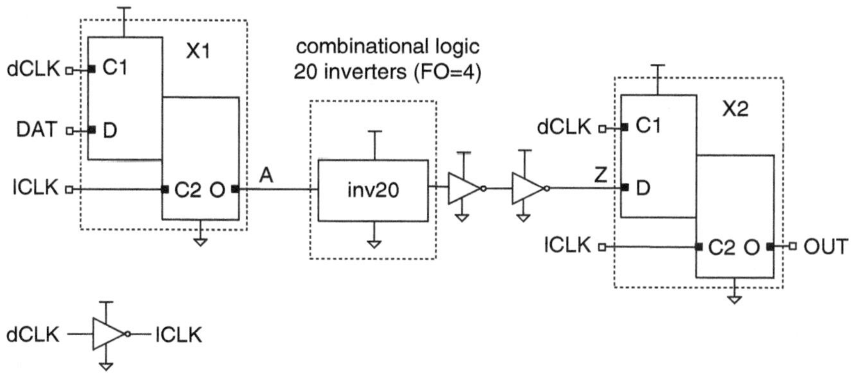

Fig. 3.32 Latch-to-latch path with a chain of 20 inverters (FO = 4)

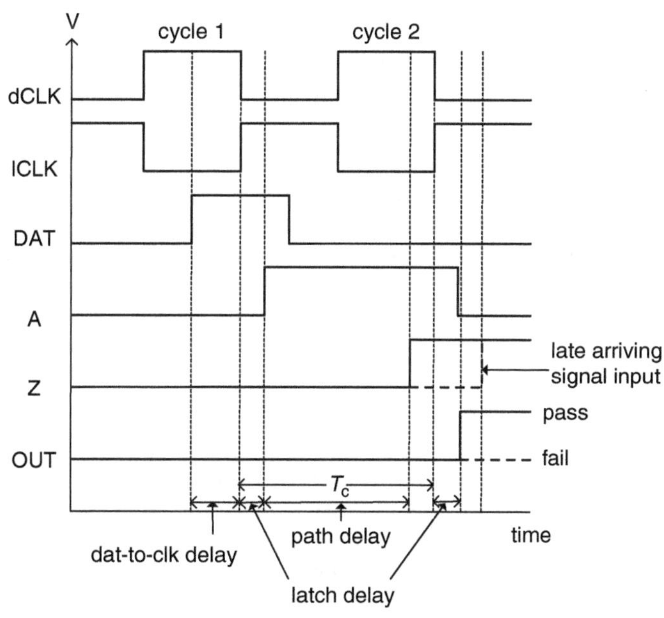

Fig. 3.33 Timing diagram for the circuit shown in Fig. 3.32. *Solid lines* for Z and OUT indicate a "pass" condition and *dashed lines* a "fail" condition

the clock in the second clock cycle, it will be propagated through X2 to OUT. This occurs following the rising edge of the lCLK at X2, with additional delay corresponding to the propagation time through the slave latch in X2. If all events occur in this sequence the circuit "passes" the timing test at the set V_{DD} and clock cycle time T_c.

If the delay across the inverter chain is too long or the clock cycle time is too short, the signal arriving at Z will not be captured by X2 in the second clock cycle. This situation is indicated by dashed lines for the signals at Z and OUT in Fig. 3.33. In such a case the signal will in general arrive at OUT in the third clock cycle, and the circuit "fails" the timing test.

The minimum cycle time T_{cmin} ($=1/f_{max}$), measured at a fixed V_{DD}, is the minimum cycle time for which a circuit passes the timing test as described above. The parameter V_{min}, measured at a fixed cycle time T_c, is the minimum V_{DD} at which the circuit passes the timing test. The pass criterion is analogous to a CMOS chip performing a function correctly, an indication that all the exercised paths are meeting the timing requirements.

SPICE simulations are carried out to determine T_{cmin} and V_{min} of this circuit. It is very convenient to be able to directly determine T_{cmin} or V_{min} in a single SPICE simulation run. This capability sets the stage to, for example, determine the distributions of T_{cmin} or V_{min} in the face of random and systematic parameter variations in an efficient and straightforward manner. The basic idea here is to perform a set of sequential pass/fail experiments as described above, while incrementing V_{DD} or T_c, and then determining the pass-fail boundary along the V_{DD} or T_c axis. The key to setting up this simulation is the specification of the data, clock, and V_{DD} waveforms. The most complex of these is the clock waveform which we implement as a concatenation of segments, each six clock cycles in length and with amplitude (V_{DD}) and cycle time (T_c) corresponding to the desired values for the corresponding pass/fail test. A single pass/fail test is executed within each six-cycle segment. The clock waveform can be constructed as a (very long) table but is more conveniently done as a set of series connected voltage sources, each producing a train of six clock pulses with an appropriate zero-offset.

Figure 3.34 shows dCLK, V_{DD} and data waveforms for determining V_{min} while Fig. 3.35 shows similar waveforms for determining T_{cmin}. In these example cases, 12 series connected voltage sources are used to generate the clock waveforms. The first voltage source initializes the two MSLs and introduces a component of the timing offset for the following 11, six-cycle segments that are generated by LTspice PULSE voltage sources. In Fig. 3.34 each of these subsequent segments steps down in amplitude by an increment of *dv* compared to the previous segment while the cycle time within all segments is the same. In Fig. 3.35 the voltage level remains constant while the period within the segments is reduced from one to the next by the same increment *dtc*. By parameterizing the voltage levels and cycle times in each source, the same series chain of voltage sources may be used for both T_{cmin} and V_{min} simulations. Once these are set up only the values of *vdd*, *tc*, *dv*, and *dtc* need to be specified in SPICE commands. The power supply (V_{DD}) waveforms that power the circuit, shown for the entire 12-segment duration in each case, can be synthesized as series connected voltage sources, but are also easily constructed as tables. Similarly, either approach can be used for the data waveforms. It is of course essential that all three waveforms be appropriately synchronized in all cases.

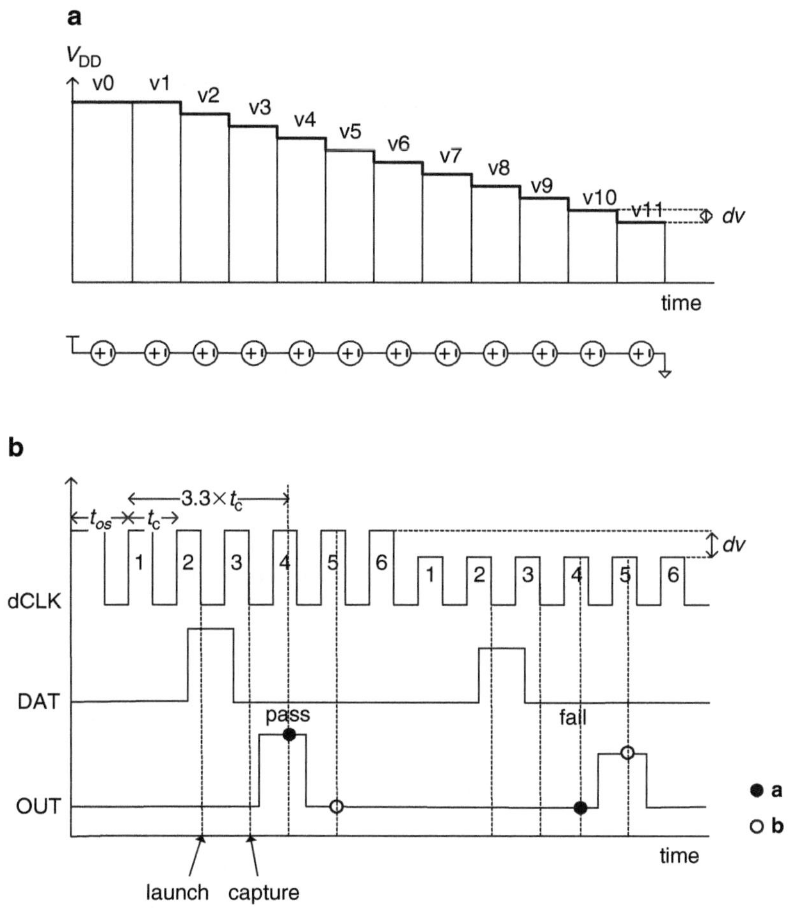

Fig. 3.34 Waveforms for V_{min} simulation tests (**a**) power supply (all segments) and (**b**) dCLK, DAT and OUT (offset plus first two 6-cycle segments). The clock waveform is created with a set of series connected voltage sources as described in the text. "*a*" (*solid circle*) and "*b*" (*open circle*) indicate voltage measurement locations on the OUT waveform

For both V_{min} and T_{cmin} tests, the OUT waveform is measured at points "a" and "b" as indicated in the figures. Each voltage that is measured is either equal to 0 or the V_{DD} of the corresponding segment. Each pair of numbers redundantly indicate whether the circuit has passed or failed the test. If the LTspice "buf" function is applied to all eleven members of the "a" and "b" OUT sets, two complementary sequences of "1"s and "0"s are obtained as indicated in Fig. 3.36. In the case of V_{min} tests, moving from left to right, the voltage corresponding to the last "1" before the transition to a "0" corresponds to V_{min} (=*vminir*) for a rising input signal at node A. Moving from right to left, the voltage corresponding to the first "0" that appears after the transition from a "1" to a "0" corresponds to V_{min} (=*vminif*) for a falling

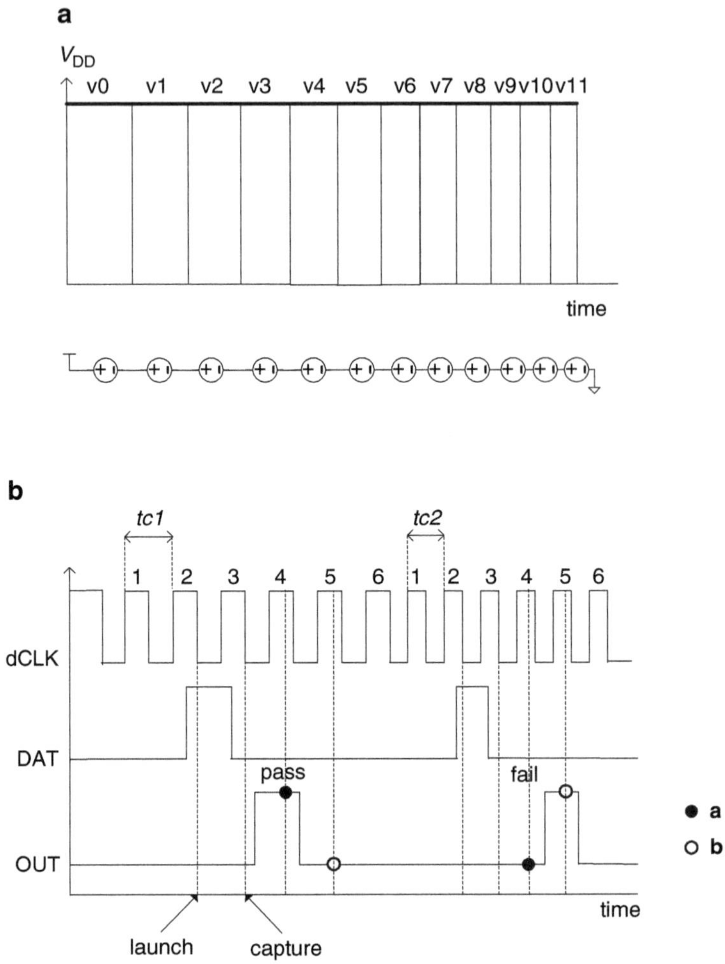

Fig. 3.35 Waveforms for T_{cmin} simulation tests (**a**) power supply (all segments) and (**b**) dCLK, DAT and OUT (offset plus first two 6-cycle segments). The clock waveform is created with a set of series connected voltage sources as described in the text. "*a*" (*solid circle*) and "*b*" (*open circle*) indicate voltage measurement locations on the OUT waveform

Fig. 3.36 Normalized measured outputs for possible regions of data path operation. 1-0 and 0-1 transitions must occur in region 2 for V_{min} or T_{cmin} to be determined

input signal at node A. In the case of T_{cmin}, moving from left to right the T_c corresponding to the last "1" before the transition to a "0" corresponds to T_{cmin} for the rising edge (=*tcnminir*). Moving from right to left the cycle time corresponding to the first "0" that appears after the transition from a "1" to a "0" corresponds to T_{cmin} for the falling edge (=*tcnminif*).

The SPICE commands and procedures for measurements are illustrated with a limited number of increments. It is initially assumed that the V_{min} or T_{cmin} are within the V_{DD} or cycle time range of the tests performed, i.e., region 2 of Fig. 3.36. If this is not the case it is important that the simulation indicates in which direction V_{DD} or cycle time must be adjusted to move the range from region 1 or region 3 into region 2.

If the path delays for rising and falling inputs are different, then $V_{minir} \neq V_{minif}$. In such cases, the true V_{min} for the path is the maximum of V_{minir} and V_{minif}. We will call this maximum value V_{minhi}. It is also useful to define V_{minlo} as the minimum of V_{minir} and V_{minif}. If $V_{minir} = V_{minif}$, then $V_{minhi} = V_{minlo} = V_{min}$. A similar nomenclature is used for T_{cmin}, with T_{cminir}, T_{cminif}, T_{cminhi}, and T_{cminlo}, correspondingly defined.

As an example consider a V_{min} simulation with 11 different V_{DD} values beginning at *v*1 and decreasing in steps of *dv* to *v*11. The first few commands to obtain the 11 voltages at the "a" measure points are given below. The circuit is initialized during the initial offset time *tos*. Since this is a V_{min} experiment the cycle time within the parameterized clock waveform is constant so $tc1 = tc2 = \cdots = tc$.

```
.measure tran v1a2 find V(out) at = tos+3.3*tc1
.measure tran v2a2 find V(out) at = tos+3.5*tc2+6*tc1
.measure tran v3a2 find V(out) at = tos+3.5*tc3+6*(tc2+tc1)
.measure tran v4a2 find V(out) at = tos+3.5*tc4+6*(tc3+tc2+tc1)
...............................................
```

These voltages are then converted to the ideal OUT voltage values with the measure commands below. The difference between the measured and the ideal voltage values should be negligible. However, note that in the case of T_{cmin} experiments, this step can be used to convert non-zero values to corresponding *tc* values.

```
.measure tran v1a1 param buf(2*v1a2)*v1
.measure tran v2a1 param buf(2*v2a2)*v2
.measure tran v3a1 param buf(2*v3a2)*v3
................................................................
```

Next, using the LTspice max function, zero-voltage values are converted to 10, while non-zero values are remain unchanged:

```
.measure tran v1a param max(v0, (0.1*(1/(v1a1+0.01))))
.measure tran v2a param max(v1, (0.1*(1/(v2a1+0.01))))
```

.measure tran v3a param max(v2, (0.1(1/(v3a1+0.01))))*

……………………………………………………

Using the LTspice min function the minimum of these 11 voltages, *vminir*1, is then determined. Note that by converting the 0 values to 10 (or any other number $\gg V_{DD}$), determining the minimum value is all that is required.

.measure tran pmin1 param min(v1a, v2a)
.measure tran pmin2 param min(v3a, pmin1)
.measure tran pmin3 param min(v4a, pmin2)

………………………………………………………

.measure tran vminir1 param min(v11a, pmin9)

Using the LTspice inv function *vminir* is then determined

*.measure tran vminir param inv(vminir1/5)*vminir1*

Note that *vminir* = *vminir1* in regions 1 and 2 of Fig. 3.36, but it is 0 in region 3, a feature that will be used later.

The corresponding *vminif* value is determined from the 11 "b" measure points using a similar methodology except that zero values need not be converted to 10 as in this case it is a maximum voltage that is determined from the set of 11 values. The set of commands in the sequence to determine *vminif* is

.measure tran pmax11 param max(v11b, v10b)
.measure tran pmax10 param max(v9b, pmax11)
.measure tran pmax9 param max(v8b, pmax10)

……………………………………………………

.measure tran pmax2 param max(v1b, pmax3)
.measure tran vminif param pmax2+dv

Next *vminhi* and *vminlo* are determined with the following two commands:

.measure tran vminlo param min(vminir, vminif)
.measure tran vminhi param max(vminir, vminif)

Provided the simulations are being done in region 2 of Fig. 3.36, *vminhi* will be the true value of V_{min} for the path, independent of whether *vminir* = *vminif* or not. If *vminir* = *vminif* then *vminlo* = *vminhi* as well. However, *vminlo* has other useful properties. If the simulations are being done in region 1 or region 3 instead of region 2, *vminlo* will have values of *dv* or 0 respectively. Thus the parameter *vminlo* at the output of the simulation either gives a value of V_{min} or informs that the simulation is being done in region 1 or in region 3 and must be re-centered. Once the V_{DD} range is centered *vminhi* can be used ensure the correct value of V_{min} in the event *vminir* ≠ *vminif*.

Using a very similar methodology, the value of T_{cmin} can be determined for a simulation in which the cycle time is varied systematically in different segments of the clock waveform as shown in Fig. 3.35b. Flags to indicate cases where the simulation has been done in region 1 or 3 instead of region 2 can similarly be introduced through T_{cminlo}.

While the methodology described above works well to determine T_{cmin} or V_{min} for the logic data path under study, there is a subtle complication that can arise that may lead to the failure of a succeeding path due to the delayed appearance of the signal at the output of X2 in an otherwise passing event. This situation is depicted in Fig. 3.37 where the rising edge is arriving at Z just before the dCLK is falling. As the arrival of the signal at Z is progressively later, the delay through the latch first increases before transmittal fails altogether, leading to a shortened output pulse with an anomalously long delay in the output of X2 with respect to the falling dCLK edge. The delay through the latch can be 2–3× nominal before transmittal ceases. This behavior is characteristic of the latch's approach to metastability as described in Sect. 3.2.3. The "a" measurement points in Figs. 3.34 and 3.35 are sufficiently delayed in the cycle that they will register such an event as a pass, provided the signal does make it through.

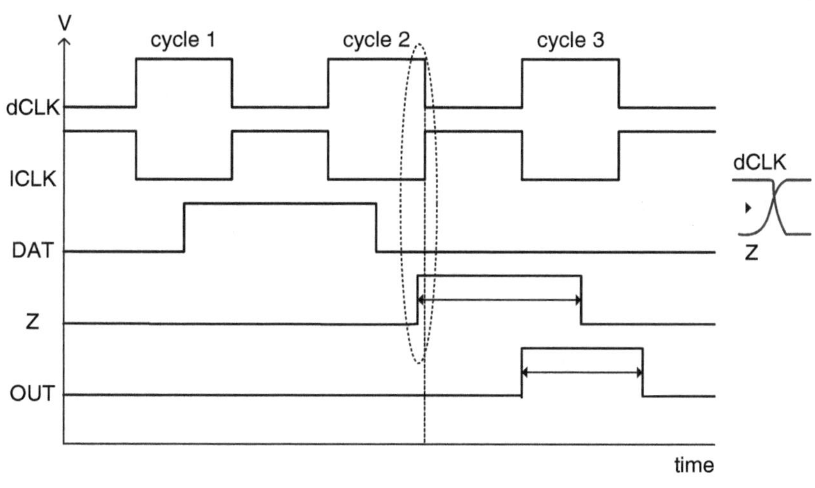

Fig. 3.37 Late arrival of signal at Z increases the delay through latch X2, and the output waveform is narrower

Note that an event like this may or may not be registered as a fail by the subsequent measurements at "b" in the next cycle depending on the path delay corresponding to the falling edge at the OUT terminal of X1. For the simulation of a single segment of data path described here, this additional delay through the latch will be of no consequence. However, for a similar path in a real machine, the delayed output of X2 could potentially cause the failure of a subsequent path, leading to a failure to meet the T_{cmin} or V_{min} criteria of the larger system.

For the schematic shown in Fig. 3.32, the path depicted comprises nominally identical series connected inverters (FO = 4). The path delay varies with the characteristics of the inverter and in turn its constituent n-FET and p-FET, resistance and capacitance of interconnecting wires, and operating power supply voltage and temperature. These inverters can be easily replaced with any other path composition of interest. For the following examples, the inverter path is used and T_{cmin}, and V_{min} correspond to *tcminhi*, and *vminhi* respectively. Note that for this special configuration with no variability in circuit parameters both the "ir" (input rising) and "if" (input falling) paths will generally have the same delay.

As a first illustration, in Fig. 3.38a the simulated values of T_{cmin} are plotted as a function of V_{DD} for 45 nm PTM HP models at 25 °C using our standard inverter (FO = 4) in the logic path. T_{cmin} decreases as V_{DD} is increased and their relationship is nonlinear. In Fig. 3.38b f_{max} (=1/T_{cmin}) is plotted as function of V_{DD}. As expected, f_{max} increases as V_{DD} is raised. A linear fit of the plot has a correlation coefficient of 0.994 and the linear equation may be used to estimate f_{max} for practical values of V_{DD}.

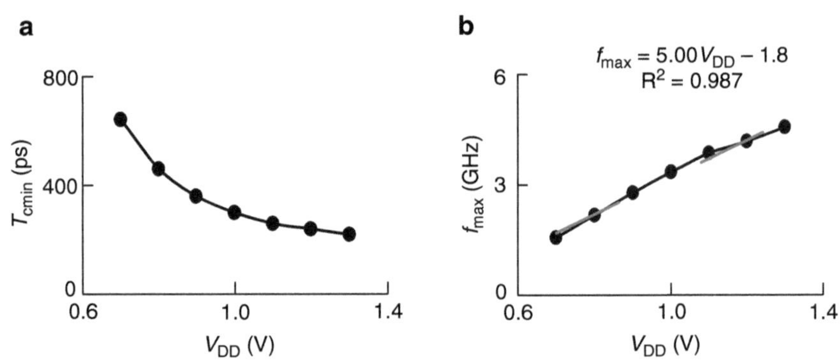

Fig. 3.38 (**a**) T_{cmin} vs. V_{DD} and (**b**) f_{max} vs. V_{DD} for the circuit in Fig. 3.32. 45 nm PTM HP models at 25 °C

Circuit simulations are next carried out to obtain the minimum operating voltage V_{min} at which the data are successfully propagated at a fixed cycle time for the same inverter (FO = 4) path. The results of these simulations for a range of cycle times are plotted in Fig. 3.39.

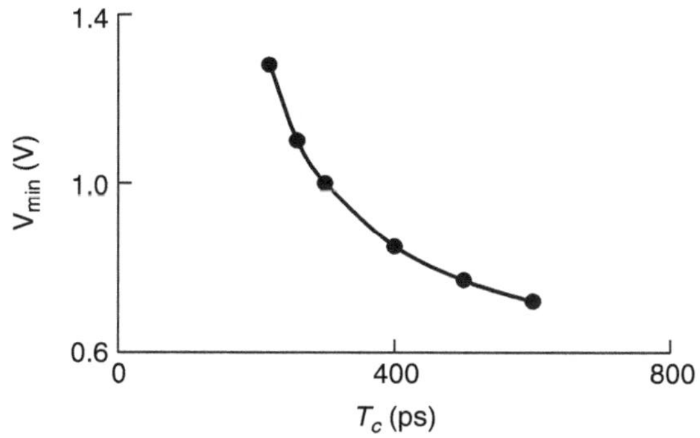

Fig. 3.39 V_{min} as a function of cycle time T_c for the circuit in Fig. 3.32. 45 nm PTM HP models at 25 °C

The accuracy with which T_{cmin} at a fixed V_{DD} is determined is dependent on the step size used in the simulations for T_c. Similarly, the accuracy of V_{min} at a fixed T_c is dependent on the step size used in the simulations for V_{DD}. Using a smaller step size gives higher accuracy but increases the simulation time for covering the same V_{DD} or T_c range.

In the chip design phase, a target cycle time (T_c) is assigned for timing the data paths. If T_{cmin} of a path is $>T_c$, it is considered to have a negative slack of ($T_c - T_{cmin}$). The circuits in such paths must be tuned so that ($T_c - T_{cmin}$) becomes positive. If, on the other hand, T_{cmin} is $>T_c$, the path has a positive slack. Circuits in such paths may be resized to reduce power while maintaining a small positive or zero slack in the worst-case corner.

When testing a CMOS chip, the procedure for obtaining T_{cmin} and V_{min} are conceptually similar to that used in the simulations described above. Accuracy of measurement in the hardware is improved by reducing the step size. The minimum step size is limited by the capability of the test equipment and the uncertainties or errors introduced by the test setup. Reducing the step size increases the test time and cost of testing, an important consideration in the overall test economics. Different sources of variation discussed in Chap. 6 tend to spread the slack distribution in the hardware, and the failing paths are often not the ones with minimum positive slack in design. Identification of failing paths in product test and relating them to circuit design is an area of major focus.

3.5 Summary and Exercises

In this chapter complex functional blocks, on-chip clock generation and distribution, clocked storage elements, and random access memory cells (SRAM and DRAM) are reviewed. Circuit schematics to determine SRAM DC characteristics and read and write times are described. Two examples of SPICE simulation setups to directly extract desired circuit parameters are described in detail. In the first example, static noise margins (SNMs) of a SRAM cell are extracted in SPICE, facilitated by a coordinate transformation of half-cell input–output characteristics. In the second example, T_{cmin} and V_{min} of a data path are directly measured in SPICE by varying the cycle time T_c or V_{DD} in predefined steps. The signal output is compared with the expected level, and T_c and V_{DD} values at which a timing error first occurs are determined. With the ability to run Monte Carlo simulations, direct measurements of SNM, T_{cmin} (f_{max}) and V_{min} facilitate relating SRAM and data path sensitivities to the properties of the devices.

Exercises 3.1–3.4 feature latch operation, setup and hold times, metastability and detection of erroneous latch outputs. Exercises 3.5 and 3.6 cover characterization of SRAM cells and sensitivity of SNM to MOSFET ratios. T_{cmin} and V_{min} for different data path compositions are investigated in Exercises 3.7–3.10.

3.1. (a) Create a schematic for a level-sensitive latch (Fig. 3.11) or select one from an existing design library. Generate a timing diagram similar to Fig. 3.12a from circuit simulations and validate the operation of the latch in the nominal corner. Include commands to measure τ_r and τ_f of the waveform at the latch output node.
(b) With data signal $\tau_r = \tau_f = 16$ ps measure latch output τ_r and τ_f.
(c) Increase the data signal $\tau_r = \tau_f = 160$ ps and measure latch output τ_r and τ_f. How do they compare with results from (a)?
(d) Keeping data signal $\tau_r = \tau_f = 16$ ps, change the clock $\tau_r = \tau_f$ to 160 ps and measure latch output τ_r and τ_f. How do they compare?
(e) What conclusion do you draw from the above simulations?

3.2. (a) Using the level-sensitive latch of Problem 3.1 with $\tau_r = \tau_f = 16$ ps at the nominal corner, measure latch delay as a function of DAT-to-CLK delay (Fig. 3.18). Determine DAT-to-CLK delay (=*tdly*) for 1.2× nominal latch delay.
(b) Add jitter to the clock signal by using a time delay (=*tdly* + *dt*) with *tdly* determined from (a) and *dt* following a Gaussian distribution with mean = 0 and standard deviation $\sigma dt = (T_c/2 - tdly)/3$. Run Monte Carlo simulations for 100 cases and record the number of failures as the latch operates in the metastable region. Adjust σdt to increase and decrease the number of failures.
(c) Determine the minimum DAT-to-CLK delay for a given σdt for zero failures in 100 simulation runs.

3.3. (a) What is the maximum V_{DD} droop at which the latch in Exercise 3.1 can function correctly in the nominal corner if the DAT-to-CLK delay is set at T_s (assume 10 % increase in latch delay at T_s).
(b) Add clock jitter = 2 % of Tc and determine maximum V_{DD} droop for this condition.

3.4. (a) Set up circuit simulation for propagating a data signal through a level-sensitive MSL with dCLK and lCLK 180° out of phase. Determine minimum DAT-to-dCLK delay such that the output of the MSL is updated.
(b) Set lCLK rising edge so that it overlaps the falling edge of dCLK by 50 ps (both dCLK and lCLK at "1"). What is the minimum DAT-to-dCLK delay for a valid output? How would you set the clocks to capture late arriving data (cycle stealing from next cycle)?
(c) Set dCLK at "1" and lCLK with a duty cycle of 20 %. Shift the lCLK to capture late arriving data.

3.5. Import any one of the SRAM cell designs from an existing design library or create an optimized design for this and the next exercise. Set up circuit simulations and measure read and write times of the SRAM cell as outlined in Sect. 3.3.1.

3.6. (a) Set up a circuit simulation to generate a butterfly curve of the SRAM cell latch in the nominal corner.
(b) Set up a circuit simulation to measure SNM of this latch using the methodology described in Sect. 3.4.1.
(c) Determine the SNM with different values of W_p/W_n of the latch inverter. Plot the corresponding butterfly curves and check for consistency.

3.7. (a) Set up a circuit to measure f_{max} and V_{min} of a data path with 20 inverters (FO = 4) as described in Sect. 3.4.2.
(b) Determine T_{cmin} in the nominal corner. Measure inverter path delay and compare it to T_{cmin}. What fraction of T_{cmin} is due to inverter path delay?
(c) Determine V_{min} at a cycle time of $1.2 \times T_{cmin}$. Determine the inverter path delay at V_{min} and compare with the cycle time. Is this result consistent with (a)?

3.8. For the circuit used in Exercise 3.7, the clock edge arrival time at the second latch is advanced by 5 % of cycle time due to jitter. What is the change in V_{min} from the case in Problem 3.7b?

3.9. (a) Simulate the circuit in Exercise 3.7 to determine V_{min} for the nominal case and for four other cases with (*pdelvto, ndelvto*) = (+0.04 V, +0.04 V), (+0.04 V, −0.04 V), (−0.04 V, +0.04 V) and (−0.04 V, −0.04 V).
(b) Compare shift in V_{min} from nominal with different *delvto* combinations.

3.10. (a) Modify the data path in Exercise 3.7 by replacing 10 of the inverters with NAND3B logic gates.
(b) Determine f_{max} of the circuit in the nominal corner.
(c) What do you need to do to get the same f_{max} as the 20 inverter path?

References

1. Henessy JL, Patterson DA (2011) The basics of logic design. In: Computer organization and design, Appendix C. Morgan Kaufmann Publishers, San Francisco
2. Rabaey JM, Chandrakasan A, Nikolic B (2003) Digital integrated circuits, 2nd edn. Prentice Hall, Upper Saddle River
3. Weste NH, Harris D (2010) CMOS VLSI design: a circuit and systems perspective, 4th edn. Addison-Wesley, Boston
4. Haraszti TP (2000) CMOS memory circuits. Kluwer Academic Publishers, Boston
5. Xanthopoulos T (ed) (2009) Clocking in modern VLSI systems. Springer, Berlin
6. Warnock J, Sigal L, Wendel D, Muller KP, Friedrich J, et al. (2010) POWER7 local clocking and clocked storage elements. In: 2010 I.E. international solid-state circuits conference, pp 178–179
7. Bhushan M, Ketchen MB, Das KK (2008) CMOS latch metastability characterization at the 65-nm-technology node. In: Proceedings of the 2008 I.E. international conference on microelectronic test structures, pp 147–151
8. Seevinck E, List FJ, Lohstroh J (1987) Static-noise margin analysis of MOS SRAM cells. IEEE J Solid-State Circuits 22:748–754

IDDQ and Power

4

Contents

4.1 Silicon Technology Scaling and Power 127
4.2 IDDQ 128
4.2.1 MOSFET Leakage Currents 129
4.2.2 IDDQ of Logic Gates and Memory Cells 130
4.2.3 IDDQ Estimation in Design and Measurements 135
4.2.4 Defect Generated IDDQ 137
4.3 Power 140
4.3.1 Measuring Power 140
4.3.2 AC Power 141
4.3.3 DC Power 146
4.4 Total Power 148
4.5 Power Management 151
4.5.1 Power Management in Chip Design 152
4.5.2 System Power Management 155
4.6 Summary and Exercises 155
References 157

Increase in MOSFET leakage currents with scaling has led to different scaling scenarios for power-performance optimization. MOSFET leakage currents and defects contribute to current drawn in the quiescent state of a CMOS chip. Measurement of this current, IDDQ, is useful in eliminating chips with gross defects early in the test flow. DC and AC components of total power in the active mode are functions of power supply voltage, switching activity and temperature. Circuit design and dynamic on-chip power management schemes help alleviate potential reliability issues and reduce overall power consumption.

Power drawn by CMOS chips in the quiescent state, standby state with minimum switching activity, and functional states is an important consideration in product design and packaging. Maximum allowable power specifications are set by external power supply, thermal, and reliability considerations. In low power and battery

M. Bhushan, M.B. Ketchen, *CMOS Test and Evaluation*,
DOI 10.1007/978-1-4939-1349-7_4

operated consumer electronics, specifications for quiescent and standby power levels are more stringent than in high performance microprocessor chips.

Typically a power supply is set to deliver a current I_{DD} at a fixed voltage V_{DD}. A CMOS chip may have one or several power supplies with different V_{DD} settings. The total power consumed at any instant is the sum of $I_{DD} \times V_{DD}$ for all power supplies. The power is dissipated as heat, raising the chip temperature. This in turn results in higher wire resistances and IR drops, and in lower circuit speeds. Aging mechanisms of circuit components are accelerated with increase in temperature causing performance degradation or catastrophic failures over long term use. Thermal runaway may occur if the power level exceeds a threshold limit. Power delivery and adequate cooling requirements add to package and product cost. Power consumption is therefore carefully specified and managed for optimum operation of a CMOS product.

When a chip is in a quiescent state and there is no AC activity, the total current drawn by the power supplies is termed IDDQ. It is the sum of leakage current contributions from all the MOSFETs on the chip. Measurement of IDDQ provides useful information on silicon technology power-performance trade-offs. AC power is a function of voltage, frequency and workload (switching capacitance) as described in Sect. 2.2.5. Total power is the sum of power consumed in switching events and background leakage power of idle circuits. Since the leakage power is wasted power, MOSFET engineering as well as circuit design methodologies are tailored to minimize it.

Additional current may be drawn by defects such as low resistance paths across power supplies or as a result of floating node voltages on MOSFET gates. In electrical testing, IDDQ measurements are useful in eliminating defective chips early in the test flow. In some CMOS products, IDDQ tests are used to detect certain types of defects.

An empirical power and thermal model generated from data collected at test is useful in determining product application range, in developing a binning methodology, and in assessing chip reliability over lifetime (Chap. 8). Manufacturing yield may be improved by customizing power supply voltages for each chip or for circuit blocks within each chip. Adaptive testing methods to enable custom specifications of V_{DD} or operating frequency to meet maximum power limits are currently in use in some high-end products. Dynamic management of on-chip power has also been put in place to reduce overall power consumption. Standby and sleep modes in CMOS chips reduce power consumption when circuit blocks or major units are not in active use.

A historical perspective on power scaling trends with reduction in CMOS feature sizes is given in Sect. 4.1. Leakage current components of MOSFETs and small circuit blocks, and IDDQ measurements for defect isolation are covered in Sect. 4.2. Measurements of power, both DC and AC components, are described in Sect. 4.3. In Sect. 4.4, a method for building an empirical power model for a CMOS chip is presented. Some methods of minimizing and managing power are discussed in Sect. 4.5.

MOSFET properties and power consumption in CMOS circuits are covered in textbooks on semiconductor devices and VLSI circuit design [1–4]. Technical publications on special topics such as defect detection and power management techniques are also cited.

4.1 Silicon Technology Scaling and Power

Through the early 1990s, MOSFET dimensions L_p, W, and t_{ox} were scaled by a scaling factor S, where $S < 1$, consistent with feature size reduction (~0.7× per technology node). The power supply voltage V_{DD} was also scaled by S and the electric-fields in both horizontal (across the channel) and vertical (across gate-dielectric) directions remained unchanged. The constant-electric field scaling rules are summarized in Table 4.1. With scaling of MOSFET dimensions, circuit density scales by a factor of $1/S^2$. Wire widths, spacings, and inter-level dielectric thicknesses are scaled by S and with increase in circuit density interconnect wire lengths can also be scaled by S. Hence MOSFET and interconnect wire capacitances both scale by S.

Table 4.1 Constant electric field scaling rules for MOSFET and circuit parameters [1]

Properties	Parameters	Scaling factor ($S < 1$)
MOSFET dimensions	t_{ox}, L_p, W, L_{ds}	S
Doping concentration	N_a, N_d	$1/S$
Power supply voltage	V_{DD}	S
Electric field	V_{DD}/t_{ox}, V_{DD}/L_p	1
Gate capacitance	$\varepsilon L_p W/t_{ox}$	S
Wire capacitance	$\varepsilon lw/h$	S
Circuit density	$\approx 1/\{W \times (L_p + 2L_{ds})\}$	$1/S^2$
Circuit delay	$\sim CV_{DD}/I$	S
AC power/circuit	P_{ac}	S^2
Power density	P_{ac}/A	1

Taking advantage of the circuit delay metric CV/I scaling by S, CMOS chip frequencies can be increased by $\sim 1/S$ per technology node. Although AC power per circuit scales as S^2, the number of circuits per unit area increases as $1/S^2$. As a result AC power density remains constant with scaling. With increase in doping concentrations as $1/S$, threshold voltage is reduced, resulting in an increase in subthreshold leakage current. However, for technology nodes down to 180 nm, leakage power remained a negligible fraction of the total power.

The idealized constant electric field scaling rules in Table 4.1 could not always be followed because of the need for a standardized V_{DD} for CMOS products manufactured in at least a few consecutive technology generations. A generalized scaling model in which V_{DD} and V_t are scaled more gradually than the feature size scaling factor S has been in use instead. The generalized scaling rules in terms of S and a second factor for electric field scaling S_k ($S_k < 1$) are listed in Table 4.2. In this approach for short channel MOSFETs, the circuit delay still scales by S allowing the clock frequency to continue to increase at the same rate. However, AC power density at constant clock frequency increases as $1/S_k^2$. Operation at higher clock frequencies further increases the power density. Even with advances in packaging and cooling technology, the general trend is continued rise in chip temperature, impacting circuit delay and reliability of CMOS products.

Table 4.2 Generalized rules for scaling for MOSFET and circuit parameters [1]

Properties	Parameters	Multiplication factor ($S < 1$, $S_k < 1$)
MOSFET dimensions	t_{ox}, L_p, W, L_{ds}	S
Gate capacitance	$\varepsilon\ L_pW/t_{ox}$	S
Power supply voltage	V_{DD}	S/S_k
Electric field	V_{DD}/t_{ox}, V_{DD}/L_p	S_k
Circuit delay	(CV/I)	S
Circuit density	$1/A = 1/\{W \times (L_p + lds)\}$	$1/S^2$
AC power/circuit	P_{ac}	$(S/S_k)^2$
Power density	P_{ac}/A	$(1/S_k)^2$

With continued reduction in V_t and exponential increase in subthreshold leakage currents, leakage power has steadily crept up to become a measurable fraction of the total power. In addition, the gate-dielectric tunneling current I_{gl} becomes significant for $t_{ox} < 3$ nm. The reduction in V_t and t_{ox} can no longer be sustained beyond the 90 nm technology node. Reduction in I_{gl} has been addressed by replacing silicon-oxide as a gate-dielectric material with a composite high dielectric constant (HK) material at the 45 nm technology node by Intel. This approach has been adopted by the silicon industry in all subsequent nodes. With further scaling of CMOS technology and the introduction of FinFETs by Intel at the 22 nm technology node, there has been a reduction in subthreshold leakage currents as well.

High-end CMOS products such as microprocessor chips could manage the increase in AC power density and leakage power, partly by expensive packaging and cooling and partly by power management techniques. Consumer electronics and battery operated products demand low power operation and sacrifice potentially higher frequency to achieve it. Both types of products benefit from scaling through manufacturing cost reduction as a result of producing more chips per wafer and packing more functions on a single chip.

Silicon foundries now offer MOSFETs optimized for either high performance (HP) or low power (LP) applications at a single technology node. In this chapter we will use 45 nm PTM models for HP and LP to illustrate the major differences in these two approaches.

4.2 IDDQ

MOSFET leakage currents and DC current paths in circuits contribute to chip IDDQ. In Sect. 4.2.1, MOSFET leakage current components are described. IDDQ of logic gates and their voltage and temperature dependencies are discussed in Sects. 4.2.2 and 4.2.3, respectively. IDDQ generated by defects is covered in Sect. 4.2.4. In 130 nm and earlier technology generations, IDDQ measurements have been successfully used to detect systematic and random defects and other yield detractors [6, 7]. In more advanced technologies, with high MOSFET leakage currents, the level of IDDQ contributions generated by a few defects is much smaller than the total chip IDDQ. In these technologies IDDQ test can only be used to reject chips with gross defects.

4.2.1 MOSFET Leakage Currents

MOSFET leakage currents are functions of gate, source, drain and body voltages. Subthreshold leakage current I_{off} is modulated by source-to-body voltage V_{bs}. In the initial discussion of I_{off} it is assumed that the body is tied to the source ($V_{bs} = 0$).

The leakage current contributions of an n-FET and a p-FET in the off-states and on-states are shown in Fig. 4.1. When the MOSFETs are in the off-state, in addition to the leakage current I_{off} across the channel, there is small component of gate-dielectric leakage current I_{gld}, in the gate-to-drain overlap region ($V_{gs} = 0$, $V_{gd} = |V_{DD}|$). In the on-state, S and D terminals are at the same potential and gate-dielectric tunneling current I_{gl} flows across G and S/D terminals. The leakage current components in the off-state are denoted by I_{offn} and I_{gldn} for an n-FET and I_{offp} and I_{gldp} for a p-FET. Similarly, in the on-state, the gate-dielectric leakage current components are I_{gln} for an n-FET and I_{glp} for a p-FET.

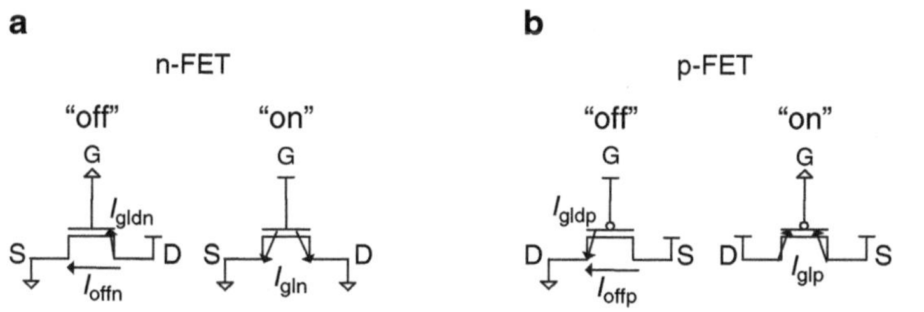

Fig. 4.1 Leakage current components in the "off" and "on" states of (**a**) n-FET and (**b**) p-FET

The voltage and temperature dependencies of I_{off} and I_{gl} per unit width are expressed in the following equations [1]:

$$I_{off} = \frac{\mu_{eff} C_{ox}}{L_p} \left\{ (\gamma - 1) \left(\frac{kT}{q} \right)^2 \exp\left(\frac{-qV_t}{\gamma kT} \right) \left(1 - \exp\left(\frac{-qV_{ds}}{kT} \right) \right) \right\}, \tag{4.1}$$

and

$$I_{gl} = c_1 \times L_p \left\{ \frac{V_{gb}}{t_{ox}} \right\} \exp\left(\frac{-c_2 t_{ox}}{V_{gb}} \right), \tag{4.2}$$

where V_{gb} is the gate-to-body bias voltage and c_1 and c_2 are constants. When the body is tied to the source, $V_{gb} = V_{gs}$. I_{off} increases with increase in V_{ds} or decrease in V_t. I_{gl} increases with increase in V_{gs} and decrease in t_{ox}. As V_{ds} follows V_{DD} in the off-state and V_{gs} follows V_{DD} in the on-state, both I_{off} and I_{gl} increase with increase in V_{DD}. The temperature dependencies of these two leakage current components are very different. With an exponential increase in carrier density with temperature, I_{off} exhibits a strong temperature dependence. The tunneling current I_{gl} is a function of t_{ox} and tunnel barrier height, and is nearly independent of temperature.

Nominal values of I_{off} and I_{gl} for an n-FET and a p-FET simulated with 45 nm PTM HP and LP models are listed in Table 4.3 at 25 and 100 °C. The simulations are carried out at the prescribed nominal V_{DD} for the 45 nm technology nodes, 1.0 V for HP and 1.1 V for LP. Note that as the temperature increases from 25 to 100 °C, I_{off} increases by >4×, while I_{gl} is nearly constant. In the HP models, the relative contribution of I_{gl} to the total leakage current is small. In LP models at 25 °C, I_{off} is comparable to I_{gl} for the p-FET and smaller than I_{gl} for the n-FET. At 100 °C, I_{off} is the dominate contributor in both HP and LP models. These trends vary at different technology nodes and with different MOSFET offerings.

Table 4.3 I_{off} and I_{gl} in nA/μm for an n-FET and a p-FET in HP (V_{DD} = 1.0 V) and LP (V_{DD} = 1.1 V) models with body tied to source. 45 nm PTM models

MOSFET	Model	V_{DD} (V)	I_{off}@25 °C (nA/μm)	I_{off}@100 °C (nA/μm)	I_{gl}@25 °C (nA/μm)	I_{gl}@100 °C (nA/μm)
n-FET	HP	1.0	20	83.7	0.62	0.79
p-FET	HP	1.0	4.5	20.4	2.52	2.63
n-FET	LP	1.1	0.024	0.29	0.111	0.122
p-FET	LP	1.1	0.020	0.22	0.031	0.034

The value of I_{gld} is <0.5 nA/μm in HP models and is nearly independent of L_p and temperature. In LP models, I_{gld} is several orders of magnitude smaller than I_{gl}. Its contribution to total leakage is henceforth ignored.

The dependencies of I_{off} and I_{gl} on MOSFET parameters L_p and V_t are also different. I_{off} is a strong function of both L_p and V_t. As an example, in the worst-case at 25 °C, with L_p and V_t at their respective -3σ values, I_{off} for an n-FET is 2,550 nA/μm, a factor of 128× its nominal value. As I_{gl} is proportional to the gate area, it increases linearly with L_p.

Because of the strong dependence of I_{off} on MOSFET process variations, CMOS circuits are designed to tolerate the worst-case (-2σ or -3σ) scenario at the highest operating temperature.

BSIM compact models for MOSFETs are typically fit to measured data from hardware with a high degree of accuracy in the saturation region. Because of the exponential increase in I_{ds} in the subthreshold region, these models may not accurately represent MOSFET behavior in this region. To circumvent this problem, silicon foundries generally offer separate models for computing subthreshold and gate-dielectric leakage currents for the sole purpose of estimating chip IDDQ.

4.2.2 IDDQ of Logic Gates and Memory Cells

The leakage currents in a logic gate depend on input voltage levels and the circuit topology. Figure 4.2 shows two stages of an inverter chain and the leakage current contributions of the constituent MOSFETs. In the first stage, with the gates at “1”, the n-FET is in the on-state and the p-FET is in the off-state. In the second stage, the n-FET is in the off-state and the p-FET is in the on-state. As described earlier, I_{off} contributions are set by MOSFETs in the off-state while I_{gl} contributions are primarily set by the MOSFETs in the on-state.

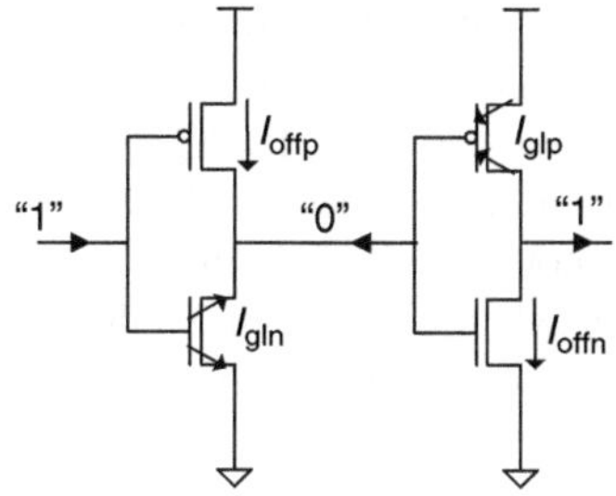

Fig. 4.2 Leakage current components of two series connected inverters

The average IDDQ/stage for the two series connected inverters is given by

$$\mathrm{IDDQ/stage} = \frac{1}{2}\left\{W_p\left(I_{offp} + I_{glp}\right) + W_n\left(I_{offn} + I_{gln}\right)\right\}. \tag{4.3}$$

In Fig. 4.3, leakage current components of two series connected NAND2T gates in a delay chain are shown. The average IDDQ/stage of this circuit is given by

$$\mathrm{IDDQ/stage} = \frac{1}{2}\left\{W_p\left(2I_{offp} + I_{glp}\right) + W_n\left(I_{offn} + 3I_{gln}\right)\right\} \tag{4.4}$$

and is higher than that of the inverter circuit in Fig. 4.2 by ~$1/2 \times (W_pI_{offp} + 2W_nI_{gln})$, assuming W_n and W_p are unchanged.

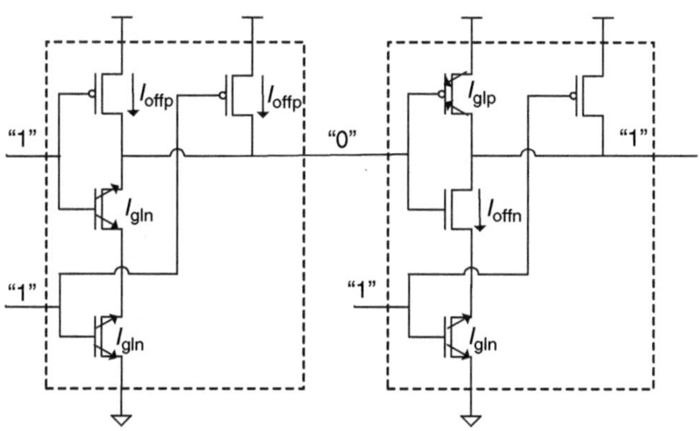

Fig. 4.3 Leakage current components of two series connected NAND2T gates

IDDQ contributions of logic gates with only bottom (B) or only top (T) input switching, or with all inputs tied together and switching simultaneously (ALL) are obtained from circuit simulations of delay chains of identical gates in their quiescent states. In the case of B or T inputs switching, all other inputs are tied to V_{DD} or GND to obtain an inverted output. The simulation results are shown in Table 4.4 for a set of combinational logic gates (FO = 1), with all MOSFETs having $W_n = 0.4$ μm and $W_p = 0.6$ μm. This width selection highlights the difference in IDDQ arising only from logic gate configurations.

From the data in Table 4.4, IDDQ increases as the stack height and the total number of MOSFETs in the logic gate increase (e.g., IDDQ (NAND3T) > IDDQ (NAND2T) > IDDQ (inverter)). However, the IDDQ per unit MOSFET width decreases with stack height as long as $I_{off} > I_{gl}$, as not all MOSFETs in a gate contribute to I_{off}. In a logic gate family, IDDQ is also higher when the bottom MOSFET in the stack remains in the on-state (top input switching), i.e., IDDQ (NAND2T) > IDDQ (NAND2B).

Table 4.4 Average IDDQ/stage of delay chains of logic gates ($W_p = 0.6$ μm, $W_n = 0.4$ μm, FO = 1). 45 nm PTM HP models @ 1.0 V, 25 °C

Logic Gate	IDDQ@25 °C (nA/μm)	Logic Gate	IDDQ@25 °C (nA/μm)	Logic Gate	IDDQ@25 °C (nA/μm)
Inverter	6.15				
NAND2B	6.45	NAND3B	7.42	NAND2ALL	4.60
NAND2T	7.42	NAND3T	9.32		
NOR2B	11.58	NOR3B	14.78	NAND3ALL	6.75
NOR2T	10.18	NOR3T	17.00		

Leakage current contributions to IDDQ of an inverter driving an n-passgate are shown in Fig. 4.4. In Fig. 4.4a, with the input at "0", D, G and S terminals of the n-passgate are at "1" and only the inverter contributes to IDDQ. In Fig. 4.4b, with the input at "1", I_{gln} of the n-passgate also contributes to IDDQ.

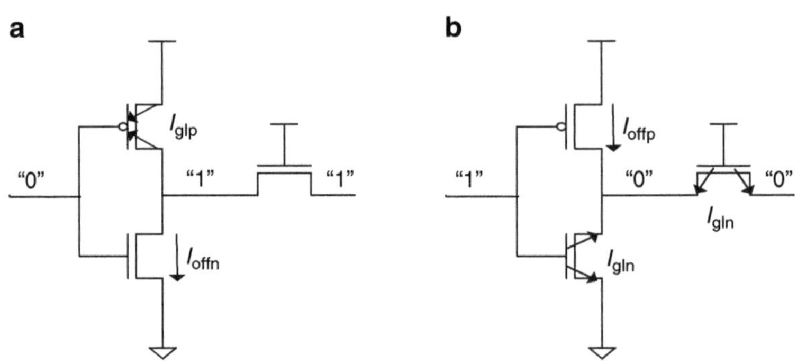

Fig. 4.4 Leakage current components of an inverter driving an n-passgate (**a**) input at "0" and (**b**) input at "1"

A 6T SRAM circuit is shown in Fig. 4.5. The word line WD is at "0" and bit lines B and B_b are at "1" when the cell is being pre-charged. Its IDDQ in this state is given by

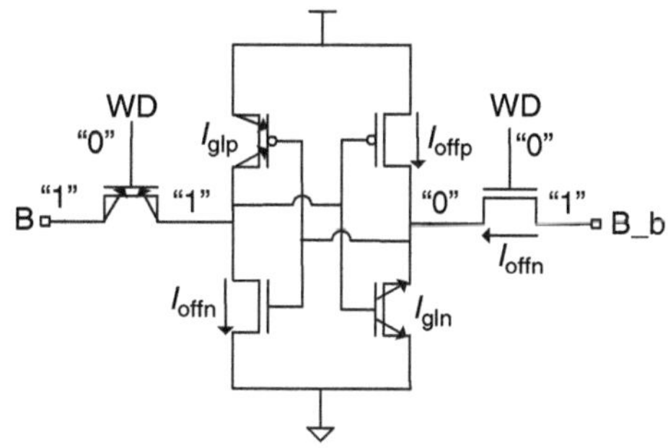

Fig. 4.5 Leakage currents in a 6T SRAM cell in the pre-charged state with word line WD at "0" and bit lines B and B_b at "1"

$$\text{IDDQ} = \{W_{\text{p}}(I_{\text{offp}} + I_{\text{glp}}) + W_{\text{n}}(I_{\text{offn}} + I_{\text{gln}})\} + W_{\text{npg}}(I_{\text{offn}} + I_{\text{gln}}). \quad (4.5)$$

The circuit schematic of a level-sensitive latch is shown in Fig. 4.6a (Sect. 3.2.1) and that of its clocked inverter in Fig. 4.6b. With CLK at "0", n-FET N1 and p-FET P1 are in on-states and n-FET N2 and p-FET P2 perform the invert function. The IDDQ in this case is that of an inverter formed by N2 and P2. With CLK at "1" both N1 and P1 are turned off and the I_{off} contribution to IDDQ is reduced due to series resistances of N1 and P1.

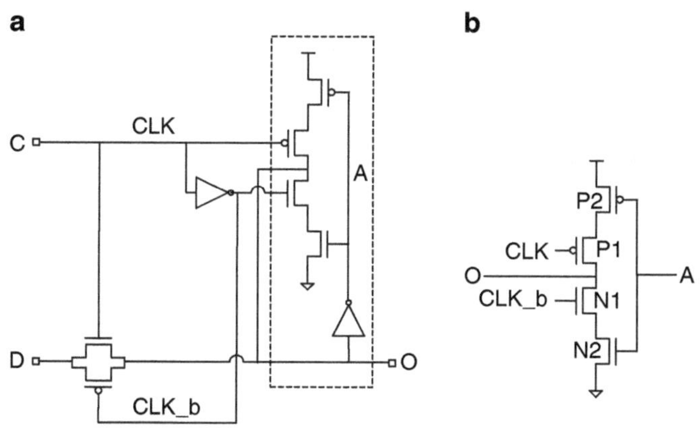

Fig. 4.6 Circuit schematic of (**a**) LSSD latch and (**b**) clocked inverter

For all circuits IDDQ varies over a wide range due to systematic process variations (Sect. 6.1.1). This spread in IDDQ is observed by running Monte Carlo simulations with $\pm 3.0\sigma$ systematic variations in L_{p} and V_{t}. In Fig. 4.7a, the IDDQ distribution of a standard inverter for 45 nm PTM HP models is shown. The distribution is highly positively skewed with a long tail because of the exponential dependence of IDDQ on V_{t}. With a log transformation shown in Fig. 4.7b, the distribution follows the characteristics of a normal distribution (Sect. 9.1.4).

The V_{t} of a MOSFET varies with body-to-source voltage, V_{bs}. The body-effect coefficient γ, which is a function of doping in the channel and t_{ox}, is given by

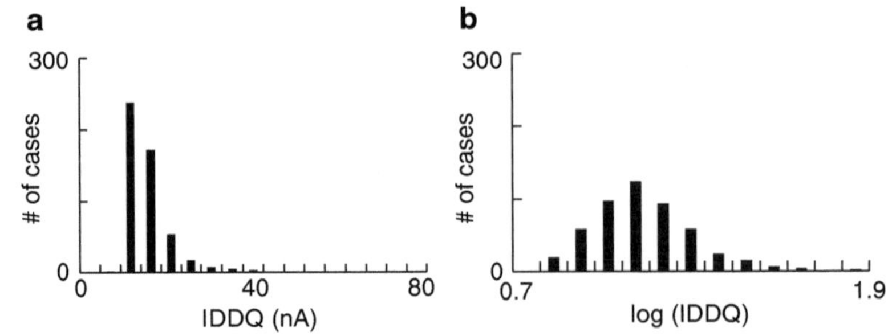

Fig. 4.7 Distribution of average IDDQ for a standard inverter with systematic L_p ($\pm 3.0\sigma$) and V_t ($\pm 3.0\sigma$) variability (500 cases): (**a**) IDDQ and (**b**) log (IDDQ). 45 nm PTM HP models @1.0 V, 25 °C

$$\gamma = 1 + \frac{dV_t}{dV_{bs}}, \tag{4.6}$$

at $V_{bs} = 0$. Typical values of γ are in the range of 1.1–1.2, corresponding to 100–200 mV change in V_t when V_{bs} changes by 1.0 V. For $V_{bs} < 0$, the body is reverse-biased with respect to the source. V_t then increases, and I_{off} is reduced.

A circuit schematic for simulating MOSFET (n-FET) characteristics as a function of V_{bs} is shown in Fig. 4.8a. In order to apply a body bias, a four terminal representation of an n-FET is used (nmos4 in LTspice). In Fig. 4.8b, c, V_{tsat} and I_{off} of an n-FET as functions of V_{bs} are plotted for 45 nm PTM HP models. The value of γ for this n-FET is 1.21, and $|dV_t/dV_{bs}|$ slightly decreases as V_{bs} becomes more negative. There is ~200× reduction in I_{off} as V_{bs} is varied from 0.0 to −1.0 V.

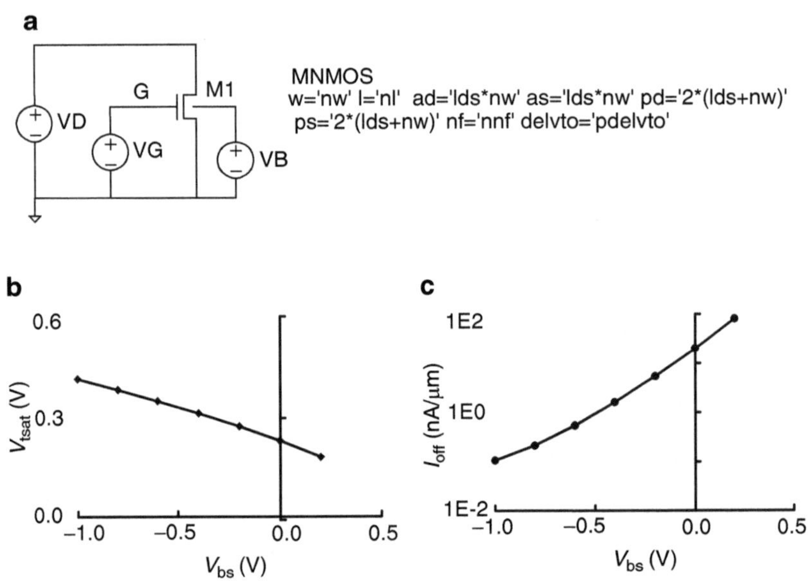

Fig. 4.8 (**a**) Circuit schematic to simulate DC characteristics of an n-FET with body bias, (**b**) V_{tsat} as a function of V_{bs}, and (**c**) I_{off} as a function of V_{bs}. 45 nm PTM HP models @1.0 V, 25 °C

The body-bias dependence of V_t may be exploited in four-terminal MOSFETs to reduce IDDQ during burn-in as discussed in Chap. 8. In bulk CMOS, a twin-well configuration is used to isolate the p-well (n-FET body) and the n-well (p-FET body) for external V_{bs} control. The twin-well configuration also provides better noise immunity in mixed signal circuits. However, because of increase in silicon process complexity and cost, this option is usually not exercised.

In PD-SOI technology, with the body of the MOSFET floating, the body potential in the MOSFET off-state is determined by a balance of charge generated by impact ionization and leakage currents between the body and the source, drain and gate terminals (Sect. 10.6). If the body becomes forward-biased ($V_{bs} > 0$), V_t is reduced and there is an increase in I_{off}. The body-potential is further modulated during switching by capacitive coupling of the gate and drain to the body. The relaxation time for the body to reach its equilibrium state can be on the order of hundreds of nanoseconds to a few milliseconds.

The floating-body effect in PD-SOI is utilized for performance gain over bulk silicon technology at the same technology node by dynamic reduction of V_t. Estimation and measurement of IDDQ in PD-SOI circuits is more complex than for bulk silicon technology. Special test structures are designed to measure IDDQ in PD-SOI circuits in the off-state and in steady-state switching configurations.

4.2.3 IDDQ Estimation in Design and Measurements

An estimation of MOSFET contributions to IDDQ of a CMOS chip in different process and operating corners can be made in early design phase. The total width for each MOSFET type on the chip is obtained from projected circuit count or with the aid of physical layout verification tools (e.g., layout to schematic verification, LVS). One approach is to assume the chip can be represented by inverters. Using Eq. 4.3, the chip IDDQ is given by the summation of IDDQ of inverters of all MOSFET pair types.

$$\mathrm{IDDQ} = \sum_i \frac{1}{2}\left\{W_{\mathrm{pi}}\left(I_{\mathrm{offpi}} + I_{\mathrm{glpi}}\right) + W_{\mathrm{ni}}\left(I_{\mathrm{offni}} + I_{\mathrm{glni}}\right)\right\}, \tag{4.7}$$

where i represents the MOSFET pair type (V_t type etc.) and W_{pi} and W_{ni} are the total widths each p-FET and n-FET type on the chip.

Typically a large fraction of the total MOSFET device width is contributed by clock buffers and other large buffers. However, as seen from Table 4.4, an inverter representation of the chip would overestimate IDDQ. Assuming the chip to comprise NAND2 gates may be more realistic. As the number of MOSFETs in a NAND2 gate is twice that of an inverter while I_{off} is only slightly higher, a NAND2 chip model drops the estimated IDDQ nearly in half.

Sophisticated EDA tools are used for power optimization and for estimating IDDQ in the chip design phase. These tools accurately determine the node voltages of each MOSFET on the chip and compute IDDQ from MOSFET leakage models.

The node voltages are determined by the inputs of logic gates which in turn vary with the inputs of circuit blocks. IDDQ may be computed with all allowed node voltage assignments for a more accurate estimation of the upper limit on IDDQ.

As I_{off} varies exponentially with V_t and temperature, it is difficult to accurately predict IDDQ of a chip comprising multiple V_t and multiple t_{ox} MOSFET offerings. Offsets among different MOSFET types in silicon manufacturing further add to this unpredictability. In addition, across chip variations in L_p, V_t, V_{DD}, and temperature are difficult to model prior to test and may vary from chip to chip (Sect. 6.1).

From a CMOS product test point of view, it is only necessary to specify limits of IDDQ. Simulations are carried out in specific application corners to obtain these limits during chip design. Three example corners for IDDQ are defined below for a chip nominal V_{DD} of 1.0 V and an operating temperature range from −25 to 75 °C.

- Nominal (NOM): Nominal L_p and V_t, $V_{DD} = 1.0$ V, temp = +25 °C
- Low IDDQ: $+3\sigma L_p$, $+3\sigma V_t$, $V_{DD} = 0.9$ V, temp = −25 °C
- High IDDQ: $-3\sigma L_p$, $-3\sigma V_t$, $V_{DD} = 1.1$ V, temp = +75 °C

Note that these corners are different from the WC and BC corners for timing in Table 2.12. Simulated values of IDDQ in these three corners for an inverter (FO = 1) for 45 nm PTM models are listed in Table 4.5.

Table 4.5 Simulated IDDQ/stage in nA/μm for a standard inverter (FO = 1) in three specified IDDQ corners. 45 nm PTM HP models

Technology	NOM	Low IDDQ	High IDDQ
45 nm HP	6.15	0.61	3,093

For the standard inverter, there is a ~5,000× increase in IDDQ from the low IDDQ to the high IDDQ corner. Meeting the requirements of the full $\pm 3\sigma$ process variation range in the design for both timing and IDDQ may be overly conservative. An upper limit for IDDQ may be set as an acceptance rule during test and chips exceeding this limit (Sect. 7.3) rejected. This is illustrated in Fig. 4.9 with IDDQ distributions for a standard inverter delay chain with systematic L_p and V_t variation ranges of $\pm 2\sigma$ and $\pm 3\sigma$. The IDDQ data are obtained from Monte Carlo simulations for 500 cases using 45 nm PTM HP models at 1.0 V, 25 °C. The maximum IDDQ at -3σ of the L_p and V_t ranges is 20× of the maximum IDDQ at -2σ, yet only 1 % of the cases fall outside the -2σ range shown as gray region in Fig. 4.9. Hence it is safe to consider only $\pm 2\sigma$ variation ranges in L_p and V_t.

The high IDDQ data points in Fig. 4.9 are attributed to short L_p and/or low V_t MOSFETs. In CMOS product chips, high fliers may also appear because of defects on the chip. It is therefore important to understand the characteristics of common defect types and separate them from silicon process related issues on non-defective chips. Measurement of chip IDDQ has been used for defect detection as described in the next section.

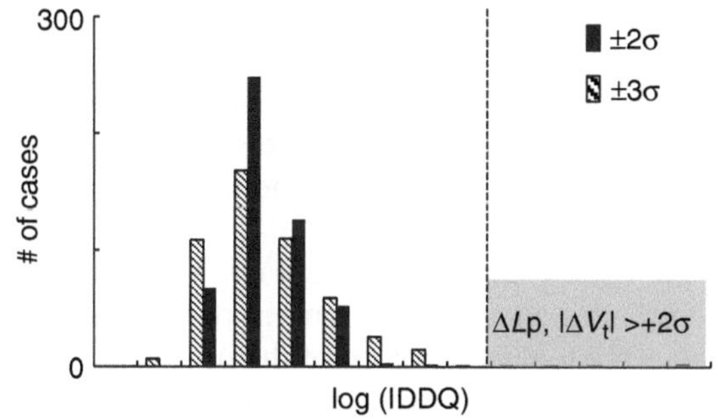

Fig. 4.9 Log (IDDQ) distributions for a chain of inverters for systematic L_p and V_t variation ranges of $\pm 2\sigma$ and $\pm 3\sigma$. 45 nm PTM HP models @1.0 V, 25 °C

4.2.4 Defect Generated IDDQ

In Fig. 4.10 three different resistive defect locations in an inverter situated in the middle of a chain of ten inverters are shown. The defect manifests itself as a resistance R_{sh} across the n-FET in Fig. 4.10a, as a resistance R_{sh} across the p-FET in Fig. 4.10b, and as a resistance R_{sh} from V_{DD} to GND in Fig. 4.10c.

In Fig. 4.10a, with the inverter input at "0" and R_{sh} of 1,000 Ω, the IDDQ of the delay chain increases by ~10,000×. The defective inverter continues to switch correctly as long as $R_{sh} \geq r_{swn}/W_n$, where r_{swn} is the normalized switching resistance of the n-FET. As R_{sh} falls below r_{swn}/W_n, such that the defective inverter's output node voltage drops to $<V_{tn}$, the inverter following the defective inverter is unable to undergo a PD transition, and IDDQ continues to increase with the inverter output stuck-at "1". When the input node of the defective inverter is at "1", there is no anomalous increase in IDDQ.

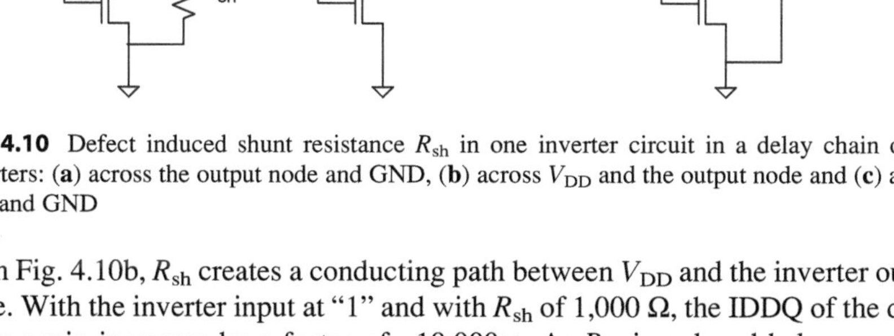

Fig. 4.10 Defect induced shunt resistance R_{sh} in one inverter circuit in a delay chain of ten inverters: (**a**) across the output node and GND, (**b**) across V_{DD} and the output node and (**c**) across V_{DD} and GND

In Fig. 4.10b, R_{sh} creates a conducting path between V_{DD} and the inverter output node. With the inverter input at "1" and with R_{sh} of 1,000 Ω, the IDDQ of the delay chain again increases by a factor of ~10,000×. As R_{sh} is reduced below r_{swp}/W_p, where r_{swp} is the normalized switching resistance of the p-FET, the output node voltage of the defective inverter eventually increases to $>(V_{DD} - V_{tp})$. Analogous to the previous case, IDDQ progressively increases and the inverter following the defective inverter in the chain is unable to undergo a PU transition with its output stuck-at "0". As in the previous case, there is no increase in IDDQ when the input node of the defective inverter is at "0".

The defect in Fig. 4.10c comprising a shunt resistance across V_{DD} and GND of the delay chain contributes an additional current of magnitude V_{DD}/R_{sh} to IDDQ, but may not influence the logical behavior of the inverter. As R_{sh} is reduced, local V_{DD} droop in the power supply may result in an increase in signal propagation delay through the inverters.

In Fig. 4.11 simulated data for IDDQ and delays of a standard inverter for 45 nm PTM HP models are plotted as functions of R_{sh} for the defect shown in Fig. 4.10c. A series resistance of $R_{pg} - 2.0\ \Omega$ is included in both the V_{DD} and GND distribution grids (Fig. 4.11a) which reduces the applied V_{DD} to the inverter as R_{sh} is decreased. The PD and PU delays increase as R_{sh} is lowered. The effective power supply voltage across the inverter, V_{DDA}, is given by

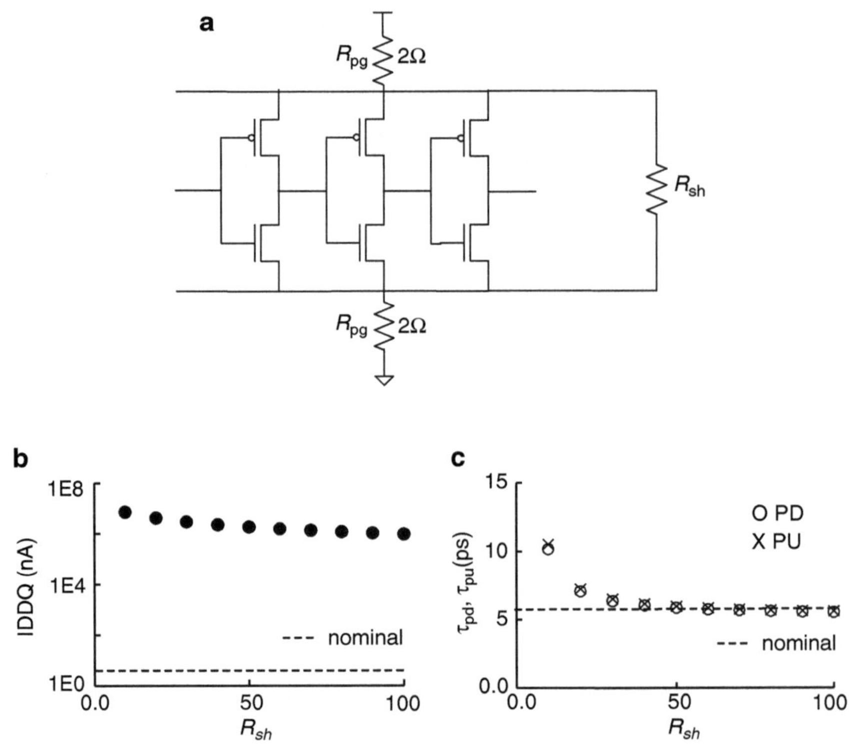

Fig. 4.11 (**a**) Circuit schematic of a segment of an inverter chain with a shunt resistance R_{sh} across V_{DD} and GND and series resistances R_{pg} (=2.0 Ω). (**b**) IDDQ and (**c**) τ_{pd} and τ_{pu} of an inverter in the chain in (**a**) as a function of R_{sh}. 45 nm PTM HP models @1.0 V, 25 °C

$$V_{DDA} = V_{DD}\frac{R_{sh}}{(R_{sh} + 2R_{pg})}. \tag{4.8}$$

With R_{sh} as low as 50 Ω, τ_p is barely affected while IDDQ increases by $10^6\times$.

IDDQ of a logic gate can also be higher when one or more of the MOSFET terminals are floating. In the case of an inverter, if the input node is floating, the output node voltage is set by the effective resistances of the n-FET and the p-FET. In a balanced inverter, the gate voltage will be ~$V_{DD}/2$, and both n-FET and p-FET will be turned "on" resulting in a large IDDQ.

In the case of an open source connection, IDDQ depends on the input voltage and the type of MOSFET. With an open in the n-FET source connection to the GND terminal, IDDQ is higher when the gate is at "1". With an open in the p-FET source connection to V_{DD}, IDDQ is higher when the gate is at "0".

Consider three different scenarios for an open drain connection. The drain of the n-FET may be disconnected from the output node, the drain of p-FET may be disconnected from the output node or the output of the inverter may be disconnected from the gate of the following inverter in the chain. When the drain connection of one of the MOSFETs to the output is open, IDDQ is input voltage dependent. With the output node floating, the input gate voltage of the next inverter in the chain is floating and the IDDQ is high.

Simulated IDDQ values for a chain of ten inverters (FO = 1) with different types of defects are shown in Table 4.6. When a connection to any of the terminals is open, the output voltages of the following gates in the chain are stuck at either "1" or "0". A more detailed discussion of stuck-at faults is included in Sect. 7.1.4.

Table 4.6 IDDQ of a chain of ten inverters (FO = 1) with one defective inverter. 45 nm PTM HP models @1.0 V, 25 °C

Defect type	R_{sh} (Ω)	Gate node @	IDDQ (A)
None		"1" or "0"	6.1E-8
Shunt across power grid	100 (R_{pg} = 2 Ω)	"1" or "0"	9.6E-2
Shunt across n-FET	2,000	"0"	5.1E-4
Shunt across p-FET	2,000	"1"	4.9E-4
Floating gate			3.8E-5
Floating source (n-FET)		"1"	6.1E-8
Floating source (p-FET)		"0"	3.3E-6
Floating drain (n-FET)		"1"	4.8E-8
Floating drain (p-FET)		"0"	3.5E-6

From Table 4.6 it is apparent that presence of a resistive short or an open can result in abnormally high IDDQ when all combinations of input vectors are exercised. In the case of an inverter, some types of defects can be detected by observing the change in IDDQ with the input of the defective inverter set at "1" and then switched to a "0". In more complex circuits, IDDQ measurements for many sets of input vectors are needed to identify defective logic gates.

The IDDQ of the inverter in Fig. 4.11a, with $R_{sh} = 50\ \Omega$ is 18 mA which is 3×10^6 times larger than the nominal IDDQ of the inverter. One such defect can be easily detected if the number of equivalent inverter circuits is ~3×10^7 (10 % increase in IDDQ) and the current measurement resolution in the power supply is adequate. If, on the other hand, a chip has one billion equivalent inverters connected to the same power supply, this defect will be undetected in IDDQ test but may cause timing failures in functional tests. The IDDQ method of detecting defects has been successfully used in manufacturing test of some CMOS products [5–7].

4.3 Power

Total power consumed by a CMOS circuit block is the sum of AC power P_{ac} and DC power P_{dc}:

$$P = P_{ac} + P_{dc}. \tag{4.9}$$

AC power is the sum of P_{sw} and P_{sc} where P_{sw} is power due to charging and discharging of nodes, and P_{sc} is short-circuit power consumed when there is a momentary direct current path between V_{DD} and GND during switching. DC power P_{dc} is the sum of leakage power drawn by circuits in the off-state P_{off} and that resulting from any direct resistive paths between V_{DD} and GND, P_{res}:

$$P_{ac} = P_{sw} + P_{sc}, \tag{4.10}$$

$$P_{dc} = P_{off} + P_{res}. \tag{4.11}$$

In the chip design phase, CMOS circuit power is determined using SPICE simulations or with EDA tools dedicated to power estimation of large circuit blocks in different operating modes. In electrical tests of CMOS hardware, only limited information can be collected on power consumption. It is therefore important to understand what data can be collected at test and how to relate it to circuit designs for model-to-hardware correlation.

4.3.1 Measuring Power

In electrical tests, power of a CMOS chip is measured as

$$P = \text{IDDA}_m \times V_{DD}, \tag{4.12}$$

where IDDA_m is the average current drawn by the power supply. In functional mode, different numbers of circuits may be switching in any given clock cycle. The measured IDDA_m is the average current over the integration time of the power supply, which is typically on the order of milliseconds (ms). With clock cycle times of several hundred ps, the integration time may be $>10^6$ clock cycles for a single

$IDDA_m$ measurement. Hence, changes in switching activity in the <ms time scale are averaged and not detectable. However, $IDDA_m$ values may change significantly over time as clocks and other major functions are turned on or off.

The measured current drawn in the quiescent state, $IDDQ_m$, corresponds to all circuits in the quiescent state. As described in Sect. 4.2.2, $IDDQ_m$ varies with the input state of circuits. The power consumed in the quiescent state is

$$P_{dc} = IDDQ_m \times V_{DD}. \tag{4.13}$$

At a fixed V_{DD}, power consumed during switching P_{ac} is obtained from $IDDA_m$ and $IDDQ_m$ measurements as

$$P_{ac} = (IDDA_m - IDDQ_m) \times V_{DD}. \tag{4.14}$$

There are several sources of variations in estimating AC and DC power consumptions. As switching activities change continuously, $IDDA_m$ is a function of time averaging. The measured $IDDQ_m$ varies with the inputs provided to each circuit block. In functional mode, background IDDQ is different than $IDDQ_m$, with only non-switching circuits contributing to IDDQ. Transients in node voltages also affect background leakage currents. Measurement of total power of CMOS chips is discussed in more detail in Sect. 4.4.

4.3.2 AC Power

Switching power P_{sw}, due to charging and discharging of nodes at a fixed rate is expressed as

$$P_{sw} = \frac{1}{2} C_{sw} V_{DD}^{n} f_m, \tag{4.15}$$

where C_{sw} is the total switching capacitance which includes MOSFET, wire and parasitic capacitances and f_m is the number of switching transitions per second. In circuit simulations power is often measured for a circuit switching at a constant rate, as in the case of ring oscillator (Sect. 2.2.5).

In Eq. 4.15, n is 2.00 for a pure capacitive load. As discussed in Sect. 2.2.2, MOSFET capacitance components have a small V_{DD} dependence. As an example, in 45 nm PTM models, n-FET gate capacitance in inversion mode increases from 1.44 fF/μm at 0.7 V to 1.52 fF at 1.3 V. Wire capacitances are independent of V_{DD}. The net effect is that for CMOS circuits, $n > 2$.

As previously mentioned, short-circuit current I_{sc} flows directly through an inverter when both the n-FET and the p-FET are momentarily "on" during a transition, resulting in a conducting path between V_{DD} and GND. The short-circuit power of an unloaded inverter is expressed as

$$P_{sc} = \beta (V_{DD} - 2V_t)^3 \frac{\tau_{ri}}{T_p}, \tag{4.16}$$

where $\tau_{ri}(=\tau_{fi})$ is the input rise (fall) time, T_p is the period of the input waveform, and β is the gain factor of the MOSFET (Eq. 2.3) [2]. Generally, this equation is expected to hold only in quasi-static situations, where τ_{ri} is long compared with output fall time τ_{fo} and similarly $\tau_{fi} \gg \tau_{ro}$. The inverter's output node voltage is then determined entirely by the relative resistance values of the n-FET and p-FET. Nevertheless, it is expected that in general P_{sc} will increase as V_{DD} or $\tau_{ri}(=\tau_{fi})$ is increased, and as T_p or V_t is lowered, although the strengths of these dependencies may be modulated significantly by capacitive effects.

To obtain further insight into short-circuit current, simulations are carried out for the circuit shown in Fig. 4.12a, where currents flowing through various terminals can be monitored with zero-voltage sources. Currents through zero-voltage sources VR and VF are monitored during PD and PU transitions of the inverter that are induced by inputs at terminal A with rise and fall times of $\tau_{ri} = \tau_{fi}$. An unloaded inverter configuration is selected so that these currents are dominated by charging and discharging of only the internal MOSFET capacitances and I_{sc}.

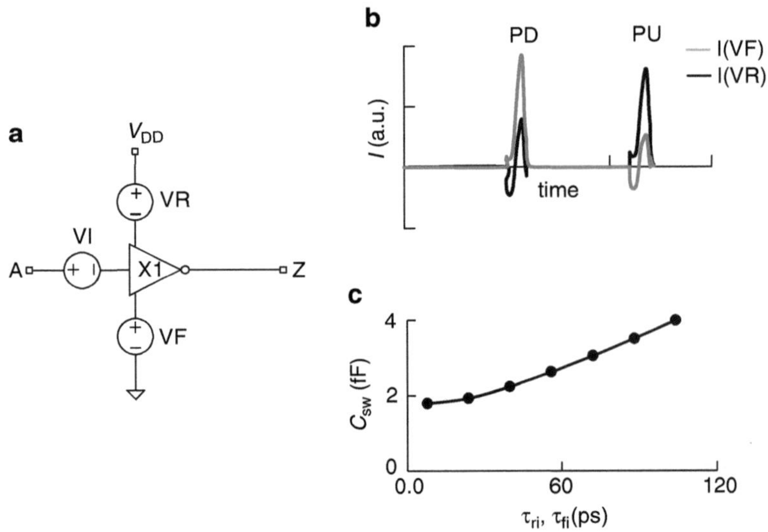

Fig. 4.12 (**a**) Circuit schematic to determine transient currents, (**b**) transient currents through VF and VR for PD and PU transitions, and (**c**) equivalent total switching capacitance C_{sw} as a function of τ_{ri} or τ_{fi} for an unloaded standard inverter @ 1.0 V

During a PD transition, current flowing through VF is associated with discharging the internal MOSFET capacitances as well as a small I_{sc} component; current flowing through VR comprises I_{sc} and a capacitive component. During a PU transition, current flowing through VR is associated with charging the internal

MOSFET capacitances along with a small I_{sc} component; current flowing through VF comprises I_{sc} and a capacitive component. These transient currents through VR and VF are shown in Fig. 4.12b.

Integrating the current I(VF) or I(VR) through one complete PU and PD cycle gives the total charge transferred from V_{DD} to GND during the cycle. Dividing that charge by V_{DD} gives an effective switching capacitance C_{sw} which comprises a pure capacitance that is charged and discharged and, in addition, an effective capacitance attributed to the two short-circuit components occurring during the PU and PD transitions. Figure 4.12c shows the simulated value of C_{sw} for this inverter as a function of $\tau_{ri} = \tau_{fi}$, as obtained from integrating the currents. The linear increase in C_{sw} above $\tau_{ri} = \tau_{fi} = 25$ ps can be attributed to I_{sc}. In this region $\tau_{ri} \gg \tau_{fo}$ and $\tau_{fi} \gg \tau_{ro}$, and a linear behavior consistent with Eq. 4.16 is expected.

In many (but not all) switching transitions the rise and fall times at the input of a logic gate are similar to those at the output, i.e., $\tau_{ri} \approx \tau_{fo}$ and $\tau_{fi} \approx \tau_{ro}$. Such is the case in a delay chain or ring oscillator with identical stages described in detail in Sect. 2.2. Instrumenting one of the central inverters with zero-voltage sources as was done for the isolated inverter in Fig. 4.12a, the value of C_{sw} can again be obtained from simulations. Figure 4.13a shows the value of C_{sw} as a function of V_{DD} over the range of 0.7–1.3 V for the standard inverter with FO = 1. C_{sw} shows a modest increase of only about 7 % over the entire range with most of that above 1.0 V. The upturn above 1.0 V can be attributed to the I_{sc} contribution which is significantly suppressed as $\tau_{ri} = \tau_{fo}$. In addition the increase in C_{in} and the diffusion capacitance components of C_{out} are also contributing to the increase in C_{sw} over the simulated range.

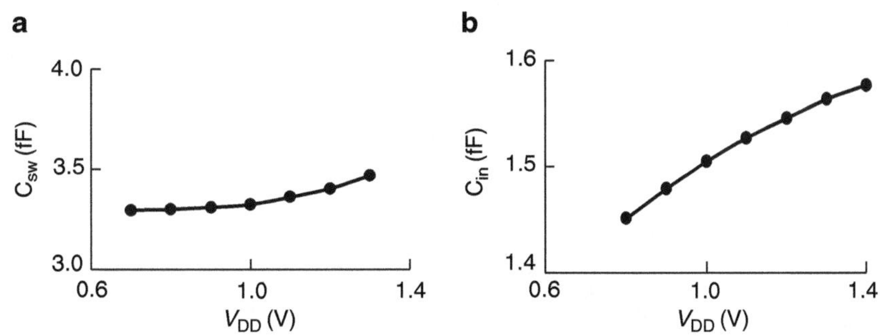

Fig. 4.13 (**a**) C_{sw} as a function of V_{DD} for PD and PU transitions in an inverter delay chain, and (**b**) C_{in} as a function of V_{DD} for a standard inverter (FO = 1), $\tau_{ri} = \tau_{fi} = 16$ ps. 45 nm PTM HP models @1.0 V, 25 °C

Figure 4.13b shows C_{in} as a function of voltage obtained by integrating the current that flows into or out of the input to the inverter, as measured through another zero-voltage source such as VI in Fig. 4.12a. Referring to Eq. 4.15, if one chooses to treat C_{sw} as a constant and absorb the additional voltage dependence from I_{sc} and C_{in} in the exponent n, both effects described above will contribute a value of $n > 2$. This situation will be next investigated in the context of a ring oscillator comprising identical stages.

The AC power of a logic gate switching at a constant frequency is simulated using a ring oscillator circuit. As described in Sect. 2.2.5, an RO is constructed with $(2\alpha + 1)$ identical inverting stages, and its frequency of oscillation is given by

$$f = \frac{1}{2\tau_p(2\alpha + 1)}. \tag{4.17}$$

From Eq. 4.15, P_{sw} is a function of f, which in turn varies inversely with $(2\alpha + 1)$.

The energy to switch a circuit,

$$E_{sw} = P_{sw}\tau_p = \frac{1}{2}C_{sw}V_{DD}^{n} \tag{4.18}$$

is independent of α. Hence, E_{sw} is independent of the number of stages in the RO and the RO frequency f. This switching energy per transition (E_{sw} = power × delay) is a useful metric for comparing different circuits, technology nodes, and different microelectronic technologies as discussed in Sect. 10.4.2.

In Fig. 4.14a, the dependence of E_{sw} on V_{DD} is shown for the standard inverter (FO = 1) in 45 nm PTM HP models at 25 °C. The FO = 1 inverter configuration is selected to give equal weightage to MOSFET gate and other internal capacitances. The simulated data are fit to a power law as suggested by Eq. 4.18, and displayed in Fig. 4.14. With an average C_{sw} of 3.38 fF over this V_{DD} range, the multiplier of 1.68 fF corresponds to ~½C_{sw} following Eq. 4.18. The value of the exponent as extracted from the fit is 2.07. This increase in n from 2.0 is as a result of increase in effective C_{sw} with V_{DD}. The exponent n increases with FO as shown in Fig. 4.14b.

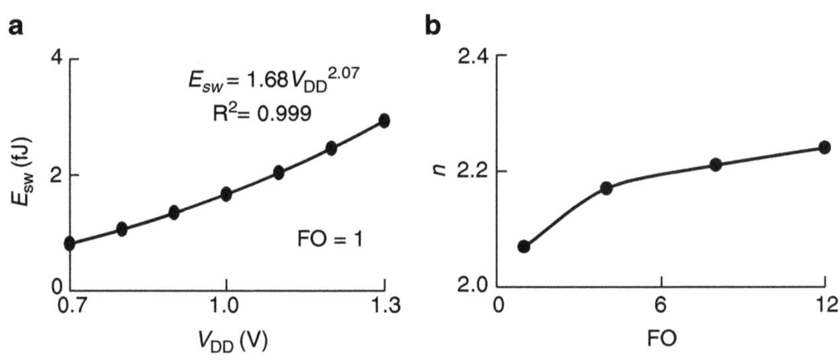

Fig. 4.14 (**a**) Inverter (FO = 1) E_{sw} (power × delay) vs. V_{DD}, and (**b**) exponent n vs. inverter FO. 45 nm PTM HP models @25 °C

C_{sw} also varies with MOSFET parameters L_p, V_t, t_{ox}, and C_{ov}. Gate capacitance increases with increase in L_p and C_{ov} and decreases with increase in V_t and with increasing t_{ox}. An RO circuit may be used to simulate process induced variations in C_{sw}. Simulated data for a standard inverter (FO = 4) in 45 nm PTM HP models are shown in Fig. 4.15. In Fig. 4.15a, C_{sw} is plotted as function of L_p, and in Fig. 4.15b as a function of $\Delta|V_t|$. In these simulations $L_{pn} = L_{pp}$ and $\Delta|V_{tn}| = \Delta|V_{tp}|$ so that both MOSFETs contribute to either an increase or a decrease in C_{sw} in the same simulation run.

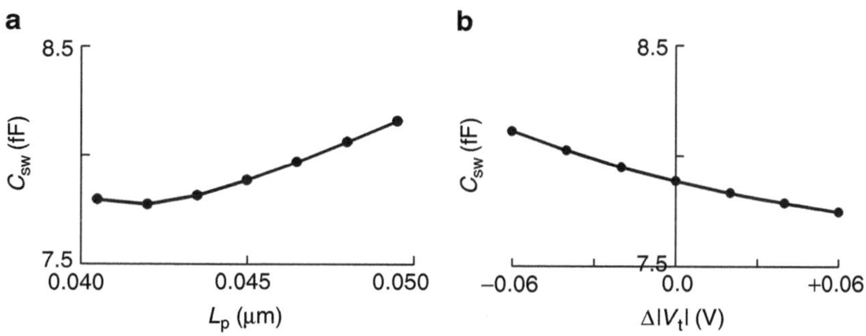

Fig. 4.15 C_{sw} for standard inverter (FO = 4) as a function of (**a**) L_p and (**b**) $\Delta|V_t|$ (n-FET and p-FET varied together in the same sense). 45 nm PTM HP models @ 1.0 V, 25 °C

In Fig. 4.15a, C_{sw} increases as L_p is increased as a result of the increase in MOSFET gate area. It levels off at very short L_p values because of V_t roll-off from short channel effect in the models. Here a capacitance increase with decrease in V_t at short channels is partly compensated by a reduction in gate-area. The relative contribution of C_{ov} to C_{sw} varies with technology node and MOSFET type.

AC power at constant frequency and V_{DD} exhibits some temperature sensitivity. Passive capacitances are dependent on layer geometry and dielectric constants of insulating layers which remain constant with temperature. MOSFET capacitances increase as the temperature rises. This is because V_t is lowered at higher temperatures. As a result MOSFETs turn on early in transitions increasing the average capacitance during switching (Sect. 2.2.2). In Fig. 4.16, the variation of C_{sw} with temperature is shown. There is a small increase in C_{sw} with rising temperature.

Simulation results for 45 nm PTM HP models for an inverter (FO = 4) indicate a ΔC_{sw} range of $<\pm 5$ % over full process and operating ranges of L_p, V_t and temperature. With the addition of fixed capacitance in physical layouts, the range of C_{sw} variation is reduced further.

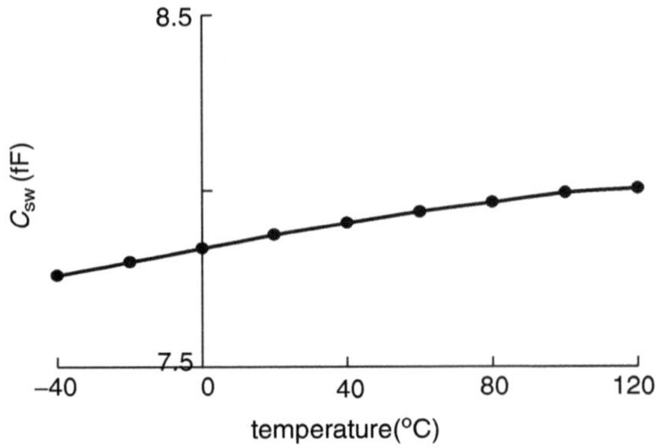

Fig. 4.16 C_{sw} for standard inverter (FO = 4) as a function of temperature. 45 nm PTM HP models @ 1.0 V

4.3.3 DC Power

DC power comprises power drawn by circuits in the quiescent state (P_{off} = IDDQ × V_{DD}) and by shunt resistances between V_{DD} and GND. The leakage power P_{off} is a function of MOSFET properties (W, L_p, V_t, and t_{ox}), V_{DD}, temperature, circuit topology and input voltages. The IDDQ/stage for a logic gate is obtained from circuit simulation of two series connected stages. The IDDQ/stage of the standard inverter (FO = 1) as a function of process variations in L_p and $\Delta|V_t|$ is shown in Fig. 4.17a, b. The data are obtained from simulation of a 51 stage RO circuit using 45 nm PTM models.

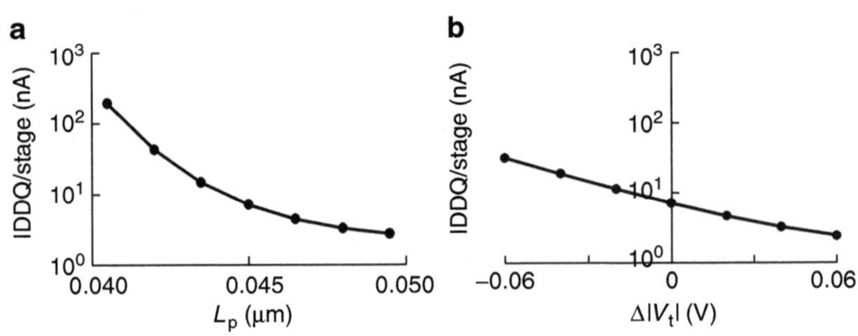

Fig. 4.17 Average IDDQ/stage of a standard inverter as a function of (**a**) L_p and (**b**) $\Delta|V_t|$ (n-FET and p-FET varied together). 45 nm PTM HP models @ 1.0 V, 25 °C

The dependencies of average IDDQ/stage on temperature and V_{DD} are shown in Fig. 4.18a, b. The temperature dependence of IDDQ at V_{DD} = 1.0 V is modeled as

$$\text{IDDQ} = c_3\, T^{5.4}, \tag{4.19}$$

where c_3 is a fitting parameter and T is the temperature in K (=temperature in °C + 273). The V_{DD} dependence of IDDQ is modeled in the same fashion using the simulated data shown in Fig. 4.18b. Note that over both the temperature and V_{DD} ranges shown, IDDQ varies by > 10×. This extreme sensitivity of IDDQ is an important factor in CMOS technology tuning for both high performance and low power applications.

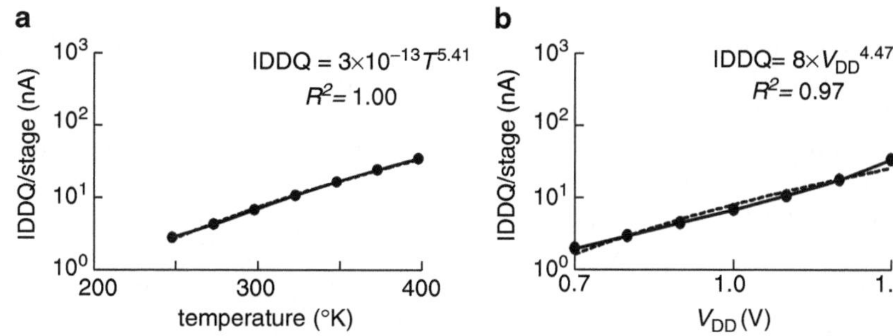

Fig. 4.18 Average IDDQ/stage of standard inverter as a function of (**a**) temperature at $V_{DD} = 1.0$ V and (**b**) V_{DD} at 25 °C. 45 nm PTM HP models

The increase in P_{off} with V_{DD} at 25 and at 100 °C for a standard inverter is shown in Fig. 4.19. The simulated data are fit to a power law with a reasonable regression coefficient over this wide V_{DD} range ($R^2 = 0.98$).

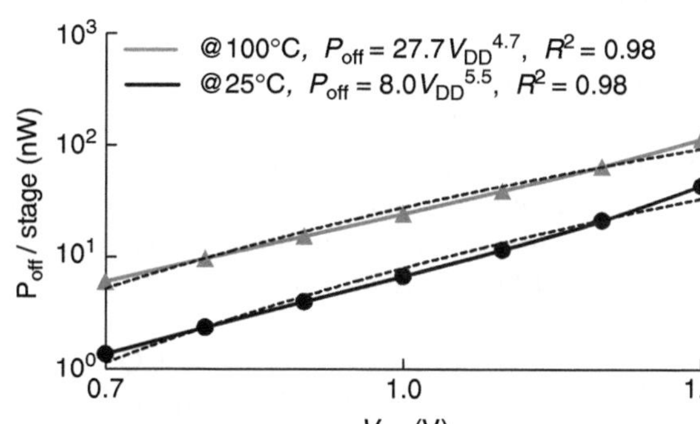

Fig. 4.19 Average P_{off}/stage for standard inverter as a function of V_{DD} at 25 and at 100 °C. Nominal 45 nm PTM HP models @ 25 °C

Based on simulation data, an empirical model of P_{off} in a temperature range of 25–100 °C, and V_{DD} range of 0.7–1.3 V is generated:

$$P_{off} = c_4 T^{5.4} V_{DD}^{5.1}, \tag{4.20}$$

where c_4 is a fitting parameter. Eq. 4.20 may be used to obtain a rough estimate of relative changes in P_{off} at different values of temperature and V_{DD}.

In addition to IDDQ, power is drawn by any shunt resistance R_{dc} between V_{DD} and GND. Power dissipated by such shunt resistance P_{res} is given by

$$P_{res} = \frac{V_{DD}^2}{R_{dc}}. \tag{4.21}$$

Metal resistances increase with temperature (TCR for Cu is ~0.30 to 0.35 %/°C) and P_{res} at constant V_{DD} decreases as the temperature is increased.

The V_{DD} and temperature dependencies of different components of total power may be compared by examining Eqs. 4.15, 4.20, and 4.21. The exponent values for V_{DD} of 5.5 at 25 °C and 4.7 at 100 °C for P_{off} are much larger than the exponent values of ~2 for P_{sw} and P_{res}. Hence with increase in V_{DD}, P_{off} increases much more rapidly than P_{sw} and P_{res}. Similarly, P_{off} has a strong dependence on temperature whereas P_{sw} is nearly constant with temperature and P_{res} decreases with increase in temperature. The ratio P_{off}/P_{total} for a circuit can vary over a wide range with variations in V_{DD} and temperature.

By design, R_{dc} is very large and P_{res} is negligible. In some chips, metal defects can create low resistance paths across the power grid. If, at very low V_{DD} values (~$0.1 \times V_{DD}$) the current drawn by a defect I_{res} is >IDDQ, the defect can be easily detected.

4.4 Total Power

It is instructive to begin with characterization of a ring oscillator in its oscillating state. In this case, total power is the sum of the switching power of a single stage and the background power of all the non-switching stages in their quiescent states. It is a simple representation of a chip in its AC mode, with a fraction of the circuits switching at any given instance and all other circuits remaining idle.

The total power P of an RO comprising $(2\alpha+1)$ stages is expressed as

$$P = \frac{1}{2} \times (2\alpha + 1)\frac{C_{sw}}{\text{stage}} \times V_{DD}^{n} \times f_{n} + 2\alpha \times \frac{\text{IDDQ}}{\text{stage}} \times V_{DD}, \tag{4.22}$$

where f_n is the number of switching transitions experienced by each stage per second. As there are two transitions (one PD and one PU) per cycle, $f_n = 2f$. Combining Eqs. 4.17 and 4.22,

$$P = 1/2\tau_p \times \frac{C_{sw}}{\text{stage}} \times V_{DD}^{n} + 2\alpha \times \frac{\text{IDDQ}}{\text{stage}} \times V_{DD}. \tag{4.23}$$

The first term in Eq. 4.23 denoting P_{ac} is independent of α and the second term denoting P_{off} increases with α. The ratio of P_{off} of the RO to total power therefore increases with the number of stages. Simulated P_{ac} and P_{off} for a standard inverter (FO = 4) RO as a function of number of stages are shown in Fig. 4.20. A small decrease in P_{ac} for $\alpha < 15$ (number of stages <31) is observed because of the increase in the relative contribution of the NAND2 delay, which is longer than the inverter delay. As a result, RO frequency decreases more slowly than the number of stages, and P_{ac} is lowered. For $\alpha > 15$, P_{ac} remains constant.

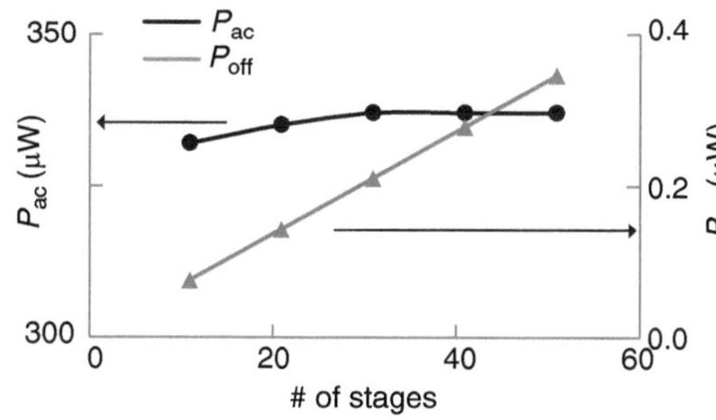

Fig. 4.20 Inverter (FO = 4) P_{ac} and P_{off} as a function of number of stages in a ring oscillator. 45 nm PTM HP models @1.0 V, 25 °C

With one stage switching at any given instant, an RO comprising nominally identical stages draws a constant current. An RO based test structure is therefore suitable for model-to-hardware correlation of total power and P_{off} of logic gates or small circuit blocks as a function of V_{DD} and temperature. The C_{sw} of the RO stage can also be extracted as described in Sect. 2.2.5 and compared with model prediction.

Unlike a ring oscillator, CMOS chip power varies over time as illustrated in Fig. 4.21. When the clocks are off, all the circuits are in a quiescent state and the chip power is low. Power increases but remains nearly constant when only the clock distribution network is on, and the clock buffers are switching twice every clock cycle. In a functional mode, with chip logic and memory activity, power fluctuates with the workload. The switching capacitance and the background P_{off} also fluctuate, changing the relative contributions of P_{ac} and P_{off}.

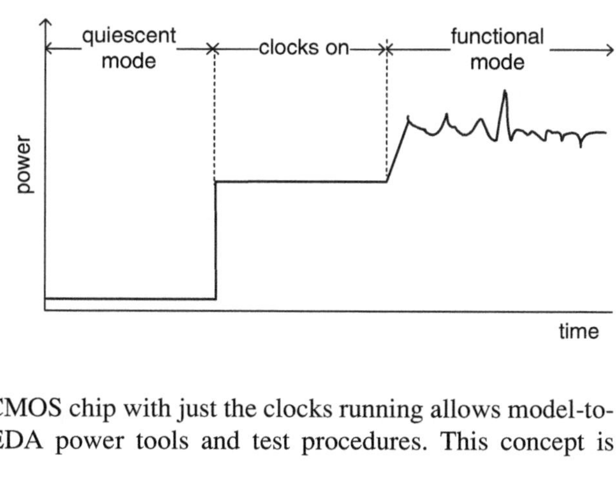

Fig. 4.21 Power as a function of time in various modes of operation

Extraction of C_{sw} of a CMOS chip with just the clocks running allows model-to-hardware correlation of EDA power tools and test procedures. This concept is illustrated by simulating a delay chain and measuring total power as a function of small frequency increments, and using Eq. 4.15 to extract C_{sw}. Unlike an RO which oscillates at fixed frequency at any given V_{DD} and temperature, the switching frequency of a delay chain may be varied independently.

A ten-stage inverter delay chain (Sect. 2.2.4) is simulated with a long input train of pulses. The frequency of the pulse train is varied, and the average value of IDDA is measured. In Fig. 4.22, total power P (=IDDA × V_{DD}) is plotted as a function of frequency f at V_{DD} = 1.0 V. From Eq. 4.15, the slope of P vs. f plot is 0.5× C_{sw}. A linear fit of the data in Fig. 4.22 gives C_{sw} = 78.8 fF. The C_{sw}/stage of a standard inverter (FO = 4) obtained from RO simulations is 7.88 fF (Table 2.15). Hence, as expected, C_{sw} of the inverter delay chain with ten stages (=78.8 fF) matches the value obtained from the slope of P vs. f plot. The intercept of the linear fit at f = 0 gives P_{off} = 0.067 μW, also consistent with simulated P_{off}.

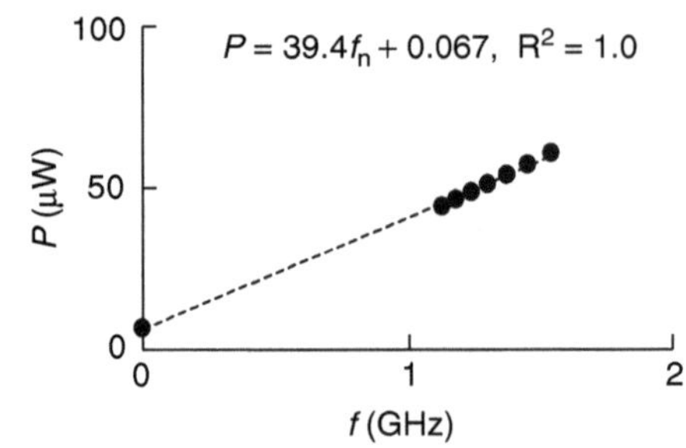

Fig. 4.22 Total power P of a standard inverter chain (FO = 4) with ten stages as a function of frequency. 45 nm PTM HP models @ 1.0 V

When multiple power supplies are used, power contributions from all the sources are summed to get the total power. If voltage sense points are available on chip, power dissipated in the board and package can be computed. As discussed in Sect. 6.1.3, this provides a means for monitoring external series resistances and *IR* drops in the wires. Any significant degradation in wire or package resistances lowering the chip V_{DD} and power serves as a warning signal. Timely corrective actions can be taken to prevent product performance degradation or possible malfunction.

Empirical models of P_{ac} and P_{off} are generated based on characterization data collected from a small subset of representative samples at early stages of manufacturing. These models are used to predict power consumption over a range of silicon process variations in different test corners.

A chip power model would include dependencies on V_{DD}, temperature, frequency, and workload. Measurement of power in the hardware may appear to be straightforward ($P = I \times V$), as all power supplies provide output voltage and current readings as a function of time. However, there are several factors to be considered in power measurements.

In the quiescent state, P_{off} (or IDDQ) is dependent on the initial states of the circuits. When power is switched on, voltage at some nodes may settle at an intermediate level between "0" and "1". Such undetermined states cause excessive current to be drawn from the power supply as discussed in Sect. 4.2.4. Prior to making an IDDQ measurement, the chip is initialized so that the internal nodes of the circuits are in definitive "1" or "0" states, and the latches are preloaded with "1"s and "0"s. This is done by running a specific set of input vector patterns. IDDQ is then measured with different initialization patterns and compared. If the differences are significant and this signature is repeated in all good chips, the highest value of IDDQ may be used for screening in IDDQ testing.

The temperature dependence of P_{off} is measured by placing the chip or package on a thermally controlled platform and incrementing the temperature over the full range of operation. Similarly, V_{DD} dependence of P_{off} is obtained from measurements made at several different values of V_{DD}. In high performance chips such as microprocessors, P_{off} may be significant, and the chip temperature may rise as V_{DD} is increased. In such cases, both V_{DD} and temperature variations must be taken into account simultaneously.

AC power on the chip has two major components: (1) power drawn by the clock distribution network and (2) power drawn by other circuits performing logic and memory functions. The clock distribution network undergoes switching every cycle and draws a steady current averaged over multiple clock periods. Total power is measured as a function of frequency at constant V_{DD} and temperature with only the clocks running. The values of C_{sw} and P_{off} are obtained from the slope and intercept of a linear fit of measured data as demonstrated with the delay chain example in Fig. 4.22. It is preferable to collect data over a small frequency range to maintain a constant chip temperature and limit P_{off} variation over this frequency range. If the silicon temperature is higher than in quiescent mode, P_{off} obtained from this method will be higher.

Power consumed by logic and memory circuits is a function of switching activity levels which can change significantly over time. As power is dependent on the functions being performed, a custom workload may be designed to set the maximum power specification for the chip during electrical testing and product qualification. This function should maintain nearly constant power over the measurement period of several milliseconds while performing maximum circuit switching activity. In this case, the silicon temperature is equilibrated over a measurement period. Typical power supply response time is of the order of a millisecond and any fluctuations in power in ns or μs time periods will not be detected.

Measures are put in place to prevent chip power from exceeding specified limits. Maximum power supply current is clamped at this limit, and the power supply is programmed to throttle by lowering V_{DD} or shutting down altogether if either the current or silicon temperature exceeds the set limits.

4.5 Power Management

Power must be carefully managed in both high performance microprocessor chips and in chips used in low power applications. Power consumption in data centers has become a great concern requiring special power delivery and cooling systems. Power reduction by lowering frequency and using systems at reduced capacity becomes necessary during peak load periods. In battery operated systems, there is a push to minimize standby power to prolong battery life. These requirements have led to different ways of managing power in different operating modes.

Power specifications for a chip are set early in the design phase. Using device models and EDA tools, the chip circuitry is designed to meet these power limits under worst-case conditions defined by the design methodology. Chips exceeding power limits are rejected during test. On-chip schemes to manage power consumption by tailoring the workload are used in both high performance and low power applications.

4.5.1 Power Management in Chip Design

On-chip power management is addressed in circuit design as well as in architecture using various design and global strategies. Based on the chip application, a high performance or low power technology option is selected. Within a technology type, circuits are optimized to consume minimum power by

- Using multiple V_t MOSFET types
- Minimizing MOSFET widths

The use of multiple V_t MOSFET pairs has been in place for many CMOS generations. Typically, MOSFET pairs with nominal V_t, and higher and lower than nominal V_t are offered for logic circuits. These will be referred to as regular V_t, low V_t and high V_t MOSFETs. The design may begin with use of only regular and high V_t MOSFETs. Logic gates with low V_t MOSFETs may then be judiciously placed in slow paths failing to meet timing requirements with regular V_t circuits. High V_t MOSFETs are predominantly used in logic paths with sufficient timing slack. The L_p and V_t values of MOSFETs for SRAM and analog applications may be independently tailored for optimum performance and power.

The differences among IDDQ values of regular, high and low V_t MOSFETs are demonstrated by making use of the *delvto* parameter to raise or lower the V_t in nominal 45 nm PTM HP models by ±0.06 V. In Fig. 4.23a, average IDDQ/stage is plotted as a function of τ_p for regular, high and low V_t inverters (FO = 4). The low V_t inverter ($\Delta|V_t| = -0.06$ V) has a 4× higher IDDQ than a regular V_t inverter and a 15 % reduction in τ_p. The high V_t inverter ($\Delta|V_t| = +0.06$ V) has 3.5× reduction in IDDQ compared with regular V_t and a 17 % increase in τ_p.

MOSFET widths are minimized to reduce both P_{off} and P_{ac}. As the widths are reduced, the current drive strength decreases and the switching resistance R_{sw} increases. The logic gate capacitances reduce in proportion to the width as described in Sect. 2.2.3. As a result, with logic gate load in the fanout, τ_p is nearly independent of MOSFET widths. This is illustrated in Fig. 4.23b for an inverter (FO = 4) with τ_p plotted as a function of ($W_n + W_p$) while maintaining $W_p/W_n = 1.5$. Also shown in Fig. 4.23b is τ_p vs. ($W_n + W_p$) for an inverter (FO = 3) and an additional fixed capacitive load $C_L = 1.5$ fF, which is ~25 % of the total load for

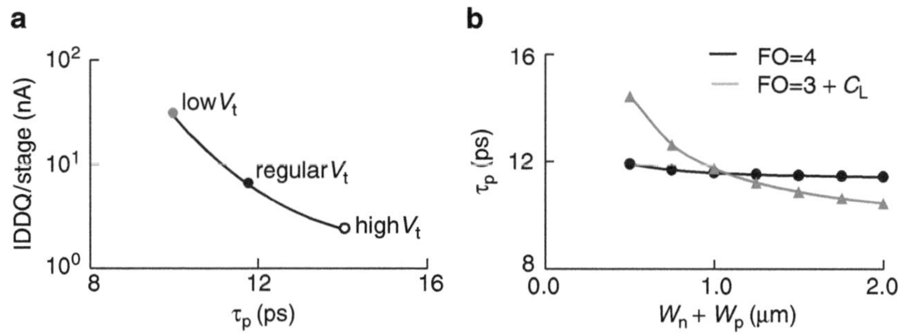

Fig. 4.23 (**a**) IDDQ/stage vs. τ_p for regular, high and low V_t inverters and (**b**) inverter τ_p vs. $(W_n + W_p)$ with FO = 4 gate load, and with FO = 3 gate load + C_L. 45 nm PTM HP models @1.0 V, 25 °C

$(W_n + W_p) = 1.0$ μm. In this case, τ_p increases with reduction in MOSFET widths. The lower limits on widths in a logic gate are determined by the load and the minimum width rules in the design kit supplied by the foundry. Sophisticated EDA tools are available for minimizing P_{off} and total power at the desired frequency of operation. The tuning process includes tailoring of device widths and a push for more high V_t usage.

Some of the global power management strategies are:

- Multiple V_{DD} domains
- Multiple frequency domains
- Power gating
- Clock gating

A CMOS chip may have several different power supply voltage domains with optimized V_{DD} for each domain. The V_{DD} for I/O circuits is set by the requirements for off-chip communications. Analog circuits such as PLLs may operate at a higher V_{DD} than the logic circuits. Logic and memory circuits may also run at different voltages. Logic circuits are sometimes partitioned to run on different power supplies. As an example, in a microprocessor chip with multiple cores, each processor core can have an independent power supply. The V_{DD} for each logic and memory block can be tuned to compensate for systematic process variations and across chip variations for overall power and performance optimization (Sect. 7.5). Separate voltage domains have the additional benefit of localizing defects through IDDQ measurements.

There are typically several different clock frequencies on the chip. Circuit blocks such as processor cores run at the highest frequency. On-chip memory circuits and peripheral logic run at an integer fraction of the core frequency and consume less power from local clock distribution and switching activity.

Power gating is used to reduce leakage currents in circuits that remain idle for sufficiently long times. The V_{DD} connection to a circuit block is controlled through a header or a footer switch shown in Fig. 4.24. A header is a p-FET in series with the connection to the V_{DD} power supply rail. The gate of the p-FET is independently controlled to turn it on or off on demand. With the header p-FET in the off-state, there is a high resistance in series with the power supply reducing the leakage current through the logic block. The p-FET is sized to minimize the *IR* drop in the on-state while providing sufficient reduction in IDDQ in the off-state. A footer switch is an n-FET in series with the GND connection of the power supply and is controlled and operated in a similar fashion as a header switch.

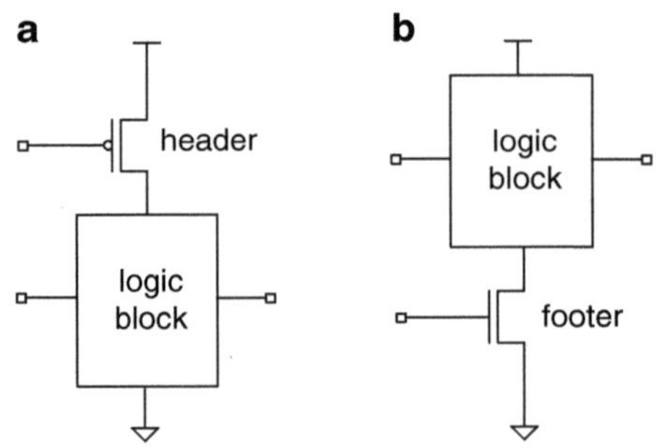

Fig. 4.24 Power gating a logic block with (**a**) p-FET header and (**b**) n-FET footer

Headers or footers may be placed in individual circuit blocks. Additional circuitry is required to turn the switches on and off and to retain critical data during the power off-state. This overhead can be made small compared with total power savings.

Clock gating is used to turn off clock delivery to unused circuit blocks. Dynamic disabling of circuit blocks on a temporary basis to reduce power and temperature may be carried out under software control. A circuit block, such as a defective core in a microprocessor chip with multiple cores, can be permanently disabled. Two representative circuit schemes that are used to enable clock gating are shown in Fig. 4.25.

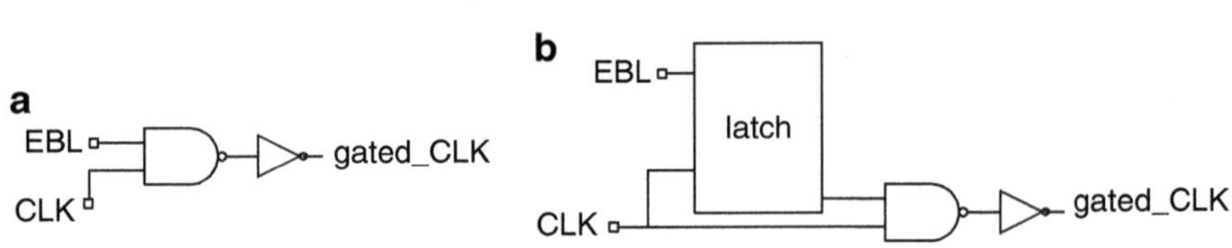

Fig. 4.25 Circuit schematics for clock gating with EBL control signal using (**a**) logic gates and (**b**) latch

A more drastic measure is to permanently turn off the power supply of unused redundant circuit blocks or defective blocks in partially good chips. This is an effective way of reducing wasted power in such chips.

4.5.2 System Power Management

Demand for energy efficiency and reliability considerations are driving power management at the system level in microprocessor chips in medium and high-end servers. These chips draw 100 W or more of power and require sophisticated cooling to keep the chip temperature within limits. Servers running under different environmental conditions have different power handling requirements. Instead of limiting workload requirements to a standard specification, each system may be dynamically managed at the customer location.

Special features may be included on the chip to enable dynamic power-performance optimization. Dynamic frequency scaling (DFS) and dynamic voltage and frequency scaling (DVFS) have been used in IBM's POWER series microprocessor chips [7]. Lowering the frequency of operation gives a linear reduction in active power in DFS. A reduction in V_{DD} with a corresponding reduction in frequency cuts down power consumption more effectively in DVFS.

Dynamic power management schemes are implemented using on-chip thermal sensors, activity counters to monitor switching activity in the processor cores, on-chip and off-chip actuators to throttle activity levels and a communication network to link on-chip power management elements to off-chip elements such as voltage regulators. This level of sophistication adds to the burden in design, software control, and qualification tests.

4.6 Summary and Exercises

The increase in leakage currents and power densities with scaling has led to different approaches for power and performance optimization in CMOS technology. MOSFET leakage current contributions to circuit IDDQ vary with circuit topologies. IDDQ measurements have been successfully used for detecting defects in chips with nominally low IDDQ levels. Measurements of DC and AC components of power at different voltage and frequency are used for building an empirical power model of the chip. Methods of power reduction in chip design, architecture and by on-chip dynamic power management are all geared towards minimizing power with optimum chip performance.

Circuit performance and power scaling parameters are evaluated in Exercise 4.1. Exercises 4.2 and 4.3 deal with IDDQ estimates for logic gates with and without defects, and Exercises 4.4 through 4.7 with total power in different scenarios. Exercises 4.8 and 4.9 cover power mitigation techniques. Reduction in IDDQ at very low temperatures (4.2 K) is considered in Exercise 4.10.

4.1. Using the data in Table A.5 in Appendix A, determine the scaling factors S_k for V_{DD}, τ_p, power and power density of an inverter (FO = 3) in transitioning from 45 to 32 nm, and 32 to 22 nm technology nodes. Use S as the scaling factor for L_p. Which parameters do not follow the scaling rules in Table 4.2?

4.2. (a) Draw circuit schematics of two series connected logic gates (inverter, NAND4T, NOR3B) and show leakage current components with input to the series connected gates as well as all unconnected inputs at "1".

(b) Estimate IDDQ values by summing leakage current components of MOSFETs, and from circuit simulation with all inputs at "1". Compare the IDDQ values obtained from the two methods.

(c) Insert a current multiplier between the two stages, and set XL = 4. Compare simulated IDDQ values with those obtained in (b). How do you explain the difference? What additional information can you extract from these simulations?

4.3. A defective chip has a shunt resistance $R_{dc} = 10\ \Omega$ across its V_{DD} and GND rails.

(a) Model the chip as a set of standard inverters and find the equivalent number of inverters N such that the total IDDQ at nominal V_{DD}, 25 °C matches the current through R_{dc} (=I_{dc}).

(b) What is the I_{dc}/IDDQ ratio at (1) 25 °C, $0.8 \times V_{DD}$ and (2) −25 °C, $1.0 \times V_{DD}$?

(c) Design a test to extract I_{dc} and IDDQ contributions for this defective chip with N inverters.

4.4. (a) Determine P_{off} and P_{ac} of a NAND3B ring oscillator with 51 stages as a function of temperature from −25 to 125 °C in steps of 25 °C, all at nominal V_{DD}.

(b) Determine the ratio P/P_{off} at each temperature.

(c) How many stages are needed in the RO for $P_{off} = 0.1 \times P_{ac}$ at each temperature step?

4.5. Total power of a chip is measured as function of V_{DD} at a constant frequency. The V_{DD} exponent on several chips is found to be 3.0.

(a) What are possible reasons for the exponent to be >2.1?

(b) What additional tests would you recommend to get to the root cause?

4.6. EDA power tool assumes a constant value of C_{in}/μm for all logic gates. Typically, the value of C_{in}/μm of a regular V_t inverter is used for this purpose.

(a) Determine C_{in}/μm for a regular V_t inverter over a $L_p \pm 3\sigma L_p$ range at nominal V_{DD} and at $1.1\times V_{DD}$.

(b) What is the maximum fractional error in estimated total switching capacitance of an inverter (FO = 4) over the range in a) if C_{in}/μm is assumed to be at its nominal value?

(c) Under what circumstances would the error introduced in the EDA power tool by assuming a fixed C_{in}/μm be significant?

4.7. A delay chain comprises ten stages of a NOR2T logic gate with $\tau_{pd} \approx \tau_{pu}$ for $(W_n + W_p) = 1$ μm.

(a) Set up a schematic to measure P_{off} and total power in the nominal corner with an input pulse train having a 50 % duty cycle.

(b) Measure total power as a function of frequency. Select an input frequency range so that total power remains approximately linear with frequency. What phenomenon sets an upper limit on the frequency range?
(c) Determine C_{sw} and P_{off} from the data acquired in (b).

4.8. Create a circuit schematic for a standard inverter with a footer device which can be turned on and off with an additional signal.
(a) Measure IDDQ as a function of footer width.
(b) By what factor is IDDQ reduced at 25 and 100 °C for a footer width = W_n of the inverter?

4.9. Assume 5 % of chip circuitry has only low V_t MOSFETs and the remaining circuits use regular V_t MOSFETs. The low V_t MOSFETs have 0.06 V lower V_t than regular V_t MOSFETs. Assume SS = 100 mV/decade.
(a) What is the relative contribution of low V_t MOSFETs to total chip IDDQ in the nominal corner?
(b) In some hardware, the V_t of low V_t p-FETs is centered 0.02 V lower than its nominal value. How does this affect IDDQ?
(c) The f_{max} of the chips turns out to be 10 % higher than target. The test team recommends converting low V_t MOSFETs to regular V_t MOSFETs. How will P_{off} be affected?
(d) An alternative to (c) is to lower V_{DD} by 10 %. Explore this option and compare with the option in (c).

4.10. It is proposed to combine CMOS circuits with superconducting circuits operating in liquid helium at 4.2 K. For this purpose, MOSFETs at the 45 nm technology node are characterized at 4.2 K.
(a) By what factor is I_{off} expected to decrease as the temperature is reduced from 298 K (25 °C) to 4.2 K?
(b) The measured IDDQ is only reduced by a factor of ~7 in cooling a circuit block from 298 K to 4.2 K. How would you explain the difference between expected I_{off} and IDDQ reduction?
(c) Would switching to 22 nm technology with high K gate-dielectric change the IDDQ reduction?

References

1. Taur Y, Ning TH (2009) CMOS device design In: Fundamentals of modern VLSI devices, 2nd edn, Chapter 4. Cambridge University Press, New York
2. Weste NH, Harris D (2010) CMOS VLSI design: a circuit and systems perspective, 4th edn. Addison-Wesley, Reading
3. Rabaey JM, Chandrakasan A, Nikolic B (2003) Digital integrated circuits, 2nd edn. Prentice Hall, Upper Saddle River
4. Bhunia S, Mukhopadhyay S (eds) (2010) Low-power variation tolerant design in nanometer silicon. Springer, New York
5. Gattiker A, Wojciech M (1998) Towards understanding "IDDQ only" fails. In: ITC 1998, pp 174–183
6. Gattiker A, Nigh P (2004) Random and systematic defect analysis using IDDQ signature analysis for understanding fails and guiding test directions. In: ITC 2004, pp 309–318
7. Floyd MS, Ghiasi S, Keller TW, Rajamani K et al (2007) System power management support in the IBM POWER6 microprocessor. IBM J Res Dev 51:733–746

Embedded PVT Monitors

5

Contents

5.1 Placement and Integration 160
5.2 Silicon Process Monitors 162
  5.2.1 MOSFETs 162
  5.2.2 Delay Chains 163
  5.2.3 Ring Oscillators 166
5.3 Power Supply Voltage and Noise Monitors 170
5.4 Critical Path Monitors 173
5.5 Temperature Monitors 174
5.6 Circuit Stages for ROs and Delay Chains 177
  5.6.1 MOSFET Parameter Extraction 182
  5.6.2 SRAM Stage Designs 186
  5.6.3 Silicon Process-Sensitive Suite 188
  5.6.4 Strengths and Limitations of RO-Based Monitors 192
5.7 Data Collection and Characterization 193
5.8 Summary and Exercises 197
References 199

On-chip monitors and sensors, hereafter referred to as monitors, play a key role in test, debug, and evaluation of CMOS chips. Silicon process monitors are useful in product yield optimization through silicon process tuning and variability reduction. Appropriately designed embedded process monitors can help bridge both upward in the hierarchy to complex circuitry and downward to the properties of the constituent components and to the silicon manufacturing line data. Such monitors may also be used for model-to-hardware correlation of circuit design and timing tools. Voltage monitors track dynamic fluctuations in on-chip local power supply voltage. Temperature monitors are used for directly measuring silicon temperature and can detect hot and cold spots on a chip. Design considerations of these process, voltage and temperature (PVT) monitors, and their applications in debug and in silicon process tuning for optimum power and performance are described.

M. Bhushan, M.B. Ketchen, *CMOS Test and Evaluation*,
DOI 10.1007/978-1-4939-1349-7_5

The use of PVT monitors extends beyond manufacturing test. Data from monitors can be collected during operation throughout the life of a chip. This data can be either stored for later inspection or used in real-time analysis and dynamic optimization of frequency, power, and workloads. Example applications include detection of frequency degradation arising from aging effects in MOSFETs, power supply voltage droop variations with workload, and of excessive rise in temperature due to insufficient cooling or degradation in package thermal resistance. Feedback from model-to-hardware correlation may be used for further circuit tuning and in next-generation designs.

PVT monitors do not directly participate in product chip functionality. However, data collected from these monitors provide valuable information to silicon manufacturing, packaging, reliability, and product application teams. In an industrial environment, these teams often have only peripheral involvement in chip design and architecture. Silicon area and added cost in design and integration, and I/O requirements of PVT monitors must therefore be justified in the early phases of chip design. Analysis and characterization methodology of the data collected should feature visualization of essential behavior in an unambiguous fashion to all participating engineering teams.

In this chapter, PVT monitors are described with focus on efficiency in design, test resources, and silicon area. Issues related to placement and integration of monitors are discussed in Sect. 5.1. Silicon process monitors including MOSFETs, delay chains, and ring oscillators are described in Sect. 5.2. A voltage noise detector is described in Sect. 5.3. The concept of a complex critical path monitoring circuit for real-time control of chip operating conditions is described in Sect. 5.4. Temperature sensing elements are covered in Sect. 5.5. Delay chain and ring oscillator circuit stage designs to bridge between silicon technology and chip performance are described in Sect. 5.6. This is followed by some examples of characterization and data visualization techniques in Sect. 5.7.

Design of test structures and the associated characterization methodology for CMOS technology evaluation form the basis of on-chip silicon process monitors [1]. References to specific technical publications on the design and applications of the PVT monitors discussed are included.

5.1 Placement and Integration

Design, placement, and efficient use of embedded PVT monitors require close collaboration among chip, package, and system design teams and the silicon manufacturing and product test teams. The number of monitors of each type and their locations on the chip are dependent on chip dimensions, silicon process variability, circuit design margins, and product specifications. There is an inherent cost associated with the use of embedded monitors on a chip. Large area, high performance microprocessor chips can take advantage of monitors sprinkled across the entire chip. Small area chips for low cost consumer applications may have a very limited number of monitor sites.

Placement schemes for monitors on different chips are illustrated in Fig. 5.1. Small area chips (<25 mm^2) are likely to have one placement location for each monitor type near the center with an option to place four more near the corners. In medium area chips (~25–250 mm^2) additional monitors may be placed near regular grid locations to get uniform coverage across chip. In large area or high performance chips (>250 mm^2), additional monitors may be placed close to critical locations such as performance gating circuit blocks or areas of high circuit activity.

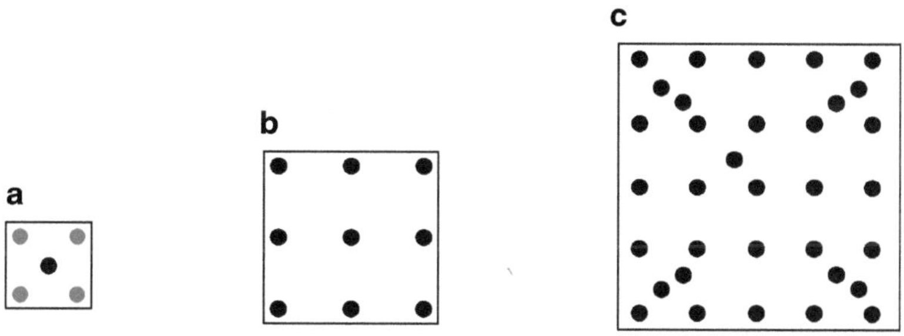

Fig. 5.1 Illustration of location and density of monitors (*solid circles*) on CMOS product chips: (**a**) small chips, (**b**) medium size chips and (**c**) large area high performance chips

Allocation of silicon area for monitors, associated control circuitry and wiring tracks are made in the early design phase. Fluctuations in temperature, voltage, and noise levels on a chip generally occur in the vicinity of high circuit switching activity. If the circuit density in these areas is near maximum, availability of silicon area for placement of monitors and utilization of wiring channels may be very limited. It is therefore prudent that monitor designs be compact with shared control circuitry and I/Os.

Test and calibration of embedded PVT monitors is carried out at wafer level testing. Provision may be made for on-chip storage and retrieval of calibration data. Data collection and characterization may continue at all test levels and during product operation. More complex monitors may incorporate a dedicated processor to generate statistical summaries of the data collected over time.

If monitors are used for feedback and dynamic control of power supply voltage or clock frequency, they must function correctly at all times. If any of the designated monitors malfunction, the chip would be rejected, impacting production yield. Hence, in addition to the requirement of compact designs, PVT monitors must be robust with their functionality validated in simulation over a wider range of parameter variations than that over which other chip circuits are validated. Appropriate documentation along with simulated targets are provided to the test engineering team to assist in model-to-hardware correlation, silicon process characterization, and chip failure diagnostics.

5.2 Silicon Process Monitors

Test structures to measure properties of MOSFETs, metal wires, resistors, and capacitors are placed in the scribe-line. Electrical data collected during silicon manufacturing are utilized to center the process within the corresponding $\pm 3\sigma$ ranges as specified in compact device models.

Variations in key MOSFET parameters (V_t and L_p) within their $\pm 3\sigma$ ranges can bring about as much as ± 50 % variation in circuit delay (Sect. 2.2.4) and associated large spreads in product chip f_{max} and power. As production continues and the silicon process is more closely centered, there may still be significant fluctuations in device parameters over time resulting in performance, power, and yield variations across lots. Process centering information from early in the silicon manufacturing cycle is therefore important for managing the CMOS chip supply line.

The sensitivities of device parameters to physical layout styles is becoming more significant with use of stress layers for MOSFET mobility enhancements, and other MOSFET performance enhancement techniques introduced with CMOS scaling. The design styles of MOSFET test structures in the scribe-line may not always be representative of the physical layout styles used in product chips. In addition, across reticle and across chip variations are not captured by scribe-line monitors (Sect. 6.1.1). In view of these sources of variability, it is important to place a set of silicon process monitors on the product chip itself.

Embedded silicon process monitors comprise individual MOSFETs for DC characterization, and delay chains and ring oscillators for measuring signal propagation delays of product representative logic and memory circuits. Designs of such monitors along with the merits of each of these monitor types are described in Sects. 5.2.1 (MOSFETs), Sect. 5.2.2 (delay chains), and Sect. 5.2.3 (ring oscillators).

5.2.1 MOSFETs

Silicon technology development and manufacturing place a strong emphasis on characterization of individual MOSFETs. *I–V* and *C–V* measurements are made on scribe-line MOSFET test structures. For ease of data analysis, a few points on *I–V* and *C–V* plots are monitored (Sect. 2.2.2). Trend charts are generated for silicon manufacturing quality control. Placement of MOSFET test structures in unused areas of a product chip itself is one way to enable a direct correlation between on-chip mean MOSFET parameter values with the scribe-line data as well as for across chip variability assessment. However, there are several limitations in this approach.

A chip may use several different MOSFET types, and many different physical layout styles may be used in circuits. This requires placement of many MOSFET test structures on a chip. In order to minimize parasitic resistances in series with their terminals, MOSFETs must be placed close to I/Os pads using low resistance wire connections. A minimum of three DC I/Os are required for characterization of an individual isolated MOSFET. There are various schemes to reduce the number of I/Os, such as shared gate terminals or shared source and drain terminals with a decoder to select the gate of any one of the MOSFETs. However, care must be exercised to minimize *IR* drops in the wires and to adequately isolate the device under test [1].

With increase in variations in narrow width MOSFET properties arising from random dopant fluctuations, many ($\geq$30) nominally identical individual MOSFETs must be characterized to get the mean and standard deviations of key parameters. Alternatively, only mean parameter values corresponding to systematic process variations may be obtained from wide MOSFETs (or narrow multi-finger MOSFETs). In wide devices, *IR* voltage drops in the wires may become significant for I_{on} and I_{eff} measurements. Limited but important information can be obtained from measurements in the subthreshold region for I_{off} and V_{t} characterization.

C–V characterization of MOSFETs placed on a product chip is even more challenging. Capacitance measurements using LCR meters require background subtraction and the test time is long compared with standard digital tests [1]. Test equipment for high speed digital and memory tests generally does not include LCR meters for capacitance measurements. A charge-based capacitance measurement (CBCM) method using an on-chip clock signal and measurement of time average DC current through a capacitor may be used instead.

I/O constraints rapidly become a bottleneck for embedded MOSFET monitors. One way to get around this is to dedicate a small number of wafers for MOSFET characterization. With this approach measurements may be obtained by sprinkling MOSFETs with product representative physical layouts in the unused space on the chip or in the fill pattern for obtaining uniform pattern densities of various layers [2]. Wafers selected for characterization are removed from the standard manufacturing flow. These wafers are processed separately using photomask layers designed to connect MOSFET terminals to metal pads for electrical probing. This method has proven to be useful for investigation of pattern density induced across chip variations originating with rapid thermal annealing and other process steps. A weakness of this approach is that process optimization is carried out on a small sample of wafers, and lot-to-lot variations cannot be tracked on an ongoing basis during production.

5.2.2 Delay Chains

Delay chains are constructed by connecting a number of logic gates (LGs) in series. A signal edge arriving at the beginning of the chain propagates through and emerges at the far end. The propagation delay is measured by comparing the arrival times of the signals at the input and output nodes. The schematic of a delay chain comprising *n* series connected LGs is shown in Fig. 5.2a. The input and output signal waveforms are shown in Fig. 5.2b. If the LGs are nominally identical, and the signal delay from node A to node Z is τ, the average delay τ_{p} through an LG is

$$\tau_{\text{p}} = \frac{\tau}{n}. \tag{5.1}$$

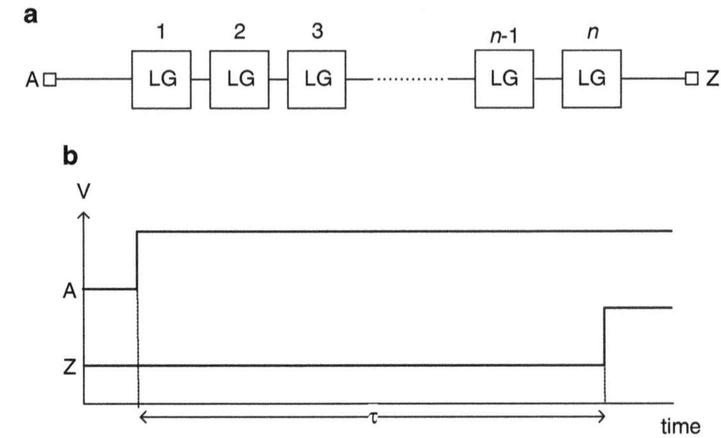

Fig. 5.2 (**a**) Circuit schematic of a delay chain comprising n inverters, and (**b**) timing diagram of rising input and output signals for even n

If the LG performs an inverting logic operation and n is even, there are equal numbers of PU and PD transitions, and τ_p represents the average of τ_{pu} and τ_{pd}. If n is odd, there is one additional PU or PD transition. In this case, the error in estimating τ_p is reduced if $\tau_{pu} \approx \tau_{pd}$ or n is large. However, increasing n increases the silicon area occupied by the delay chain.

On-chip measurement of τ across a delay chain requires additional circuitry. The principle of operation of one such scheme utilizing a time-to-digital converter is shown in Fig. 5.3a [3]. A signal is launched at the input of the delay chain comprising even number of inverting stages. This signal and the signal through

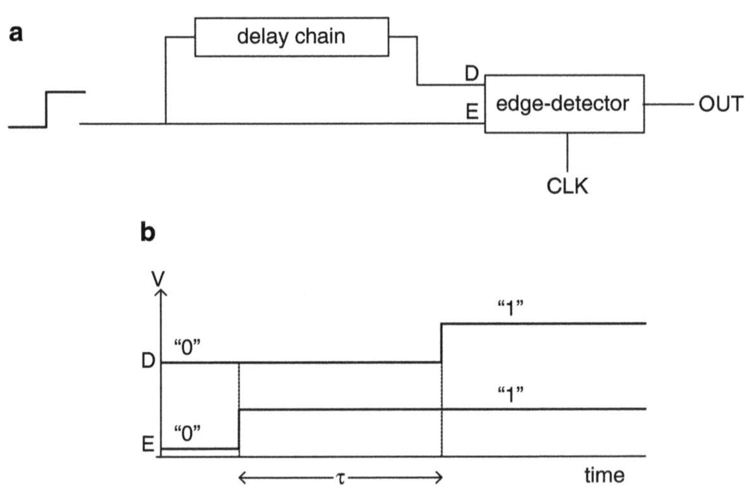

Fig. 5.3 A scheme to measure signal propagation delay τ through a delay chain: (**a**) schematic and (**b**) timing diagram of signals at D and E inputs of the edge detector

the delay chain, after a delay τ, arrive at the edge detector inputs E and D, respectively. The timing diagram of the signals at nodes E and D is shown in Fig. 5.3b. As the rising signal edge travels through the delay chain, node voltages at (E, D) transition from ("0", "0") to ("1", "0"), and after a time delay τ, to ("1", "1"). The time difference between the arrival of the rising edge at node E and the arrival of the rising edge at node D is measured with the edge detector. The measurement may be repeated with a falling signal edge.

A section of an edge detector circuit schematic suitable for this application is shown in Fig. 5.4. It comprises a chain of lightly loaded inverters (FO ~ 2) and a series of edge-triggered latches together with XOR2 and XNOR2 gates. The output of each inverter feeds into an edge-triggered latch. A snapshot of the signal level at the output node of each inverter is captured by the edge-triggered latches asserted with a time-adjustable clock edge. The outputs of the latches and the signal E go through alternating XOR2 and XNOR2 gates, generating a string of "1"s and "0"s.

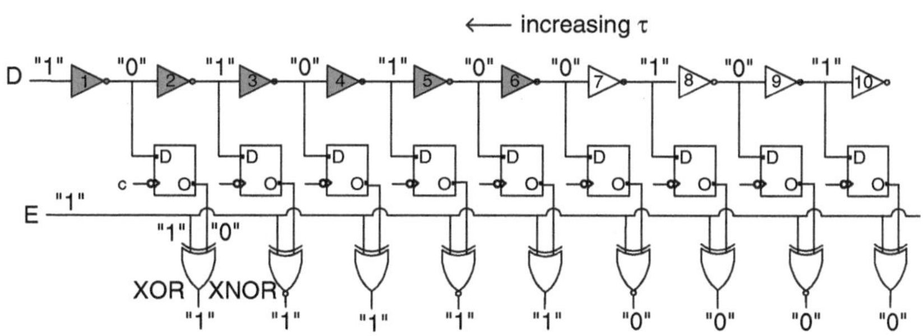

Fig. 5.4 Circuit schematic of an edge detector

In Fig. 5.4, the rising signal D from the output of the delay chain under test is launched at the input of the inverter chain, while node E has already transitioned to a "1". A snapshot of the outputs of XOR2 and XNOR2 gates is taken when this edge has propagated through the first five inverters, giving a string "111110000." The transition from a string of "1"s to a string of "0"s indicates the position of the rising signal edge at D at the "1" → "0" transition. In order to measure τ, the two rising edge locations for D and E shown in Fig. 5.3b are captured by making the inverter chain in the edge detector longer. An output string of the type "11111000000000000000011" is captured, giving $\tau = 16 \times$ inverter (FO ~ 2) delay. As τ increases, the transition "1" → "0", indicating the signal edge at D, moves to the left in the output string.

The delay chain and the edge detection circuit in Fig. 5.3 are integrated with the chip logic. The outputs of the XOR2 and XNOR2 gates are either streamed through for instantaneous measurement and real-time feedback, or scanned out later. The time resolution of this edge detector is ~ two inverter stage (FO = 2) delay. At the 45 nm technology node and beyond, this is of the order of 15 ps or less. To achieve

1 % accuracy in τ, the delay chain should be sufficiently long with τ of ~1,500 ps, or with equivalent delay of 200 inverter (FO = 2) stages. Also the number of stages in the edge detector itself must be >200 to observe both edges. Although the edge detector may be shared among several delay chains, the delay chain itself cannot be compacted without losing measurement accuracy.

If one or more capture latches enter the metastability region (Sect. 3.2.3) during the snapshot, the output stream may contain erroneous values such as "111**01**00000000000" instead of "1111100000000000." In such cases, a "010" sequence with a single "1" is replaced with "110" during post-processing of the data, or the erroneous data filtered from the analysis.

The delay chain under test may comprise series connected nominally identical logic gates or any combinational logic circuit block. The output of the chain may be an inverted or a non-inverted waveform with respect to the input signal. The control and peripheral circuitry may be shared among a large number of chains. This allows measurements on different logic gates and circuits while maintaining an acceptable footprint for the process monitor site. Multiple copies of such monitors may be placed on the chip as described in Sect. 5.1.

A delay chain can be used to study properties associated with the floating-body effect in PD-SOI circuits (Sect. 10.6). In one such application the switching history effect is measured. The first signal edge measures 1SW delay if there has been no signal activity for a long time (millisecond time scale). If a second signal is launched within a few nanoseconds of the first edge, 2SW delay is measured. These delays may be different from each other and also from the steady-state delay measured with an RO. In another configuration, the floating-body effect in either only p-FETs or only n-FETs in PD-SOI technology is investigated. Body-contacted MOSFETs in PD-SOI do not exhibit floating-body effects. With body-contacted p-FETs and floating-body n-FETs in an inverter, the history effect in n-FETs can be measured. Similarly, with body-contacted n-FETs and floating-body p-FETs, the history effect in p-FETs is measured.

5.2.3 Ring Oscillators

A brief description of ring oscillator operation is given in Sect. 2.2.2. A representative ring oscillator circuit comprising $(2\alpha + 1)$ identical inverting logic gates (LGs) is shown in Fig. 5.5a. With the switch "S" closed, the voltage at any node oscillates with a period T_p as shown in Fig. 5.5b. Oscillations are suspended when "S" is open. For sustained oscillations, a signal edge at node A should arrive with a complementary level at A after travelling through the loop once. For the RO circuit in Fig. 5.5a, this criterion is met when the number of inverting stages is odd.

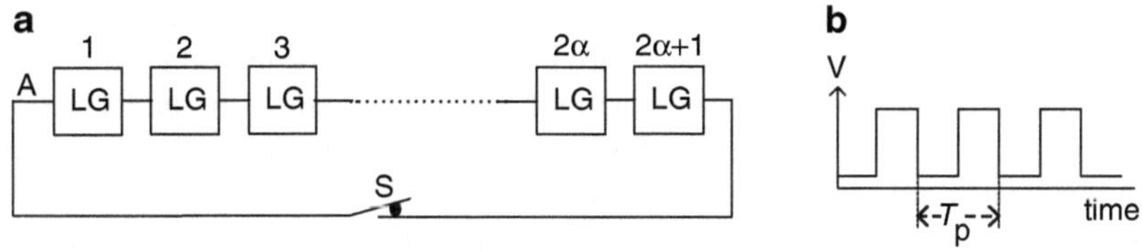

Fig. 5.5 (**a**) Ring oscillator circuit with $(2\alpha + 1)$ inverting logic gate (LG) stages, and (**b**) output voltage waveform at any node as a function of time

The period of the RO corresponds to the time for a signal to travel twice around the loop, with one PU and one PD transition at each node. The period T_p is given by

$$T_p = \left(\tau_{pd1} + \tau_{pu1} + \tau_{pd2} + \tau_{pu2} \ldots + \tau_{pd(2\alpha+1)} + \tau_{pu(2\alpha+1)}\right). \tag{5.2}$$

Here τ_{pd1} and τ_{pu1} are the PD and PU delays of the first LG and this naming convention is followed for all $(2\alpha + 1)$ LGs. The average of the PU and PD delays, τ_p, of all the LGs in the RO is given by

$$\tau_p = \frac{T_p}{2(2\alpha + 1)} = \frac{1}{2f(2\alpha + 1)}, \tag{5.3}$$

where f is the frequency of oscillation. By measuring either the period or the frequency of oscillation, τ_p of any set of identical logic gates in an RO can be determined. The LGs in Fig. 5.5a may be replaced by combinational logic representative of any data path producing an inverted output signal.

Switch "S" in Fig. 5.5a may be implemented with a logic gate, a latch or another suitable circuit scheme. Examples of switch "S" configurations are shown in Fig. 5.6. In Fig. 5.6a, one of the two inputs of a NAND2 closes the RO loop and the second input EBL is held at V_{DD} to enable the oscillations. The RO oscillations are disabled when the EBL connects to GND. In a similar scheme shown in Fig. 5.6b, the NAND2 is replaced by a NOR2 gate. The EBL input is held at GND to enable the oscillations and at V_{DD} to disable the oscillations. With a level-sensitive latch switch in Fig. 5.6c, the clock input (EBL) is held at V_{DD} to enable the oscillations and at GND for disabling the RO. An inverter is added to the output of the latch to maintain an odd number of inverting stages in the ring.

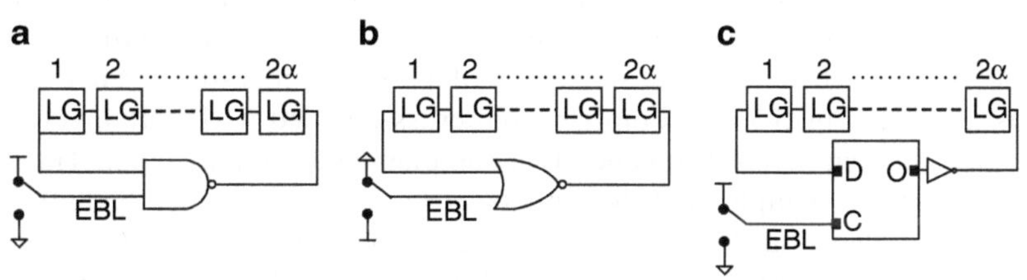

Fig. 5.6 Switch "S" configurations for ROs with EBL input signal set to enable oscillations: (**a**) NAND gate, (**b**) NOR gate and (**c**) clocked latch

The delay through the switch is generally different than the delay through the other 2α logic gates in the RO. This may introduce an error in computing τ_p using Eq. 5.3. The error is reduced by increasing the number of stages in the RO ($\alpha \geq 50$). However, for multiple placements of ROs across the chip, it is preferable to have a small physical footprint, and hence a small value of α (≤ 25).

A complete set of measurements to characterize an RO comprises determination of frequency of oscillation or period, and current in the quiescent and active states, IDDQ and IDDA, respectively (Sect. 2.2.5). The measured data are used for determining power drawn in the off- and on-states, signal propagation delay, switching capacitance, and switching resistance per stage.

RO circuits in a scribe-line may have an independent power supply, separate from the power supply of the control circuitry, to permit IDDQ and IDDA measurements. If a number of ROs share the same power supply, the measured IDDQ is the sum of the IDDQs of all ROs. By turning on only one RO at a time, and subtracting the background IDDQ of all ROs, one can determine (IDDA – IDDQ) and thereby C_{sw} and R_{sw} of each stage of the active RO.

On a product chip, if an RO has its own independent V_{DD}, the same analysis is applied as with a scribe-line RO. However, if there are a large number of embedded ROs, to reduce the number of I/Os, ROs are typically tied to the chip power supply. The IDDQ of the chip is generally much higher than the IDDA of a single RO which is of the order of ~1 mA. Switching activity in the control and readout circuits add to the background current during RO measurements. In this case, accurate determination of IDDA of any single RO and extraction of C_{sw} and R_{sw} are not feasible. Circuit characterization is based on measured frequencies of ROs with different circuit topologies as described in Sect. 5.6.

RO frequency or period can be measured on-chip or with off-chip equipment. An on-chip counter circuit using the chip clock as a time reference can be integrated with the RO circuit. The RO period is computed from the number of RO cycles in one clock cycle. Off-chip measurements of RO cycle time are carried out using the tester (ATE) clock as a reference. Alternatively, an off-the-shelf frequency counter may be used. ATE clocks and frequency counters typically have an upper frequency limit of a few 100 MHz whereas RO frequencies may reach several GHz. Also low frequency signals have a higher immunity to noise and can travel over longer distances, requiring fewer buffer insertions to maintain signal integrity. Generally, it is highly desirable to lower the RO frequency to ~1–10 MHz by using a frequency divider circuit before delivering the signal for off-chip measurements.

A frequency divider function is implemented with a flip-flop unit or a master–slave latch that divides the input frequency by a factor of two per unit with a high degree of precision. The circuit schematic of a positive edge-triggered T-flip-flop unit is shown in Fig. 5.7a. It comprises four cross-coupled AND-OR-Invert (AOI) logic gates and an inverter to generate a complementary input signal. The signal frequency at the complementary outputs at Q and Q_b is exactly half of the frequency of the input CLK signal. In Fig. 5.7b, four divide-by-two units are connected in series to divide the input frequency by 2^4 (=16). In this chain, outputs Q and Q_b of the previous unit are connected to the complementary inputs CLK and CLK_b of the following unit. With n such units, the output frequency is divided by 2^n.

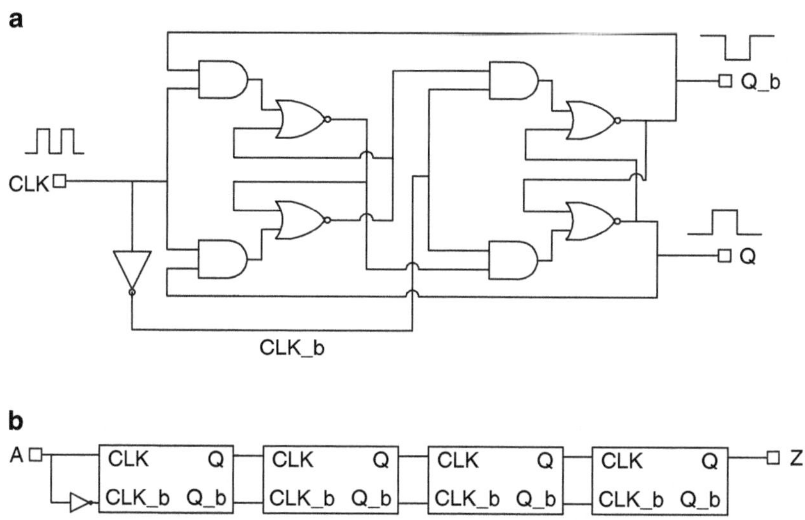

Fig. 5.7 Circuit schematic of (**a**) divide-by-two flip-flop unit and (**b**) frequency divider with a chain of four flip-flop units to divide by 16

A scheme for sharing the inputs and outputs of multiple ROs on a chip is shown in Fig. 5.8. A decoder selects and enables an RO and a multiplexer or a wide OR propagates the oscillating signal from the selected RO to OUT while all other ROs are in quiescent states. A frequency divider circuit at the output of each RO lowers the frequency to a few 100 MHz. Buffers are placed at suitable intervals to maintain signal integrity if the ROs are distributed across chip. Additional divide-by-two units may be added at the common output to lower the frequency further for connecting to an ATE or off-the-shelf frequency counter.

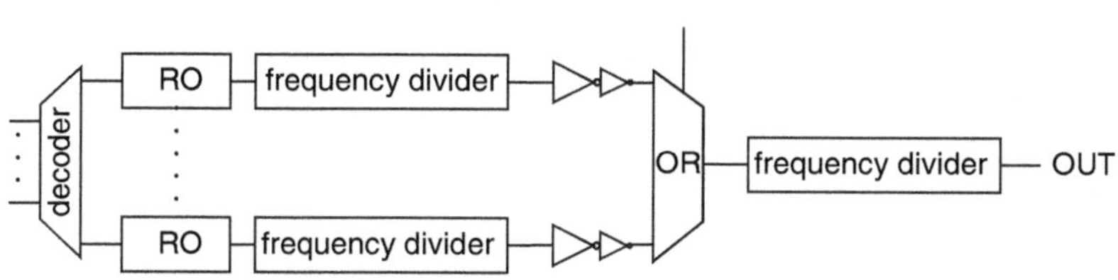

Fig. 5.8 Scheme for sharing control circuitry to enable and readout multiple ROs on a chip

ROs form a closed loop and are generally not compatible with chip timing tools. In such situations, RO loops need to be "black-boxed" with only inputs and outputs visible to the timing tools. Each RO loop design must be manually checked to prevent errors resulting in a dysfunctional monitor.

Measurement accuracy of RO frequencies is typically better than ±1 %. The output waveform may be corrupted if higher order harmonics of the fundamental frequency are generated. This occurs due to noise or with a sudden change in power supply voltage as the RO is being enabled, resulting in multiple signal edges propagating in the loop simultaneously. Only odd multiples of f ($3f$, $5f$, ...) can be sustained in such situations, and the most common observed is the third harmonic. Even multiples of the fundamental frequency can also be recorded when glitches or spikes in the output waveforms are not properly filtered. If the range of expected RO frequency is known, such erroneous data can be easily rejected during post-processing.

5.3 Power Supply Voltage and Noise Monitors

Signal propagation delays through logic circuits vary with the power supply voltage, and as a result T_{cmin} (f_{max}) varies with V_{DD} (Fig. 3.38). Local variations in V_{DD} may occur because of *IR* drops associated with series resistance in the power grid. Temporal variations in local V_{DD} values occur as circuit switching activities and the current drawn fluctuate over time. Such variations in V_{DD} are tracked over time with on-chip monitors, as a dip in V_{DD} may lower f_{max} and be the root cause of a timing fail.

Chip timing, T_{cmin} and f_{max} are affected by local clock skew, variations in duty cycle and jitter at the clock edges (Sect. 3.1.3). Any change in V_{DD} may also impact clock path delays. These affect the timing of the clock signals for launch and capture at clocked storage elements such as latches and flip-flops. Signal-to-signal coupling may also affect waveform shapes and signal propagation delays.

It is difficult to separate and quantify each of these effects by location on the chip and time of occurrence. Monitors are designed to capture the net effect of power supply, clock, and noise in critical locations on the chip. One such implementation is described below.

The clock skew and jitter, SKITTER (SKew + jITTER) monitor comprises a tapped delay chain configured to function as an edge detector [4]. A ten inverter section of such a delay chain is shown in Fig. 5.9. A clock edge is launched at input A. At time $= t_1$, input at A is at “0” and all the inverter outputs along the line alternate between “1” and “0”. At time $= t_2$, the rising edge of the CLK signal has propagated through the first two stages, with the outputs of stages 2 and 3 both at “1”. At time $= t_3$, the falling edge of the CLK signal has also been launched at A. The rising edge is now at inverter 7 and the falling edge at inverter 2. The inverter outputs have a logic bit pattern of 0**11**0101**00**1. The two consecutive “1”s and “0”s give the locations of the clock edges in the delay chain at t_3.

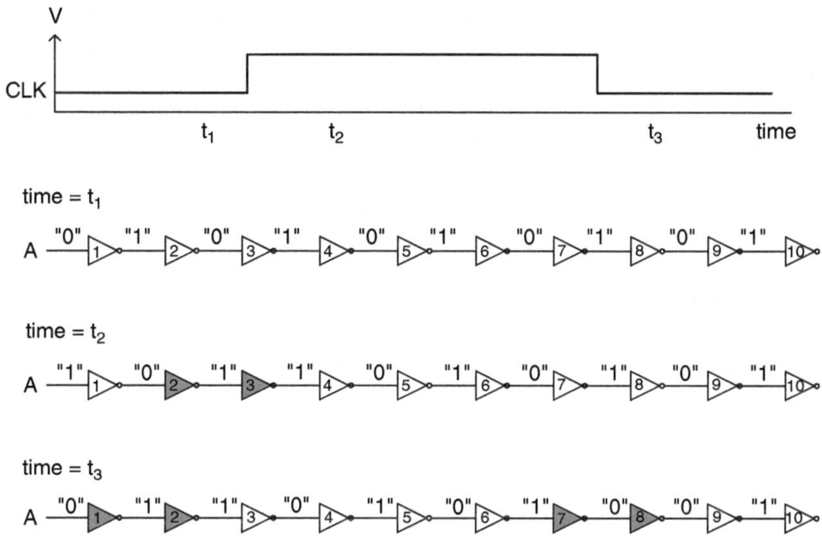

Fig. 5.9 Timing diagram of a clock signal along a delay chain. The locations of the signal rising and falling edges are indicated by consecutive "1"s or "0"s at the inverter output nodes

The output nodes of the delay line are tapped and fed into edge-triggered latches as shown in Fig. 5.10. The outputs of neighboring latches drive the XNOR2s as indicated in the figure. Two consecutive "1"s or "0"s in the delay line result in a "1" from the corresponding XNOR2. The output bit pattern of 10000010 indicates two edges in the delay chain at the time the snapshot was taken via the latches triggered by input clock signal C. The outputs may be read in real time or stored in register files to be scanned out at a later time.

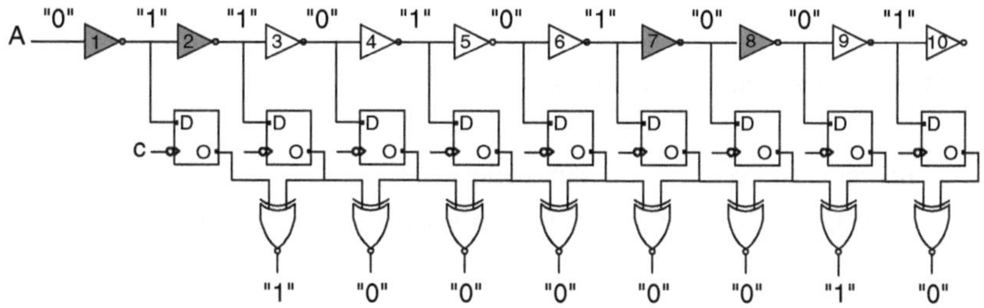

Fig. 5.10 Tapped delay chain to capture clock rising and falling edges indicated by XNOR2 outputs at "1"

With a sufficiently long delay chain, three clock edges may be captured and the clock duty cycle estimated from the output bit pattern. The accuracy of clock edge location in time is on the order of an inverter delay. With an inverter delay of 7 ps and a clock cycle time of 250 ps (4 GHz), the worst case error estimate for a full cycle is 6 %.

Noise in the clock distribution tree, PLL jitter, and other variations cause jitter in the clock edges. With additional circuitry, the movement of the clock edges over time may be captured. One implementation of this idea is shown in Fig. 5.11 [4]. The outputs of XNOR2s and an additional control signal, sticky, go through AOI logic and are captured in a second set of edge-triggered latches. With the sticky bit = "1", the second set of latches is updated if the XNOR2 output is a "1". If the clock edge moves by more than one inverter delay at any time, the event is captured and the total range of clock edge movement recorded as a set of "1"s. The output bit patterns for an ideal clock signal with 50 % duty cycle, with 45 % duty cycle and with jitter are shown in Fig. 5.12.

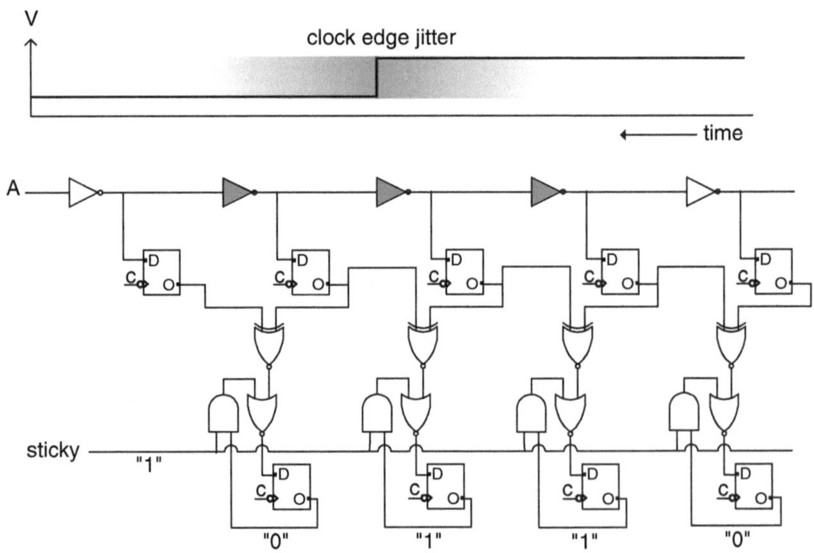

Fig. 5.11 SKITTER with sticky mode enabled

normal clock cycle duty cycle = 50%

1000000000000000000010000000000000000000100000000000000000001

duty cycle = 45%

1000000000000000001000000000000000000000001000000000000000000100

duty cycle = 50% sticky mode on

1000000000000000000011100000000000000001110000000000000011111

Fig. 5.12 SKITTER output bits with normal clock cycle, with 45 % duty cycle and with jitter

The time calibration of the SKITTER is accomplished by sending a clock signal while the chip is in an inactive state. The clock cycle time divided by the number of bits in the cycle from the SKITTER gives the inverter delay per stage in picoseconds. This is the time resolution per bit. Next, the inverter delay using the above method is measured over a range of V_{DD} values, centered about the nominal V_{DD}. A linear fit of the data gives ΔV_{DD} in units of number of bits. The shift in clock edge due to a V_{DD} droop can then be used to estimate the magnitude of the shift in V_{DD}.

5.4 Critical Path Monitors

A critical path monitor (CPM) combines the functions of silicon process monitors with those of voltage, jitter, and noise monitors in a single unit. Dynamic measurements of path delays on a chip and comparisons of the measured delays with initial calibration data stored on the chip provide information on delay changes with time due to temperature, voltage, noise, and aging effects. This information is used to manage power and clock frequency for optimum performance at all times.

The basic operation of a CPM implemented in IBM's POWER6 microprocessor is shown in Fig. 5.13 [5]. The simplified block diagram includes a signal generator to launch a clock pulse on the delay path. The edge detector circuit converts the path delay to digital bits as described in Sect. 5.2.2. Initial measurements on the path delay are made with minimum switching activity on the chip and the data are stored for tracking changes at later times. The path delay is then measured at predetermined intervals when the chip is in functional mode. The dynamic path delay measurements are compared with the initial calibration and any significant changes are used for adjusting system operating parameters for optimum power and frequency.

The circuit concept of a CPM in Fig. 5.13 is expanded to include several delay paths configured to represent different circuit topologies or circuit blocks with high sensitivity to f_{max} (T_{cmin}) for a specific chip design. The delay path configuration provides flexibility to accommodate variety of such paths. Unique paths may be created by emulating a hybridized combinational circuit block.

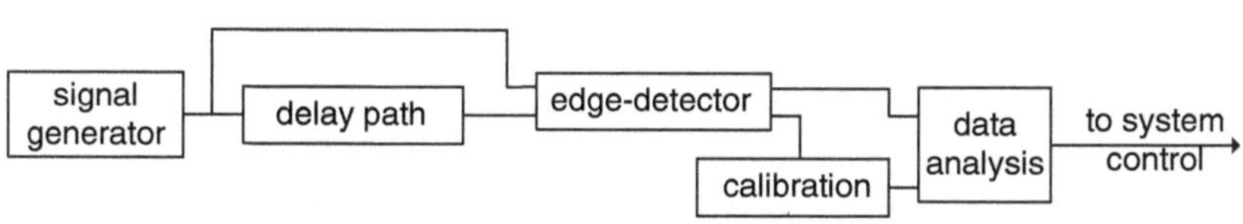

Fig. 5.13 Simplified schematic of a critical path monitor

There are clear advantages of deploying CPMs with analysis capability and feedback control on CMOS chips. Three key advantages of CPMs are listed below:

- Customized path configurations mimic critical path and f_{max} behavior.
- Dynamic read out enables tracking circuit performance during functional operation and over the lifetime of a CMOS product.
- Real-time feedback from the CPM can be used to tune application conditions (V_{DD}, T_c) for optimum power/performance trade-off over time.

There are also limitations of CPMs which must be considered in selecting the CPM design and its overlap with other monitors on chip. EDA timing tools used for present day chip designs try to equalize the delay of all paths by appropriately sizing the transistors. If a path has positive slack (fast path), the transistor widths of the gates in the path are tuned to decrease power while still meeting the cycle time. An outcome of this approach is that no single path dominates T_{cmin} and it is difficult to identify a single representative critical path. As there is considerable overhead in the design, integration, silicon area, and test of on-chip CPMs, the associated cost needs to be justified in view of the gains. High-end microprocessor chips, where performance is critical to the product, benefit the most from CPM implementation.

5.5 Temperature Monitors

The ambient temperature range over which CMOS chips are specified to operate varies with product application and may be as wide as −40 °C to 120 °C. Power dissipation on the chip raises the average silicon surface temperature T_j above ambient. Local variations in power density result in nonuniform temperature distribution across chip, producing hot spots in regions of high switching activity. The locations of hot spots and the local rise in temperature may change over time with changes in workload. In high performance CMOS chips with high IDDQ, an increase in temperature may be observed even in the quiescent state. Variations in chip-to-package thermal resistance and external cooling system efficiency also impact T_j (Sect. 6.1.4).

Since MOSFET parameters and interconnect wire resistances vary with temperature, electrical tests must cover the temperature range over which chips are expected to operate. At wafer and package level testing, the test platform is held at a fixed temperature. One or more temperature monitors are embedded on chip to enable T_j measurements. If the chip area is small and the power density is low, a single embedded temperature monitor may be adequate. High performance chips generally require placement of multiple on-chip temperature monitors. Thermal analysis of the chip in the design phase is useful in identifying potential hot spots when operating in functional mode. If possible, temperature monitors should be placed in the vicinity of such areas in addition to other cooler locations.

On-chip temperature monitors are based on temperature sensitivity of device or circuit parameters. As circuit delay varies linearly with temperature, one can make use of the infrastructure already in place for silicon process monitors. A delay chain or RO with a circuit stage design comprising an inverter driving MOSFET gate loads is used as a reference. A second delay chain or RO with a circuit stage design comprising and an inverter driving a long metal wire serves as the temperature sensor. Circuit schematics of these two stage designs are shown in Fig. 5.14a, b.

Typically, delay varies by <0.12 % per 1 °C change in temperature for the reference stage. The temperature coefficient of copper metal wire is ~0.35 %/°C, higher than R_{sw} of the inverter. The ratio of inverter R_{sw} to wire $R_w l$ for the circuit in Fig. 5.14b is selected to get a higher temperature sensitivity than the reference stage in Fig. 5.14a. The % changes in τ_p, $(\delta\tau_p/\tau_p)$, for the two stages as a function of temperature are shown in Fig. 5.14c for $R_w l/R_{sw} = 1.2$ and $C_{wl}/C_{in} = 1$ for 45 nm PTM HP models. Such calibration plots are generated by measuring the delays at two known temperatures indicated by solid white circles in the plot and a linear fit of the data obtained. The temperature can then be determined from the relative differences in measured $\delta\tau_p/\tau_p$ values shown as gray circles. With this differencing scheme the impact of silicon process variations and local temporal variations in V_{DD} are minimized.

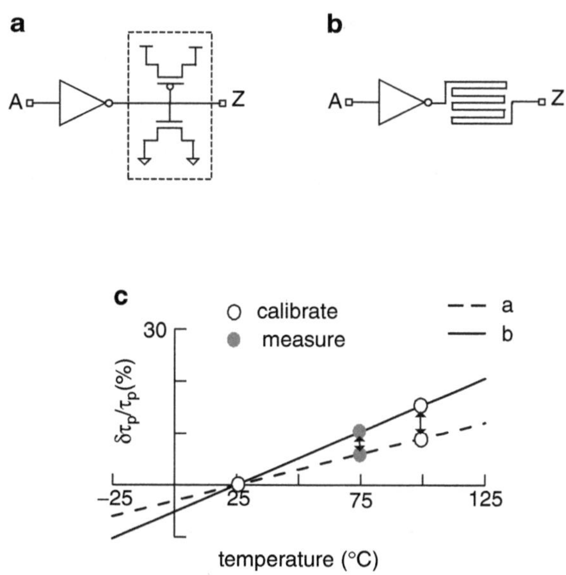

Fig. 5.14 Circuit schematics of delay chain/RO stages for sensing temperature differences: (**a**) inverter driving a MOSFET gate load and (**b**) inverter driving a wire *RC* load. (**c**) $\delta\tau_p/\tau_p$ (%) as a function of temperature for stage schematics in (**a**) and (**b**)

It is generally desirable to measure T_j prior to electrical testing using stand-alone temperature monitors. This is to ensure that there is good thermal contact between the wafer and the chuck or between the chip and the package. Properties of p/n junction diodes are exploited for constructing temperature monitors which can be configured either as stand-alone or integrated with on-chip circuitry to provide a digital output. At a constant current, the forward voltage V_F of a p/n junction diode varies linearly with temperature. Typically, a constant current of 100 μA is passed through the diode biased in the forward direction. The voltage V_F is in the 0.7–0.9 V range and can be easily measured. The power dissipation in the diode is small and the error in T_j due to self-heating of the diode is negligible.

The equations that describe the *I–V* characteristics of a p/n junction diode are

$$I = I_0 \exp\left(\frac{qV_F}{kT_j}\right) \tag{5.4}$$

and

$$I_0 = c_1 T_1^n \exp\left(\frac{-qV_{gp}}{kT_j}\right), \tag{5.5}$$

where c_1 and n (~3.5 for silicon) are constants, V_{gp} is the bandgap voltage (1.14 eV for silicon), k is the Boltzmann constant (8.617×10^{-5} eV/ °K), and T_j is the p/n junction temperature in °K.

A four terminal configuration for measuring voltage across a diode, and its *I–V* characteristics at two different temperatures T_{j1} and T_{j2} ($T_{j2} > T_{j1}$) are shown in Fig. 5.15. Each diode requires two current and two voltage leads for accurate measurements. A constant current is supplied by an external power supply, and the voltage V_F is recorded by the tester. A calibration curve is obtained from a linear fit of V_F vs. T_j plot. If the expected temperature range of T_j is very wide (>100 °C), different calibration coefficients may be used in different temperature ranges. Alternatively, a more complex fitting routine may be used by taking measurements at three different known temperatures for calibration. Typically, the uncertainty in temperature measurement is of the order of ±5 °C.

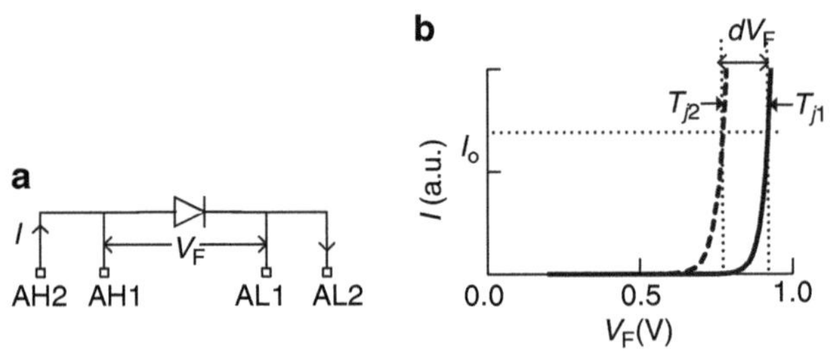

Fig. 5.15 p/n junction diode: (**a**) four terminal measurement scheme and (**b**) *I–V* characteristics at temperatures T_{j1} and T_{j2}

Calibration of temperature monitors is performed with minimal on-chip power dissipation to ensure T_j remains close to the temperature of the thermally controlled block. The calibration coefficients are stored in the tester memory or in on-chip ROM for later use at board and system level. The diode voltages may be digitized with the use of a comparator circuit. The digital output can then be read using chip logic at any time during operation. There is silicon area overhead associated with additional circuitry in this scheme, but dedicated I/Os are not required. The digital temperature monitor implementation is suitable for large area chips with nonuniform T_j distributions. Multiple monitors may be embedded for thermal profiling with the option of tracking T_j in the field.

Thermal time constants for heat flow are of the order of hundreds of microseconds to milliseconds. Unlike SKITTER, the temperature monitors can only respond to changes in local activity over a relatively long time scale.

5.6 Circuit Stages for ROs and Delay Chains

Delay chain and ring oscillator configurations and the associated circuitry for measuring signal propagation delays through series connected logic gates or small circuit blocks are described in Sects. 5.2.2 and 5.2.3. Their primary application is, however, for measuring delays through a chain of nominally identical circuit stages to obtain the average delay per stage τ_p. With a measurement accuracy of ~1 %, τ_p for our standard inverter (FO = 4) can be measured with a precision of better than 0.1 ps. This facilitates fairly accurate model-to-hardware correlation by comparing the measured τ_p values of different circuits with those obtained from simulations using device models and EDA tools.

The nominally identical circuit stages, henceforth referred to as ckt_stgs, may comprise a single logic gate, a logic gate driving additional capacitive or *RC* load, a small logic circuit block, or a memory cell. Relative differences in τ_p values among different ckt_stgs are employed for characterization. In order to implement this differencing scheme, ckt_stgs and their physical layouts need to follow the guidelines described below.

In Fig. 5.16, four ckt_stgs, each with an inverter as the driving logic gate, are shown. Fig. 5.16a shows an inverter (FO = 1). The inverter in Fig. 5.16b drives a MOSFET gate load in addition to the inverter in the following stage. The gate load C_{gXL} comprises an n-FET of width $W_{ncg} = (XL - 1) \times W_n$ and a p-FET of width $W_{pcg} = (XL - 1) \times W_p$, where W_n and W_p are the n-FET and p-FET widths of the inverter, and XL = FO. The ckt_stg schematic in Fig. 5.16c is close to a real FO = 3 with the inverter driving two load inverters in addition to the inverter of the next stage. Ideally, the two load inverters should also be in a FO = 3 configuration. In order to limit the physical dimensions of the ckt_stg, the two load inverters are terminated with C_{gXL} loads, having three times the widths of the n-FET and p-FET in the driver inverter. In Fig. 5.16d, a current multiplier circuit with XL = FO described in Sect. 2.2.4 is inserted for simulations with different FO values. This scheme is included for comparing simulation results with designs which can be realized in silicon.

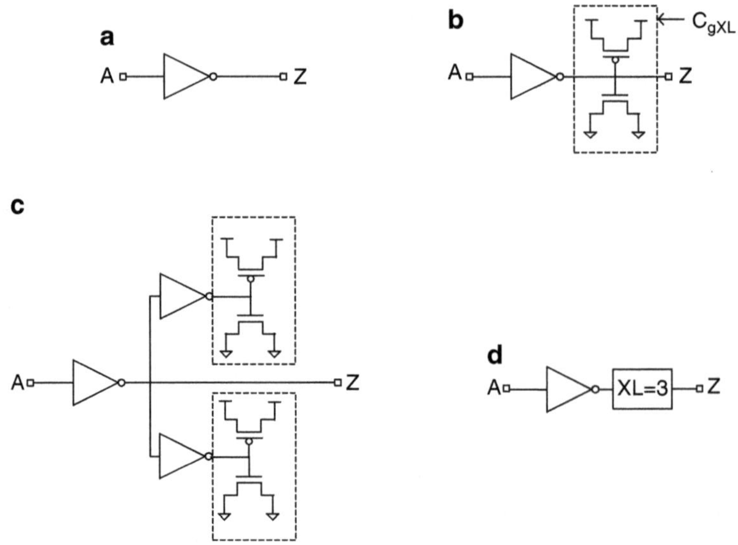

Fig. 5.16 Circuit schematics of ckt_stgs: (**a**) inverter (FO = 1), (**b**) inverter with MOSFET gate load C_{gXL}, (**c**) inverter (FO = 3), and (**d**) inverter with a current multiplier XL

In Sect. 2.2.5, a methodology to determine C_{sw} and R_{sw} of an RO stage from measured values of frequency, and active and quiescent currents, IDDA and IDDQ, is described. Using this procedure, τ_p, C_{sw}, and R_{sw} for the four ckt_stgs in Fig. 5.16 are obtained from circuit simulations of ring oscillators. These values for our standard inverter in 45 nm PTM HP models at 1.0 V, 25 °C are listed in Table 5.1. Because of the additional currents through the load inverters in the ckt_stg in Fig. 5.16c, the R_{sw}, C_{sw} and IDDQ values do not represent the inverter (FO = 3) circuit. This is a limitation of an RO monitor of this design using the measured frequency, IDDA and IDDQ methodology.

Table 5.1 Inverter τ_p, R_{sw}, and C_{sw} derived from measured *f*, IDDA and IDDQ for ckt_stg schematics with our standard inverter shown in Fig. 5.16. 45 nm PTM HP models @1.0 V, 25 °C

Circuit schematic in Fig. 5.16	FO	τ_p (ps)	R_{sw} (Ω)	C_{sw} (*fF*)	IDDQ (nA)
(a)	1	5.27	1,582	3.33	6.15
(b)	3 (XL = 3)	9.21	1,600	5.75	7.91
(c)	3	9.64	–	–	–
(d)	3 (XL = 3)	9.55	1,498	6.37	6.57

The ckt_stg design in Fig. 5.16b for an inverter (FO = 3) is more compact, with ~5 % smaller delay, than the design in Fig. 5.16c. It is often more convenient to use this design in RO monitors placed in the scribe-line where, with an independent power supply, IDDA and IDDQ measurements can be used to extract C_{sw} and R_{sw}. This RO design has the additional advantage that it can be wired with one metal level (M1) to get an early readout during manufacturing. The true FO = 3 design in Fig. 5.16c is used in embedded silicon process monitors, relying on relative τ_p values for tracking performance. A true FO = 4 design, used in circuit simulations is a less practical choice on silicon because of the larger area requirement. It can be easily demonstrated that FO = 3 and FO = 4 track well and only one such design is sufficient to monitor silicon technology performance.

Ckt_stg delays τ_{pa}, τ_{pb}, and τ_{pc} for the designs in Fig. 5.16a–c are expressed as

$$\tau_{pa} = R_{sw}\left(C_{in} + C_{out} + C_{p1}\right) \tag{5.6}$$

$$\tau_{pb} = R_{sw}\left(C_{in} + C_{gXL} + C_{out} + C_{p2}\right) \tag{5.7}$$

$$\tau_{pc} = R_{sw}\left(3C_{in} + C_{out} + C_{p3}\right) \tag{5.8}$$

where C_{p1}, C_{p2}, and C_{p3} are the respective parasitic capacitances. If $C_{p1} \approx C_{p2} \approx C_{p3}$, various comparisons involving differences and ratios of τ_p values provide clear and more accurate results.

A standardized ckt_stg physical layout template is followed to keep the ckt_stg parasitics nearly the same. The height and widths of ckt_stg layouts are fixed for all stage designs. The height matches the standard power grid on the chip and the width is selected to accommodate the largest design in the set of delay chains or ROs to be embedded. Example physical layouts of FO = 1 and FO = 3 with C_{gXL} load are shown in Fig. 5.17a, b. M1 wire lengths for connecting to the input and output terminals of the stages (A and Z) and the M1 power busses are identical in the two designs. Similarly, M2 and higher level metal interconnect wire shapes should be matched as closely as possible. Dummy PS shapes are placed to minimize line-width variations by providing a uniform PS density for the optical lithography and etching tools in silicon processing.

Schematics of the type shown in Fig. 5.16b, c may be implemented in different V_t and L_p offerings from the silicon foundry. For different V_t offerings, C_{sw} values are nearly the same, and τ_p ratios indicate the difference in R_{sw} values or current drive strengths of different V_t MOSFET pairs. Ckt_stg designs with identical schematics but different physical layout styles of MOSFETs (Fig. 10.2) are useful for determining the impact of parasitic RC on τ_p. Such designs may be used for model-to-hardware correlation of the layout extraction tools and for optimization of standard cell layouts. Another variation is in MOSFET widths and number of PS fingers for the same width. Such designs are utilized to track MOSFET width bias relative to model predictions.

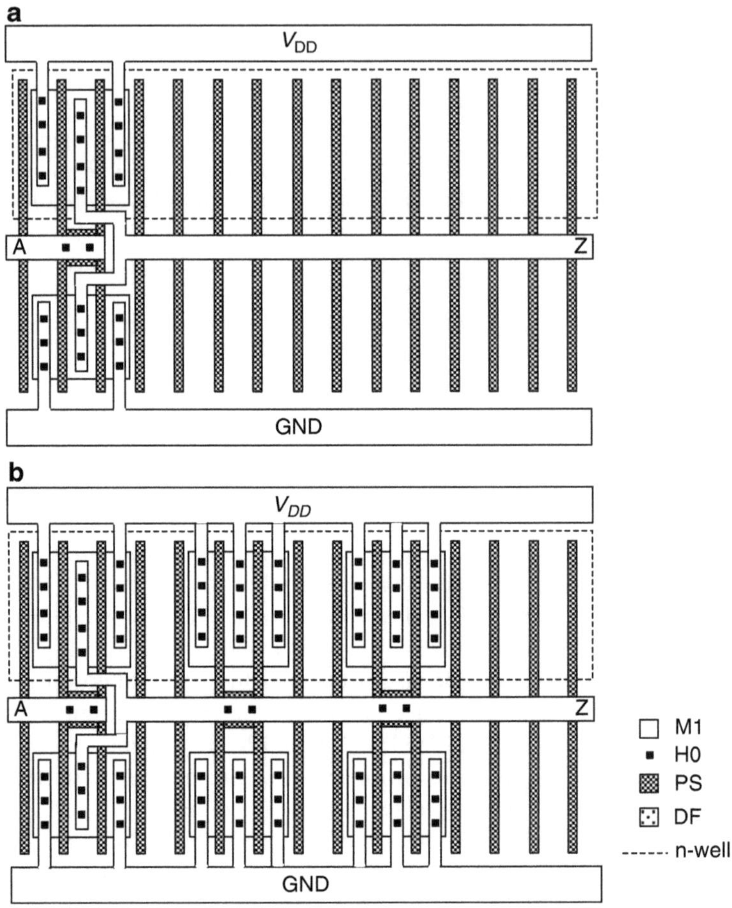

Fig. 5.17 Physical layouts of stages with fixed dimensions, (**a**) inverter (FO = 1), and (**b**) inverter with equivalent FO = 3 gate load corresponding to the circuit schematic in Fig. 5.16b

In Fig. 5.18a, b, ckt_stg schematics for tracking overlap capacitance C_{ov} and diffusion region capacitances C_{ds} and C_{dd} are shown. In Fig. 5.18a, MOSFET loads are configured such that their switching capacitance is dominated by C_{ov}. In Fig. 5.18b, the inverter drives diffusion region capacitive loads of an n-FET and a

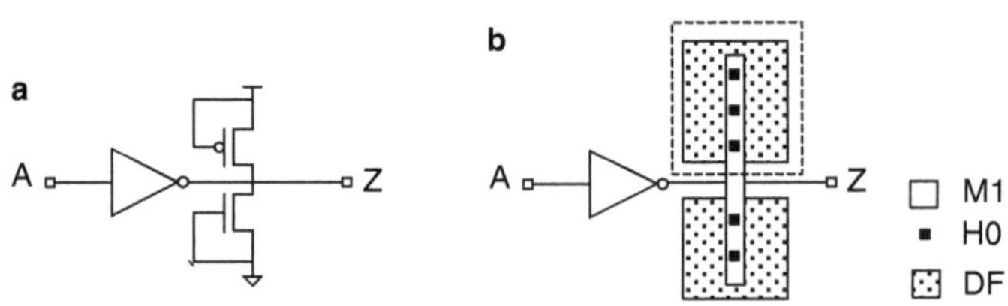

Fig. 5.18 Schematics of ckt_stg to determine (**a**) C_{ov} and (**b**) C_{ds}, C_{dd}

p-FET, depicted by their respective physical layouts. By varying the perimeter to area ratios of these regions in a set of ckt_stgs, relative values of area and perimeter components C_{js} and C_{jsws} can be extracted.

Designs of ckt_stgs with inverters driving wire capacitance loads are illustrated in Fig. 5.19a, b. In Fig. 5.19a, M2 signal (S) wire capacitance of a comb structure includes C_{up}, C_{down}, C_{left}, and C_{right} as the neighboring M2 wires are held at GND potential. In Fig. 5.19b, only C_{up} and C_{down} contribute to C_{sw} since all M2 wire segments are at the same potential. In both designs M1 and M3 wire planes with 50 % density shown in Fig. 5.19c serve as GND planes below and above the M2 signal wires. The wire resistance $R_{w}l$ is $\ll R_{sw}$, and the $R_{sw}C_{w}l$ component dominates the wire contribution to delay. The wire length is selected to give $C_{w}l \sim 0.15 \times (C_{in} + C_{out})$ so that $C_{w}l$ contributions are easily detected while meeting the constraints of standardized physical template width for ckt_stgs.

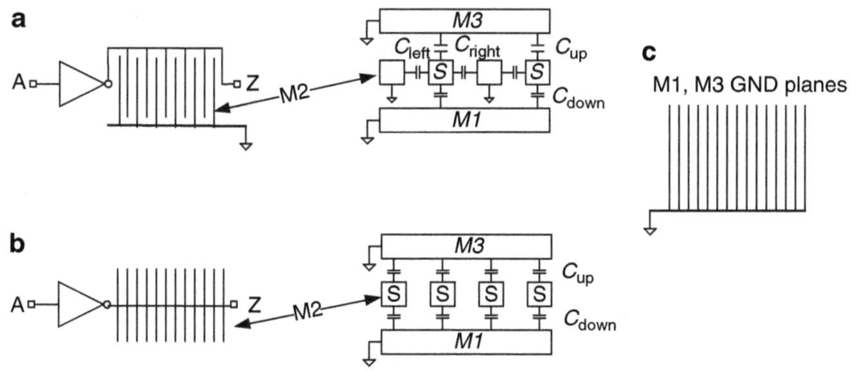

Fig. 5.19 Schematics of ckt_stg to determine (**a**) $C_w = C_{up} + C_{down} + C_{left} + C_{right}$, and (**b**) $C_w = C_{up} + C_{down}$ of signal wires in M2 layer. A schematic layout of M1 and M3 GND planes is shown in (**c**)

RC loads comprising interconnect wires and precision resistors are shown in Fig. 5.20a, b. The inverter is designed to be wider such that $R_{sw} < 2 \times R_{w}l$. This assures that the wire *RC* delay is a significant fraction of the stage delay. As minimum pitch of higher level interconnect layers increases, these designs cannot be accommodated in a standard template shown in Fig. 5.17.

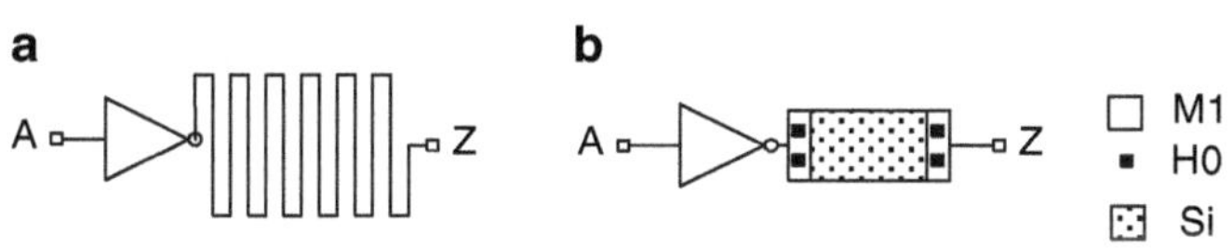

Fig. 5.20 Schematics of ckt_stg: (**a**) inverter driving an interconnect wire *RC* load and (**b**) inverter driving a precision resistor load

For measurement of τ_p of other logic gates with FO = 3, the C_{gXL} load gives a more compact design for the standard template shown in Fig. 5.17. Circuit schematics for a NAND3B and a NOR3T with C_{gXL} loads are shown in Fig. 5.21. The widths of the MOSFETs in the C_{gXL} load are (FO − 1) × the width of the switching MOSFETs in the NAND3B and NOR3T. These types of ckt_stg designs may be used to characterize a variety of logic gates in the standard cell library.

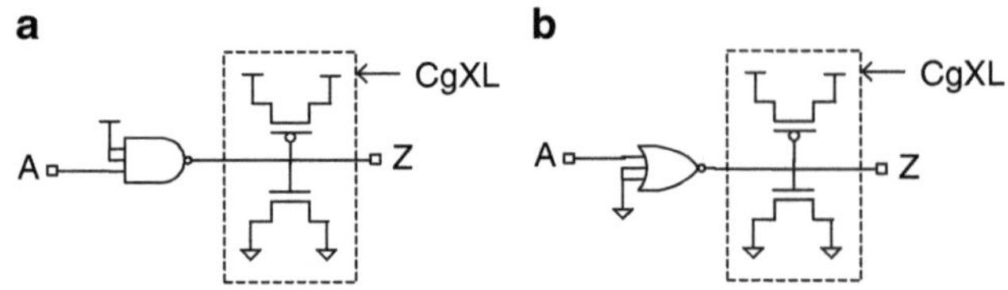

Fig. 5.21 Circuit schematic of FO = 3 equivalent ckt_stg: (**a**) NAND3B and (**b**) NOR3T

5.6.1 MOSFET Parameter Extraction

Additional information on MOSFET parameters and current drive in different regions of the I_{ds}–V_{ds} trajectory can be extracted from delays of appropriately designed ckt_stgs. The ckt_stg designs for monitoring silicon technology variations are skewed to enhance sensitivity to device parameters and do not necessarily represent circuits in typical data paths. Ring oscillators or delay chains comprising such topologies (e.g. skewed stacked logic gates, passgates) give an early readout of silicon process centering during test. Data from embedded monitors can be collected even on partially defective chips. Furthermore, with a high resolution mapping of data, across chip, across wafer, and wafer-to-wafer variations of critical MOSFET parameters may be obtained (Chap. 6).

The first set of ckt_stg designs described in this section is used for extracting the characteristics of MOSFETs in stacked gates. Modified circuit schematics of an inverter, a NAND2T and a NAND3T are shown in Fig. 5.22. The n-FETs with non-switching inputs are shown as resistors R(N2) in the NAND2T and R(N2) and R(N3) in the NAND3T.

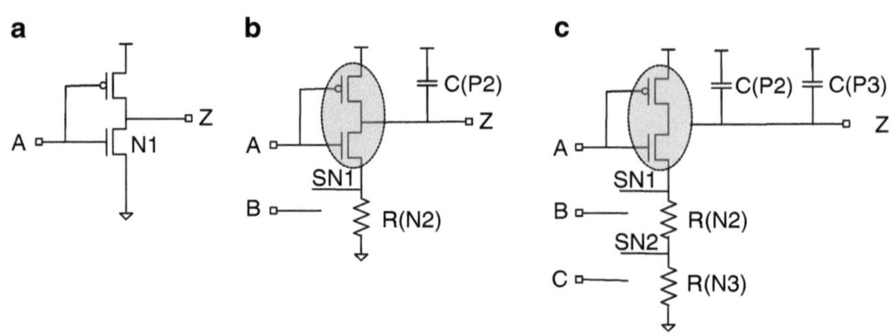

Fig. 5.22 (**a**) Circuit schematic of an inverter. Circuit schematic with equivalent R and C of additional MOSFETs in (**b**) NAND2T and (**c**) NAND3T

I_{ds}–V_{ds} trajectories of the n-FETs in the stack are shown in Fig. 5.23. The switching n-FET N1 in all three logic gates traverses the full range of V_{ds}. The n-FETs N2 in the NAND2T, and N2 and N3 in the NAND3T, operate in the linear region and act as series resistors, consistent with the schematics shown in Fig. 5.22. The net effect is an increase in R_{sw} as the stack height S_h is increased. This increase per n-FET gives a measure of the resistance in the linear region of the n-FET.

The C_{sw} also increases with stack height from the additional p-FETs in parallel with the switching p-FET. The drain terminals of all p-FETs switch between "0" and "1" for PD and PU transitions and the diffusion region capacitance associated with the additional p-FETs increases with S_h. These p-FETs with non-switching inputs are represented as C(P2) in the NAND2T and C(P2) and C(P3) in the NAND3T. The net increase in C_{sw} is from the contributions of junction and overlap capacitances of these p-FETs.

Simulated values of τ_p, R_{sw}, and C_{sw} for an inverter, NAND2T, NAND3T, and NAND4T with FO = 1 for 45 nm PTM HP models @1.0 V, 25 °C are listed in Table 5.2. The n-FET and p-FET widths are the same in all four gates ($W_n = 0.4$ μm, $W_p = 0.6$ μm) to maintain a constant C_{in}. All three delay parameters, τ_p, R_{sw}, and C_{sw}, increase with S_h and the increments with each S_h value, $\delta C_{sw}/S_h$ and $\delta R_{sw}/S_h$ are similar.

Table 5.2 Logic gate (FO = 1) τ_p, R_{sw}, and C_{sw} and increase in R_{sw} and C_{sw} with S_h. $W_n = 0.4$ μm, $W_p = 0.6$ μm for all gates. 45 nm PTM HP models @1.0 V, 25 °C

Logic gate	S_h	τ_p (ps)	C_{sw} (fF)	$\delta C_{sw}/S_h$ (fF)	R_{sw} (Ω)	$\delta R_{sw}/S_h$ (Ω)
Inverter	1	5.27	3.33		1,582	
NAND2T	2	8.62	4.21	0.877	2,048	466
NAND3T	3	12.66	5.08	0.866	2,494	447
NAND4T	4	17.35	5.94	0.867	2,919	425

Estimates of R_{sw} and C_{sw} for the inverter and NAND gates are made using BSIM model parameters and DC characteristics of the MOSFETs. The BSIM model parameters for junction and overlap capacitances and drain-source resistance *rdsw* are listed in Table 5.3. These components are functions of V_{gs} and V_{ds} and their integrated value during switching can be obtained from circuit simulations. However, a rough estimate of δR_{sw} and δC_{sw} is made by using the fixed model values applied to the MOSFET geometry. For $W_p = 0.6$ μm, C_{ds} from Table 5.3 is 0.636 fF. Using zero-bias $C_{ov} = cgdo + cgdl$, the p-FET C_{ov} is 0.225 fF. The total increase in capacitance with S_h is then 0.861 fF, comparable to the value of ~0.87 fF. in Table 5.2.

Table 5.3 BSIM model values for zero-bias junction and overlap capacitance components and drain/source resistance for 45 nm PTM HP models (Appendix B)

	c_{jd} (fF/μm²)	*cjswd* (fF/μm)	*cjwgd* (fF/μm)	*cgdo* (fF/μm)	*cgdl* (fF/μm)	*rdsw* (Ω-μm)
n-FET	0.50	0.50	0.30	0.11	0.2653	210
p-FET	0.50	0.50	0.30	0.11	0.2653	250

The contribution to R_{sw} with n-FET stack height comes from the series resistances of the bottom n-FETs as can be seen in the trajectories in Fig. 5.23, and from the reduction in the drive current of the top n-FET (N1) due to reverse body-bias and higher source voltage during switching. From the linear part of the $I-V$ trajectory, the series resistance of each n-FET (N2 and N3) in the stack is ~650 Ω, comparable to the series resistance of 525 Ω ($W_n = 0.4$ μm) in the I_{ds}–V_{ds} plot of an n-FET. This series resistance contribution is only for the PD transition and hence the net increase in average R_{sw} for PD and PU transitions is ~325 Ω compared with the observed $\delta R_{sw}/S_h$ of ~450 Ω.

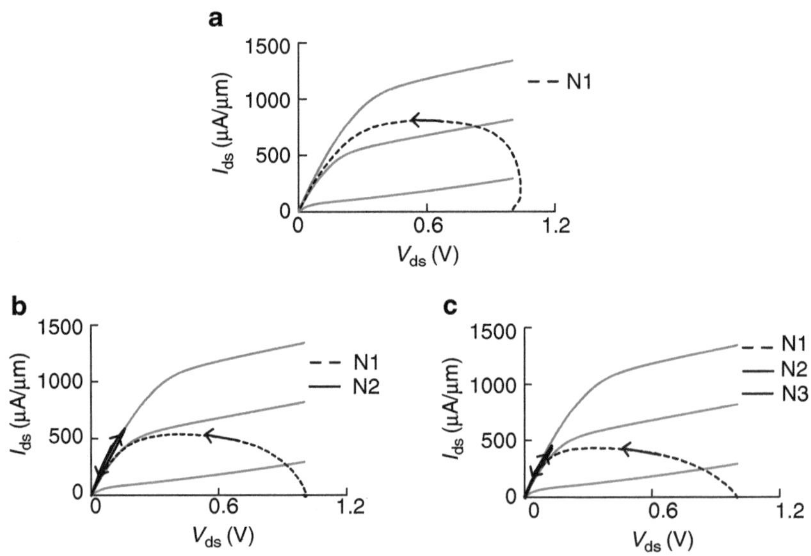

Fig. 5.23 I_{ds}–V_{ds} trajectories of n-FETs in a PD transition superimposed on DC I–V characteristics for (**a**) inverter, (**b**) NAND2T and (**c**) NAND3T. I_{ds} is normalized to W_n. 45 nm PTM HP models @1.0 V, 25 °C

The second contribution to δR_{sw} comes from the increase in switching resistance of the top n-FET in the stack. During the transition, the voltage at the source of n-FET N1 increases as shown in Fig. 5.24a for a NAND2T. Its gate overdrive V_{gs} is reduced resulting in a higher series resistance. The body of the n-FET is tied to its source in the NMOS model in LTspice. In a design with the body terminal of the n-FET at GND, its V_{bs} will be negative during the transition along with a reduction in V_{ds}, further reducing its current drive. The source terminal voltages for a NAND3T are shown in Fig. 5.24b. In this case both N1 and N2 in the stack are affected.

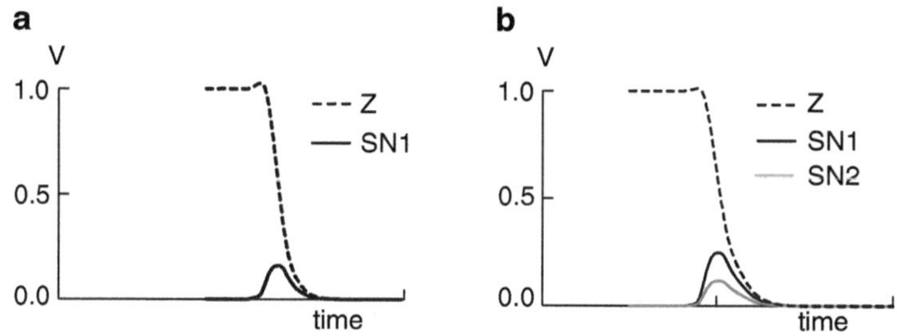

Fig. 5.24 Voltages during a switching transition at nodes Z, and at (**a**) SN1 (NAND2T) and (**b**) SN1 and SN2 (NAND3T)

The increase in R_{sw} and τ_p with stack height in NAND gates is modulated by the V_t of the n-FETs. Similarly, the increase in R_{sw} and τ_p with stack height in NOR gates is modulated by the V_t of the p-FETs. Delay sensitivities to V_t can be enhanced by designing tapered gates, with narrow width non-switching MOSFETs.

Ckt_stgs comprising n-passgates or p-passgates provide enhanced sensitivities to variations in V_t. Circuit schematics for one such configuration for an n-passgate and a p-passgate are shown in Fig. 5.25. Other schemes and their relative merits are discussed in Sect. 5.6.3.

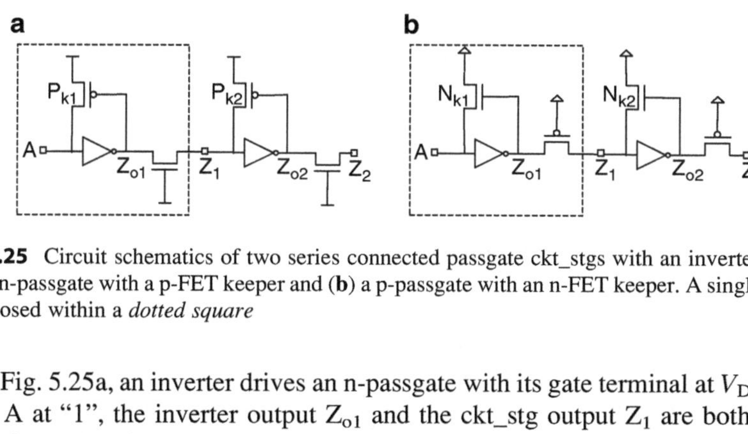

Fig. 5.25 Circuit schematics of two series connected passgate ckt_stgs with an inverter driving (**a**) an n-passgate with a p-FET keeper and (**b**) a p-passgate with an n-FET keeper. A single ctk_stg is enclosed within a *dotted square*

In Fig. 5.25a, an inverter drives an n-passgate with its gate terminal at V_{DD}. With input A at "1", the inverter output Z_{o1} and the ckt_stg output Z_1 are both at "0". When input A switches to a "0", Z_{o1} rises to a "1" and node Z_1 will be at V_{DD}–V_{tn}, where V_{tn} is the threshold voltage of the n-passgate. Instead, when node Z_{o2} in the next stage switches to a "0", its p-FET keeper P_{k2} turns on and pulls node Z_1 to V_{DD}.

In Fig. 5.25b, an inverter drives a p-passgate with its gate terminal at GND and the keeper is an n-FET. With input A at "0", the p-passgate passes a "1" to Z_1. When input node A switches to a "0", the n-FET keeper N_{k2} of the next stage assists in pulling down node Z_1 to GND.

I_{ds}–V_{ds} trajectories of the n-passage in Fig. 5.25a, superimposed on the corresponding DC I_{ds}–V_{ds} characteristics are shown in Fig. 5.26. The n-passgate traverses its saturation region at the beginning of the PD transition and its linear region at the beginning of the PU transition. Its V_{gs} and V_{ds} remain $<0.5 \times V_{DD}$ and its switching resistance R_{swnpg} is more sensitive to V_t than the switching resistance R_{swi} of the driving inverter. The impact on τ_p is increased by selecting the n-passgate width W_{npg} to be less than W_n of the inverter ($R_{swnpg} > R_{swi}$). The p-FET keeper width is selected to be very small to limit its capacitive load.

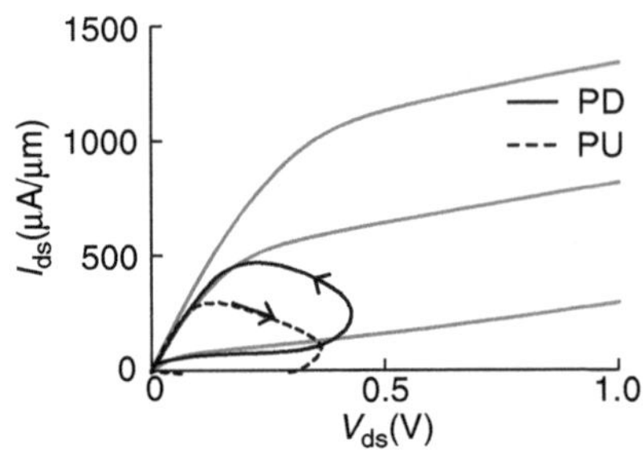

Fig. 5.26 I_{ds}–V_{ds} trajectories of the n-passage (superimposed on its DC characteristics) in the circuit schematic in Fig. 5.25a during PD and PU transitions. I_{ds} is normalized to W_n

Although the ckt_stg designs in Fig. 5.25 are not representative of typical circuits used in CMOS chips, the higher delay sensitivity to V_{tn} (n-passgate) and V_{tp} (p-passgate) than in standard logic gates makes them more favorable candidates for tracking systematic variations in relative p/n strengths (beta ratio). In another application, n-FET and p-FET degradation over time due to bias temperature instability (BTI) is tracked with ring oscillators comprising n-passgate and p-passgate circuits, respectively (Sect. 8.2.1).

5.6.2 SRAM Stage Designs

Memory cells and their components can also be characterized by constructing ckt_stgs for delay chains and ring oscillator monitors. The circuit schematic of a 6T SRAM cell is shown in Fig. 5.27a. The cell schematic is perfectly symmetric but physical layout differences and line-width biases in layer shapes during silicon processing as well as random V_t variations can produce systematic offsets between the two half-cells. Three ckt_stg designs using components of an SRAM cell are shown in Fig. 5.27b, d. In these ckt_stg, inverter INR, INR driving inverter INL, and INR driving access n-FET NPR are extracted from the cell. Similar ckt_stgs are designed using the inverter INL. Delays per stage with inverter INR are compared with the corresponding designs with inverter INL. Systematic differences in delays of left and right inverter pairs can be used to quantify cell asymmetry.

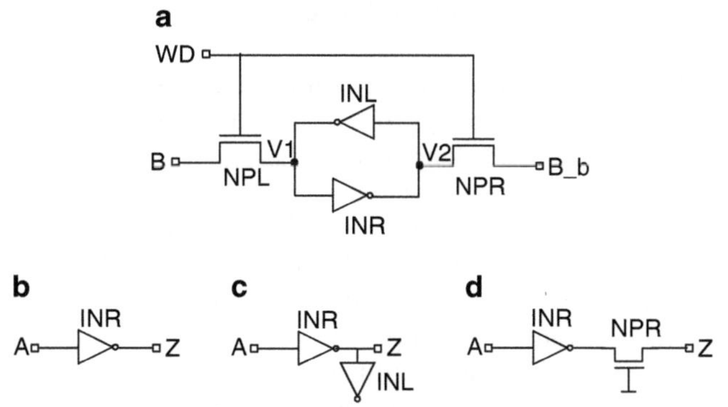

Fig. 5.27 (**a**) 6T SRAM cell schematic. Ckt_stg schematics with SRAM cell components: (**b**) inverter IR, (**c**) inverter IR driving inverter IL and (**d**) inverter IR driving access n-FET NPR

An SRAM cell can be reconfigured to form a ckt_stg that provides a measure of cell read/write sensitivity without the use of any peripheral circuitry. The cell is rewired as shown in Fig. 5.28a. Word line WD terminals in Fig. 5.27a are separated and connected to the inputs of the latch inverters INL and INR. A ring oscillator comprising the modified SRAM cell is shown in Fig. 5.27b. The output node Z1 is connected to node A2 of the following ckt_stg and similarly node Z2 is connected to node A1 of the following ckt_stg. The circuit is initialized with EBL = "0" which turns N1 on and N2 off, and node a1 is set to GND. Node A2 of the first stage is at "0" and A1 is at "1". Outputs Z1 and Z2 of the first stage are then at "1" and "0", respectively. With EBL set at "1", n-FET N1 is turned off and N2 is turned on, closing the loop and enabling oscillations. This ckt_stg can also be placed in a delay chain configuration.

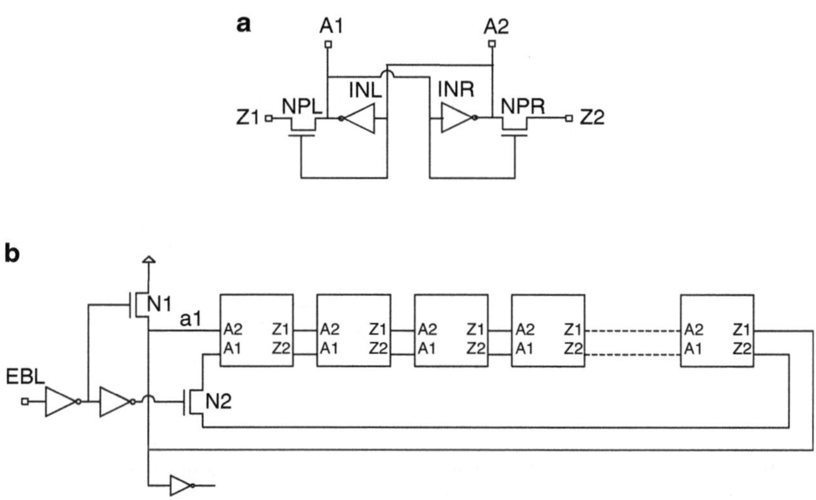

Fig. 5.28 (**a**) SRAM cell rewired to track cell performance, and (**b**) SRAM ring oscillator circuit

5.6.3 Silicon Process-Sensitive Suite

The number of individual ckt_stgs to be characterized can become unmanageably large as different MOSFET offerings, physical layout styles, logic gate topologies, and memory cells are included along with ckt_stg designs for MOSFET parameter extraction. Full characterization of circuit delays is possible only on test chips dedicated to silicon technology monitoring and evaluation. A practical approach to embedding monitors on a product chip is to judiciously select a small suite of ckt_stgs. The selection is based on sensitivity to the most critical MOSFET and interconnect parameters and to the chip T_{cmin}. An even smaller suite comprising one to four ckt_stg designs may be made sufficiently compact so that multiple copies of this group of monitors can be distributed across the chip. This small group is then useful for tracking across chip circuit delay variations arising from variability in silicon process, V_{DD}, and temperature.

The design of such a group of ckt_stgs is tuned by performing a sensitivity analysis to the parameters of interest [6]. Here we have selected three critical MOSFET parameters V_{tn}, V_{tp}, and L_p. The sensitivity analysis is demonstrated with a standard inverter (FO = 3) design and then applied to a small set of ckt_stgs. A more detailed description of sensitivity analysis and use of condition number CN in the selection of ckt_stgs is described in Sect. 9.3.2.

Circuit simulations are carried out to measure τ_p from delay chains comprising ten standard inverter (FO = 3) stages (Sect. 2.2.4). Each of the parameters V_{tn}, V_{tp}, and L_p are individually varied over their full $\pm 3\sigma$ range, and τ_p measured for each case.

The variations in inverter (FO = 3) delay are plotted in normalized units in Fig. 5.29a. The V_t of all the p-FETs, V_{tp}, in the delay chain is varied over its $\pm 3\sigma$ (± 0.06 V) range. The normalized value of τ_p, τ_p/τ_{p0}, where τ_{p0} is the average stage delay under nominal conditions, is plotted as a function of $\Delta|V_{tp}|/\sigma V_{tp}$ in Fig. 5.29a. A linear fit of the data is carried out to get the sensitivity of normalized τ_p per σV_{tp}.

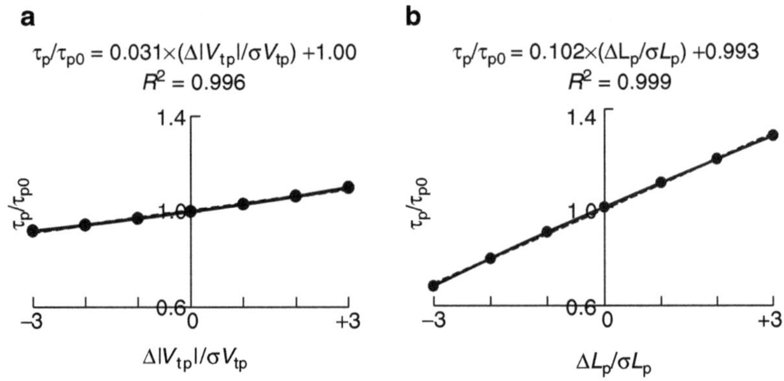

Fig. 5.29 Normalized τ_p (τ_p/τ_{p0}) for a standard inverter (FO = 3) as a function of (**a**) $\Delta|V_{tp}|/\sigma V_{tp}$ and (**b**) $\Delta L_p/\sigma L_p$. *Dashed lines* show linear fits to the simulated data. 45 nm PTM HP models @1.0 V, 25 °C

The slope of the linearly fit line is 0.031, hence there is 3.1 % change in τ_p per σV_{tp} (=0.02 V). Over the full $\pm 3\sigma V_{tp}$ range, τ_p is expected to vary by ±9.3 %. Sensitivity to σV_{tn} is obtained in a similar fashion by varying the V_t of all the n-FETs in the delay chain over the full $\pm 3\sigma$ (±0.06 V) range.

In Fig. 5.29b, τ_p/τ_{p0} is plotted as a function of $\Delta L_p/\sigma L_p$. It is assumed that L_p values for p-FETs and n-FETs track together ($L_{pp} = L_{pn}$), as is typically the case in silicon processing. From a linear fit of the data, the sensitivity of τ_p is 10.2 % per σL_p, and τ_p varies by ±30.6 % over the full range of L_p variation.

The fractional change in τ_p arising from each of the three parameters V_{tn}, V_{tp}, and L_p is expressed as

$$\frac{\Delta\tau_p}{\tau_{p0}} = m_{vn}\frac{\Delta|V_{tn}|}{\sigma V_{tn}} + m_{vp}\frac{\Delta|V_{tp}|}{\sigma V_{tp}} + m_l\frac{\Delta L_p}{\sigma L_p} \tag{5.9}$$

where m_{vn}, m_{vp}, and m_l are the sensitivity coefficients for V_{tn}, V_{tp}, and L_p, respectively. The values of m_{vn}, m_{vp}, and m_l are determined from the slopes of the plots shown in Fig. 5.29.

Following the procedure described above, sensitivity coefficients are determined for a variety of ckt_stg designs shown in Fig. 5.30a–j. Ckt_stgs (a) through (d) are logic gates: inverter (FO = 3), inverter driving a transmission gate (TG), NAND3B, and NOR3T. In these circuits, the FO = 3 configuration reflects true fanout but a C_{gXL} gate load may be used in compact designs. In Fig. 5.30e–j, there are three sets of passgate ckt_stgs. The first set, NPG_1 and PPG_1, is the same as discussed in Sect. 5.6.1. In the second set in Fig. 5.30f, g, NPG_2, and PPG-2, the n-passgate is in series with the gate terminal of the inverter's n-FET, and the p-passgate is in series with the gate terminal of the inverter's p-FET. The third set in Fig. 5.30i, j, NPG_3, and PPG_3 is similar to the first set except the keeper MOSFETs are not present.

Sensitivity coefficients for the ckt_stgs in Fig. 5.30 along with the MOSFETs widths for each of the designs are listed in Table 5.4. The values of W_p/W_n in NAND3B_F and NOR3T_F designs are fixed to be the same as in the inverter. In NAND3B_B and NOR3T_B, W_p/W_n values are balanced to give equal PU and PD delays, ($\tau_{pu} = \tau_{pd}$) while keeping $(W_n + W_p) = 1.0$ μm to maintain the same load capacitance as for NAND3B_F and NOR3T_F. Simulations are carried out using 45 PTM HP models at 1.0 V, 25 °C.

The models show the general trend in sensitivity coefficients for the ckt_stgs. Balanced logic gates (inverter, NAND, and NOR) and inverter driving a TG with $\tau_{pu} = \tau_{pd}$ have m_{vn} and m_{vp} values in the range of 0.027 to 0.044, and m_l values in the range of 0.098 to 0.104. For these four ckt_stgs, m_l values are ~3× larger than the m_{vn} and m_{vp} values. Also m_{vn} values are nearly equal to m_{vp} values for balanced gates. The m_{vn} value is slightly higher than the m_{vp} value in a NAND3B_F gate and slightly lower in a NOR3T_F gate as expected.

For the passgate ckt_stgs, values of m_l are generally higher than for logic gates. A more distinguishing feature of the NPG stages is that m_{vn} is 4× to 10× larger than m_{vp}. Similarly for the PPG stages, m_{vp} is 4× to 10× larger than m_{vn}. Hence the

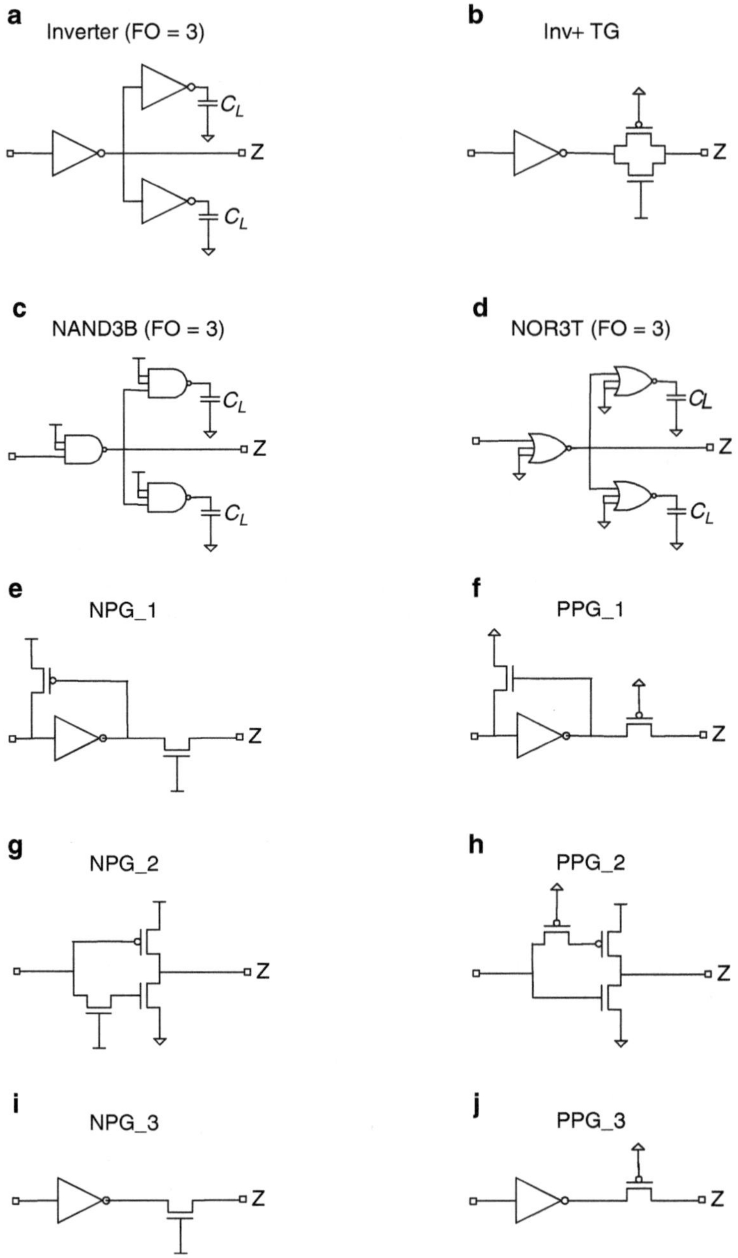

Fig. 5.30 Ckt_stg designs for determining sensitivity coefficients for V_{tn}, V_{tp}, and L_p : (**a**) Inverter (FO =3), (**b**) Inv + TG, (**c**) NAND3B (FO =3), (**d**) NOR3T (FO =3), (**e**) NPG_1, (**f**) PPG_1, (**g**) NPG_2, (**h**) PPG_2, (**i**) NPG_3 and (**j**) PPG_3

NPG and PPG ckt_stgs are well suited for separating V_{tn} variations from V_{tp} variations. In the simulation data using PTM models, different NPG and PPG designs have different sensitivity coefficients. To some extent the sensitivities can be enhanced by tuning the device widths.

Table 5.4 Sensitivity coefficients for RO stage designs. 45 nm PTM HP models @ 1.0 V, 25 °C

	Driver	Passgate	Keeper				
Circuit	W_p/W_n (μm)	W (μm)	W (μm)	FO	m_{vn}	m_{vp}	m_l
Inverter	0.60/0.40	–	–	3	0.027	0.031	0.102
Inv + TG	0.60/0.40	0.40/0.40		1	0.034	0.032	0.100
NAND3B_F	0.60/0.40	–	–	3	0.037	0.029	0.095
NAND3B_B	0.46/0.54	–	–	3	0.034	0.030	0.098
NOR3T_F	0.60/0.40	–	–	3	0.027	0.044	0.104
NOR3T_B	0.80/0.20	–	–	3	0.028	0.037	0.098
Inv + NPG_1	0.60/0.40	0.30	0.10	1	0.117	0.012	0.117
Inv + NPG_2	0.60/0.40	0.30	–	1	0.085	0.025	0.125
Inv + NPG_3	0.60/0.40	0.30	–	1	0.114	0.011	0.119
Inv + PPG_1	0.60/0.40	0.60	0.20	1	0.023	0.128	0.165
Inv + PPG_2	0.60/0.40	0.60	–	1	0.028	0.106	0.160
Inv + PPG_3	0.60/0.40	0.60	–	1	0.001	0.169	0.108

With the known sensitivity coefficients obtained from circuit simulations, changes in τ_p as a result of shifts in V_{tn}, V_{tp}, and L_p can be computed using Eq. 5.9. In the case of data collected from embedded process monitors, it is highly desirable to track $\Delta|V_{tn}|$, $\Delta|V_{tp}|$, and ΔL_p from the measured τ_p values of ckt_stgs. At least three ckt_stg designs are needed to extract these three unknowns. This requires solving an equation for parameters Δp of the type

$$\mathbf{\Delta p} = \mathbf{M}^{-1} \times \Delta\boldsymbol{\tau} \tag{5.10}$$

$$\text{where } \mathbf{M} = \begin{vmatrix} m_{vn1} & m_{vp1} & m_{l1} \\ m_{vn2} & m_{vp2} & m_{l2} \\ m_{vn3} & m_{vp3} & m_{l3} \end{vmatrix}.$$

As the condition number CN of the matrix **M** is reduced, errors in estimating $\Delta|V_{tn}|$, $\Delta|V_{tp}|$, and ΔL_p are minimized as explained in Sect. 9.3.2.

Six different sets each comprising three ckt_stgs are considered as potential candidates for distributed monitors. These sets are listed in Table 5.5 along with their condition numbers. The first ckt_stg in each of the first five sets is an inverter (FO = 3) which tracks well with most of the combinational logic circuits on the chip, but is equally sensitive to variations in V_{tn} and V_{tp}. The other two ckt_stgs are selected to detect variations in V_{tn} and V_{tp}. In the sixth set, an inverter + TG ckt_stg replaces the inverter (FO = 3) stage.

Table 5.5 Condition numbers of each set of three ckt_stg designs. 45 nm PTM HP models @ 1.0 V, 25 °C

set #	ckt_stg1	ckt_stg2	ckt_stg3	CN
1	Inverter	NAND3B_F	NOR3T_F	29.73
2	Inverter	NAND3B_B	NOR3T_B	40.00
3	Inverter	Inv + NPG_1	Inv + PPG_1	13.72
4	Inverter	Inv + NPG_2	Inv + PPG_2	17.41
5	Inverter	Inv + NPG_3	Inv + PPG_3	9.15
6	Inv + TG	Inv + NPG_1	Inv + PPG_1	17.60

From Table 5.5, set #3, #4, #5, and #6 have lower CNs than set #1 and set #2. Set #3 is preferred over set #5 with the lowest CN, as node voltages in PPG_3 and NPG_3 designs without keeper MOSFETs do not reach full power rail voltages. This is illustrated in Fig. 5.31 for NPG_1 and NPG_3 designs. The output voltage for the PU transition reaches full power supply voltage for NPG_1 as the p-FET keeper turns on. In the NPG_3 design, the output voltage for the PU transition reaches only 0.8 V. As V_{DD} is lowered NPG_3 fails to function altogether well before NPG_1 does. Hence, set #3 is a sound and robust choice. The standard inverter (FO = 3) serves as a reference for all static CMOS logic gates to track variations in L_p, while NPG_1 and PPG_1 track variations in V_{tn} and V_{tp}.

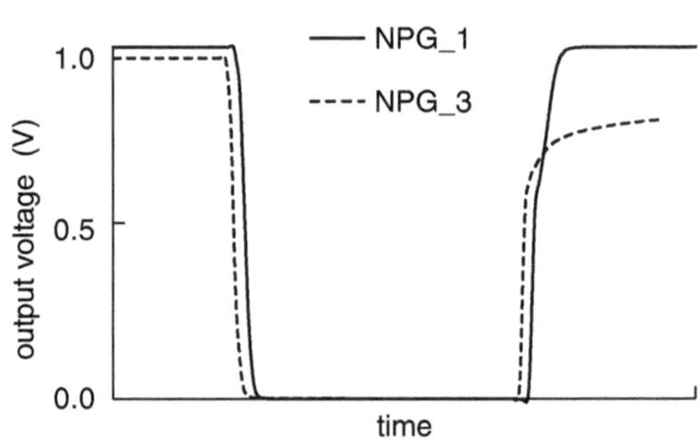

Fig. 5.31 RO stage output waveforms for NPG_1 and NPG_3 designs. 45 nm PTM HP models @1.0 V, 25 °C

5.6.4 Strengths and Limitations of RO-Based Monitors

The ckt_stg designs described above are suitable for placement in a delay chain or ring oscillator monitor configuration. There are some distinct differences in the utilities of these two types of monitors. Some of the advantages of RO-based monitors are listed below:

- RO frequency is independent of circuit delays external to the RO itself.
- By lowering the RO frequency with a frequency divider circuit, timing constraints on control circuits are relaxed.

- Use of decoders/multiplexers allows shared control and output circuits, including frequency divider, for multiple ROs distributed across chip.
- RO frequency or period can be directly measured with external test equipment with an accuracy of better than 1 % even with the chip clocks off. With a uniform temperature distribution across chip, only silicon process variations are observed.
- Delay measurements can be directly correlated with data collected from scribe-line measurements on similar ROs measured after the delineation of the first metal level in silicon manufacturing. This provides early readout of chip power/performance.
- RO designs can be compact with 11 stages per RO. Delay chains have to be larger with 100 or more stages to achieve 1 % measurement accuracy.

In general, RO monitors are easier to design and require only DC inputs to enable. There are, however, some limitations of using only RO-based embedded monitors.

- Chip timing tools often cannot handle a closed loop in which case some degree of manual checking or modification of tools is needed. Delay chain monitor designs are compatible with EDA tools.
- Circuit simulation times for ROs are longer than for delay chains. In some cases the simulator may fail to converge on a DC operating point.
- In PD-SOI technology, the delay of a logic gate depends on its switching history. ROs in PD-SOI provide only steady-state delays whereas delay chains can provide 1SW, 2SW, and steady-state delays (Sect. 10.6).

5.7 Data Collection and Characterization

Data are collected from the embedded monitors at each stage of testing and in some high-end products even during operation in the field. The analog nature of the data requires more resources for characterization than digital pass/fail data. The data volume can be very large: for twelve ring oscillators per chip measured at three different V_{DD} settings, with 400 chips per wafer and 20 wafers per lot, each lot would generate ~3×10^5 frequency measurements at one test stop alone. Efficient and effective characterization methods are needed to reduce the cost of data analysis, easing the burden on the test team.

Nominal target values of delay/stage for each ckt_stg design are generated from circuit simulations at the test V_{DD} and temperature. Measured data are expected to fall within an allowed spread around the nominal target value as illustrated with a histogram in Fig. 5.32. A pass/fail criterion may be set such that chips with all the monitors within the target range pass while chips with any monitor outside the range fail. This approach works well if the number of failing chips is a very small fraction of the sample population.

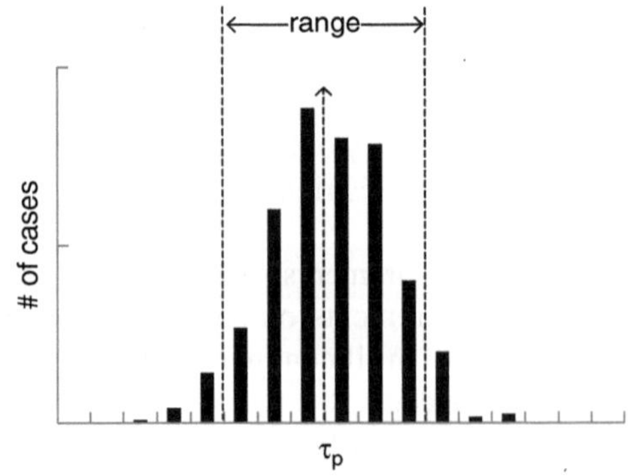

Fig. 5.32 Histogram showing target and acceptable range for a monitor parameter value

Often it is necessary to gather more information from monitors for tuning the silicon process, for debugging circuit design issues and for test diagnostics and failure analysis. Characterization methods for variability, correlation to product f_{max}, V_{min} and power, and statistical analysis are covered in Chaps. 6, 7, and 9, respectively. Here we describe methods for model-to-hardware correlation using silicon process monitors and graphical visualization techniques for straightforward assimilation of the results.

Target values of τ_{po} are obtained from circuit simulations using nominal silicon process as defined in circuit design tools, and nominal operating V_{DD} and temperature. Further circuit simulations are carried out to determine values of τ_p while varying key process parameters, for example, L_p over $\pm 3\sigma$ ranges. Delay at a given value of L_p, normalized to the nominal target value, τ_p/τ_{po}, is a convenient parameter for expressing the change in delay with shift in L_p. The normalized delay values τ_p/τ_{po} for a ckt_stg are plotted vs. those of a reference ckt_stg.

The normalized delays of an inverter (FO = 3) and a NAND3B (FO = 3) are simulated while varying L_p in 1σ steps. In Fig. 5.33a, τ_p/τ_{po} for NAND3B is plotted against τ_p/τ_{po} of our reference inverter at nominal V_{DD}, 25 °C for 45 nm PTM HP models. A diagonal line shows 1:1 correspondence for the two ckt_stgs. Data points

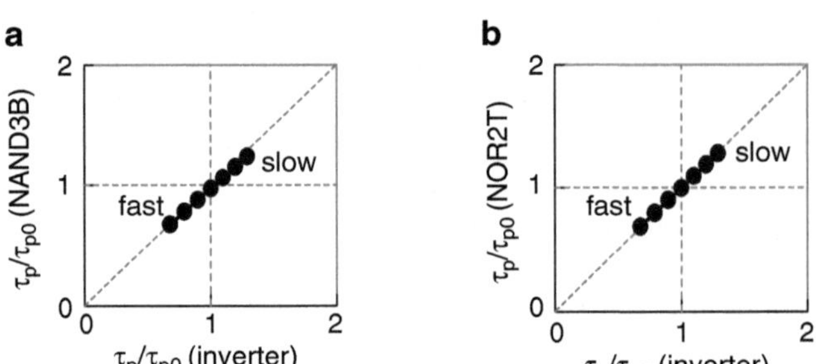

Fig. 5.33 Normalized circuit delays τ_p/τ_{po} as a function of normalized standard inverter (FO = 3) delay over $\pm 3\sigma$ range of L_p: (**a**) NAND3B (FO = 3) and (**b**) NOR2T (FO = 3). 45 nm PTM HP models @ 1.0 V, 25 °C

in the top right quadrant represent longer delays than nominal and those in the bottom left quadrant represent shorter delays than nominal. An important message here is that the two ckt_stgs track with each other, slowing down or speeding up with the same fractional changes with variation in L_p. If a CMOS chip comprises only inverters (FO = 3) and NAND3Bs (FO = 3), all data paths will speed up or slow down by the same amount with L_p variation and cycle limiting paths will remain the same for all the chips. In Fig. 5.33b, τ_p/τ_{po} for a NOR2T is plotted against τ_p/τ_{po} of the reference inverter and has the same characteristics as a NAND3B. All three logic gates (inverter, NAND3B, and NOR2T) in these examples are designed to have $\tau_{pu} = \tau_{pd}$. Generally all balanced static logic gates track with each other as L_p is varied.

The graphical representation methodology shown in Fig. 5.33 is next extended to the three selected ckt_stgs in set # 3 of Table 5.5, namely inverter (FO = 3), NPG_1, and PPG_1. The plots in Fig. 5.34 cover four possible scenarios for variations of V_{tn} and V_{tp} at constant L_p. Only systematic variations in V_{tn} and V_{tp} over a range of ±0.06 V are considered. Circuit simulations are carried out at nominal L_p of 0.045 μm, 1.0 V, and 25 °C using 45 nm PTM HP models.

In Fig. 5.34a, V_{tn} is varied which affects the inverter n-FET and n-passgate in NPG_1. Variations in V_{tn} of the keeper n-FET in PPG_1 have negligible impact on its τ_p. It is apparent that τ_p/τ_{po} of the NPG_1 shows a large range compared to the inverter and PPG_1. In Fig. 5.34b, V_{tp} is varied and has a strong impact on τ_p/τ_{po} of PPG_1 while the inverter and NPG_1 show smaller variations.

In Fig. 5.34c, both V_{tn} and V_{tp} are varied simultaneously and in the same sense, either making both n-FET and p-FET stronger or both weaker. In this case both NPG_1 and PPG_1 show a large variation in τ_p/τ_{po} compared with the inverter, confirming their strong sensitivities to variations in V_{tn} or V_{tp}. In Fig. 5.34d, V_{tn} and V_{tp} are simultaneously moved in opposite directions such that $|V_{tn}|$ increases while $|V_{tp}|$ decreases. Here when the n-FET is weaker, the p-FET is stronger. For the inverter, τ_p/τ_{po} is nearly unchanged as τ_{pu} and τ_{pd} change in opposite sense. Again NPG_1 and PPG_1 vary over a wide range.

With appropriately designed ckt_stgs in a delay chain or ring oscillator configuration and delay targets obtained from circuit simulations in the chip design environment, having netlists generated using parasitic extraction models, graphical displays of the type shown in Figs. 5.33 and 5.34 are extremely useful for getting a quick assessment of average L_p, V_{tn}, and V_{tp} on every CMOS chip during test. The graphical display technique described here can also be effectively used for examining variations in other process parameters such as C_{ov}, C_{ds}, R_{ds}, R_w, and C_w. A large number of plots can be accommodated on a single page and only deviations from the dotted diagonal line noted for further investigation. A single chart may thus be used for tracking process variations from a circuit operation perspective. This information should, however, be validated with the measurements made on scribe-line test structures in the silicon manufacturing line for obtaining consensus on conclusions drawn from in-line and chip test data.

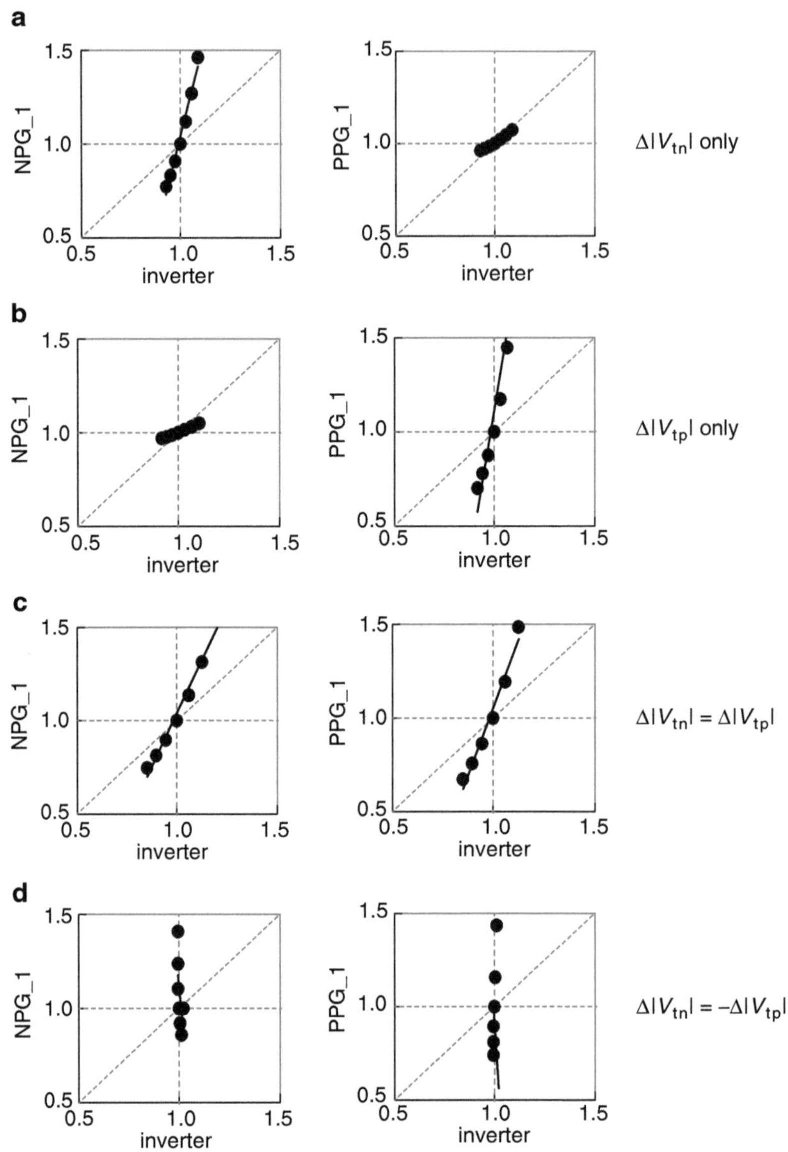

Fig. 5.34 τ_p/τ_{po} of NPG_1 and PPG_1 ckt_stgs vs, inverter (FO = 3) with variation in: (**a**) V_{tn} only, (**b**) V_{tp} only, (**c**) $|V_{tn}|$ and $|V_{tp}|$ tracking together and (**d**) $|V_{tn}|$ and $|V_{tp}|$ changing in opposite sense. 45 nm PTM HP models @ 1.0 V, 25 °C

Data collected from voltage and temperature monitors can also be compared to target values. This is illustrated in Fig. 5.35 with a hypothetical CMOS chip while running five different applications. Because of difference in workload and local switching activities, the monitors indicate different ranges of excursions from a nominal target for each of the applications (apps). These variations can cause f_{max}, V_{min}, and P_{ac} differences in different customer work environments. It is important to ensure that product qualification tests are carried out mimicking worst-case applications.

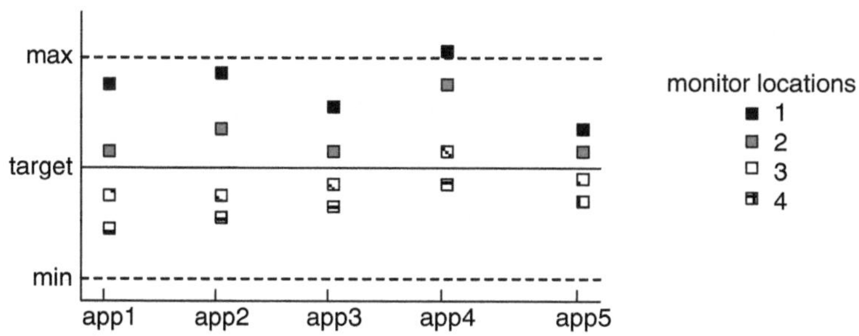

Fig. 5.35 Hypothetical data showing tracking of local V_{DD} or temperature for five different apps with four monitors distributed across chip

Finer spatial resolution in silicon process parameters, and in V_{DD} and temperature excursions, is obtained by placing more monitors on the CMOS chip. Temporal variations are obtained by repeated measurements over suitable time intervals.

5.8 Summary and Exercises

Design principles and utilization of on-chip silicon PVT monitors are described. Delay chains, ring oscillators, and CPMs provide accurate measurements of average logic gate delays. Compact designs, representative of on-chip circuitry, may be distributed across chip for routine monitoring of silicon process and for model-to-hardware correlation. The sensitivities of delays to changes in MOSFET and interconnect properties are exemplified with circuit simulations. A procedure to select a subset of designs with higher sensitivities to individual MOSFET parameters is demonstrated. These designs coupled with differencing schemes in data analysis extend the value of the monitors in process tuning and debug beyond just delay measurements.

Exercises 5.1–5.3 relate to delay chain and RO designs. Exercises 5.4 and 5.5 involve voltage and temperature monitors. Exercises 5.6 through 5.9 involve parameter extractions from process monitors. Sensitivity analysis is explored in Exercise 5.10.

5.1 The delay chain monitor circuit described in Sect. 5.2.2 is designed to measure delay/stage τ_p of logic gates. The edge detector circuit comprises inverter (FO = 2) gates.

(a) What is the minimum number of inverter (FO = 2) stages required in the delay chain to measure τ_p with an accuracy of ±1 %?
(b) How many NAND3T (FO = 4) stages are needed to measure τ_p with an accuracy of ±1 %?

5.2 What is the minimum required number of stages in the edge detector circuit for the inverter (FO = 2) and NAND3T (FO = 4) delay chains in Exercise 5.1, at nominal V_{DD} and at $0.8 \times V_{DD}$.

5.3 ROs are distributed across the chip and share a common frequency counter in the center of the chip. The chip operating frequency is 1 GHz and it is recommended that the signals from the RO be at a frequency lower than the clock frequency.
(a) What is the minimum number of RO stages for an inverter FO = 3 design required to maintain the RO output frequency lower than 1 GHz, at $1.2 \times V_{DD}$?
(b) If an inverter FO = 1 design is used instead, can the number of stages remain the same?
(c) The length of signal wire travelling from the RO output to the on-chip frequency counter is 5 mm. Does this impact the accuracy of frequency measurements? If so, what measures can be taken to get accurate measurements?

5.4 Simulate the circuit below to generate an initial V_{DD} dip > 10 % by selecting appropriate values of L and C. Connect an inverter (FO = 3) delay chain to V_{DD} and adjust L and C to observe change in τ_p as a function of time.

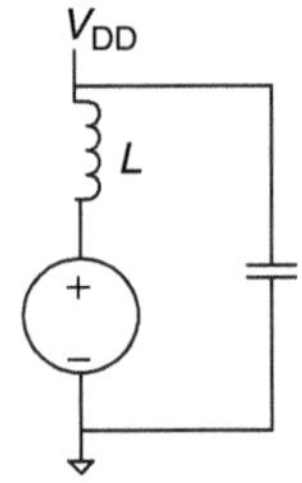

5.5 One out of twelve temperature monitors on a microprocessor chip records a temperature 20 °C above the others during operation. How would you verify that the temperature is truly higher?

5.6 RO frequency values of nine distributed ROs are averaged for each chip and a histogram of the average values is plotted to show the distribution. Some chips have a much higher average and the data points lie more than $+4\sigma$ away from the mean.
(a) How would you check the data to ensure these outliers are real?
(b) What are possible fixes to get accurate data?

5.7 Plot I_{ds}–V_{ds} trajectories for the n-FET and p-FET of a transmission gate driven by a standard inverter for both PU and PD transitions, along with that of the inverter n-FET and p-FET.
(a) What changes in the design are needed to change the delay sensitivities to V_{tn} and V_{tp} of the MOSFETs?

5.8 The silicon foundry proposes a new low K inter-level dielectric for thin metal layers based on 30 % reduction in C_w from scribe-line data. From the data collected on embedded silicon process monitors, the ckt_stg designs in Fig. 5.19a show much less improvement while results from the design in Fig. 5.19b are consistent with the foundry data. How can these differences be explained?

5.9 (a) Using a standard SRAM cell offered in the technology node, design and simulate a full-cell SRAM RO. Measure read and write time of the SRAM cell (Sect. 3.3.1).

(b) Compare SRAM RO delay with read and write times for different beta ratios. What trends do you find?

5.10 (a) Determine sensitivity coefficients for R_w and C_w and V_{DD} for inverters with $(W_n + W_p) = 1.0$ μm driving a wire of length 100 μm. Let $R_w = 1.0$ Ω/μm and $C_w = 0.2$ fF/μm. Assume 3σ of 50 % for R_w and C_w each, and 30 % for V_{DD}.

(b) Repeat above with an inverter of width $(W_n + W_p) = 60$ μm driving the same 100 μm long wire.

(c) How different are the sensitivity coefficients for the two inverter designs? Can you explain the differences based on an *RC* model of the circuit?

References

1. Bhushan M, Ketchen MB (2011) Microelectronic test structures for CMOS technology. Springer, New York
2. Yu X, Gluschenkov O, Zamdmer ND, Deng J, Goplenet BA, et al. (2011) Chip-level power-performance optimization through thermally-driven across-chip variation (ACV) reduction. IEDM pp 25.3.1–25.3.4
3. Drake AJ, Senger RM, Singh H, Carpenter GD, James N (2008) Dynamic measurement of critical path timing. IEEE international conference of integrated circuit design and technology and tutorial ICICDT 2008, pp 249–252
4. Franch R, Restle P, James N, Huott W, Freidrich J, Dixon R. et al (2008) On-chip timing uncertainty measurements on IBM microprocessors. Proceedings of the IEEE international test conference, ITC'08, pp 1–7
5. Drake A, Senger R, Deogun H, Carpenter G, Ghiasi S, Nguyen T, et al. (2007) A distributed critical-path timing monitor for a 65 nm high-performance microprocessor, IEEE international solid state circuits conference, ITC'07, pp 398–399
6. Mahfuzul IAKM, Tsuchiya A, Kobayashi K, Onodera H (2012) Variation-sensitive monitor circuits for estimation of global process parameter variation. IEEE Trans Semicond Manuf 25:571–580

Variability 6

Contents

6.1 Sources and Impact of Variations 202
6.1.1 Silicon Process Variations 205
6.1.2 Random Variations 211
6.1.3 Voltage Variations 212
6.1.4 Temperature Variations 214
6.2 Variability Characterization 215
6.2.1 Silicon Manufacturing Tests 216
6.2.2 On-Chip Embedded PVT Monitors 216
6.2.3 Functional Parameters 218
6.2.4 Optical Imaging 218
6.2.5 Thermal Imaging 220
6.3 Minimizing Variations 220
6.3.1 Chip Design and Floorplanning 221
6.3.2 Reticle and Wafer Assembly 222
6.3.3 Silicon Process Improvements 222
6.4 Accommodating Variability in Circuit Design 223
6.4.1 Simulation Corners 225
6.4.2 Impact of Random Variability on Circuits 227
6.5 Summary and Exercises 235
References 239

CMOS chips are designed to function over the published tolerances of physical parameters of circuit components and over a specified range of environmental conditions. Electrical tests are defined to cover the range of operating conditions such as power supply voltage and temperature over which any chip may need to function. Metrology and electrical test data are collected and analyzed to isolate factors influencing chip yield and performance. Characterization of variability and understanding the underlying sources of variations are important components of

M. Bhushan, M.B. Ketchen, *CMOS Test and Evaluation*,
DOI 10.1007/978-1-4939-1349-7_6

electrical test and analysis. Efforts are made to maximize yield by accommodating anticipated variations and by minimizing variability with continuous improvements in manufacturing processes.

Variations occur in the properties of circuit components and in the chip operating conditions. Silicon processes drift over time, and quality control is put in place to maintain key circuit component parameters within their specified $\pm 3\sigma$ limits. Random statistical variations in some properties of circuit components tend to increase as feature dimensions are reduced. Power supply voltages may be affected by variations in the current drawn during operation and by series resistances in the power grid network. The operating temperature of the chip varies with local power density, ambient temperature, and the efficiency of the cooling system.

Typically, variations in process, voltage, and temperature (PVT) are taken into account in circuit design margins. However, larger than anticipated excursions may sometimes be observed arising from lack of process quality control or process modifications, package degradation, or tester malfunction. Understanding the ranges of different variations is important in determining test conditions and in failure diagnostics. This knowledge is further applied in guard-banding and in speed binning of chips to improve yield and profit margins.

Sources of variability in silicon process, models, and operating conditions are introduced in Sect. 6.1. Variability characterization through electrical test data analysis and imaging techniques is described in Sect. 6.2. Some examples of variability reduction methods are given in Sect. 6.3. Accommodation of variability in circuit design together with the impact of systematic and random variations are covered in Sect. 6.4.

Variability in CMOS circuits has been given limited coverage in textbooks on semiconductor devices and circuits, although a few recent books devote a chapter or more to this topic [1–3]. Books on silicon manufacturing quality control and statistical analysis cover process-related variations [4, 5]. There has been an increasing focus on the impact of variability on CMOS circuits in recent years, and there are many publications on different aspects of variability in technical journals.

6.1 Sources and Impact of Variations

There has been considerable focus in the technical literature on the increase in variability in CMOS device properties with scaling. As device dimensions continue to be scaled to the sub-100 nm regime, and with the introduction of FinFETs, variability among nominally identical devices due to statistical fluctuations in the number of dopant atoms and line-edge roughness continues to increase. These sources of variability are fundamental in nature and their impact is most significant on analog circuits requiring device matching and on SRAM cells.

There are many other sources of variations inherent in the manufacturing process with hundreds of processing steps, and batch production of wafers tiled with chips.

Operational and environmental factors and degradation in device and package characteristics over time contribute further. These variations are accounted for in design and by guard-banding the chip operating conditions (Sect. 8.3.3). Model-to-hardware offsets introduce another source of variability seldom discussed. With increasing complexity in MOSFET characteristics and physical layout dependencies, and a proliferation of device offerings at each technology node, device models used in circuit design can sometimes contain errors or may not reflect the current manufacturing process. EDA tools used in circuit design use some level of abstraction in device characteristics, compromising the model accuracy, and are not immune to errors.

A taxonomy of sources of variations in CMOS technology, circuit design tools and models, chip manufacturing, and operating conditions of CMOS products is shown in Fig. 6.1a. The nature of factors contributing to variability can also be categorized as systematic, random, spatial, and temporal as indicated in Fig. 6.1b.

Variations arising from inherent technology definition include impact of feature sizes, pattern densities, and aging effects. Efforts are made to minimize these variations through process and device engineering early in the technology development cycle. It is typically more challenging to reduce variability in a MOSFET due to random dopant fluctuations (RDF) and line-edge roughness (LER) as the feature dimensions are reduced. However, the net impact of random variations averaged over many devices in a logic circuit block is less than that on circuits where matching of device characteristics is required.

The variations in silicon technology and manufacturing are typically accounted for in the models used for chip design, timing, and power analysis. In complex designs with hundreds of millions to several billion devices, simplified versions of compact device models are generated for use with timing tools. The simplification and abstraction of the inherent nonlinear behavior of MOSFETs introduces offsets between models and hardware which may vary with circuit topologies and device widths.

Variations occur between identical circuits in batch processing such as among chips on a wafer and among chips from different wafers. As manufacturing processes and tools drift over time, the mean values and distributions of parameters in wafer lots comprising 5–25 wafers tend to fluctuate. Variations in device parameters as well as in relative strengths of p-FETs and n-FETs and in differences among various V_t type MOSFETs have an impact on the operating margins of chips.

Packaged chips experience variability from sources both internal and external to the chip. Power supply voltages may vary over time with changes in workload activity. Local temperature variations due to fluctuations in power densities introduce time-dependent spatial variations in circuit parameters over and above those introduced in silicon manufacturing. Systems in different customer locations may be operating with different ambient temperatures and cooling capabilities, affecting global chip temperature. As aging effects on MOSFETs and wires are temperature-dependent, circuits in hot spots and entire chips with less efficient cooling tend to degrade faster than circuits located in cooler locations.

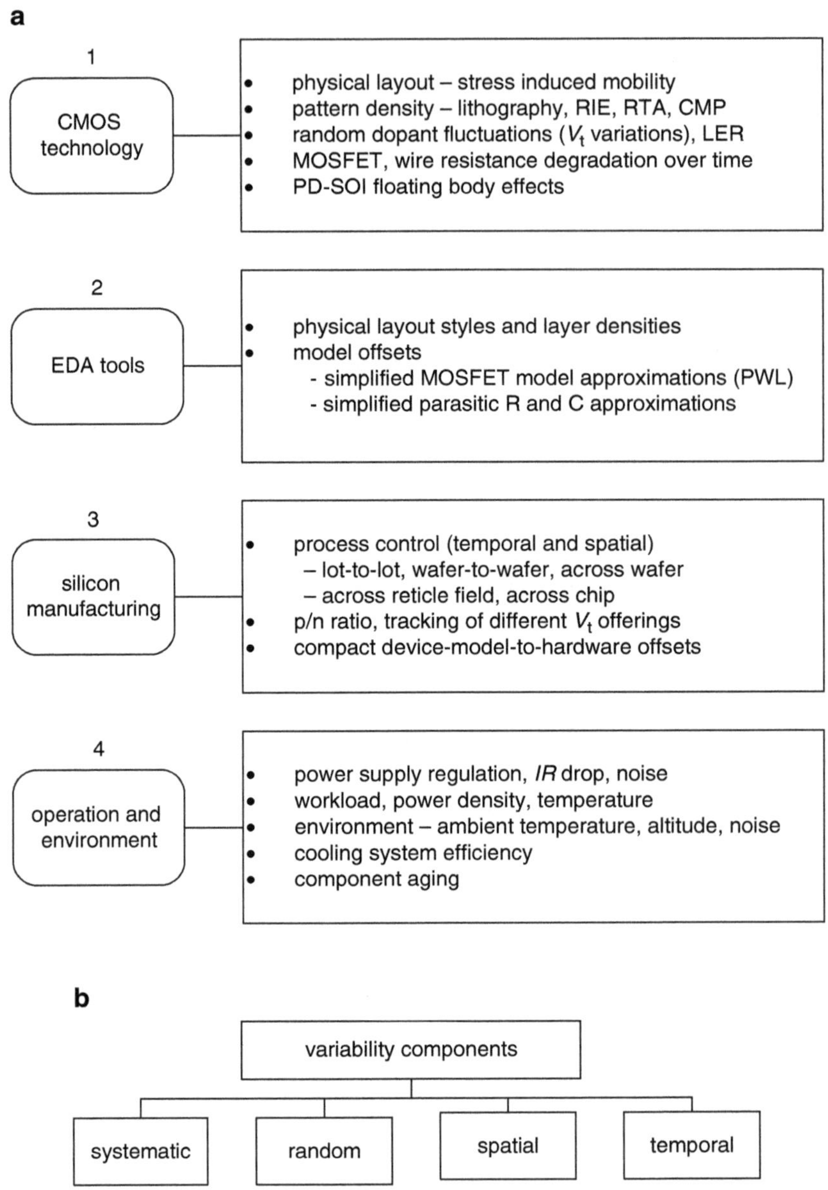

Fig. 6.1 (**a**) Taxonomy of sources of variations in (1) CMOS technology, (2) EDA tools, (3) silicon manufacturing, and (4) operation and environment, and (**b**) variability components

Management of variability is depicted as a triangle in Fig. 6.2, with three focus areas comprising characterization, minimization, and accommodation. Each of these areas is addressed in the following sections with examples spanning silicon technology, circuit design, and test.

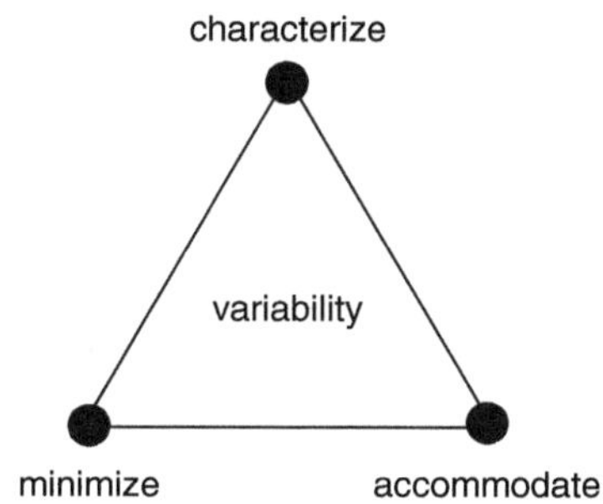

Fig. 6.2 Aspects of managing variability

6.1.1 Silicon Process Variations

Process-induced systematic and random variations are included in the device and parasitic models. As described in Chap. 2, these models include nominal values and the statistical distributions of all key device parameters so that the designs can be verified to work correctly over the published range of variations. In a mature technology, silicon manufacturing process is well calibrated to the models and the process is continually tuned to stay within the specified limits. In the early stages of a technology, manufacturing processes may undergo changes after the models have been released to the circuit design team. It is often the case that the early hardware is off-center, with some key parameters outside the published range. Even in a mature technology, there is a continuous push towards reducing variability to increase yield. An understanding of the sources of process variations and detection of any significant deviations in the hardware from design assumptions are helpful in electrical tests and diagnostics, in improving the overall economics of manufacturing, and in providing timely feedback to future designs.

Wafer level processing in silicon manufacturing allows many chips to be fabricated simultaneously. The physical layout of each processing layer is printed on a glass reticle known as a photomask. A scaled down version of the pattern on the photomask is then transferred to silicon wafers coated with a photosensitive material (photoresist) by a photolithographic process. After the various processing steps have been completed the photoresist is removed in preparation for the subsequent step. In multilayer processing, each layer is accurately aligned to preceding layers on the wafer to faithfully render the physical design. With traditional steppers, the entire reticle field is exposed and the process repeated after stepping the wafer in X or Y directions to a new location until the entire wafer surface has been exposed.

Step-and-scan exposure tools have made it possible to use larger reticle areas without sacrificing image resolution and alignment accuracy. The schematic of a step-and-scan optical exposure tool is shown in Fig. 6.3. The photomask and wafer move in opposite directions during exposure. Instead of exposing the entire reticle, only a fraction of the area is exposed through a slit which is as wide as the reticle in one direction and narrow in the orthogonal direction. The image of the photomask is scanned through the slit onto the wafer. There may be small differences in the widths of exposed lines in the direction of the scan compared with those in the orthogonal direction. To reduce the impact of such variations, in advanced technologies, a preferred direction of orientation is specified for narrow lines on critical layers such as the PS layer that defines MOSFET gate-lengths.

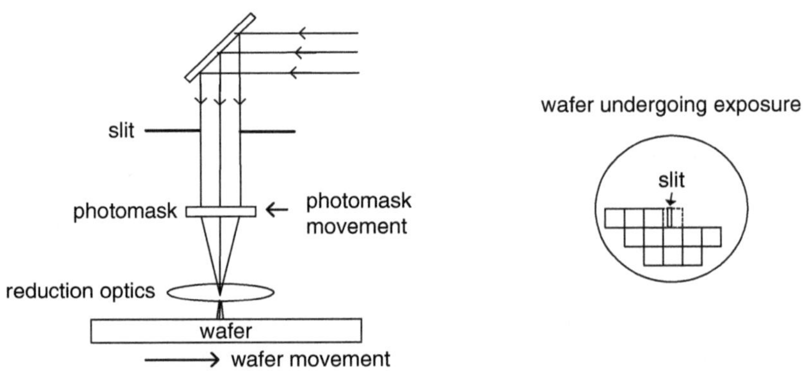

Fig. 6.3 Schematic of a step-and-scan photolithography exposure tool and a partially exposed wafer

The silicon wafer map in Fig. 6.4a shows rows and columns of exposed reticle locations covering the surface area of the wafer. Each reticle location may comprise one or more individual chips of identical or different designs as shown in Fig. 6.4b. The chips are separated from each other by a scribe-line or kerf ~30 to 100 μm in width. This scribe-line, with no product circuit content, is filled with test structures to monitor the manufacturing process. After completing silicon processing and initial electrical testing to identify good chips, a diamond saw or laser is used to dice the wafer into individual chips for packaging.

Many of the process steps introduce parameter variations across the wafer such that chips in different locations on the wafer may differ in their characteristics from other chips on the same wafer. Since in a typical foundry there are multiple tools installed for each process step, the across wafer variation may itself be different on different wafers, and the signatures may change with time due to process drift, intentional adjustments in process recipes, aging of tools, or introduction of new tools.

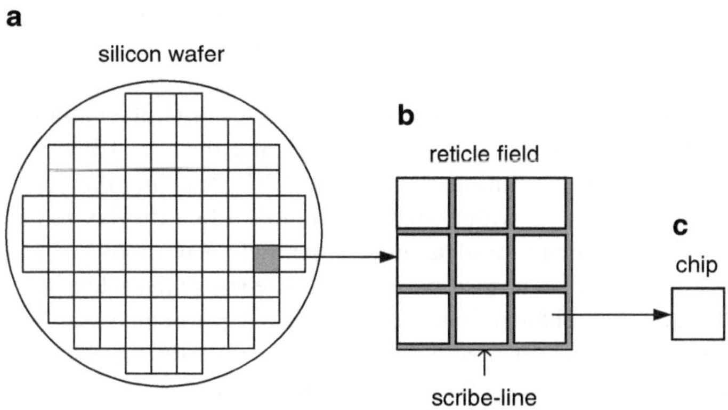

Fig. 6.4 (**a**) Silicon wafer map showing reticle exposure areas, (**b**) photomask reticle with nine chips, and scribe-lines, and (**c**) single chip after dicing the wafer

Across wafer variations arise from the inherent nature of various process steps and the discontinuity at the wafer edge. Some representative process steps are shown in Fig. 6.5. In reactive-ion-etching (RIE) to remove material after patterning photoresist, etch rate variations across wafer result from gas flow patterns and loading effects. In photoresist coating and in chemical–mechanical polishing (CMP) for interconnects, wafer rotation causes layer thickness variations. Thermal gradients during baking and annealing may introduce variations in line-widths and doping profiles. Changes in reflectivity with layer pattern density can cause temperature variations during rapid thermal annealing (RTA), resulting in spatial variations of V_t. Line-width and layer thickness variations affect circuit component properties such as resistances, capacitances, and MOSFET channel lengths.

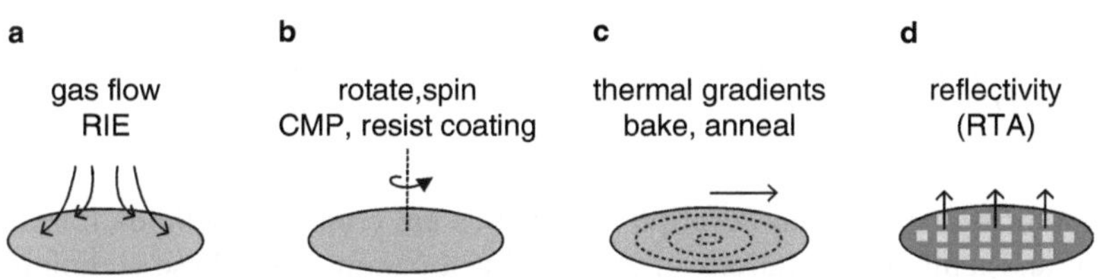

Fig. 6.5 Sources of nonuniformity from wafer level processing: (**a**) gas flow in reactive-ion-etching (RIE), (**b**) wafer movement (CMP, photoresist coating), (**c**) thermal gradients (baking, annealing), and (**d**) reflectivity (RTA)

In silicon manufacturing wafers are processed in batches of 5–25. Each batch, commonly known as a lot, is processed with the same recipes but different wafers may go through different tools or be placed in different locations in the same tool. Tools and processes may drift with time, and process recipes may be modified to compensate for the drift and for recalibration. Process recipes may also be fine-tuned to improve yield and performance. Hence variations occur among wafers processed in the same lot and from one lot to another.

A classification of variations associated with batch processing is shown in Fig. 6.6. Manufacturing quality control tracks lot-to-lot (L2L) variations to detect changes in the mean values of key parameters and their distributions over time. Wafer-to-wafer (W2W) variations are observable by tracking the wafer mean and standard deviation. Across wafer (AcW) variations often display radial symmetry as well as *X* and *Y* gradients. With a multi-chip reticle, variations across reticle-field (ACR) may be present because of optical exposure and pattern density variations in different designs and field locations on the wafer. Chip-to-chip (C2C) variations are associated with chip locations on wafer and also within reticle. Across chip (AcC) variations arise from layer pattern density variations and are also carried over from AcW and AcR variations, especially near the wafer edge.

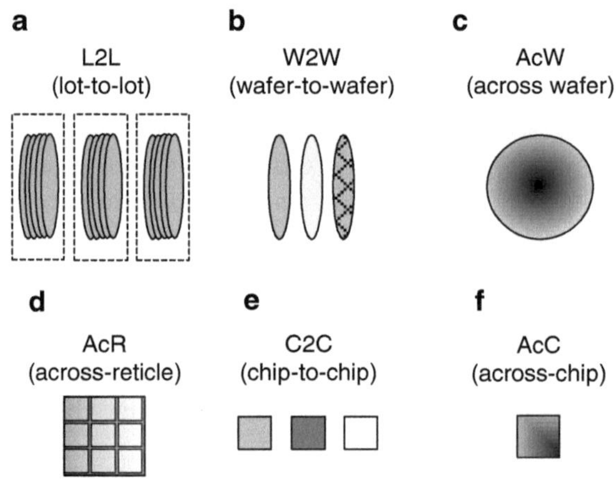

Fig. 6.6 Variation classifications associated with batch processing, (**a**) L2L, (**b**) W2W, (**c**) AcW, (**d**) AcR, (**e**) C2C, and (**f**) AcC

Graphical displays and descriptive statistics are common methods for viewing spatial and temporal variations in electrical test parameters. These techniques are described in more detail in Chap. 9. Here a few examples of visualizing variability from data collected at test are shown. Trend charts for different parameters show L2L and W2W variations. An example trend chart in Fig. 6.7 (box and whiskers plot) shows the range of variation of parameter p with a nominal target mean value of $\overline{p}$ and a standard deviation of σp. When the parameter drifts out of range, a correction to the recipe to re-center p closer to $\overline{p}$ is made. In the meantime the yield of lots already past the faulty processing step may be impacted because of p being out of range on some chips.

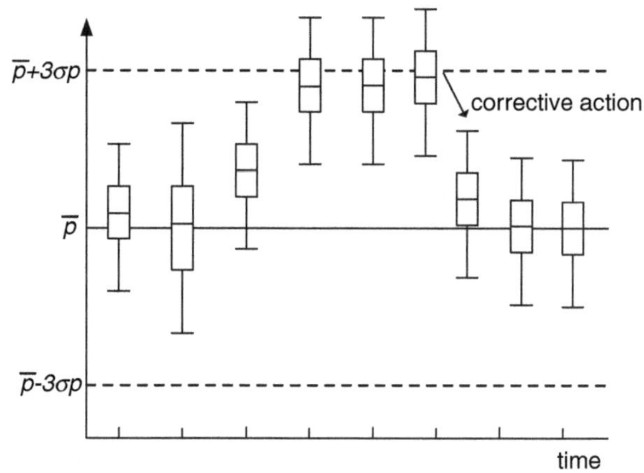

Fig. 6.7 Box and whiskers plot showing L2L trend of parameter p

Electrical test data on CMOS chips become available only after completion of the process. If all the chips on all the wafers are tested, high resolution wafer maps of key electrical parameters can be generated. A pictorial representation of AcW variation for a parameter p is shown in Fig. 6.8a. The parameter has a radial variation with dotted circles indicating regions of different ranges of p values. The number of chips near the outside edge of the wafer is greater than near the center. Nearly 35 % of the chips out of a total of 212 on this wafer lie in the edge zone. Based on the observed AcW patterns, a wafer may be divided into different zones and the parameter distributions in each zone tracked for monitoring zone-to-zone variability.

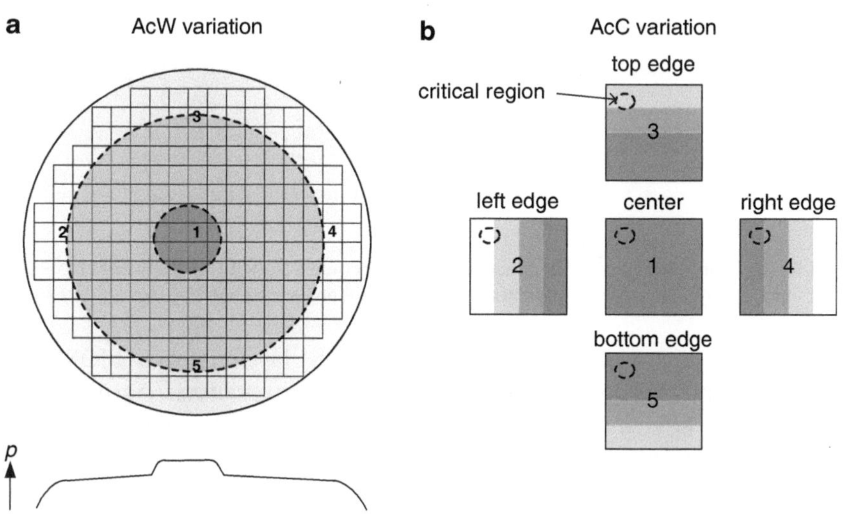

Fig. 6.8 (**a**) Wafer map showing variation of parameter p in gray scale across wafer. *Dotted lines* indicate boundaries of different p value regions, (**b**) AcC variations induced by AcW variation on five chips selected from the wafer map in (**a**). *Dotted circles* show location of a critical region on the chip

In Fig. 6.8b, values of parameter p are mapped in gray scale across five selected chips numbered 1–5 on the wafer map in Fig. 6.8a. Dashed circles in the top left corner of each chip indicate locations of critical circuits. Because of the radial dependence at the edges of the wafer, the value of p in the critical area is different on each of the edge chips. Chips for detailed electrical characterization must be carefully selected after examining such data, as individual chips may have unique signatures.

In Fig. 6.9a, AcC variation of a single parameter is shown in shades of gray. Systematic variation of this type may be present in the photomask and hence repeated on every chip. At advanced technology nodes, sensitivity to local pattern density may also introduce such systematic variations across all chips. The chip in Fig. 6.9a has six identical critical circuit areas, such as cores in a microprocessor chip. If the parameter p affects circuit delays, these cores of identical design and physical layout may have different values of f_{max} and V_{min}. Another consequence of AcC variations indicated in Fig. 6.9a is that while chip f_{max} is determined by the slowest core, the leakage current contributions of faster cores are disproportionately higher.

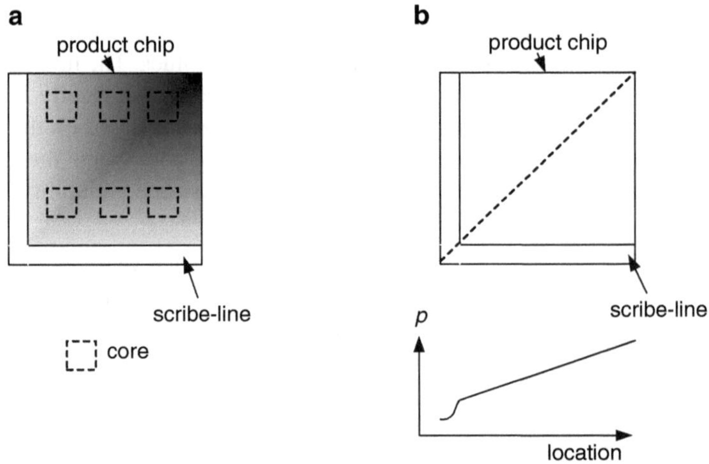

Fig. 6.9 (**a**) Spatial variation of parameter p across a reticle field with six identical circuit blocks on the chip, and (**b**) variation of parameter p across chip and scribe-line (along *dotted line*)

Systematic line-width variations of the PS layer impact MOSFET L_p and consequently MOSFET drive strength and capacitances. As discussed in Sect. 3.4. L_p variations affect f_{max}, V_{min}, IDDQ, and total power. Line-width variations in narrow interconnect wires are more significant than in thick, wide wires of the metal stack. Line-width narrowing along wire-length or at corners and bends can reduce the current carrying capability of the wires. A wire path can become highly resistive or even non-conducting due to electromigration. Similarly, widening of narrow wires may cause low-resistance paths (shorts) between adjacent metal lines.

Systematic variations in line-widths may be classified by their frequency of occurrence over time (L2L) and by the spatial distance over which they are observed (AcW, AcC). This classification is useful in detecting and correctly identifying the sources of observed variations. Systematic variations, once identified, may often be corrected by lithographic compensation techniques.

Variations in circuit properties continue to occur with use over product lifetime. These include changes in MOSFET properties arising from bias temperature instability (BTI), hot carrier injection (HCI), and time-dependent dielectric breakdown (TDDB). Interconnect wire and via resistances may increase over time due to electromigration, which may even result in catastrophic failures. The underlying mechanisms are functions of voltage, temperature, and current density. The extent of the degradation from these mechanisms may vary locally on a chip and from chip-to-chip based on product use. Guard-bands are applied in product qualification to cover the impact of these changes. Circuit reliability and strategies for applying guard-bands are discussed in more detail in Chap. 8.

6.1.2 Random Variations

Random variations in device parameters are fundamental in nature and their magnitude increases as feature dimensions are reduced with CMOS scaling. The most significant of these is random dopant fluctuation (RDF) in the MOSFET channel region. This effect results in random variations in the subthreshold characteristics of nominally identical MOSFETs. The resultant statistical distribution in V_t has a standard deviation expressed as

$$\sigma V_{\mathrm{tr}} = \frac{A_{\mathrm{vt}}}{\sqrt{WL_{\mathrm{p}}}}, \tag{6.1}$$

where A_{vt} is a constant for a given technology. Typical values of A_{vt} are in the range of 0.003–0.005 V μm and may be different for n-FETs and p-FETs. The variation increases with decreasing MOSFET area, or with decreasing width W for fixed L_{p}. In multi-finger devices, W represents the sum of the widths of all fingers. As an example, at the 45 technology node ($L_{\mathrm{p}} = 0.045$ μm) with $A_{\mathrm{vt}} = 0.004$ V μm, σV_{tr} is 0.027 V for a 0.5 μm wide device, and 0.042 V for a 0.2 μm wide device. Measures are taken during technology development to reduce random variability in MOSFETs.

Random variations also occur in gate-dielectric thickness, further modulating MOSFET properties. Line-width variations from line-edge roughness in the gate (PS layer) result in variations in electrical properties along a MOSFET's width. Because of V_t roll-off in short-channel MOSFETs, V_t and I_{off} can be strongly affected by the narrow width regions.

Although random variations among identical MOSFETs may be large, the net effect is less significant in combinational logic where circuit delays are typically averaged over many gates in a path (Sect. 6.4.2). SRAM cells, analog circuits, and single-ended narrow pass-transistors in logic paths are more susceptible to random

V_t variations. In a marginal design, random variations may result in timing issues in some test corners or cause reliability fails with aging.

Occasional process irregularities, unpredictable malfunction of a process tool, or unintentional failures to meet process recipe specifications constitute another class of random events. Such events can produce a wafer or an entire lot with uniquely different characteristics. Most cases like this are detected during in-line testing and the wafers eliminated from the flow. However, since scribe-line testing does not cover all chips and all product representative circuit components, some random excursions of this nature may be detected only with rigorous testing of product chips.

6.1.3 Voltage Variations

Standard ATEs are equipped with precision DC voltage and current source measure units (SMUs). The precision with which power supply voltage is applied to the board/package or chip I/Os is determined by contact and cable resistances, capacitances, inductances, and average and instantaneous currents drawn by circuitry under test. Internally, the power supply voltage V_{DD} may vary across the chip. These local variations in V_{DD} are dependent on the properties of the internal power grid, contact and wire resistances connecting the power grid to MOSFETs, and time-varying switching activities. The voltage variations discussed in this section pertain to differences between the power supply set-point and the actual on-chip V_{DD} at MOSFET terminals.

An ideal power supply is set to deliver a constant value of V_{DD}. In practice, the output voltage of the power supply varies with the load or current drawn. The load-line and the ideal current load range to best match the output to the set-point are included in the power supply specifications. The average V_{DD} applied at the chip I/Os is reduced by the *IR* drops in the series resistances between the power supply and the chip. This is partially compensated by sensing the voltage at points physically closer to the chip I/O and using a feedback loop to adjust the external power supply set-point.

A schematic of the test setup with resistances, inductances, and capacitances in the V_{DD} and GND network is shown in Fig. 6.10. Here R_{svb}, R_{svp}, and R_{svc} are the resistances in the V_{DD} lines on the board, package and chip respectively; R_{sgb}, R_{sgp}, and R_{sgc} are the corresponding resistances in the GND lines. If there are multiple power supply sources with a common GND, the current in the GND line is the sum of the currents through all the power supplies. In the absence of any dynamic feedback based on sense point data, the difference between the average applied voltage V_{DDi} of the *i*th power supply and the voltage at the chip is given by

$$\delta V_{DDi} = -\left\{ I_{dci} \times \left(R_{svb} + R_{svp} + R_{svc} \right) + \sum\nolimits_{i=1}^{n} I_{dci} \times \left(R_{sgb} + R_{sgp} + R_{sgc} \right) \right\}, \tag{6.2}$$

where n is the number of independent power supplies and I_{dci} is the average current drawn by the *i*th power supply. Variations in the series resistances in Eq. 6.2 can result in V_{DD} variations from chip-to-chip. Any degradation in contact or wire resistances over time adds to the V_{DD} variability, which in turn affects the chip performance and power.

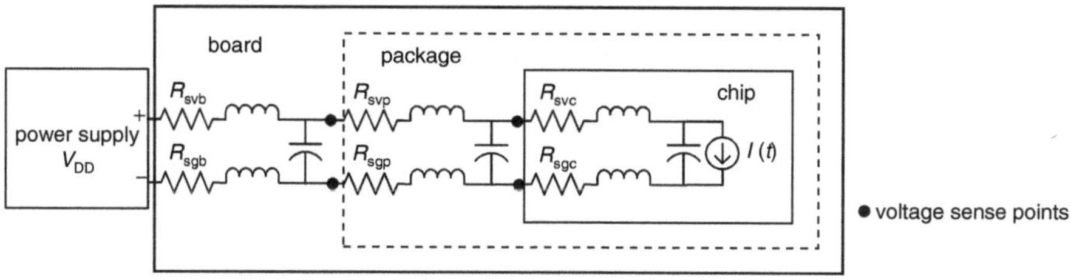

Fig. 6.10 Schematic showing series resistances, inductances, and capacitances in the V_{DD} and GND power distribution network of the board, package, and chip

Voltage variations on the chip may also arise from the power grid design. Power is distributed from the chip I/Os to the MOSFETs through the metal interconnect layers. The power grid resistance is designed to be small, however, parasitic series resistances of thin wires and vias connecting the power grid to the MOSFET terminals may vary. Local heating in dense areas of the chip also affects the resistances in series with the power supply.

In the standby or off-state, the chip draws a small DC current. As IDDQ is small, the *IR* drop is much smaller in low activity states. It is still worthwhile to consider the effect of V_{DD} variation on IDDQ of our standard inverter (FO = 4) circuit. In Fig. 4.18, IDDQ vs. V_{DD} data obtained from circuit simulations using 45 nm PTM HP models at 25 °C are plotted. From the exponential trend line, an increase in V_{DD} (=1.0 V) by 1 % increases the inverter IDDQ by 4.8 %.

Once the chip clock distribution network is turned on, both DC and AC currents flow in the power grid. The current drawn by switching of the clock tree is averaged over a time much longer than the clock cycle time and is nearly constant. In the absence of clock gating, the average AC component of this current can be effectively treated as a DC component for the purpose of estimating *IR* drops in the power supply distribution grid.

Any sudden change in switching activity on the chip causes a change in voltage. This change is governed by the expression

$$\delta V_{DD} \approx -L\frac{di}{dt}, \tag{6.3}$$

where L is the inductance of the wire network. The change in voltage can be large with large changes in switching activity within a few clock cycles (small dt). This situation arises when clock gating or power gating features are used to save power as described in Sect. 4.5. It can also happen during chip initialization if all of the circuitry is turned on at once. The Ldi/dt effect can be reduced in several different ways:

- in chip design: by ensuring there is sufficient decoupling capacitance on chip
- in board and package design: by placing adequate decoupling capacitance and minimizing R and L in the power distribution network
- in test codes and in microarchitecture: by increasing the number of cycles over which large changes in switching activities take place

A sudden increase in switching activity resulting in a dip in V_{DD} increases circuit delays and T_{cmin}, and may cause a timing failure. On the other hand, a sudden decrease in activity generates a spike in V_{DD} reducing the path delays, and may exacerbate race conditions in fast paths.

For rigorous testing of the chip, running actual field applications is the only way to fully exercise V_{DD} swings. Chips must be tested in functional mode at high V_{DD} and temperature corners to ensure adequate margins are present in the design. Although di/dt at the low-voltage corner is smaller, any nonlinear V_{DD} dependence may enhance changes in path delays. Special test programs can be generated to imitate worst-case situations in customer workloads and apps. If test time budget or equipment limitations at wafer and package level testing do not allow for full functional tests of this kind, such tests can be run on selected chips fabricated in different process corners.

The SKITTER macro described in Sect. 5.3 may be embedded on the chip in critical locations to calibrate and monitor the effect of V_{DD} variations arising from IR drops, Ldi/dt effects, and noise. Tracking V_{DD} variations in critical areas of the chip is very useful in optimizing workloads and power management techniques, and in diagnosing intermittent failures.

6.1.4 Temperature Variations

Average temperature of the circuitry on the chip T_j changes with power consumption. Variations in T_j among chips during test are determined by on-chip power dissipation, chip-to-package thermal resistance, and temperature set-point precision. Ambient temperature in the field and effectiveness of the external cooling system, such as a fan or circulating water or refrigerant, also vary from system-to-system and over time. Temperature variations of as much as 20–30 °C within chip are induced by local power density variations creating hot and cold spots. The locations of these hot and cold spots may change with time as different circuit blocks or units are exercised during operation.

In manufacturing test, a silicon wafer or a packaged chip is placed on a thermally controlled block or heat sink. The temperature of this block is held constant by a temperature controller. The average chip temperature, as determined by the thermal resistance of the chip to the heat sink and the total on-chip power dissipation, may be higher than the set-point.

Thermal resistance R_{th}, of the chip to the thermal platform expressed in °C/W is defined as

$$R_{th} = \frac{dT}{dP} = \frac{T_{j2} - T_{j1}}{P_2 - P_1}, \tag{6.4}$$

where P_1 and P_2 are the powers drawn at temperatures T_{j1} and T_{j2} respectively. The heat flow directions from the chip to the heat sink are illustrated in Fig. 6.11, with uniform power and with one hot spot on the chip. When hot spots develop on the chip, the heat flows in both lateral and vertical directions to cooler locations of silicon. In this case, the measured value of R_{th} using a temperature monitor in the vicinity of the hot spot will be higher than if the monitor is in a cooler location.

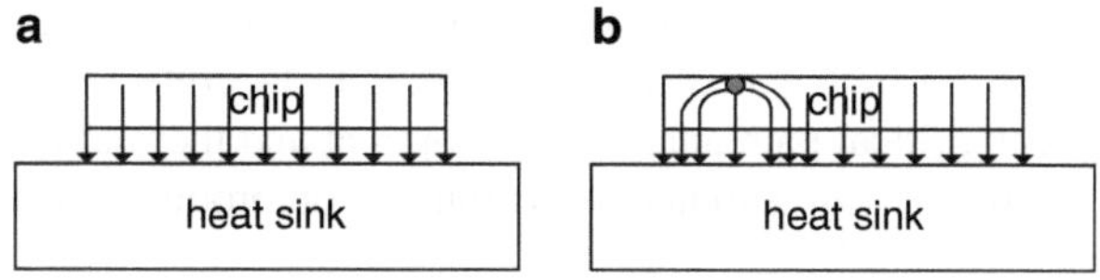

Fig. 6.11 Heat flow from chip to package or heat sink (**a**) uniform flow and (**b**) with hot spot on the chip

For $R_{th} = 0.05$ °C/W, with a uniform power dissipation of 100 W, T_j increases by 5 °C above the set-point of the heat sink. Except in high power/performance chips, power dissipation is typically lower than 100 W, and temperature variations on the average are not significant. However, poor quality control of the thermal interface material (TIM) between the heat sink and the chip may result in significant variations in R_{th}. Also, R_{th} may increase with degradation of the TIM in response to thermal cycling and over time. Corrective actions, such as TIM replacement or re-test on a different tester, are taken when R_{th} exceeds the acceptable limit.

Temperature variations have significant impact on MOSFET characteristics in the subthreshold region and on circuit reliability. Leakage current varies exponentially with temperature (Fig. 4.18). High T_j regions of the chip add disproportionately to the background leakage power in the active state. The temperature dependence of MOSFET characteristics in the saturation region can be tailored by process engineering and may be negligible at some technology nodes. Interconnect wire resistances increase with temperature. For logic gates driving signals over long wires, the wire *RC* delay may be a significant fraction of the total delay. The temperature dependence of circuit delays therefore varies with the mix of MOSFET gate and wire loads, and may have different signatures at different CMOS technology nodes.

The NBTI and PBTI degradation mechanisms in MOSFETs are thermally activated as described in Sect. 8.3.1. Circuits in a hot spot of the chip degrade faster than in a cold spot, increasing the signal propagation delays in regions of highest activity. With inadequate design margins, the chip may develop timing issues with aging. Another reliability issue is electromigration in metal wires (Sect. 8.2). Metal agglomeration occurs in the direction of current flow, creating metal deficit areas, thereby increasing wire resistance over time. A catastrophic failure may occur if the wire becomes non-conducting. Electromigration is also a function of T_j, and hot regions of the chip are more strongly affected.

Finally, a combination of inadequate cooling, high-circuit switching activity or shorts developed within the power grid may cause thermal runaway. In this case instantaneous data collected from thermal sensors on the chip or package can be used to shut the power off to prevent a meltdown.

6.2 Variability Characterization

Variability characterization begins during CMOS technology development. Dedicated test chips with a large variety of electrical and metrology test structures provide early feedback for minimizing variations induced by different process steps.

However, variations in product chips, specific to a chip design, can only be characterized when hardware becomes available. Data collected from embedded monitors along with critical path analysis, and optical and thermal imaging are some of the techniques for analyzing sources of variations on product chips.

6.2.1 Silicon Manufacturing Tests

Electrical tests are conducted in the manufacturing line on scribe-line test structures. Generally, a subset of wafers from a lot and only a few selected locations on each wafer undergo full characterization. Average values of the parameters measured on these sites, or "chip mean" values, are used for centering the process. The parameter statistics for AcW, W2W, and L2L variability are analyzed for model-to-hardware correlation. This is described in more detail in Sect. 7.4.1.

Relative positions of a product chip and its scribe-line are shown in Fig. 6.9a. Mismatch among circuit component properties in the scribe-line and on the product chip may occur for the following reasons:

- differences in designs and physical layouts
- spatial separation between chip and scribe-line in the presence of across reticle variation
- test V_{DD} and temperature

Test structures for optical inspection and line-width measurements for metrology are typically placed within the product chip to measure AcC variation. Probing of electrical test structures within chip may be conducted only on selected wafers with additional photomasking levels as described in Sect. 5.2.1. With the availability of noncontact electrical test structures, such as ring oscillators with wireless power and signal delivery features, it is now feasible to obtain on-chip measurements during manufacturing, although cost considerations may preclude routine implementation.

6.2.2 On-Chip Embedded PVT Monitors

Electrical test data collected on embedded process, voltage and temperature (PVT) monitors described in Chap. 5 are extremely useful for characterizing C2C and AcC variations. Delay chains, ROs or critical path monitors track the impact of silicon process variations on circuit delays. Circuit stage designs for tracking MOSFET parameter variations are described in Sect. 5.6.

In order to separate silicon process variations from V_{DD} and temperature variations, it is best to measure RO frequencies with an external frequency counter, initially with the on-chip clock distribution off. In this case, the power dissipation is small and any voltage droop in the power grid is minimized. Also, T_j is uniform across chip and close to the set-point on the heat sink. If measurements are made at a lower V_{DD} value, the magnitudes of voltage droops and T_j variations are further reduced. Also the sensitivity to variations in V_t is greater, providing a larger signal (% change in delay). By repeating these measurements at the operating V_{DD} with the clock distribution on, any significant V_{DD} droops can be detected.

Data collected from ROs are mapped for visualizing spatial variations. In Fig. 6.12a, locations of 16 ROs uniformly distributed across a product chip are shown. The measured RO frequencies are mapped for one wafer in Fig. 6.12b. The product chip is depicted as a square box within scribe-lines and small squares representing relative RO locations are coded in a gray scale to indicate their frequency range. It is immediately apparent that the RO frequencies vary across chip on all chips. Chips located within the dotted circle have similar across chip variation, with ROs in the lower part of the chip being the slowest. The AcC variation outside the dotted circle is different for chips at the top, bottom, left and right sides of the wafer. Silicon process recipes, if not adequately compensated at the wafer edges, can often result in such variations.

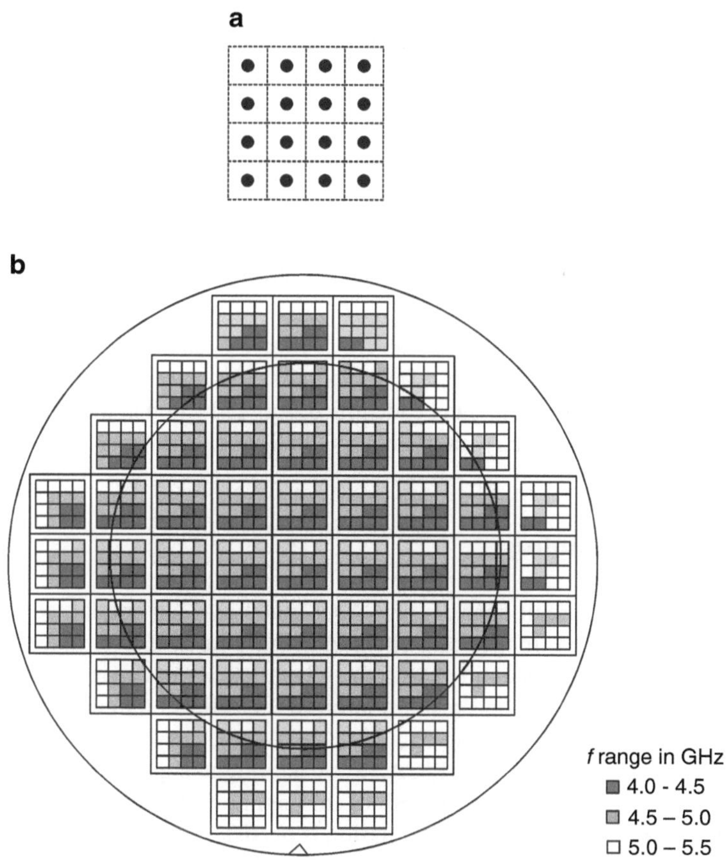

Fig. 6.12 (**a**) Locations of 16 ROs uniformly distributed across a chip, and (**b**) RO frequencies in a gray scale showing across-chip (AcC) and across-wafer (AcW) variations. Chips within *inner circle* have similar AcC variation while edge chips differ

6.2.3 Functional Parameters

Variations in f_{max}, V_{min}, P_{off} (IDDQ), and total power measurements on each chip may also be mapped across wafer. A wafer map of f_{max} is shown in Fig. 6.13. This map has less resolution than the map in Fig. 6.12 as there is only one f_{max} measurement per chip. However, it represents one of the critical outcomes of variability. If the map shows a uniform distribution with f_{max} at or above the target, no further analysis is required. A higher resolution may be obtained, as for example, in a microprocessor chip having multiple cores, with independent measurements of f_{max} for each core.

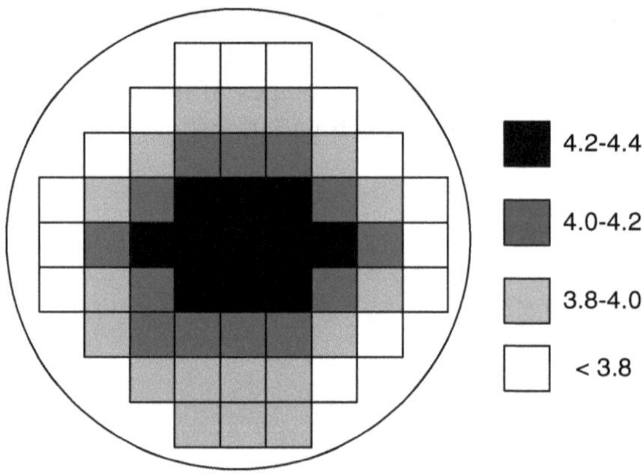

Fig. 6.13 Wafer map of chip f_{max} showing radial dependence across wafer

The AcW variations in chip parameters f_{max}, V_{min}, P_{off}, and P_{ac} are compared with the AcW variations of RO and MOSFET monitors. If a correlation exists, the most significant parameter variations can be identified. Any reduction in the spread of these parameters can then be directly translated into yield or performance improvements.

6.2.4 Optical Imaging

Picosecond imaging circuit analysis (PICA) is a noninvasive optical technique for obtaining time resolved images of switching activity in CMOS circuits. During a switching transition light is emitted with the relaxation of hot carriers in MOSFETs in saturation mode. Picosecond resolution capability allows diagnostics of timing related failures on CMOS chips (Sect. 7.3).

In a different application of PICA, a weaker light signal is observed from MOSFETs in the off-state. The strength of this signal in the infrared region of the spectrum is proportional to I_{off}. This emission is persistent unlike the emission from hot carrier relaxation, which occurs only during switching. At the 180 nm technology node and beyond, with increasing electric field in the channel region, the emitted infrared signal strength becomes sufficiently strong for mapping across chip variations in I_{off} [6]. I_{off} increases with reduction in channel length L_p and V_t as discussed in Sect. 4.2.1. Qualitative measurements of variations in L_p and V_t across chip can be accomplished by viewing the PICA image. Quantitative measurements via this technique are, however, very time consuming.

With thinner gate-dielectrics, there is an additional light emission component from the gate tunneling current. In high performance MOSFETs, this contribution is negligible compared with emission from the channel region. Also, gate-dielectric thickness variation occurs from wafer-to-wafer and is insignificant across chip, further diminishing the diagnostic potential of this component.

To avoid interference from metallization, light emission for PICA is observed from the backside of the silicon substrate after it is thinned to ~50 μm as shown in Fig. 6.14a. The back surface is polished to give an optical finish to reduce light loss from surface scattering. Light emission is recorded using an infrared camera, equipped with a liquid nitrogen-cooled HgCdTe focal plane array detector. Dark current subtraction is carried out to correct for any nonuniformities in the detector imaging system. It may be necessary to stitch several images to cover the entire chip area.

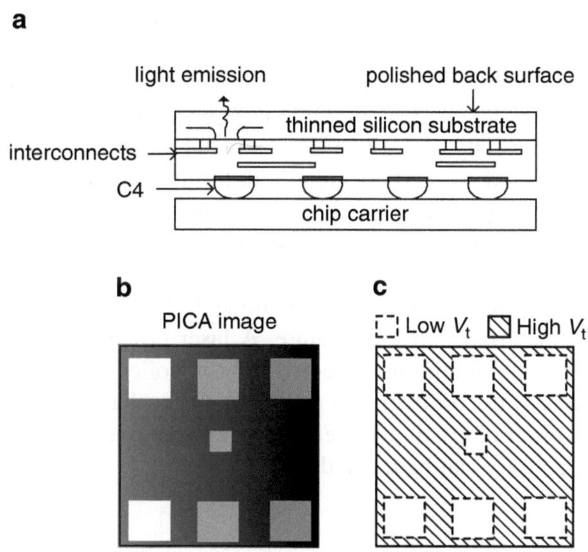

Fig. 6.14 (**a**) Cross section of PICA sample with thinned substrate for backside light emission, (**b**) PICA image showing light emission variation across chip, and (**c**) low and high V_t usage across on this chip

Prior to imaging, the chip is initialized to ensure that all latches are reset and that there are no floating gates generating excessive leakage currents. Data collected are post-processed and overlaid on a front side optical image of the chip or physical layout design data to facilitate identification of areas of high or low device strengths. Rendering of a PICA image is shown in Fig. 6.14b. The six core circuits of identical design exhibit different emission intensities, corresponding to the image in Fig. 6.14c with AcC variation in intensity from left to right. Interpretation of the image requires knowledge of V_t device usage across chip. The region surrounding the cores uses higher V_t devices and appears darker with the exception of one small square area of low V_t devices at the center in brighter shade. The AcC variation may result from a gradient in L_p or V_t across chip and should correlate with data from distributed embedded monitors (Sect. 5.6.3). With high-resolution imaging, this technique is most useful in identifying areas of nonuniformity not revealed by mapping ROs or CPMs.

Although PICA is noninvasive, sample preparation is time consuming and expensive. This limits the number of samples on which PICA images can be obtained. Data from mapping RO frequencies or delay/stage give the first indication of significant variability. In order to get a higher resolution map, a few representative chips from different locations on the reticle and wafer may be selected for PICA imaging.

6.2.5 Thermal Imaging

Thermal imaging of a chip complements temperature profiling obtained from on-chip thermal monitors. A high-resolution thermal map of a chip is obtained with an infrared camera. The spatial resolution is limited by the wavelength of the radiation to 3–10 μm. Intensity of emission is dependent on black-body radiation as well as thermal emissivity of the surface. As metal areas have a higher emissivity than oxide, background correction is required. Infrared camera detectors are liquid nitrogen-cooled to reduce background noise. A special setup is required for imaging and only a few representative chips may be sampled.

The spatial resolution of thermal images may be further improved with thermoreflectance imaging [7]. This technique is based on the changes in the reflectivity of a material with temperature. A laser probe in the visible spectrum is used as a light source and the reflected light is collected with a CCD camera. Time resolution is of the order of 100 ns or less, faster than thermal time constants which are of the order of ms, and movement of hot spots on the chip with changes in local switching activity can be tracked.

6.3 Minimizing Variations

Efforts to reduce variability begin with chip architecture and floorplanning and follow through circuit design, physical layout, and reticle design for photomasks. In the first pass of a chip design, information provided by the silicon foundry and learning from other products at the same technology node or from previous technology nodes, if applicable, is useful in mitigating the effect of anticipated variations. Once silicon hardware is available, feedback from electrical test data early in the manufacturing cycle is critical for making improvements in silicon processes to reduce variations (Sect. 6.3.3). Systematic offsets in EDA tools may also surface if some of the cycle limiting paths cannot be accounted for by timing margins in design together with observed AcC variations. There may be an opportunity to retune these paths if there is a second design pass. However, close interactions among silicon technology, chip design, and test teams are important in validating the design assumptions with measured hardware data.

6.3.1 Chip Design and Floorplanning

Some of the unavoidable variability challenges should be considered in chip architecture, floorplanning, and design. Smaller area chips with uniform circuit density (such as gate arrays) have the least amount of AcC variations. Large area microprocessor chips with multiple cores and large on-chip memory banks are likely to suffer more from AcC variations.

Critical circuits should not be placed near the chip edges, especially in large chips occupying 40 % of the reticle area or more (one or two chips per reticle). Generally, there may be some L_p and V_t variations near the edges of the reticle, producing an even larger variation for chips located at the edges of the wafer.

The power grid should be uniformly robust to reduce its resistance and consequently *IR* drops, particularly in the most active and dense circuit areas. Clock distribution should be well designed with adequate buffering to reduce clock edge uncertainty. Critical circuits with high-power density should be placed near the center of the chip to prevent excessive heating at the chip edges where the thermal conduction to the package may not be as good as at the center.

Standardization of physical layout styles and placement of small features on a regular gridded pattern reduce optical proximity correction (OPC) errors and lithography-induced variations. PS shapes are placed on a fixed pitch. At the 32 nm technology node and beyond the gridded approach is being applied to H0 vias and M1 wires in critical areas. Some examples of non-gridded and gridded physical layouts are shown in Fig. 6.15. MOSFET physical layouts in Fig. 6.15a, with PS layers in orthogonal directions and different PS to M1 spacing, are representative of older CMOS technologies. In Fig. 6.15b, minimum PS pitch for defining MOSFETs and minimum M1 pitch for making contacts to their source and drain regions are fixed. Also H0 vias are placed on a fixed grid in both *X* and *Y* directions.

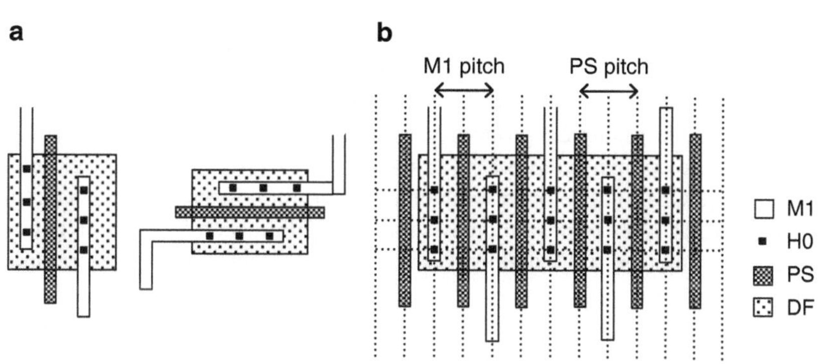

Fig. 6.15 Physical layout styles: (**a**) M1 and H0 off-grid and orthogonal PS shapes, and (**b**) PS, M1, and H0 on grid

6.3.2 Reticle and Wafer Assembly

The typical reticle field area at the 45 nm technology generation is ~850 mm^2. A large reticle may contain multiple instances of a chip design or multiple instances of different chip designs. Even though efforts are made to provide nearly uniform layer pattern densities within chip, proximity to the scribe-line test structures with large metal pads for probing and low-density CMOS circuitry creates nonuniformity near the chip edges. Circuits at the edges of the reticle and bordering scribe-line area are likely to have larger process-induced variability.

The scribe-line test structures for monitoring the manufacturing line process should be representative of the designs on the chip to ensure that L2L and W2W variability control in the manufacturing line translates into a similar control on the product chip.

Manufacturing process variations at the edges of the wafer are quantified on test chips designed for technology development. Based on the data collected, dimensions of an exclusion ring at the wafer edge are defined. All reticle areas on the wafer are placed inside the exclusion ring. This comes at a cost of reducing the number of chips on a wafer. As the number of chips near the edge of the wafer is large, the chip yield may be improved with proper exclusion limits at the cost of fewer chips per wafer.

6.3.3 Silicon Process Improvements

Timely feedback from test data is key to making silicon process improvements for variability reduction. This type of feedback is best provided during the early characterization phase of the chip. Ideally data collected from on-chip embedded delay chains or ROs are mapped across wafer, reticle field, and chip. If possible this is carried out for each operational chip on every wafer, even if the chip is not fully functional or does not meet the required frequency and power criteria (Sect. 7.3.2). A dense wafer map of circuit delays is thus created from which systematic spatial variations can be de-convolved [8]. Analysis of these variations with the silicon technology team and getting to the source of variability is essential in identifying root causes.

Typically L_p variations arise from photolithography and etching processes. Across reticle variation may originate with the photomask or result from exposure dose variations. Measurement of critical dimensions (CDs) on the PS photomask is the first step in addressing L_p variability. Any variations caused by the exposure or etch tools can be compensated for by adjusting the exposure dose. One example is dose mapper (DOMA) developed for such compensation in step-and-scan exposure tools. With DOMA, the exposure dose is mapped within each reticle area to correct for AcC variations on all locations on the wafer. Other techniques continue to evolve.

Variations in V_t may be partly due to L_p variations (short-channel effect), and partly due to other processes such as, for example, rapid thermal annealing (RTA) at the 45 nm technology node. RTA-induced variations arise from the changes in the reflectivity of the silicon surface based on the pattern density. If the range of the pattern density variations is large, a fill pattern to provide a uniform pattern density can be adopted. In more advanced technology nodes, the sources of variation may be different, and each technology node has to be individually evaluated.

If the chip frequency is dominated by long wire paths, variations in the interconnect wire widths and thicknesses may also have a measurable effect. Variations in material compositions of thin metal and inter-level dielectric layers impact their resistances and capacitances. These variations are more difficult to quantify using delay chains and ROs because of number of metal layers and silicon area requirements. Use of metrology tools to measure metal line-widths and thicknesses on the chip, and scribe-line measurements on interconnect test structures during fabrication are more practical methods for assessing variations in interconnect resistances and capacitances.

6.4 Accommodating Variability in Circuit Design

In this section, circuit design practices for including systematic and random variations of device parameters are described. Individual logic gates, circuit blocks, and functional units are designed and characterized with a common set of models and tools. Systematic variations in device parameters, and V_{DD} and temperature values over which the chip design is specified to function correctly span a wide range. Monte Carlo analysis to cover this space is time consuming and can only be efficiently implemented for a subset of critical circuit blocks and memory cells. The bulk of the designs are validated in a few simulation corners, exercising the nominal and extreme ranges of device properties and application conditions.

Random variations in device properties are most significant in narrow width MOSFETs. In Sect. 6.4.2 we show that in combinational logic paths, these variations are averaged over many gates and their net impact is relatively small. Analog circuits requiring matching devices, as for example in current mirrors, as well as memory cells must be designed with sufficient margins to accommodate random variations.

AcC variations in identical circuit blocks, placed in different locations on the chip, are usually not considered in design as there is no a-priori knowledge of these variations. This is further complicated by the fact that AcC variation itself may be dependent on the location of the chip on the reticle and the wafer. Also from a design point-of-view, replicating copies of circuit blocks, functional units, and microprocessor cores across chip reduces the cost of circuit design. In a robust design, process-induced AcC variations may be included in global systematic parameter distributions. AcC variations are then characterized in the hardware and silicon processing tuned to minimize them.

CMOS chip design methodology is established to account for all the sources of variability along with expected yield, performance, and power. A practical approach is to design for worst-case scenarios for frequency, power, and noise.

These worst-case scenarios are defined based on parameter spreads in the device models, estimated variations in V_{DD}, temperature and noise levels, and projected degradation of device properties over time.

As an example, consider only the variation in L_p and its impact on the frequency of operation. This variation is represented by a normal distribution in Fig. 6.16a, with a mean value of L_{po} and a standard deviation of σL_p. A CMOS chip design is initially carried out to meet a single frequency target while accommodating a $\pm 3.0\sigma$ range of L_p. A chip fabricated with a channel length for all MOSFETs of $(L_{po} + 3.0\sigma L_p)$ will have longer circuit delays and hence a lower frequency of operation. By design, it will meet the frequency target, while most of the other chips will operate at a frequency lower than their potential. The probability of finding a chip with $L_p > (L_{po} + 3.0\sigma L_p)$ in a large population is 0.015 %. Hence, such a design is overly pessimistic with a wide design margin for frequency. By setting a lower frequency target to meet this L_p constraint, the product may be less profitable in the marketplace.

If instead, chip design is carried out to accommodate a range of $(L_{po} \pm 2.0\sigma L_p)$, the frequency target can be raised over the previous $(L_{po} \pm 3.0\sigma L_p)$ design. For our standard inverter (FO = 4) delay in 45 nm PTM HP models, this corresponds to 8 % increase in the frequency target. About 2.3 % of the chips are expected to fail the frequency test as indicated in Fig. 6.16b. These chips may be selectively binned for lower frequency, or specified to operate at a higher V_{DD} to meet the frequency target.

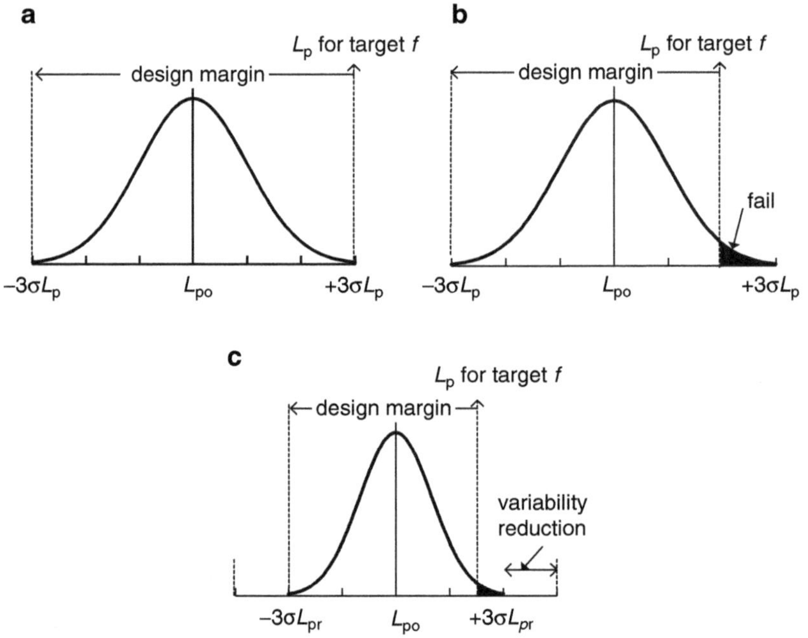

Fig. 6.16 L_p distribution and target chip frequency corresponding to: (**a**) $L_p + 3\sigma L_p$, (**b**) $L_p + 2\sigma L_p$ with 8 % higher frequency and 2.3 % yield loss, and (**c**) $L_p + 2\sigma L_{pr}$ ($\sigma L_{pr} = 2/3\sigma L_p$) with 13 % higher frequency and 2.3 % yield loss

In a third scenario, the manufacturing process is improved to reduce σL_p by 1/3 of its original value to σL_{pr}. The new distribution with $\sigma L_{pr} = 0.66\sigma L_p$ is shown in Fig. 6.16c. The target frequency can now be increased by 13 % over the first design in our example and approximately 2.3 % of the chips are either rejected or binned to operate at a higher V_{DD}.

With many sources of variation, accommodating a $\pm 3.0\sigma$ range for all variables becomes even more pessimistic than for a single variable (Sect. 9.3.1). The trade-offs between wide design margins and manufacturing yield need to be established with all aspects of manufacturing, test, product specifications, and profitability in mind. Establishing the worst-case design assumptions prior to having any hardware in hand requires good physical insight into the sources of variations, data from other products or from previous technology generations, and engineering acumen.

As the manufacturing process matures, there is a constant push to reduce variability and improve yield. This approach requires modeling the sources of variations in each product and mitigating the most significant ones. Because of the differences in chip dimensions, packaging, operating conditions, workload, and environment of different CMOS products, post-design variability reduction is a complex exercise, customized to each chip design.

6.4.1 Simulation Corners

Key parameters affecting circuit delays are listed in Table 6.1. The last three columns in the table indicate the impact on delay τ, P_{off}, and P_{ac} with increase in each of the corresponding parameters at constant V_{DD}, temperature, and frequency. An upward pointing arrow ↑ indicates an increase in τ, P_{off}, or P_{ac} and a downward pointing arrow ↓ a decrease. In general, these parameters vary independently, but some of them have second-order interdependencies such as decrease in V_t at shorter channel lengths.

Table 6.1 Impact of increase in major parameters on circuit delay and power: ↑ (higher), ↓ (lower), – (unchanged)

Component	Parameter	τ	P_{off}	P_{ac}
MOSFET	V_t	↑	↓	↓
	L_p	↑	↓	↑
	t_{ox}	↑	↓	↓
	R_{ds}	↑	–	–
	μ	↓	↑	–
	C_{ov}	↑	–	↑
	C_j	↑	–	↑
Interconnect wires	R_w	↑	–	–
	C_w	↑	–	↑

Assuming optimally engineered MOSFETs with good control on stress-induced mobility, variations in V_t and L_p are the major contributors to circuit delay variations. With a single exposure and etch of the PS layer for n-FETs and p-FETs, L_{pn} and L_{pp}

generally track together and a single value of L_p ($=L_{pn}=L_{pp}$) is assigned to all MOSFETs with identical drawn PS lengths. Threshold voltages, tailored by implant and rapid thermal annealing (RTA) processes, are controlled separately for the n-FETs and the p-FETs. Hence, with independent variations in V_{tn} and V_{tp}, the p/n ratio may vary from L2L, W2W as well as AcW and sometimes even AcC.

In order to account for changes in p/n ratio, five different process corners are selected using different combinations of high, low, or nominal values of V_ts. These simulation corners are shown in Fig. 6.17 and cover the full range of p/n ratios over which the design must pass. The corners shown here are designated based on MOSFET intrinsic delays: slow (s), fast (f), and nominal (nom), also referred to as typical. As an example, a ff corner denotes a fast n-FET (lower V_{tn}) and a fast p-FET (lower V_{tp}).

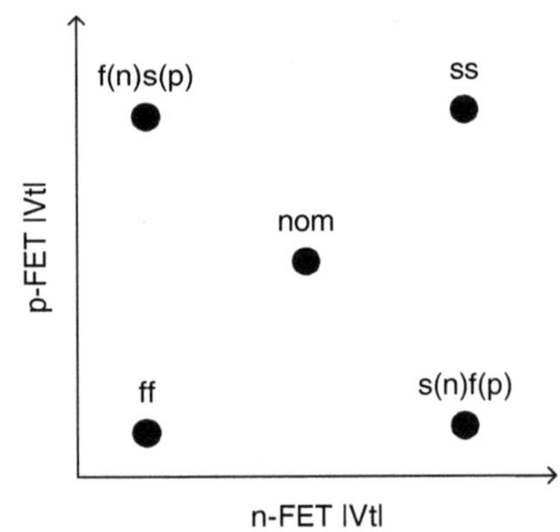

Fig. 6.17 Simulation corners to account for n-FET and p-FET V_t variations: nominal (nom), slow/slow (ss), fast/fast (ff), slow/fast s(n)f(p), and fast/slow f(n)s(p)

The corners described in Fig. 6.17 are limited to V_t alone. Four commonly used worst-case simulation corners to include both timing and power are listed in Table 6.2. To meet chip cycle time, circuit delays are longest in the slow timing corner. Race conditions and early arrival of data are tested in the fast timing corner. P_{off} is maximum in its worst corner with minimum V_t and L_p and at maximum V_{DD} and temperature. The worst-case corner for AC power P_{ac} is defined at maximum V_{DD}, temperature, and L_p and at lowest V_t to maximize C_{sw}. Worst-case for wire resistances, capacitances, and other properties may also be included in corner definitions.

Table 6.2 Simulation corner definitions

Simulation corner	L_p	V_{tn}, V_{tp}	V_{DD}	Temperature
Timing (slow)	Maximum	High	Minimum	Maximum
Timing (fast)	Minimum	Low	Maximum	Minimum
P_{off} (worst)	Minimum	Low	Maximum	Maximum
P_{ac} (worst)	Maximum	Low	Maximum	Maximum

The values assigned to L_p, V_t, V_{DD}, and temperature in each of the four corners listed in Table 6.2 are based on trade-offs between higher design margins, chip specifications, and yield. Understanding these trade-offs is important for defining test corners, and for isolating design weaknesses from silicon process or other test-related excursions. Global penalties on timing to accommodate noise, clock jitter, and aging effects are added to the design margins.

6.4.2 Impact of Random Variability on Circuits

In addition to the process simulation corners described in the previous section, the impact of random variations in advanced technologies is included in circuit design tools. Random variations in V_t, governed by Eq. 6.1, are the most significant random variations in narrow-width MOSFETs. As described below, such variations can be accounted for by using the device models to generate a delay multiplier $m_{vtr}(W)$, applicable to all combinational logic gate topologies.

To get a physical feel for the influence of V_t variations on signal propagation delay, it is instructive to consider an ideal hypothetical case of a long delay chain of nominally identical circuit stages as shown in Fig. 6.18a. Each stage has a nominal delay τ, with a random variation that follows a normal distribution of standard deviation $\sigma\tau$. The stage can be non-inverting or inverting and τ and $\sigma\tau$ are the same for a rising or a falling input signal.

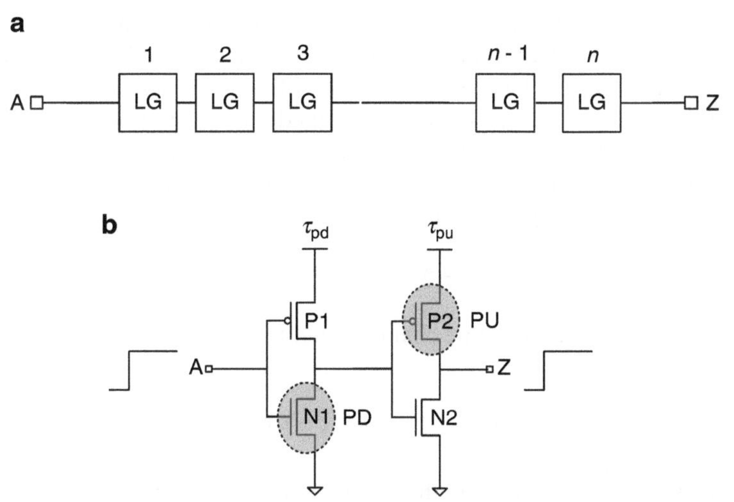

Fig. 6.18 (**a**) Delay chain of n logic gates, and (**b**) two-stage segment of an inverter chain indicating n-FET N1 and p-FET P2 driving the PD and PU transitions for a rising input

The nominal delay of an n-stage segment of the chain is given by

$$\tau(n) = \tau + \tau + \tau + \cdots + \tau = n \times \tau, \tag{6.5}$$

while the standard deviation of $\tau(n)$ is

$$\sigma\tau(n) = \sqrt{(\sigma\tau^2 + \sigma\tau^2 + \sigma\tau^2 + \cdots + \sigma\tau^2)} = \sqrt{n \times \sigma\tau^2} = \sqrt{n} \times \sigma\tau. \tag{6.6}$$

The corresponding fractional delay variation can be characterized as

$$\frac{\sigma\tau(n)}{\tau(n)} = \frac{\sqrt{n} \times \sigma\tau}{n \times \tau} = \frac{1}{\sqrt{n}} \times \frac{\sigma\tau}{\tau}. \tag{6.7}$$

Thus while the delay of an $n = 100$-stage segment is 100× that of a single stage, from Eq. 6.6, $\sigma\tau(100)$ only increases by 10×. The fractional variation in delay $\sigma\tau(100)/\tau(100)$ from Eq. 6.7 is just 0.1× that of a single stage. The net impact of random variations on delay is expected to get smaller as the length n of the chain increases.

Next, consider the more realistic case of an inverter chain comprising standard inverters with $W_n = 0.4$ μm and $W_p = 0.6$ μm. In 45 nm PTM models, this inverter design has $\tau_{pd} \approx \tau_{pu}$, although not precisely identical. The PD transition is governed by n-FET current drive strength, and the PU transition by p-FET current drive strength. Hence, with different σV_{tn} and σV_{tp}, $\sigma\tau_{pu}$ may not be equal to $\sigma\tau_{pd}$.

A two-stage segment of an inverter chain is shown in Fig. 6.18b. With a rising input, the PD and PU delays are mainly governed by the current drive strengths of n-FET N1 and p-FET P2. Similarly, with a falling input, the PU and PD delays are mainly governed by p-FET P1 and n-FET N2. Although the two n-FETs and the two p-FETs are nominally identical by design, random V_t variations will cause their properties to be different. Hence, τ_{pd} and τ_{pu} vary in different two-stage segments in the chain, resulting in a variation in the delay across two stages $\tau(2)$.

The signal delay going through the two stages is

$$\tau(2) = \tau_{pd} + \tau_{pu}, \tag{6.8}$$

while the variation in $\tau(2)$ can be characterized by the sum of the variances,

$$\sigma\tau(2) = \sqrt{\left(\sigma\tau_{pd}^2 + \sigma\tau_{pu}^2\right)}. \tag{6.9}$$

Note that these expressions are the same for either a rising or a falling edge at the input of the two-stage segment. Furthermore, Eq. 6.9 holds independent of the relative magnitudes of τ_{pd} and τ_{pu}. This two-stage segment well fits the description of a single stage of the ideal hypothetical chain of nominally identical logic gates previously discussed. To facilitate investigation of variability in chains of inverting stages of length n, it is convenient to rewrite Eqs. 6.5–6.7 in terms of $\tau(2)$ and $\sigma\tau(2)$.

$$\tau(n) = \frac{n}{2} \times \tau(2), \tag{6.10}$$

$$\sigma\tau(n) = \sqrt{\frac{n}{2}} \times \sigma\tau(2), \tag{6.11}$$

$$\frac{\sigma\tau(n)}{\tau(n)} = \sqrt{\frac{2}{n}} \times \frac{\sigma\tau(2)}{\tau(2)}. \tag{6.12}$$

Equations 6.10–6.12 above are valid for all even values of n, and for odd values of n in the limit of large n.

The average delay per stage is given by $\tau(2)/2$. This is differentiated from the average delay of nominally identical stages τ_p used elsewhere in this book, as τ_p could be an average across any number of stages. In the following discussion on random variability for a path with n stages, the average delay/stage is estimated as $\tau(n)/n$, where the variable n denotes the number of stages.

Circuit simulations of delay chains of various identical logic gates are carried out to investigate the magnitude of the effects of random variability. The results are compared with the expressions in Eqs. 6.8–6.12. Four different delay chains are investigated, comprised of standard inverters, NOR3T, NAND3T, and NAND3B logic gates. Circuit schematics of a NOR3T, NAND3T, and NAND3B indicating the driving MOSFETs for PD and PU transitions are shown in Fig. 6.19a–c.

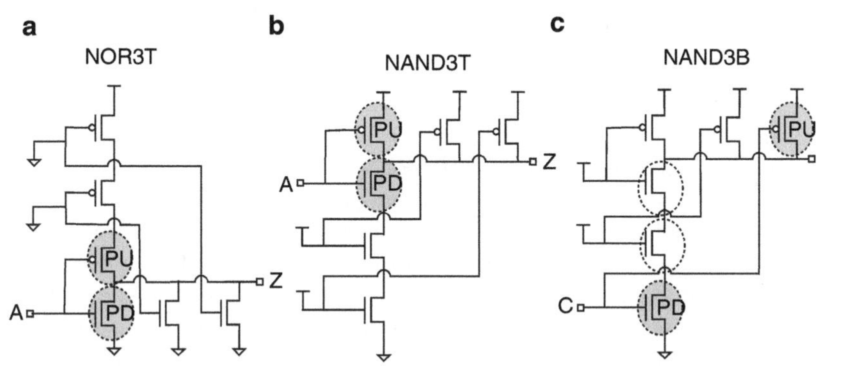

Fig. 6.19 MOSFETs in signal path for PU and PD transitions in (**a**) NOR3T, (**b**) NAND3T, and (**c**) NAND3B

The ratio W_p/W_n in each logic gate is selected to give $\tau_{pu} \approx \tau_{pd}$ while keeping $(W_n + W_p) = 1.0$ μm. The ratio W_p/W_n is 1.5 for an inverter, 4.0 for a NOR3, and 0.85

for a NAND3. Each stage has a FO = 4 current multiplier at its output node. Monte Carlo simulations for 500 cases are carried out with random variations only in V_t as described in Chap. 2.2.7, setting *delvtonom* = 0 and *delvtosigma* = 0.004/√WL.

For each chain, the PD and PU delays, τ_{pd} and τ_{pu}, are measured across two sequential single logic gates. The delay across two stages in series with input signal rising or falling gives $\tau(2)$ and thus $\tau(2)/2$, the average delay per stage. The delays follow normal distributions, and $\sigma\tau_{pu}/\tau_{pu}$, $\sigma\tau_{pd}/\tau_{pd}$ and $\sigma\tau(2)/\tau(2)$ are the fractional standard deviations of these distributions (Sect. 9.3.1). The results of these simulations are summarized in Table 6.3.

Table 6.3 Signal delays per stage and sigma/mean (%) for τ_{pu}, τ_{pd}, and $\tau(2)/2$ for a two-stage segment of a logic gate chain (FO = 4). 45 nm PTM HP models @ 1.0 V, 25 °C, 500 cases

Logic gate	W_p (μm)	W_n (μm)	τ_{pd} (ps)	τ_{pu} (ps)	$\tau(2)/2$ (ps)	$\sigma\tau_{pd}/\tau_{pd}$ (%)	$\sigma\tau_{pu}/\tau_{pu}$ (%)	$\sigma\tau(2)/\tau(2)$ (%)
Inverter	0.60	0.40	11.90	11.34	11.63	7.79	6.30	5.44
NOR3T	0.80	0.20	23.64	24.46	24.06	11.17	7.47	7.25
NAND3T	0.46	0.56	20.77	19.44	20.16	7.27	8.43	5.59
NAND3B	0.46	0.56	21.35	21.72	21.58	9.53	8.49	6.38

In NOR3T and NAND3T, similar to an inverter, the dominant effect on $\sigma\tau_{pu}$, and $\sigma\tau_{pd}$ is that of the switching p-FET or n-FET as indicated in Fig. 6.19a, b. The $\sigma\tau_{pu}/\tau_{pu}$ and $\sigma\tau_{pd}/\tau_{pd}$ qualitatively track with the MOSFET widths, i.e., the variation is larger for the narrower dominant MOSFETs in the transition. This trend is reversed in the NAND3B case where the non-switching n-FETs in the stack contribute significantly to the delay: the result is $\sigma\tau_{pd} > \sigma\tau_{pu}$, even though $W_n > W_p$.

An important observation from Table 6.3 is that although the average delay across the gates varies by nearly 2×, $\sigma\tau(2)/\tau(2)$ varies over a small range (5–7 %) and is nearly independent of the gate type.

In order to see the averaging effect in random variations in longer delay chains, Monte Carlo simulations are carried out to measure delays for 2, 8, and 18 stages. The average delay $\tau(n)/n$ and $\sigma\tau(n)/\tau(n)$ for the four different logic gates are listed in Table 6.4. Although $\tau(n)/n$ is independent of the number of two-stage segments, $\sigma\tau(n)/\tau(n)$ decreases by a factor of ~2 in four two-stage segments and by a factor of ~3 in 9 two-stage segments, as anticipated from Eq. 6.12. The net effect in a typical combinational path of 20 FO = 4 stages is a fractional standard deviation of ~2–2.5 %.

Table 6.4 Signal delays per stage $\tau(n)/n$, and $\sigma\tau(n)/\tau(n)$ (%) for $n = 2$, 8, and 18 stages (FO = 4). 45 nm PTM HP models @ 1.0 V, 25 °C, 500 cases

Logic gate (n)	W_p (μm)	W_n (μm)	$\tau(n)/n$ (ps)	$\sigma\tau(n)/\tau(n)$ (%)
Inverter (2)	0.60	0.40	11.63	5.44
Inverter (8)	0.60	0.40	11.62	2.87
Inverter (18)	0.60	0.40	11.62	2.01
NOR3T (2)	0.80	0.20	20.14	7.25
NOR3T (8)	0.80	0.20	20.18	3.70
NOR3T (18)	0.80	0.20	20.65	2.65
NAND3T (2)	0.46	0.56	20.16	5.59
NAND3T (8)	0.46	0.56	20.13	2.91
NAND3T (18)	0.46	0.56	24.15	1.99
NAND3B (2)	0.46	0.56	21.58	6.38
NAND3B (8)	0.46	0.56	21.61	3.21
NAND3B (18)	0.46	0.56	21.60	2.17

In Fig. 6.20, $\tau(2)/2$ and $\sigma\tau(2)/\tau(2)$ for an inverter are plotted as a function of $1/\sqrt{(W_n + W_p)}$. As expected $\tau(2)/2$ is nearly constant and $\sigma\tau(2)/\tau(2)$ varies linearly with $1/\sqrt{(W_n + W_p)}$.

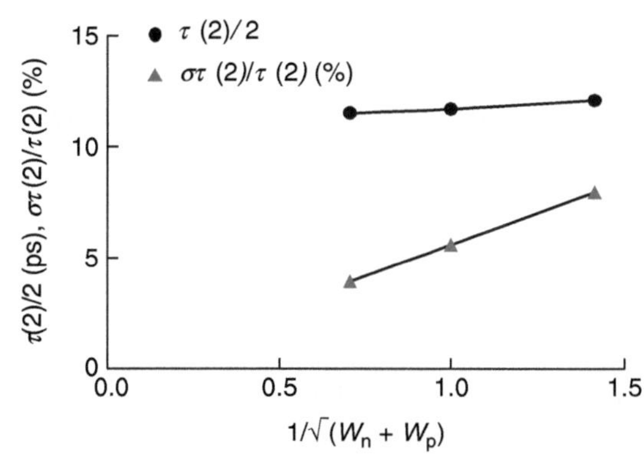

Fig. 6.20 Inverter (FO = 4) $\tau(2)/2$ and $\sigma\tau(2)/\tau(2)$ as a function of $1/\sqrt{(W_n + W_p)}$. 45 nm PTM HP models @ 1.0 V, 25 °C

The effect of random V_t variations in the presence of systematic V_t variations is studied by running Monte Carlo simulations in the ff and ss corners. The simulation results for a standard inverter, NOR3T and NAND3B are listed in Table 6.5. The systematic shift in V_t of ±0.06 V has a bigger impact on $\tau(2)/2$ than on $\sigma\tau(2)/\tau(2)$.

The net result is that $\sigma\tau/\tau$ varies as $1/\sqrt{(W_n + W_p)}$ and is a slow varying function of systematic V_t shifts. The spread in τ of a typical logic path due to random V_t variations is small (2.0–2.5 %) and relatively independent of the mix of logic gate types. In timing tools, fixed delay penalties using a delay multiplier m_{vtr} may be applied to all logic gates based on switching MOSFET device widths. As an

Table 6.5 Logic gate τ(2)/2 and στ(2)/τ(2) for inverter, NOR3T and NAND3B with systematic $|V_t|$ offsets of ±0.06 V. 45 nm PTM HP models at 1.0 V, 25 °C

Logic gate	W_p (μm)	W_n (μm)	Corner	τ(2)/2 (ps)	στ(2)/τ(2) (%)
Inverter (nominal)	0.60	0.40	nom	11.65	5.44
Inverter ($\Delta|V_t| = -0.06$ V)			ff	9.94	5.28
Inverter ($\Delta|V_t| = +0.06$ V)			ss	14.04	5.93
NAND3B (nominal)	0.46	0.54	nom	21.58	6.38
NAND3B ($\Delta|V_t| = -0.06$ V)			ff	18.06	6.30
NAND3B ($\Delta|V_t| = +0.06$ V)			ss	26.50	6.82
NOR3T (nominal)	0.80	0.20	nom	24.14	7.25
NOR3T ($\Delta|V_t| = -0.06$ V)			ff	20.22	7.05
NOR3T ($\Delta|V_t| = +0.06$ V)			ss	29.77	7.84

example, a value of 1.06 for m_{vtr} may be selected corresponding to a 6 % increase in delay due to random variations alone.

Measurement of variability in logic gates in silicon hardware can be accomplished by constructing and characterizing ROs with different numbers of stages. The period or frequency of an RO will vary with the number of stages $(2\alpha + 1)$ and the W_n and W_p of the switching MOSFETs. As shown in Fig. 6.21, as a signal edge

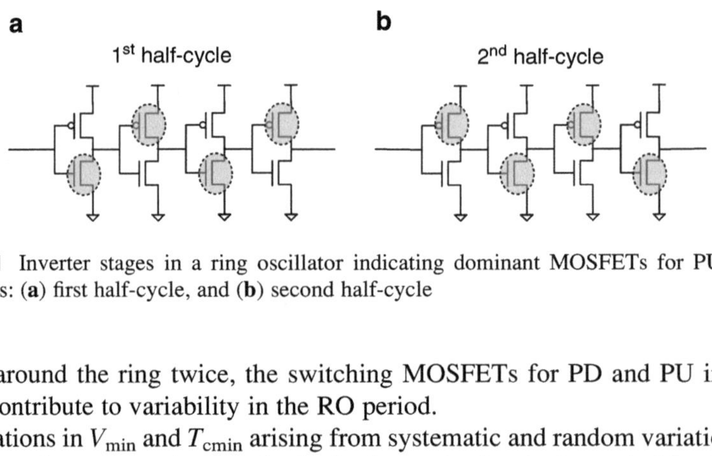

Fig. 6.21 Inverter stages in a ring oscillator indicating dominant MOSFETs for PU and PD transitions: (**a**) first half-cycle, and (**b**) second half-cycle

travels around the ring twice, the switching MOSFETs for PD and PU in all the stages contribute to variability in the RO period.

Variations in V_{min} and T_{cmin} arising from systematic and random variations in V_t and L_p are demonstrated using the logic data path circuit shown in Fig. 3.32 and reproduced here in Fig. 6.22 for reference. A signal launched by latch X1 propagates through a chain of 20 inverters (FO = 4) and arrives at the input of the second latch X2 after the signal edge is sharpened by two additional unloaded inverters in the path. If the signal at Z arrives before the falling edge of the dCLK signal in the second cycle, it will propagate to the node OUT. If the signal at Z arrives later, it will propagate to OUT in the third dCLK cycle. Implementation of this circuit to measure V_{min} and T_{cmin} in LTspice is described in detail in Sect. 3.4.2.

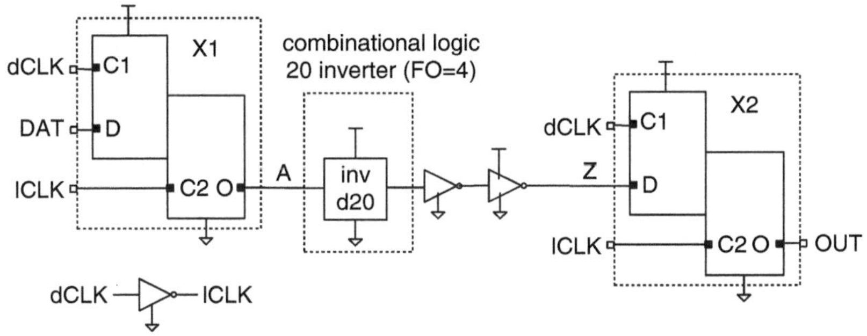

Fig. 6.22 Logic data path with a chain of 20 inverters (FO = 4)

The variations in V_{min} and T_{cmin} are simulated in three different scenarios:

(a) Simultaneous and identical variation of V_t or L_p in all n-FETs and p-FETs
(b) Independent systematic variations of V_t in n-FETs and in p-FETs
(c) Independent and random variations of V_t in all MOSFETs

The variation in T_{cmin} with V_t or L_p for case (a) where all n-FETs and all p-FETs get the same value of $\Delta|V_t|$ or L_p are shown in Fig. 6.23a, b. T_{cmin} increases with increase in $\Delta|V_t|$ or L_p as expected from the longer path delay through the inverter chain. A linear fit of the data removes some of the "wiggles" because of the $\Delta|V_t|$ or L_p step sizes in the SPICE simulations.

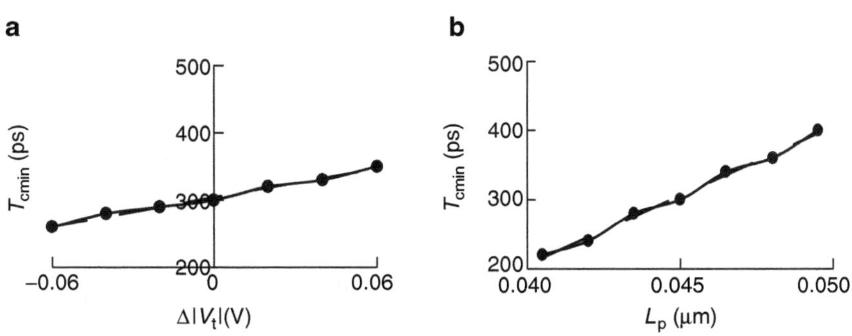

Fig. 6.23 T_{cmin} for the circuit in Fig. 6.22 as a function of simultaneous and identical change in (**a**) V_t and (**b**) L_p of all n-FETs and p-FETs in the combinational logic circuit. 45 nm PTM HP models @ 1.0 V, 25 °C

Monte Carlo simulations are conducted for cases (b) and (c) to obtain V_{min} and T_{cmin} distributions for systematic and random variations in V_t. The V_{min} distributions for 100 cases are shown in Fig. 6.24a, b for 45 nm PTM HP models. The clock cycle time is increased to 500 ps, lowering V_{min} into the 0.7–0.8 V range.

The σV_{min} is 0.017 V with $\sigma V_t = 0.02$ V for case (b) in Fig. 6.24a where all n-FETs in the inverter chain get the same $\Delta|V_t|$ and all of the p-FETs also get the same $\Delta|V_t|$ but independent of that of the n-FETs. This is a typical scenario for systematic V_t variations as discussed previously.

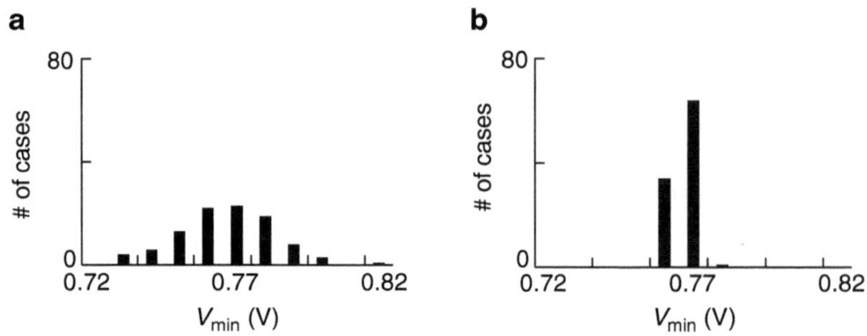

Fig. 6.24 V_{min} distributions for the circuit in Fig. 6.22: (**a**) case (b) for systematic variations with $\sigma V_t = 0.02$ V, and (**b**) case (c) for random variations alone. Monte Carlo simulations for 100 cases, 45 nm PTM HP models at $T_c = 500$ ps, 25 °C. Note that the X–Y scales and bin sizes are identical in both plots for visual comparison

As shown in Fig. 6.24b, the V_{min} distribution for case (c) is significantly narrower when random V_t values are assigned to each n-FET and each p-FET in the inverter chain.

In small analog circuits where matching characteristics of MOSFETs pairs are critical to functionality, Monte Carlo simulations may be carried out by varying the key parameters of each device individually. One such example is the current mirror circuit shown in Fig. 6.25, where the current through n-FET N2 is matched to the reference current through N1.

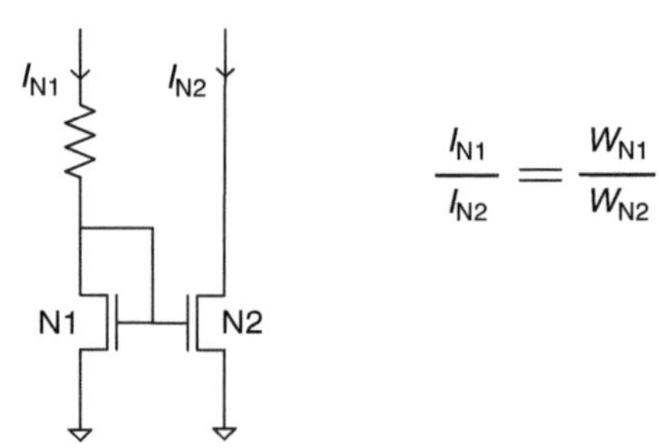

Fig. 6.25 Current mirror circuit

The V_t mismatch $\Delta|V_t|$ between the two n-FETs N1 and N2 in Fig. 6.25 is a function of the difference in their V_t values due to random variations, ΔV_{tr}. The distribution of $\Delta|V_t|$ is characterized by

$$\sigma\Delta|V_t| = \sqrt{\left(\sigma V_{\rm tr}^2 + \sigma V_{\rm tr}^2\right)} = \sqrt{2} \times \sigma V_{\rm tr}, \qquad (6.13)$$

and is $\sqrt{2}\times$ of that of a single device.

SRAM and DRAM memory cells use narrow width MOSFETs which are more prone to random V_t variations. The operation of a 6 T SRAM cell is described in Sect. 3.3.1. In determining the nominal static noise margins, it is assumed that the cell is perfectly symmetric and the MOSFETs on the left half of the cell are identical to the corresponding MOSFETs on the right half of the cell.

Random V_t variations in all the MOSFETs cause the 6T SRAM cell to naturally deviate from symmetry even if the physical layout is perfectly symmetric. The result is reduction in the static noise margins (SNMs) of the cell from the nominal case. As a reminder, the circuit schematic of a 6T SRAM cell is shown in Fig. 6.26a. LTspice simulations are carried out using methodology described in Sect. 3.4.1. Monte Carlo simulations are performed at the nominal corner with random V_t variations for all MOSFETs in the cell. The distribution of the SNM for hold is shown in Fig. 6.26b, illustrating reduction from the nominal value.

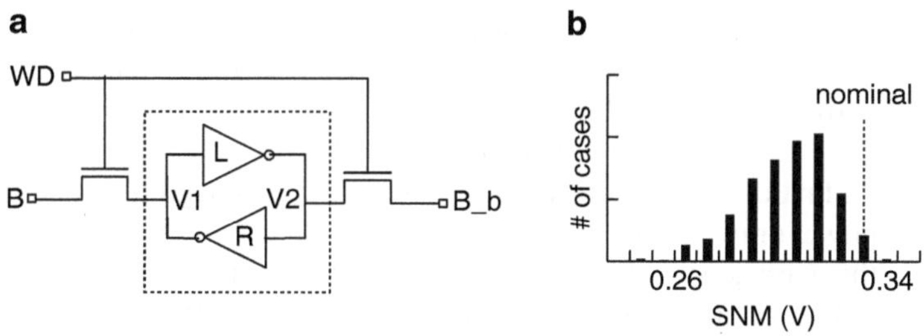

Fig. 6.26 (**a**) Schematic of an SRAM cell showing the *L* and *R* inverters of the latch, and (**b**) variation in static noise margin (SNM) of the latch in response to random V_t variations. $W_p = 0.1$ μm, $W_n = 0.3$ μm, 45 nm PTM HP models @ 1.0 V, 25 °C

Including random variability in timing tools may be accomplished by assigning a fixed penalty or guardband as a % of cycle time. In performance critical CMOS chips, the penalty may be tuned to MOSFET widths so that narrow width devices get a higher delay penalty than wider devices. Analog and memory circuits are simulated in many additional corners and Monte Carlo simulations of small logic blocks are carried out for a robust design.

6.5 Summary and Exercises

Sources of variations in silicon technology, circuit design, and operating conditions of CMOS chips are described. Characterization methods for mapping variability induced by these different sources are discussed. Variations, although unavoidable, can be minimized by careful design and by silicon process improvements based on feedback from electrical test data. Circuit design practices accommodate both systematic and random components of variability through design margins.

Exercises 6.1–6.3 investigate power supply voltage and temperature variations. Exercise 6.4 is based on variability mapping through imaging analysis. An optimized wafer map is generated in Exercise 6.5. Exercises 6.6–6.10 involve variability assessment using circuit simulations.

6.1. Variation in power supply voltage translates directly into variation in circuit delay. One component of this variation is the static response of the power distribution system to variation in circuit activity.
   (a) Create a simplified circuit diagram of a power grid with series resistances R_s in both the V_{DD} and GND busses. With the power supply voltage set at 1.0 V, the chip draws 2 A in the quiescent state and 80 A with clocks running. What is the value of R_s for a 5 % V_{DD} droop with the clock on?
   (b) What is the change in measured τ_p of an inverter (FO = 3) with and without clocks running?

6.2. Another source of power supply voltage variation is very dynamic in nature and associated with turning large circuit blocks on or off. Consider the following circuit diagram for power delivery to the circuitry on a 4-core microprocessor chip. Initially assume that L is very large (>1 H), $C_1 = C_2 = C_3 = R_s = R_{sc} = 0$ and $R_{L2} = 10^9\ \Omega$.
   (a) At first the current drawn is V_{DD}/R_{L1}. If P_{off} is negligible and the active power drawn by the chip is CV_{DD}^2f, calculate the value of R_{L1} in terms of the average switching capacitance C and the frequency f.
   (b) If $V_{DD} = 1.0$ V, and the initial total power is 50 W with only two cores on, what is the numerical value of R_{L2}, and how much current is flowing through R_{L2}?
   (c) After a long period of time, a third core is turned on over a few machine cycles changing the value of R_{L2} from $10^9\ \Omega$ to $2\times\ R_{L1}$. Simulate this event and generate a plot of V_{ch} vs. time (Hint: Model variable resistor R_{L2} as a behavioral current source with parameters generated by an auxiliary circuit).
   (d) Set $C_2 = 0.5\ \mu F$, representing the device and wire capacitance across the power grid. Re-simulate and determine how long the stored charge can keep $V_{ch} > 0.9$ V.
   (e) Add $C_3 = 0.5\ \mu F$ and $R_{sc} = 2 \times 10^{-3}\ \Omega$ to represent additional decoupling capacitance added across the chip power grid and re-simulate the circuit. Compare with results from (d).
   (f) Assuming that the power supply provides V_{DD} independent of the current drawn with a 1.0 ms time constant, come up with reasonable values of package components L, C_1 and R_s that will minimize the V_{ch} disturbance as the third core is turned on.

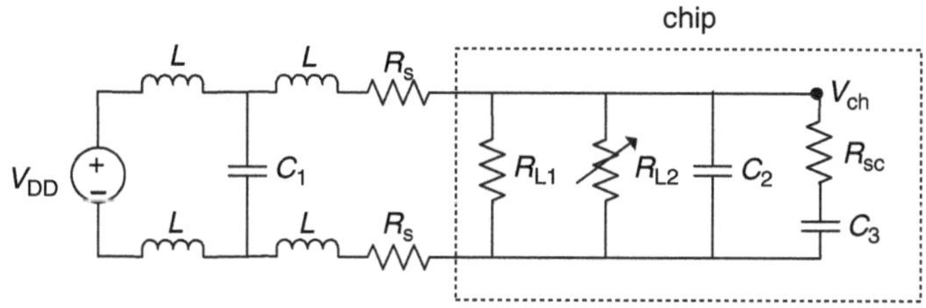

6.3. A temperature monitor on a high-performance chip is calibrated at set-points of 25 and 75 °C with the chip in the quiescent state. The average R_{th} is 0.1 °C/W and $P_{off} = 20$ W at 25 °C.
   (a) What is the error in temperature measurement at 25 °C?
   (b) What is the error in temperature measurement at 75 °C with higher P_{off}? Hint: Use the P_{off} model in Chap. 4.

6.4. A PICA image of chip leakage current shows different average brightness in four quadrants as indicated in the figure below. The intensity indicates relative leakage currents as shown. There is one inverter (FO = 3) RO in the middle of each quadrant.

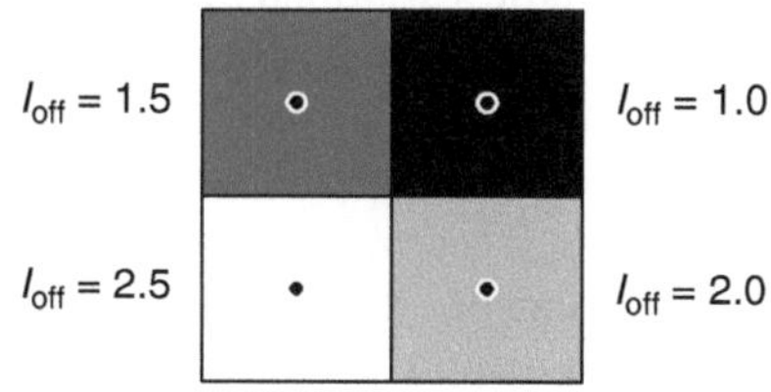

   (a) Based on the I_{off} ratios, make a quantitative assessment of the relative frequencies of the ROs in each of the four quadrants. Hint: assume SS = 100 mV/decade.
   (b) All the chips on a wafer display this behavior. What is at least one possible root cause?
   (c) Chips in the center of the wafer are relatively uniform while edge chips show various AcC variation patterns. Give at least one explanation for this.

6.5. The silicon foundry recommends increasing the wafer edge exclusion zone from 4 to 6 mm for the second pass design.
   (a) Estimate the number of chips on a 300 mm wafer with 4 and 6 mm edge exclusions for chip sizes of (1) 5 mm × 5 mm, and (2) 20 mm × 15 mm. Assume a 0.1 mm wide scribe-line between chips.
   (b) What will be the % reduction in chip count for the two chip sizes in increasing the edge exclusion from 4 to 6 mm?
   (c) What electrical test data from the first pass design would you need to show that increasing edge exclusion to 6 mm has no measurable benefit?

6.6. Set up a 10-stage delay chain simulation in the nominal corner using a standard inverter (FO = 4) stage.

(a) Measuring across two central inverters determine the value of $\tau(2)$ for values of V_{DD} ranging from $1.3 \times V_{DD}$ down to $0.7 \times V_{DD}$ in steps of $0.1 \times V_{DD}$.

(b) Run Monte Carlo simulations for 100 cases at each V_{DD} in which the V_ts of all MOSFETs vary randomly following Eq. 6.1 with $A_{vt} = 0.004$ V μm. At each voltage determine $\sigma\tau(2)$ and $\sigma\tau(2)/\tau(2)$ and compare the values. What do you conclude?

6.7. A logic path simulation was first described in Sect. 3.4.2 and has subsequently been a component of several exercises. The corresponding circuit diagram is redrawn in Fig. 6.22. Simulation results in the face of various variations are shown in Figs. 6.23 and 6.24.

(a) Set up a similar simulation with combinational logic consisting of 20 standard inverters (FO = 4) and determine *vminir*, *vminif*, *vminlo*, and *vminhi* using 10 mV voltage steps. Center the voltage search range on V_{min}.

(b) With the configuration set up in (a) run Monte Carlo simulations for 100 cases with systematic variations in V_t over the $\pm 3\sigma V_t$ range.

(c) Generate histograms (analogous to Fig. 6.24a) of *vminir*, *vminif*, *vminlo*, and *vminhi*.

(d) Examine the data used to create the histograms carefully and explain any differences.

6.8. The delay of a circuit path can be altered by circuit asymmetry. The impact of random variations can also display asymmetry effects.

(a) Reconfigure the logic path circuit used in Problem 6.7 by replacing one of the standard inverters in the middle of the path with another having $W_n = 0.1$ μm and $W_p = 0.9$ μm. Run simulations with a voltage step of 5 mV, center V_{min} in the search range and determine *vminir*, *vminif*, *vminlo*, and *vminhi*. Explain how the results differ from those in Exercise 6.7a.

(b) Swap the skewed inverter used in (a) with a standard inverter. With a 5 mV step size, run Monte Carlo simulations for 100 cases, this time with random values of V_t for all MOSFETs varying independently of each other.

(c) Again generate histograms of *vminir*, *vminif*, *vminlo*, and *vminhi*.

(d) Explain the differences among the histograms in (c). Which one represents the true V_{min} of this path?

6.9. The impact of V_t variations on an SRAM cell is very different than for a logic path. A methodology for determining the static noise margins (SNM) of an SRAM cell was introduced in Sect. 3.4.1 and was used to generate the histogram shown in Fig. 6.26b. This figure illustrates the variation in the SNM of the cell's latch in the face on random variations in V_t values of all of the MOSFETs.

(a) Set up a similar Monte Carlo simulation for an SRAM cell and recreate the histogram shown in Fig. 6.26b with 100 cases.
(b) Examine the simulation files and determine the V_t values that gave the worst observed SNM for the latch. Qualitatively explain why this set of V_t values leads to a very low SNM.
(c) Re-simulate the cell with systematic rather than random variations in V_t values. Plot a histogram of the data similar to the one created in (a). Compare and contrast the results.

6.10. An SRAM cell can be more prone to failure during read and write operations. SRAM read and write static noise margins are described in Sect. 3.2.1
(a) Reconfigure the SRAM latch static noise margin simulation to determine the read static noise margin (RSNM).
(b) With this new arrangement run Monte Caro simulations for 100 cases varying all V_t values randomly and independently. Plot a histogram of the resultant RSNM values and determine the range over which the RSNM varies.
(c) Repeat the above exercise for the write static noise margin (WSNM) of the same SRAM cell.

References

1. Bansal A, Rao RM (2010) Variations: sources and characterization. In: Bhunia S, Mukhopadhyay S (eds) Low-power variation tolerant design in nanometer silicon, Chapter 1. Springer, Berlin
2. Orshansky M, Nassif S, Boning D (2007) Test structures for variability. In: Design for manufacturability and statistical design: a constructive approach. Springer, Berlin
3. Weste NH, Harris D (2010) CMOS VLSI design: a circuit and systems perspective, 4th edn. Addison-Wesley, Boston
4. Jaeger RC (2001) Introduction to microelectronic fabrication, vol 5, 2nd edn, Modular series on solid state devices. Prentice Hall, Upper Saddle River
5. Campbell SA (2007) Fabrication engineering at the micro and nanoscale. Oxford University Press, Oxford
6. Polonsky S, Bhushan M, Gattiker A, Weger A, Song P (2005) Photon emission microscopy of inter/intra chip device performance variations. Microelectron Reliab 45:1471–1475
7. Yazawa K, Kendig D, Christofferson J, Marconnet A, Shakouri A (2012) Fast transient and steady state thermal imaging of CMOS integrated circuit chips considering package thermal boundaries. In: IEEE 13th ITHERM conference, pp 1405–1411
8. Gattiker A (2008) Unravelling variability for process/product improvement. In: Proceedings of the international test conference, ITC'08, pp 1–9
8. Die per wafer calculator. http://mrhackerott.org/semiconductor-informatics/informatics/toolz/DPWCalculator/Input.html. Accessed 21 Jul 2014

Electrical Tests and Characterization in Manufacturing

7

Contents

7.1 Digital CMOS Chip Tests 242
7.1.1 Test Flow 243
7.1.2 Test Equipment 245
7.1.3 DC and AC Parametric Tests 246
7.1.4 Structural Faults and ATPG 246
7.1.5 IDDQ Tests 251
7.1.6 DFT and Diagnostics 254
7.1.7 Scan Design 255
7.1.8 Built-in Self-Test 257
7.1.9 Boundary Scan 258
7.1.10 Measurements of T_{cmin}, V_{min}, and AC Power 259
7.2 Yield 260
7.2.1 Defect Limited Yield 261
7.2.2 Cycle Time Limited Yield 263
7.3 Failure Analysis 265
7.4 Product Chip Characterization 267
7.4.1 Silicon Manufacturing Line Tests 267
7.4.2 Silicon Process-Split Hardware 269
7.4.3 Embedded Process Monitors 270
7.4.4 Aggregate Behavior 277
7.4.5 Silicon Manufacturing Process Window 278
7.5 Adaptive Testing and Binning 278
7.6 Summary and Exercises 281
References 284

Electrical tests are conducted during manufacturing for verification of all functions of individual CMOS chips and systems. In logic testing, test vectors are applied to inputs, and output responses are compared with expected results. Memory tests are conducted by writing and reading each individual cell in all arrays. Design for testability (DFT) features incorporated in the chip design help improve test efficiency and facilitate debug. Scribe-line tests of circuit components during silicon

M. Bhushan, M.B. Ketchen, *CMOS Test and Evaluation*,
DOI 10.1007/978-1-4939-1349-7_7

manufacturing, on-chip process, voltage and temperature monitors, and characterization and modeling of the aggregate behavior of the chip provide physical insight and assist rapid failure diagnostics and resolution. Adaptive testing methods for managing silicon process variations and yield enhancement are becoming increasingly important with shrinking design and profit margins for chips manufactured in advanced CMOS technologies.

Electrical testing continues through various steps in CMOS chip and product manufacturing as outlined in Chap. 1 (Fig. 1.3). Essential supporting activities related to test include DFT, test-code generation, test equipment maintenance, characterization, and debug. Rejecting defective chips early in the manufacturing cycle with minimum impact on throughput is an essential part of the overall test strategy. Getting to the root causes of failures and finding workable resolutions in a timely fashion minimizes further production of defective hardware. Test errors that may result in throwing away good chips or passing bad chips which may fail in the hands of the customer must be eliminated early in the production cycle. Hierarchical testing and model-to-hardware correlation help build a robust test strategy for maximum yield.

This chapter focuses on test methods for digital CMOS circuits and characterization techniques to bring together the learning from CMOS circuit components, on-chip monitors, and functional tests. An introduction to parametric, structural, and functional testing of CMOS chips along with DFT features is given in Sect. 7.1. Yield loss and modeling are discussed in Sect. 7.2. Methods of failure analysis are covered in Sect. 7.3. Chip characterization and correlating scribe-line test data with that from on-chip monitors and parametric tests are described in Sect. 7.4. Adaptive test methods and binning strategies to improve overall yield are reviewed in Sect. 7.5.

Detailed descriptions of VLSI testing can be found in textbooks [1–3]. There are two books that specifically focus on design and test of scribe-line test structures for device and circuit components [4, 5]. Several references to journal articles reviewing CMOS test techniques are also cited in the chapter.

7.1 Digital CMOS Chip Tests

Logic and memory functions are verified through simulations prior to shipping chip design data to the silicon foundry. Functional verification is carried out in early hardware to ensure that the design has been faithfully reproduced in silicon. At this early stage, if applicable, verification is extended to board and system level to catch any design issues in integrating the chips into the final product. In the following discussions we will assume that the chip logic is correct by design, and that test fails in manufacturing result from defects and variability in silicon processing or from test errors.

A generic test setup is shown in Fig. 7.1. An automated test equipment (ATE) unit supplies power, clock signals, and input test vectors for digital testing to the I/O pins on the chip. Measured data are fed back to the ATE. DC currents and voltages are measured at selected pins on the chip. The digital output signals are compared with expected results for each set of input test vectors. A test fail is detected when either a DC level exceeds a predetermined threshold or a miscompare in the digital output signature occurs. Standard testing is aborted at the first fail. Further investigations of failing chips may be carried out for design, silicon process, and test debug.

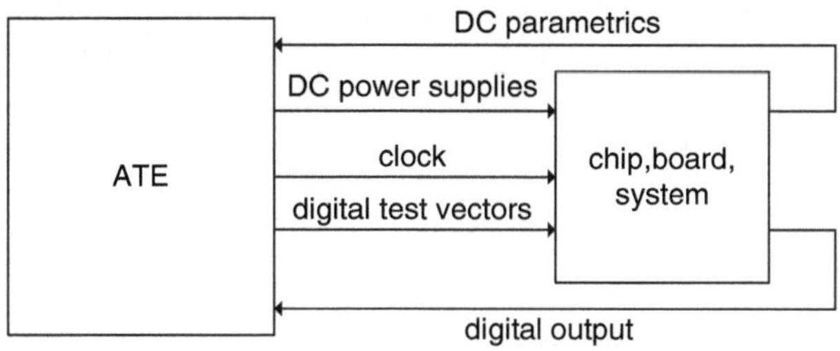

Fig. 7.1 Schematic of a setup for DC parametric and digital testing

7.1.1 Test Flow

Manufacturing test flow is shown in Fig. 7.2. Electrical tests begin during silicon processing. Test structures in the scribe-line are measured at different stages of the manufacturing sequence starting with delineation of the first metal layer when contacts can be made to MOSFET terminals. Product chips are first tested on the wafer after completion of processing, including that of I/O pads or solder balls for flip-chip bonding. The wafer is diced to separate the chips, and only those chips passing all of the initial tests are packaged. Packaged chips must pass additional tests before moving on to board or system assembly and acceptance testing by the customer, if applicable. For some product applications, chips may be subjected to accelerated stress testing (burn-in) to improve reliability in the field (Sect. 8.3.2)

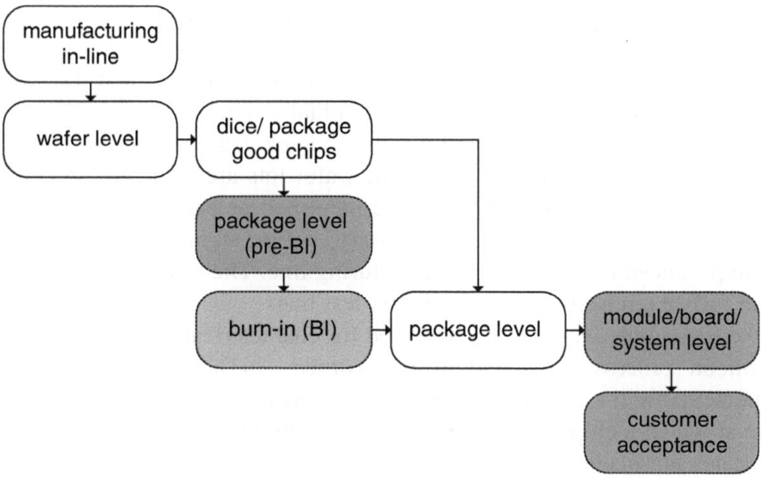

Fig. 7.2 CMOS product chip test flow. Shaded rectangles indicate optional tests for medium to high cost chips

The sequence of tests conducted at each stage is shown in Fig. 7.3. DC parametric tests are carried out to identify chips with catastrophic failures such as opens and shorts in the power grid and I/O pins. Input clock signals are then applied to activate the logic. Essential elements include scannable clocked storage elements and register files, I/O drivers and receivers, on-chip clock generators, and selected embedded monitors. These and other nonredundant elements must pass before commencing full digital testing.

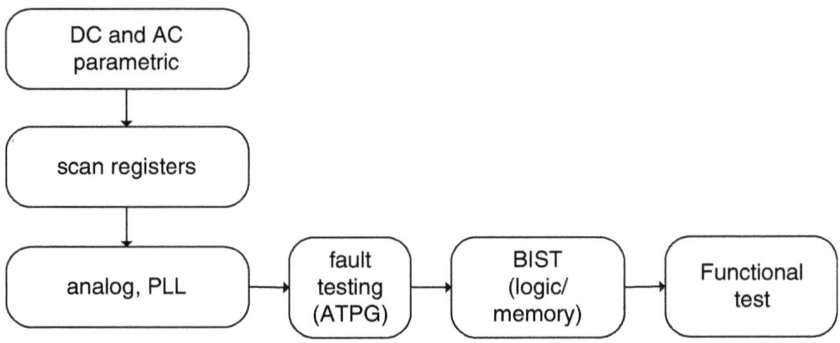

Fig. 7.3 Test sequence of major test types

Structural tests using automated test pattern generation (ATPG) algorithms check the node voltages at the closest observable location, typically the output of a clocked storage element. Functional testing verifies the chip logic, analogous to logic verification in design. Built-in self-test (BIST) allows test results to be compared on-chip. BIST for on-chip memory exercises the address sequence and read and write operations of each cell in the array. Error correction is applied based on test data to disable defective cells and replace them with good ones. Full functional tests may be carried out at chip and board or system level. The extent of tests in each of these categories varies with the stage at which they are being conducted, and is customized for each individual product.

The test focus evolves as production begins and transitions to full-scale manufacturing. Chips from the first set of wafer lots are dedicated to chip design verification. Major design flaws must be identified as early as possible and new corrected photomasks obtained, if necessary, before significant volumes of hardware are produced in the silicon manufacturing line. The test codes and procedures are also fully exercised to identify and fix test bugs.

In the second phase, tests are conducted for characterization of the chip behavior with silicon process splits and at test corners. In addition to pass/fail tests, measurements of the maximum frequency of operation f_{max}, minimum operating voltage at which a chip remains functional V_{min}, and power in the off, standby and active states are made. The silicon process is tuned for optimum yield and performance. The data may be correlated to model predictions for learning and providing feedback for essential design modifications, if needed, in present and future designs.

In preparation for full-scale production tests, test corner definitions are established to include adequate margins for variations in test equipment, environmental conditions, and degradation in the properties of circuit components over time. A discussion of these margins, referred to as guard-bands, is included in Sect. 8.3.3. In the final phase, a subset of the tests is used in production, and chips are sorted as pass, fail, or in some cases as partially good.

7.1.2 Test Equipment

The schematic of a test station is shown in Fig. 7.4. A state-of-the-art test station is a very complex and expensive system. The ATE houses multiple power supplies, a reference clock, as many as 2,048 high performance digital channels for driving and receiving signals, and large buffers for data storage and retrieval. The voltage levels and signal timing in each channel connected to a chip or package I/O can be controlled independently. A set of input test vectors defined in the test program along with corresponding predicted output responses is loaded in the tester memory for pass/fail determination on the fly.

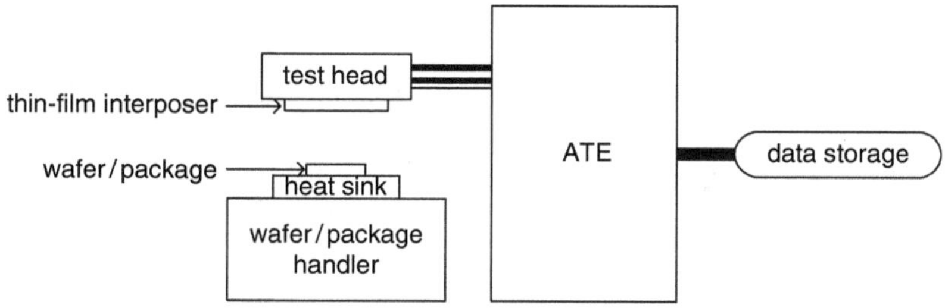

Fig. 7.4 Schematic of a test setup for wafer or package level test

The ATE is coupled to the chip through a test head in conjunction with a wafer or package handler. A wafer handler automatically loads a wafer from a cassette onto a thermally controlled chuck. The associated test head contains a thin-film interposer for making contact to the chip I/Os and electronics for connecting to the ATE. An optical alignment system positions the wafer with the required degree of precision with respect to the thin-film interposer in the test head. For testing of packaged chips, a handler loads and unloads the packages, and there is a different test head for connecting to the ATE.

The locations of known good die (KGD) on a wafer are either marked with ink or stored in a file for sorting good chips after dicing. Chip identification (chip id) by wafer and (x, y) location may be stored in on-chip ROM during wafer test. The electronic chip id and other vital information such as calibration data are retrieved in all subsequent test stages for tracking and characterization.

7.1.3 DC and AC Parametric Tests

DC parametric tests are conducted to eliminate chips with gross defects. A short in a power supply grid in the quiescent state results in a current in excess of the maximum IDDQ limit. A large current may overheat and ultimately melt wires. In order to prevent possible damage to the thin-film interposer and package I/O connections, initially a small voltage (~100 mV) is applied to the power grid. A short is detected if the IDDQ exceeds a preset limit, and the chip is rejected. In addition maximum current drawn on each pin is clamped to prevent overheating of contacts and wires, and to prevent thermal runaway.

All I/O pins are tested for opens and shorts by forcing a current out of the pad and comparing the measured voltage to preset limits. As an example, for the circuit shown in Fig. 3.2a, the measured voltage is determined by the forward bias voltage of the ESD diode (~0.7 V) and the *IR* drop in the input resistor. A smaller than expected voltage indicates a short and a larger than expected voltage indicates an open. I/O pins are screened for leakage currents between V_{DD} and GND and also from pin-to-pin. The signal I/Os are tested for output current drive capability and output voltage levels in the low and high states.

7.1.4 Structural Faults and ATPG

Structural testing is carried out to verify the logic structure of the design. It relies on fault models to represent physical defects introduced in silicon processing. These fault models are used in circuit simulations to generate test patterns for fault detection. An advantage of structural tests is that test patterns can be generated automatically using the design data for any of a wide range of chip designs with subsequent analysis of node voltages at the gate level. ATPG algorithms are used for detecting a large variety of structural faults. For technologies with low MOSFET leakage currents, IDDQ testing has also been used to detect the presence of faulty circuits as described in Sect. 7.1.5.

There are a number of different kinds of faults that may occur in circuits due to silicon process-induced defects. One example of a fault model is "stuck-at" in which a fixed voltage is assigned to a signal node in a circuit. A "stuck-at" fault occurs when an input or output signal level of a logic gate is stuck at a low or a high voltage: stuck-at-0 (s-a-0) is a fault where the voltage level stays fixed at "0" and stuck-at-1 (s-a-1) is a fault when the voltage stays fixed at "1". The response of the circuit is simulated for all combinations of s-a-0 and s-a-1 at each input and output node of a gate in combinational logic. The input combinations that result in a logic error being propagated to an observable node are identified and used for generating test vectors. The observable node is typically the output of a clocked storage element.

The main categories of faults and their possible impacts are summarized in Table 7.1. A stuck-at fault as described above may cause a logic error. A resistive short between signal wires may result in node voltages at intermediate levels between "0" and "1". High-resistance wires increase signal propagation delay and rise and fall times causing timing errors. Shorts in MOSFET gate-oxide add to the leakage current. Examples of these fault types in a NAND2 and in a small combinational logic block are subsequently discussed.

Table 7.1 Types of defects and their impact

Defect	Possible impact
Logic gate input/output stuck-at	Logic malfunction
MOSFET stuck-open or stuck-short	Logic malfunction, higher IDDQ
Bridging (resistive shorts)	Logic malfunction, higher IDDQ
High-resistance wires or vias	Increased delay and transition times
Gate-oxide shorts	Higher IDDQ

In Fig. 7.5a, input A of the NAND2 is s-a-1. The truth table shows output Z for a correctly functioning NAND2 and one with input A s-a-1. The fault is detected only with inputs A = "0", B = "1". In Fig. 7.5b, input C of a small combinational logic block is s-a-1. The internal node Y2 is then s-a-0. From the truth table, the fault is propagated to output Z for 5 out of 8 input vectors.

a

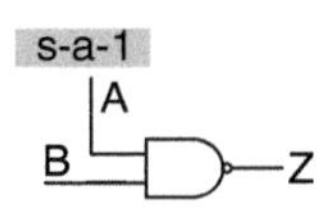

$Z = \overline{A \bullet B}$

			A s-a-1
A	B	Z	Z*
0	0	1	1
0	1	1	0
1	0	1	1
1	1	0	0

b

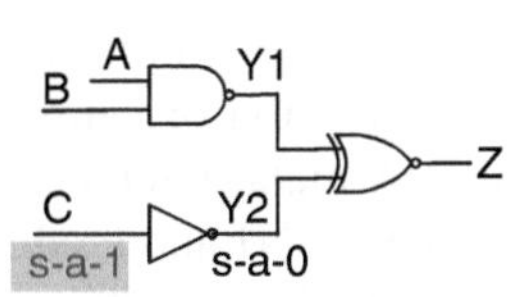

$Z = \overline{(\overline{A \bullet B}) \oplus \overline{C}}$

						C s-a-1	
A	B	C	Y1	Y2	Z	Y2*	Z*
0	0	0	1	1	1	0	0
0	0	1	1	0	0	0	0
0	1	0	1	1	1	0	0
0	1	1	1	0	0	0	0
1	0	0	1	1	1	0	0
1	0	1	1	0	0	0	0
1	1	0	0	1	0	0	1
1	1	1	0	1	0	0	1

Fig. 7.5 Circuit schematic and truth table for (**a**) NAND2 input A s-a-1 and (**b**) small combinational circuit block with input C s-a-1

A stuck-short or stuck-open fault occurs when a MOSFET is permanently in an on- or off-state. This type of fault may cause a node voltage to float to an intermediate level between a "1" and a "0". A stuck-short fault in a NAND2 gate is shown in Fig. 7.6. The gate of the top n-FET, N1, is disconnected from the input pin A and stuck at "1" and N1 stays on at all times. A fault may occur when with A = "0" and B = "1", the output Z does not match its expected value of "1". In this case, p-FET P1 and n-FETs N1 and N2 are all in the on-state. The voltage level of the output Z depends on the on-state resistance ratio R(P1)/{R(N1) + R(N2)}. For our standard NAND2 ($W_p = 0.5$ μm, $W_n = 0.5$ μm), the voltage level at Z is 0.56 V when $V_{DD} = 1.0$ V. This voltage level may be sufficient to switch the following logic gate in a path at an increased signal delay, and the fault may not be propagated to an observable location at long-cycle times.

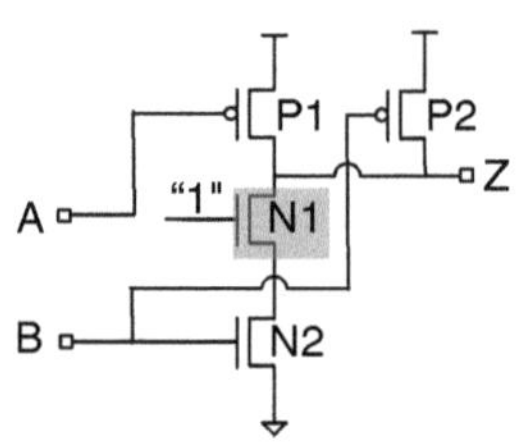

$Z = \overline{A \bullet B}$

A	B	Z	N1 stuck-on Z*
0	0	1	1
0	1	1	?
1	0	1	1
1	1	0	0

Fig. 7.6 Defect in n-FET N1 of a NAND2 gate turning it permanently "on" (stuck-on, or stuck-short) and truth table

In Fig. 7.7 three different physical defects in NAND2 gates and the impact on logic functions of the gate are shown. In Fig. 7.7a, a metal puddle shorts the source and drain of the top p-FET, P1, in the NAND2. This connects the output Z directly to the V_{DD} terminal and Z is s-a-1. The fault is detected with A = "1" and B = "1".

In Fig. 7.7b, c two different H0 via induced defects are shown. The H0 vias are the smallest dimension vias and incomplete etching of the via hole can increase the via resistance outside of the normal resistance distribution or cause an electrical open. A missing via in Fig. 7.7b disconnects input A metal interconnect M1 from the gate nodes of p-FET P1 and n-FET N1. With gates of N1 and P1 floating, IDDQ is very high when input B = "1". The fault is detected with A = "0" and B = "1" as node *X* floats to a voltage >$V_{DD}/2$. In Fig. 7.7c, the vias on the source side of the bottom n-FET N2 have a much higher resistance than nominal. The added series resistance to N2 lowers its current drive strength causing an increase in the signal propagation delay when, for example, with A = "1", B transitions from a "0" to a "1" turning N2 on.

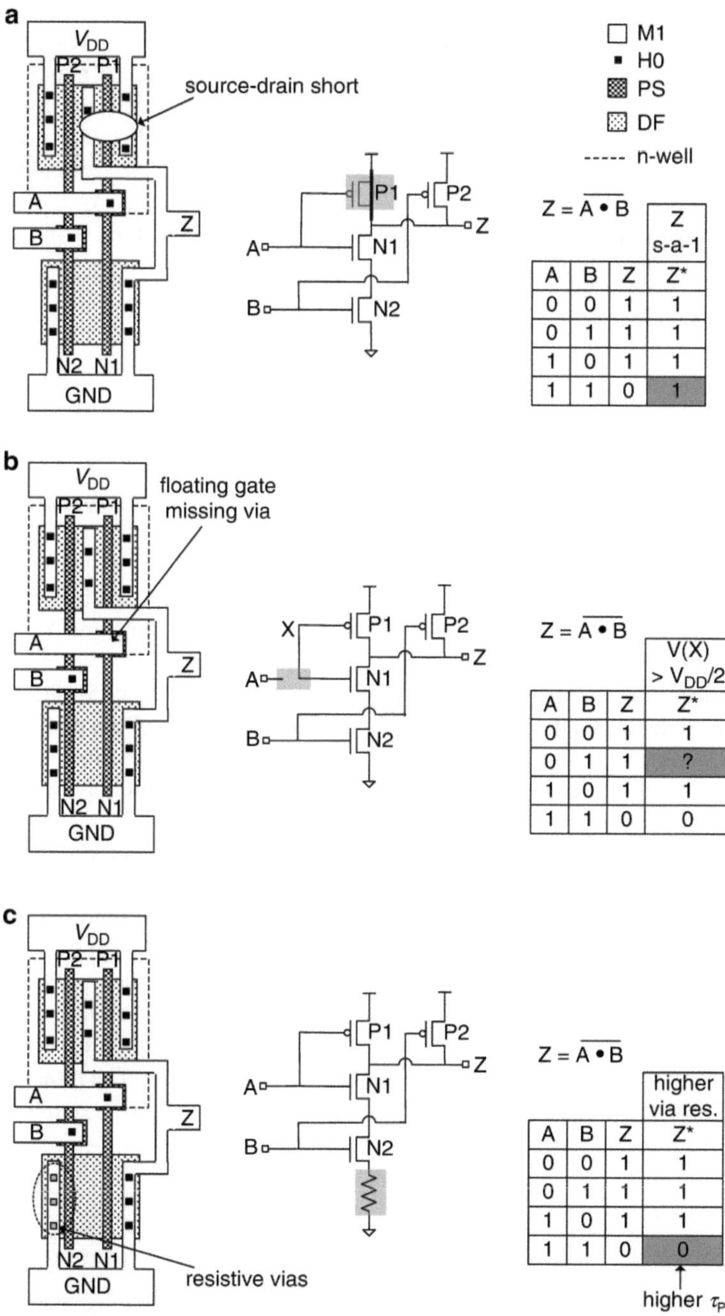

Fig. 7.7 Three types of defects in a NAND2 gate: (**a**) metal short, (**b**) floating MOSFET gates, and (**c**) resistive vias

Defects in signal lines may act as resistive shorts or as high-resistance interconnects. In Fig. 7.8a, the outputs of two NAND2 gates have a resistive bridge between them. This type of fault is detected when the logical outputs of the two gates, Z1 and Z2 are complementary. One such combination is A = "1", B = "1", C = "0", and D = "0", giving Z1 = "0" and Z2 = "1" in a fault-free circuit. In the presence of a shunt resistance R_{sh}, the voltage levels at Z1 and Z2 are determined by the value of R_{sh} and MOSFET strengths in the two NAND2 gates. With $R_{sh} = 0$, Z1 and Z2 are at ~$V_{DD}/2$ for equal MOSFET strengths and the fault may or may not be propagated to the subsequent stages.

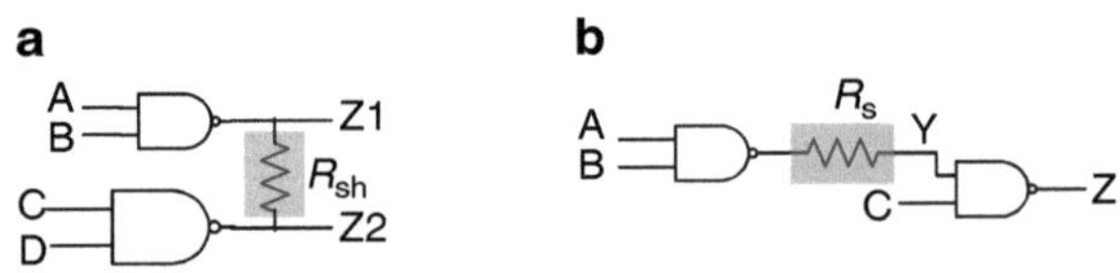

Fig. 7.8 (**a**) Low-resistance R_{sh} between outputs of two NAND2 gates, and (**b**) series resistance R_s at node Y

In Fig. 7.8b, additional series resistance R_s in the signal path increases the propagation delay and transition times of the voltage signal at node Y. This type of fault may only be detected when a timing failure occurs.

A short or pinhole in the thin gate-dielectric increases the gate leakage current. Such gate-dielectric faults are "soft" defects. If the resistance of the pinhole short is high (>10^6 Ω), it has little impact on logic functionality or performance. Often these defects degrade over time causing higher IDDQ and ultimately, with the gate shorted to source, drain, or body, a logic error may occur.

The total number of possible faults required to build a fault model is a function of the number of external and internal signal lines or nodes in a logic block. In Fig. 7.5a, the NAND2 gate has six possible stuck-at faults with inputs A and B and output Z each s-a-0 or s-a-1. The logic block in Fig. 7.5b with inputs A, B, and C, intermediate nodes Y1 and Y2, and output Z has 12 possible stuck-at faults. As the size of circuit block increases, the number of input vectors to build a fault model also increases rapidly. In a logic circuit block, such as the 64 bit adder in Fig. 7.9a, there are 129 (64 + 64 + 1) unique inputs and 65 unique outputs. The total number of input patterns is then 2^{129} (~10^{40}) generating a total of 2^{65} (~10^{20}) output patterns.

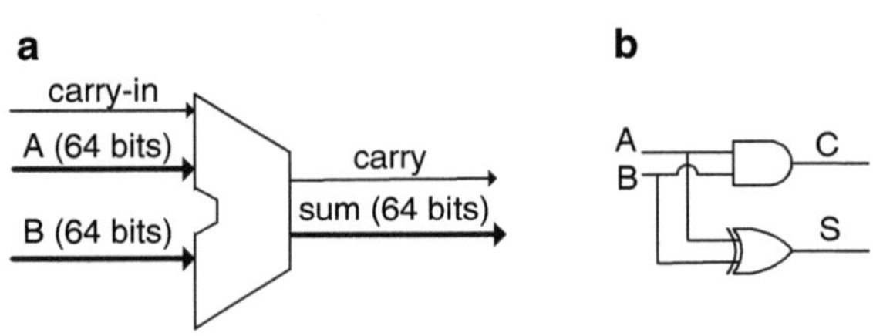

A	B	C	S
0	0	0	0
0	1	0	1
1	0	0	1
1	1	1	0

Fig. 7.9 (**a**) 64 bit adder symbol, and (**b**) circuit schematic of a half-adder slice with inputs A and B and outputs C (carry) and S (sum), and the corresponding truth table

Testing for all the combinations becomes prohibitive. Fortunately, fault models provide a way to significantly reduce the number of input patterns required for detecting faults. While not as comprehensive as an exhaustive functional test, fault detection is carried out with a high degree of coverage using fault modeling to generate the test patterns.

Fault models determine fault equivalence and fault dominance of all observable faults to collapse the number of test vectors. Faults are considered equivalent if each one of the faults generates the same output pattern. Fault dominance occurs when the tests for one fault always detect a second dominating fault. While this collapse typically reduces the number of test vectors for a small logic block by only about a factor of 2, the methodology described here provides insight into how fault models work.

A reduction in input patterns for detecting faults is illustrated with a half-adder circuit shown in Fig. 7.9b. For this circuit, each input and output may be s-a-0 or s-a-1, and there are eight possible single faults. From the truth table, a fault with A s-a-0 or a fault with B s-a-0 is detected with erroneous S and C outputs, and these two faults are considered equivalent. Considering the complete set of possible faults, A s-a-0, B s-a-0, C s-a-0, and S s-a-1 are all detected with inputs A = B = "1". Faults A s-a-1, B s-a-1, C s-a-1, and S s-a-1 are all detected with A = B = "0". Fault S s-a-0 is detected with either A = "0", B = "1", or A = "1", B = "0". Hence, only three input combinations are required for detecting any one of the eight faults.

The test routine determines how a fault is detected and propagates the fault to an observable point, which is typically a scannable CSE. Test patterns for detecting structural faults are generated using ATPG. There are several different algorithms in use to generate ATPG tests. Most ATPG tools operate on combinational logic and are not easily usable with sequential logic. Scan designs as described in Sect. 7.1.7 can convert sequential logic into combinational logic for test purposes.

ATPG routines may not be applied to all the logic circuits on the chip or may not cover all possible faults. Fault coverage for test patterns is defined as

$$\text{Fault coverage} = \frac{\text{Number of detected faults}}{\text{Total number of possible faults}}$$

It is desirable to have a fault coverage of 98 % or higher for single stuck-at faults and 100 % for interconnect faults.

7.1.5 IDDQ Tests

IDDQ tests are routinely used for identifying chips with gross defects such as shorts in the power grid. In a CMOS product chip, IDDQ of each voltage island with an independent power supply is measured separately to increase the fraction of defect-induced IDDQ over "good" IDDQ. This is the first step towards isolating the physical location of defects.

IDDQ tests may also be used for detecting systematic and random defects in circuits [6]. This type of testing is applicable to CMOS chips with worst-case IDDQ in the mA range, so that an additional current contribution of a few 100 μA from a defect can be resolved.

The contributions of different defect types to IDDQ are described in Sect. 4.2.4. Floating source, drain, or gate terminals of MOSFETs caused by opens in wire connections exhibit high IDDQ. The IDDQ is time dependent if the time to charge the floating node to its equilibrium value is longer than the time it takes to make an IDDQ measurement. The time dependence is detected by repeatedly measuring the IDDQ after initializing the chip.

Defect generated IDDQ arising from resistive bridges is dependent on node voltages. As an example, the metal short ($R_{sh} = 0$) in Fig. 7.10a creates a direct low-resistance path between V_{DD} and GND when outputs Z1 and Z2 are complimentary. This path is shown in Fig. 7.10b for the case of A = "0", B = "0", and C = "1", turning on MOSFETs P1, P2, and N3. This circuit draws a few hundred μA (for a nominal design with 45 nm PTM HP models). Such shorts are more easily detected at low temperatures with lower background I_{off} values.

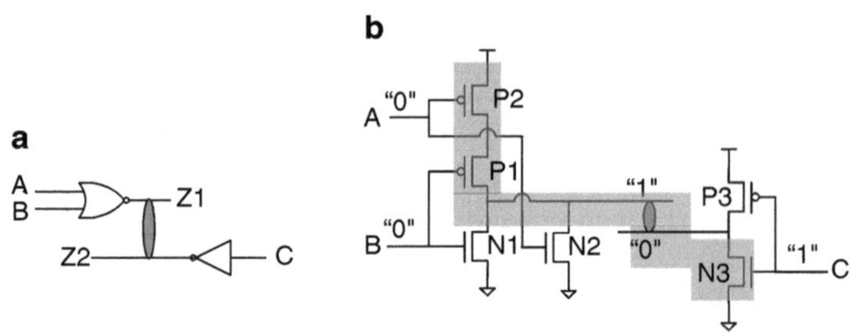

Fig. 7.10 (**a**) Circuit schematic showing metal shorting of inverter and NOR2 outputs, and (**b**) V_{DD} to GND path with A = "0", B = "0", and C = "1" for the circuit in (**a**)

In Fig. 7.11, IDDQ of 1,000 "good" inverters is compared with an inverter having a resistive short across the power grid and a second inverter with a resistive shunt across the source and drain of its p-FET while its input is at a "1". Simulations are carried out for the standard inverter designs in 45 nm PTM HP models at 1.0 V, 25 °C. In this small sample, the defects clearly stand out.

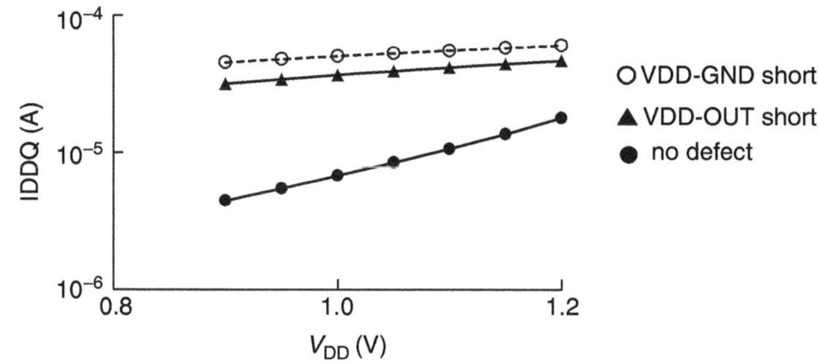

Fig. 7.11 IDDQ of an inverter (FO = 1) vs. V_{DD} with 1) 1,000 Ω shunt resistance across the power grid, and 2) 1,000 Ω shunt resistance across p-FET source and drain, and 3) a chain of 1,000 inverters with no defects. 45 nm PTM HP models @1.0 V, 25 °C

Anomalies in defect-generated IDDQ are dependent on node voltages which can be manipulated by running different input test vectors and measuring IDDQ for each test. Measured values of IDDQ vs. test sequence with different input vector patterns are shown in Fig. 7.12a. Steps in IDDQ are indicative of defects on the chip.

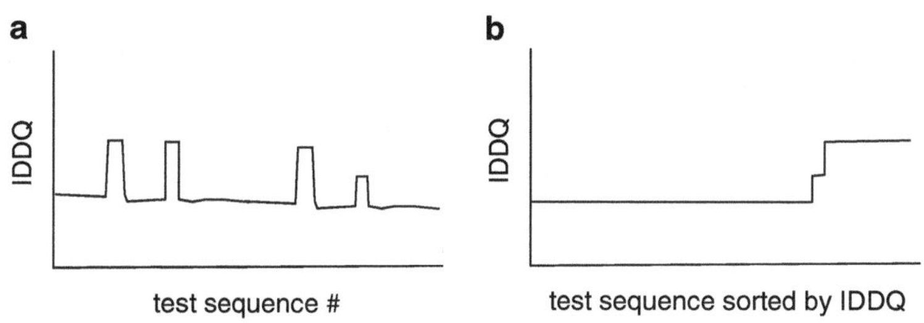

Fig. 7.12 (**a**) IDDQ vs. test pattern vector sequence, and (**b**) test sequence ordered by ascending IDDQ

The data are plotted in Fig. 7.12b by sorting IDDQ values in an ascending order to check for systematic vs. random defects. If steps are present and the pattern is repeated on all chips for the same input test vectors, there may be systematic defects originating from the design, photomask, or process-related problems (i.e., shorting between wires or floating nodes). If the pattern is repeated only on some chips, it may be related to the location of the chip on a multi-chip reticle or a signature of across-wafer systematic process variations.

If the steps occur in random test sequence on different chips, then the defect locations are random and may be caused by particulates on the wafer. The density of such particulates is mapped using microscopic imaging tools in silicon manufacturing. Images from inspection tools can identify chips with a large number of optical irregularities which can be correlated with electrical test data. This is discussed in more detail in Sect. 7.3.

For IDDQ diagnostics, the power supply current meter should have sufficient resolution to detect changes in current in the μA range. It takes 1–10 ms to make an IDDQ measurement, and there can be significant impact on total test time when running hundreds of patterns.

7.1.6 DFT and Diagnostics

DFT and diagnostics is a methodology in which features in the chip circuit design are added to assist in testing, failure diagnostics, and in some cases to validate proposed fixes. DFT includes additional I/O pins for observation and circuitry that enable scan tests, BIST, and boundary scan tests. Ease in test pattern generation, test time reduction, higher test coverage, and rapid debug are all facilitated by DFT. Additional features that can aid in characterization and diagnostics are silicon process, voltage and temperature (PVT) monitors, programmable clock buffers, electronic fuses, and spare logic gates.

The schematic of a CMOS chip with minimum required input and output pins is shown in Fig. 7.13a. DC pins supply voltages and currents. AC signal pins carry signals to and from the chip circuitry and enable communication to other electronics on a board or system. For test and debug, additional I/Os pins may be added for diagnostics as indicated in Fig. 7.13b. As an example, voltage sense pins can track within chip power supply excursions. Internal probe pads may be added for observing critical locations on the chip not accessible externally.

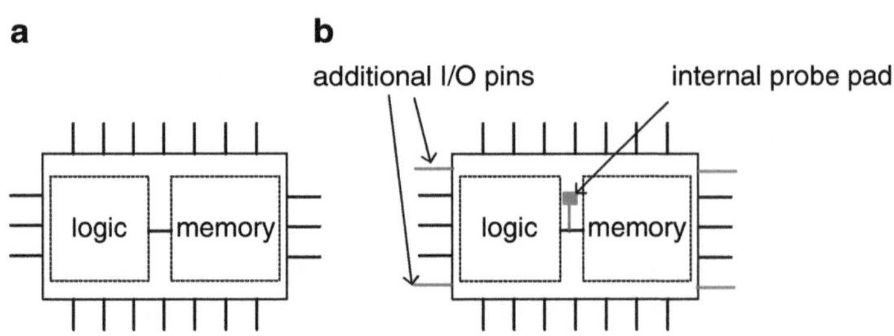

Fig. 7.13 Schematic of CMOS chip with (**a**) minimum required I/O pins, (**b**) additional I/O pins and internal probe pads for test and debug

Programmable electronic fuses (eFuses) are used for storing vital information on the chip which can be read out during test. Storing the chip id which identifies the parent wafer and (*x*, *y*) location of the chip on the wafer is extremely useful in tracking and correlating test data collected at different stages of test.

Spare logic gates are sprinkled in unused spaces so that design fixes can be incorporated with new photomasks for only a few metal layers. These photomasks cost less than those for delineating MOSFETs, and the fixes can be implemented in wafers that have already gone through process steps for defining MOSFETs.

Circuits to monitor chip temperature, silicon process, and local V_{DD} variations are embedded in the chip design. Temperature monitors are calibrated during test and the calibration coefficients stored for reference (Sect. 5.5). A suite of silicon process monitors comprises ring oscillators, delay chains, and CPMs. Voltage droops during switching in critical areas may also be monitored in high-performance chips. The design and implementation of embedded PVT monitors is covered in Chap. 5.

Local clock buffers are configured to enable corrections of timing errors in different clock domains. In these locating critical path (LCP) buffers, the timing of the clock edges can be adjusted during test to compensate for clock skew or longer delay through the combinational logic [7]. Referring to Fig. 7.14, if the signal arrives late at node Z, the clock edge at C2 is delayed. If, on the other hand, the signal arrives early at C1 with respect to C2, the clock edge at C1 is delayed. Such adjustments can be made for a limited number of critical paths on a chip-by-chip basis during test. In pulsed latches, the pulse width and the arrival time can both be adjusted with LCP buffer control signals.

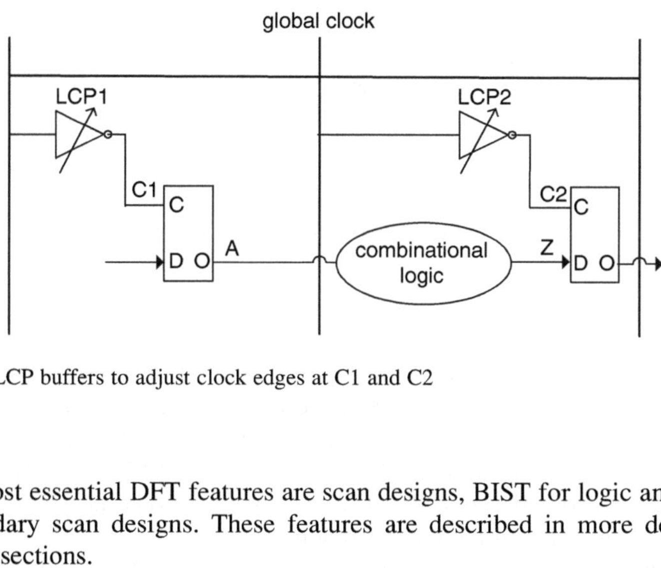

Fig. 7.14 LCP buffers to adjust clock edges at C1 and C2

The most essential DFT features are scan designs, BIST for logic and memory, and boundary scan designs. These features are described in more detail in the following sections.

7.1.7 Scan Design

Scannable level-sensitive latch and register file designs allow testing the functionality of all CSEs independent of the combinational logic circuit paths between them. An input mux selects the scan mode in which the combinational logic is bypassed so that the CSEs form a serial shift register as shown in Fig. 7.15. In this mode, the output of one CSE directly feeds into the input of the following CSE. During scan test, sclk is enabled, lclk is held at "1" and dclk is at "0". Scan-in input "01" is scanned out as "01" from the second CSE.

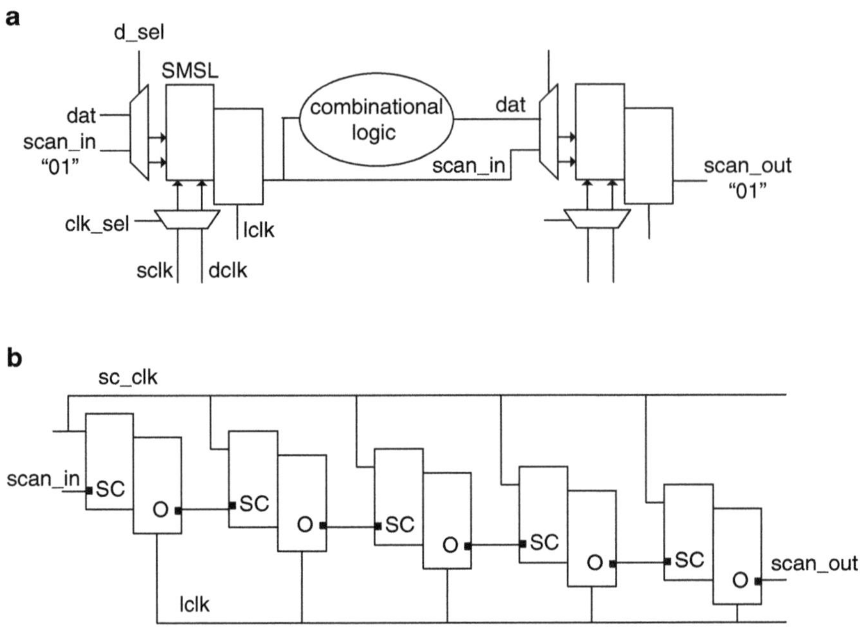

Fig. 7.15 Circuit schematic for scan tests: (**a**) input mux to select scan or data paths in two CSEs, and (**b**) scan design for test

The scan-in signal may, for example, be a set of "1"s, or a set of "0"s to load either all "1"s or all "0"s into the latches. Loaded bits are then shifted out with a regular clock signal to be observed externally. With a "00110011…" sequence of scan-in bits, all four transitions "0" → "0", "0" → "1", "1" → "1", "1" → "0" are exercised in a pair of adjacent CSEs. This test configuration is also used for loading "1"s and "0"s in CSEs for chip initialization.

In operational mode, scan_in input and sclk are disabled and data input is selected with dclk and lclk enabled. Logic functions are performed through the combinational logic circuits. At any desired point, the outputs of CSEs may be shifted out as in scan test, new bits shifted in, and again the state of each CSE after a few more clock cycles can be observed. The number of clock cycles required to read out the state of all the CSEs in a chain is equal to the number of CSEs in a chain.

As the length of the scan chain increases, it takes longer to read the outputs. This issue is addressed by using a random access scan approach. The CSEs are connected to form an array of rows and columns similar to a random access memory array. With decoders to select rows and columns, rows can be written and read out in parallel.

Scan tests are conducted at a low frequency to separate structural faults from timing or delay errors. Testing the functionality of all CSEs prior to logic or memory tests is extremely useful in debugging. Chips failing scan tests are rejected and removed from the test flow.

7.1.8 Built-in Self-Test

BIST techniques allow testing for correct operation within the chip itself. There is an overhead in design, but test efficiency is greatly improved by BIST for memory as well as for logic. BIST can also be run at the board or system level, checking the state of the chip periodically or at reboot.

The block schematic for a memory BIST (MBIST) scheme, also known as array BIST (ABIST), is shown in Fig. 7.16. The MBIST engine addresses all the cells and writes a checkerboard pattern of "1"s and "0"s in the array. The cells are read and checked against the expected results. This is followed by a second test with an inverse checkerboard pattern and all cells are again read out and validated.

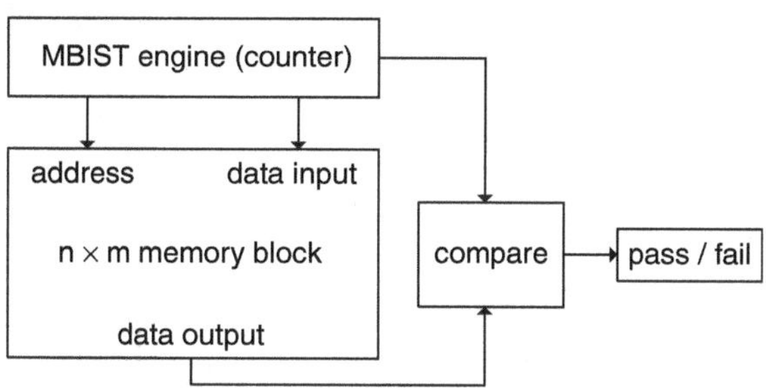

Fig. 7.16 Memory built-in self-test (MBIST) scheme

BIST implementation for logic circuits (LBIST) is illustrated in Fig. 7.17. A pseudo random pattern generator (PRPG) is used for generating input patterns for combinational logic circuits. The response of the logic circuit is analyzed and compacted into a signature which can be compared with that expected from a fault-free circuit, as determined from simulation or as obtained from KGDs.

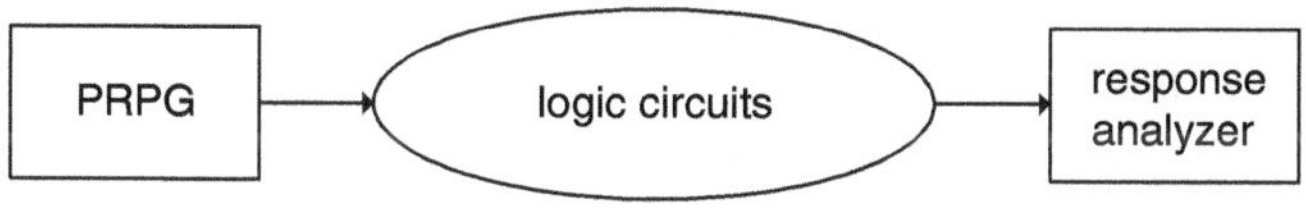

Fig. 7.17 BIST implementation with a PRSG and a signature analyzer

There are various schemes to minimize the additional logic required and the test time for LBIST. In one such scheme, a linear feedback shift register (LFSR) is configured with flip-flops and XOR gates to generate a pseudo random pattern sequence. A 3-bit LFSR along with the sequence of random numbers it generates are shown in Fig. 7.18a, b. A LFSR can also be configured as an output response compactor to reduce the number of bits in the circuit response. A multiple-input shift register (MISR) is used for response compaction to reduce the number of flip-flops and circuit overhead.

a

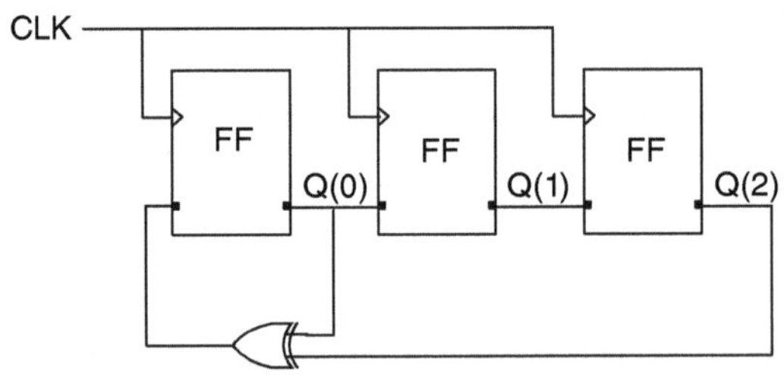

b

step	output
0	111
1	110
2	101
3	010
4	100
5	001
6	011
7	111 (repeats)

Fig. 7.18 (**a**) Circuit schematic of a pseudo random sequence generator implemented with flip-flops and an XOR2 gate, and (**b**) output responses

There are sophisticated EDA tools available for generating LBIST compatible circuit designs with a high degree of test coverage and for generating test code. The final outcome of LBIST is again a pass/fail decision.

7.1.9 Boundary Scan

Boundary scan design provides a way to connect the chip I/Os for test independent of the internal chip circuitry. Each I/O pin is connected to a boundary scan register cell. These register cells are connected in a serial fashion to form a boundary scan register. The connections from I/O pins to the internal circuitry are made through the boundary scan register cells. Input data is serially loaded into these registers and read out in a serial fashion during boundary scan test.

Boundary scan was originally developed by the Joint Test Access Group and is generally known as JTAG. A schematic of the boundary scan scheme based on IEEE standard 1,149.1 is shown in Fig. 7.19. There are a number of other registers such as instruction and bypass registers connected to the boundary scan register. The TDI pin is used to serially load input data into the boundary scan registers and the TDO pin provides serial output from the boundary scan registers. The TCK (test clock) pin is used for the test clock which is operated asynchronously from the chip clock. The TMS (test mode select) pin is used for selecting different test modes. A test access port (TAP) is used for remotely observing and controlling each I/O pin through boundary scan test hardware.

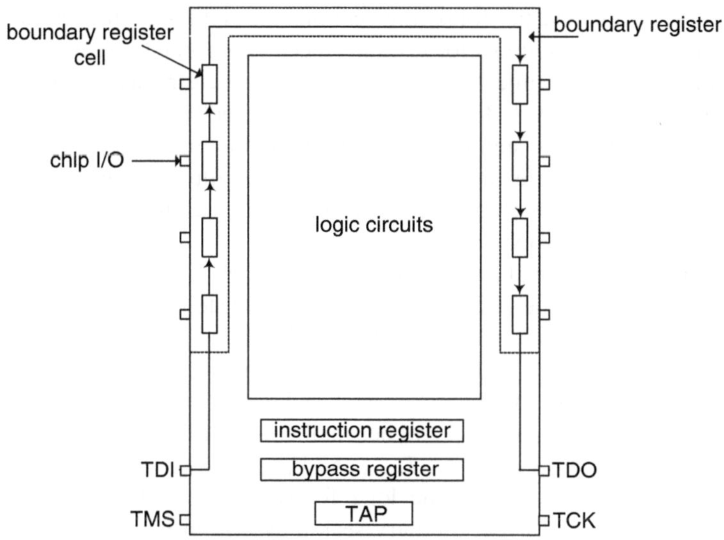

Fig. 7.19 High level schematic of boundary scan logic

Boundary scan hardware can be operated in several different modes. In normal mode, the standard operation of chip circuit functions takes place with transparent connection to the chip I/O through boundary scan register cells. In scan mode depicted in Fig. 7.20a, the boundary scan registers are scanned to read the state of the chip circuitry. In bypass mode shown in Fig. 7.20b, chip Y is bypassed.

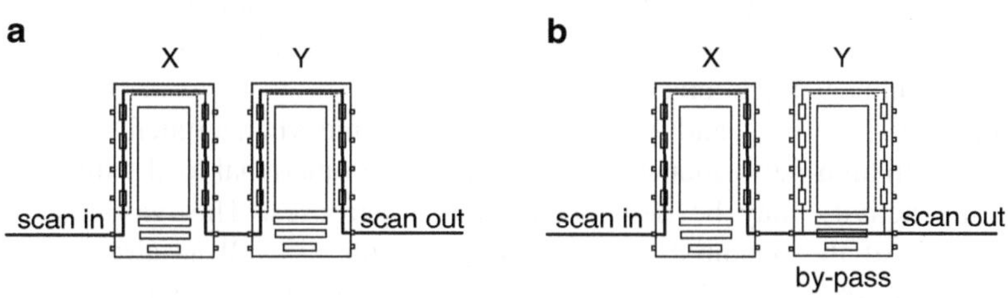

Fig. 7.20 Boundary scan register connections in chip X and Y in (**a**) scan mode, and (**b**) bypass mode for chip Y

7.1.10 Measurements of T_{cmin}, V_{min}, and AC Power

Full operation of chips is validated by running functional applications at specified cycle time, power supply voltage, and temperature. The test settings are based on worse-case product specifications and include guard-banding as discussed in Sect. 8.3.3. Typically, tests are conducted at the shortest specified cycle time (highest frequency), lowest specified V_{DD}, and at either one or both extremes of specified temperature. In this and the following sections on yield and characterization, we will assume that these settings for test have been appropriately selected.

The minimum cycle time T_{cmin} at a fixed V_{DD} is determined by first running a functional pattern at a cycle time T_c at which all defect-free chips are expected to pass. The cycle time is stepped down in small increments and the functional patterns run at each cycle time decrement until a failure occurs. The minimum cycle time at which the chip passes the test is T_{cmin}, and its reciprocal is the maximum frequency of operation f_{max}.

The minimum voltage at a fixed frequency at which the chip passes a functional test is V_{min}. It is determined by lowering V_{DD} in small steps until a failure occurs, in a manner similar to the measurement of T_{cmin} described above.

Measurement of T_{cmin} and V_{min} in the hardware is similar to the simulation example of a logic path described in Sect. 3.4.2. For accurate measurements of these parameters, the step size should be small (~1 % of the target T_c or V_{DD}). As a result of silicon process variations, T_{cmin} and V_{min} can vary over a wide range and the measurement time can become long for some chips. The test time can be significantly reduced by using a binary search algorithm within a selected range. The first test is conducted at the mid-point of the range. If it passes, the set-point is lowered to a value half-way between the mid-point and the lower extreme and if it fails, the set-point is raised to a value half-way between the mid-point and the upper extreme. This procedure is continued until a failure is recorded within a step size increment that gives the desired measurement accuracy. As examples, the desired measurement accuracy for T_{cmin} may be 1 % of the nominal cycle time and 10 mV for V_{min} at a nominal V_{DD} of 1.0 V.

AC power measurement is carried out by measuring the average current drawn during a functional test at the prescribed settings. If there are multiple power supplies, the total power is the sum of the power consumed by all the power supplies. AC power may vary with switching activity, and the highest total power must be within specifications.

Often the T_{cmin}, V_{min}, and AC power vary with time when a functional workload is running. In high-performance chips, a worst-case functional load with maximum switching activity may be developed for test purposes. This workload should exercise all of the potential cycle limiting circuits as well as circuits in high-power density regions of the chip. A high level of switching activity may increase the chip temperature and noise level. Voltage droops and higher silicon temperature tend to increase T_{cmin} as well as V_{min}. In addition, increase in IDDQ with temperature adds to the total power.

7.2 Yield

Once the test procedures are fully debugged and the test codes have been validated in simulations as well as in hardware, the focus in test turns to improving yield. The overall yield of a product is defined as

$$Y = \frac{\text{Number of good chips}}{\text{Number of manufactured chips}}. \tag{7.1}$$

The object is to maximize yield for maximum profit. Hypothetical yield trends for two different product chips are shown in Fig. 7.21. The yield is lower in the beginning of the manufacturing cycle and improves with time in response to design fixes, silicon process tuning, and lower defect density. When fabricated at the same technology node, small area ASIC chips typically have a higher yield than large area microprocessor chips. Yield also varies with circuit density, minimum feature sizes, level of built-in redundancy, and design margins.

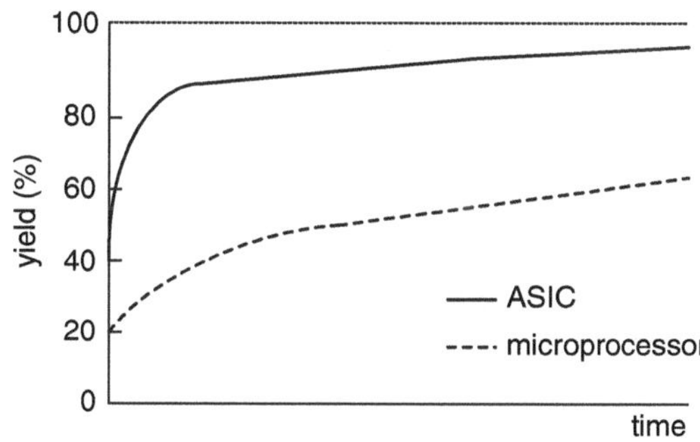

Fig. 7.21 Hypothetical yield as a function of time in the product manufacturing cycle for a commodity ASIC chip, and a large area high-performance microprocessor chip

There are several factors contributing to yield loss. Defects introduced during silicon processing result in malfunction due to opens and shorts. The source of these defects may be particulates, metal residuals, or shape definitions such as corners in wires and via holes. Some chips, although free of structural defects, may fail to meet product specifications, such as operating frequency or power level, and are rejected. Defect limited yield and cycle time limited yield issues are covered in Sects. 7.2.1 and 7.2.2 respectively.

In order to improve product yield, silicon foundries set guidelines for design for manufacturability (DFM). Some of the general guidelines include use of redundant vias, wider metal overlap on via borders, increased spacing for greater than minimum wire widths, and regular or gridded patterns for minimum features at lower levels. Following these guidelines in chip design is strongly recommended to promote higher yield.

7.2.1 Defect Limited Yield

At wafer level testing, DC and AC parametric, structural, and functional tests conducted at low frequencies identify most of the chips with electrical faults or anomalies that are a consequence of physical defects. Some resistive defects can only be detected at the operating frequency in packaged chips. Yield loss in memory arrays is minimized by exploiting built-in redundancy and repair capability features. A single defect in random logic, however, can cause a chip to fail.

The focus in silicon processing has been primarily on visual defects, such as metal residues, missing metal, via opens, and line-width variations exemplified by cases shown in Fig. 7.7. There are, in addition, nonvisual fault-causing defects arising from organic, metallic, or surface contamination. Electric charge damage to dielectric layers during processing is another class of non-visual defects.

Physical defects caused by particulates are observed with the aid of photo limited yield (PLY) tools, such as high-resolution scanning electron microscopes. A defect in unused area in a layer, indicated by a dark region in Fig. 7.22a, is of little consequence. The small defect in Fig. 7.22b is not likely to result in a fault whereas the defect in Fig. 7.22c is very likely to cause a fault. A critical area is defined as an area where the center of a particle of a certain size must fall to result in a fault-causing defect. Hence, the defect locations are compared with the circuit layer definitions to isolate fault-causing defects in the PLY data analysis.

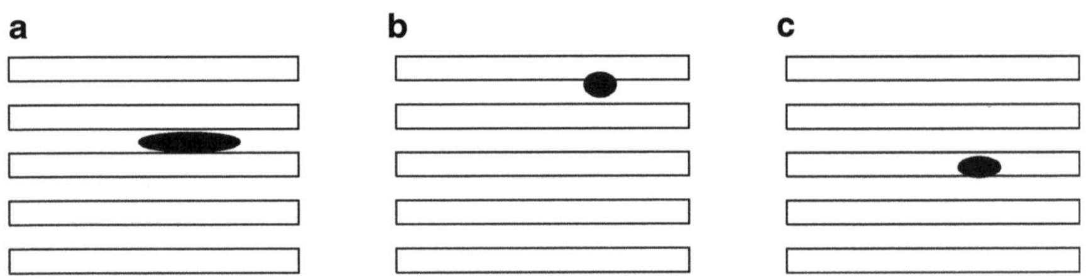

Fig. 7.22 Location of defects (dark regions) in an area filled with parallel interconnect wires: (**a**) defect in unused space, (**b**) noncritical area defect, and (**c**) critical area defect (indicated by dots)

The number of defects per unit area, or defect density, is monitored in the silicon manufacturing line. The defects are classified by their size and impact. Locations of each defect type are mapped on the wafer at major processing steps. A wafer map of random defects obtained from a PLY tool is illustrated in Fig. 7.23a. Defects typically occur in clusters as shown in Fig. 7.23b: the defect density is nonuniform, with a higher probability of a defect occurring in the vicinity of other defects.

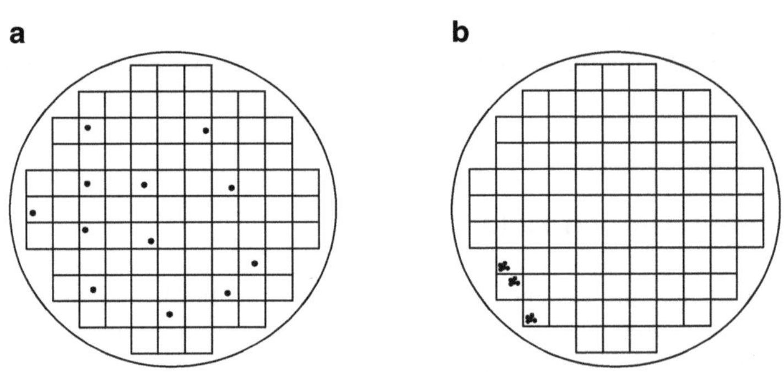

Fig. 7.23 Wafer with (**a**) randomly distributed defects, and (**b**) clustered defects (indicated by dots)

The yield from one type of defect at a level is given by

$$\mathrm{PLY}_j = \left(1 - \mathrm{PF}_j\right) \times 100, \tag{7.2}$$

where PF_j is the probability of fail from defect j.

The probability of failure from n defect types at level i is

$$\mathrm{PLY}_i = \prod_{j=1}^{n} \mathrm{PLY}_j, \tag{7.3}$$

which is the product of the yield factors. The net yield for all the processing levels is the product of the yields at each level, assuming the faults caused by the defects are independent of the level

$$\mathrm{PLY} = \prod_{i=1}^{n} \mathrm{PLY}_i. \tag{7.4}$$

The above equations highlight the increasing impact of defects with increase in the number of processing steps and photomask levels.

If defects are truly randomly distributed, having a fault-causing defect density DD, the yield of a chip of area A is given by Poisson distribution

$$Y = e^{-(A \times DD)}, \tag{7.5}$$

assuming a single fault causes the chip to fail. The yield plummets as $A < 1/\mathrm{DD}$.

The yield for clustered defects is modeled as

$$Y = \left(1 + \frac{A \times DD}{\alpha_c}\right)^{-\alpha_c}, \tag{7.6}$$

where α_c is a cluster parameter. A very small value of α_c (~0) indicates extreme clustering. Typical values of α_c are between 0.3 and 3.

Some fault-causing defects can be "healed" by applying a high voltage. Resistive heating of metal bridge shorts can accelerate electromigration and create an open circuit. Testing at elevated voltages for a short time or application of high-voltage pulses are some of the common voltage screening methods to heal defects. Burn-in at high temperature and voltage is another technique to accelerate evolution of fault-causing defects and may even heal some of them prior to shipping the product as described in Sect. 8.3.2.

7.2.2 Cycle Time Limited Yield

A chip may pass all structural and functional tests at low frequencies but may fail to operate at the target frequency. Such fails fall in the category of cycle time limited yield (CLY) detractors. Other operational yield detractors are V_{min} above target, IDDQ (P_{off}), or total power P exceeding acceptance limits, and noise-related fails which may be intermittent. As power and frequency can both be adjusted by turning the V_{DD} knob, in the following discussion we will use the term CLY to include all such fails.

Ideally, in a well-centered silicon process, CLY should be 100 % with all fails attributed to silicon processing defects. However, some physical defects such as those that lead to high-series resistances can reduce CLY. If chips failing cycle time tests are randomly distributed in the population and the related yield loss is relatively small, they may be ignored. If the fails are larger in number and traced to a small subset of circuit faults, a design fix such as wider wires or redundant vias may be implemented.

CLY is reduced as the silicon process drifts off-center or when the design margins are small. A V_{DD}–T_{cmin} shmoo plot may be generated during test to map the operating region of a chip. Such a shmoo plot for a good chip is shown in Fig. 7.24a. An open square indicates a pass and a gray square a fail. A dark circle near the center of the plot is placed at the nominal target (V_{DD}, T_{cmin}) location. The boundary of the pass and fail region shows T_{cmin} as a function of V_{DD}, similar to Fig. 3.38a obtained from simulating a 20 inverter (FO = 4) data path. The shmoo plot in Fig. 7.24b has a similar shape except that this chip fails at the target T_{cmin}. Such fails are usually caused by weak MOSFETs or resistive signal wires. In such cases, correlation with embedded process monitors (Sect. 7.4.3) helps get to the root cause.

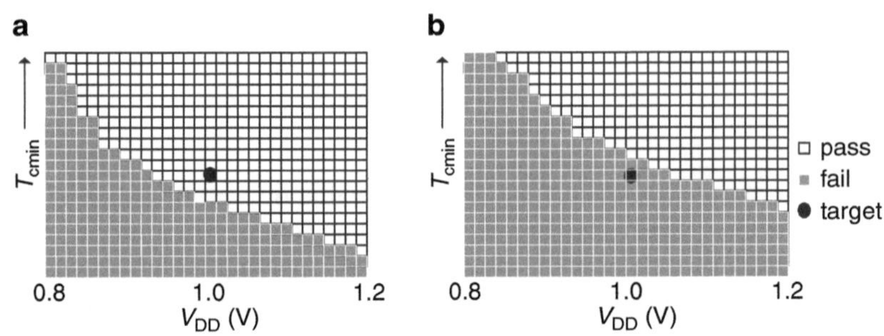

Fig. 7.24 (V_{DD}–T_{cmin}) shmoo plot: (**a**) good chip passing at the (V_{DD}–T_{cmin}) target point, and (**b**) chip failing at (V_{DD}–T_{cmin}) target point

Two examples of shmoo plots of failing chips with unusual behavior are shown in Fig. 7.25. In Fig. 7.25a, additional fails occur in the passing region indicative of noise coupling at higher V_{DD}. In Fig. 7.25b, there are no passing chips for $V_{DD} > 0.96$ V, possibly as result of a race condition.

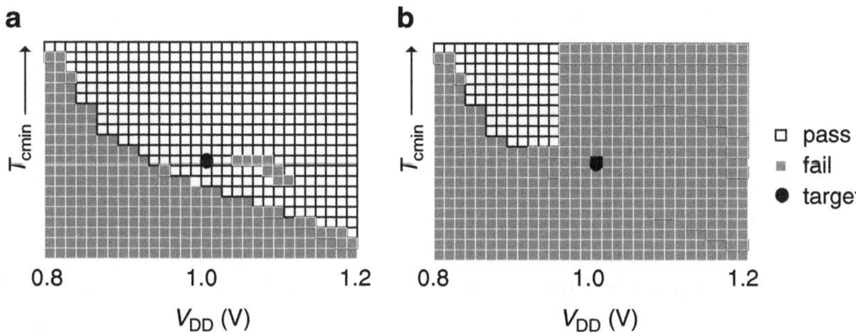

Fig. 7.25 T_{cmin}–V_{DD} shmoo plot: (**a**) additional fails at higher V_{DD} indicating noise coupling, and (**b**) abnormal T_{cmin} response at higher V_{DD} possibly due to a race condition

CLY improvement may be achieved by re-centering the silicon process to get optimum MOSFET performance, by customizing V_{DD} for failing chips or by speed binning. These topics are covered in Sect. 7.4.

7.3 Failure Analysis

Fault models are used for automated diagnostics of chips failing ATPG tests. A fault is detected when the output pattern fails to match the expected pattern. The first failing observable CSE is identified and fault models are applied to all the logic that feeds into that latch. The faults that do not generate the failing pattern are removed from the list. The remaining faults are possible candidates for defect detection. Automated software tools for fault analysis can trace the cone of logic that feeds into the failing CSE and inject faults using circuit simulations to match the failing pattern. These tools can also relate the possible fault locations directly to the physical layout. Inspection of the physical layout may lead to areas more susceptible to process variations such as bends in wires and nonredundant via contacts.

When the same failure mode is repeated on a statistically significant population, sophisticated failure analysis methods can help get to the root cause, leading to a course of action to eliminate such defects in the future. The defect must be localized using standard electrical test data coupled with DFT and design data. Sample preparation and analysis are expensive and time consuming and the use of physical failure analysis tools and techniques is recommended only when systematic fails become significant yield detractors. Some of the commonly used techniques for failure analysis are described here.

High-resolution scanning electron microscopy provides images of chip surfaces for detecting silicon process-induced defects such as metal residues, particulate contamination, and extraneous deposits. Special delayering techniques are used to expose defects in lower metal interconnects or in silicon. Chemical composition of the defect area can be obtained from electron emission spectra or secondary ion mass spectroscopy. Transmission electron microscopy is used for cross-sectional analysis. The spatial resolution is of the order of a few tenths of a nm or less, but sample preparation is time consuming.

Electron microscopes are also used for imaging voltage levels in circuits with a voltage contrast (VC) technique. The number of secondary electrons collected from areas at lower voltages is higher than from areas at higher voltages, creating a contrast in intensity. This image intensity map of a circuit area, when compared with a defect-free sample, can help localize opens and shorts.

MOSFET and gate terminals can be directly contacted using pico-probes with $<1\ \mu m^2$ contact area for measurement of *I–V* characteristics in failing circuits. Only a small number of probes may be used simultaneously and the current carrying capability of the probes is limited to <1 mA.

Switching activity and timing errors can be detected in the active mode using picosecond imaging circuit analysis (PICA). This is a noninvasive technique for obtaining time-resolved optical images of switching activity in CMOS circuits [8]. In CMOS circuits when a MOSFET switches between the on-state and off-state, a transient current flows through it producing a corresponding transient light emission signal from the relaxation of hot carriers. The intensity of light emission is proportional to the current and is much higher than that arising from light emission in the subthreshold region described in Sect. 6.2.4. A photon detector coupled with time-correlated photon counting techniques can track a signal as it passes through a circuit path. By overlaying the light signal on an optical image of the circuit and physical design data, the exact location and timing of signal propagation is recorded. Sample preparation requires thinning the substrate as shown in Fig. 6.14.

A temporary repair of a systematic fault-causing defect may be made on selected chips to confirm the findings prior to committing resources for an expensive fix such as a new high-definition photomask. Any recommended changes in the silicon process parameters have to be carefully evaluated to avoid the possibility of introducing a different yield detractor, such as a timing error in another path. MOSFET current drive strengths can be increased or reduced under the conditions at which a fail occurs with a laser-assisted device alteration (LADA) tool. An infrared laser beam of wavelength 1.3 μm, incident on the front or back of the chip raises the local temperature. MOSFET drive currents are reduced, increasing signal delays within a small area. The laser beam-induced heating also increases wire resistances and impacts delays in wire loaded paths. Alternatively, a 1.06 μm laser with photons of energy greater than the silicon bandgap of 1.1 eV generates electron-hole pairs. The result is an increase in drive currents of n-FETs and p-FETs.

Faulty circuits may be repaired using a focused ion beam (FIB) technique. A beam of Ga+ ions is focused to a diameter of ~10 nm on the sample surface. The beam moving in a raster pattern can cut through metal, polysilicon and dielectric layers. Ion milling provides access to layers buried below the sample surface. Metal may also be removed and deposited to make new connections as shown in Fig. 7.26 where a metal wire is cut and a new wire connection made to bypass a buffer to increase timing slack in a path. Spare logic gates may be added to change the logic or to add delay. Validating the fix using FIB prior to ordering a new photomask can help save considerable time and money.

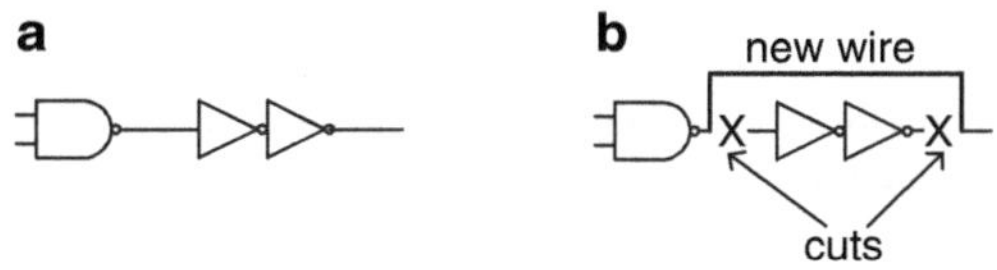

Fig. 7.26 (**a**) Sub-section of a larger circuit with a timing error, and (**b**) delay reduction with a new wire path that bypasses the two inverters in (**a**), implemented with FIB techniques

7.4 Product Chip Characterization

A comprehensive chip characterization plan early in the product manufacturing cycle is necessary for estimating silicon technology and design weaknesses. Chip yield, performance, and power are correlated with design model predictions and with data collected from embedded monitors and scribe-line test structures. Test programs are reviewed to resolve any inconsistencies. Process-split hardware and data collected at different test corners are used for tuning the silicon process and to determine test conditions for routine production testing.

7.4.1 Silicon Manufacturing Line Tests

The earliest readout of silicon process is obtained from electrical measurements of test structures placed in the scribe-lines (Fig. 7.27). A detailed account of test structure design and test can be found in the literature [4, 5]. Circuit components are tested, and the data are correlated with model predictions. A subset of device parameters are monitored on selected sites for a few wafers in all lots, where each lot comprises 5–20 wafers. Any deviations from the published $\pm 3\sigma$ parameter limits are flagged to alert the manufacturing process team. The silicon process is recentered to ensure future lots will stay within the six sigma range.

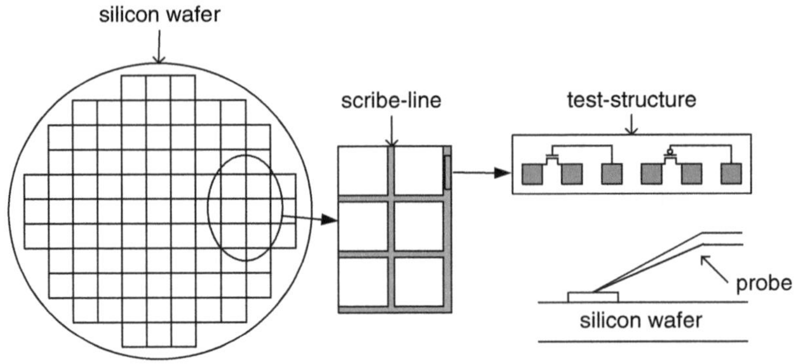

Fig. 7.27 Wafer showing location of chips and placement of test structures with metal pads for landing probes in the scribe-line

MOSFETs can be fully tested after the completion of the first metal wiring layer M1. Contacts to MOSFETs are made by landing probe needles on metal pads connected to MOSFET terminals (Fig. 7.27). *I-V* characteristics are evaluated by measuring I_{ds} at fixed V_{ds} and V_{gs} bias voltages or by making voltage sweeps with small step sizes and then extracting key parameters (I_{on}, I_{eff}, I_{off}, V_t) from the data. MOSFET physical layouts for scribe-line test structures may be extracted from logic circuit layouts used in the product chip. Such product representative test structures improve the correlation between scribe-line tests and product chip test data. SRAM cells are modified to access the terminals of each MOSFET in the cell while maintaining the basic cell layout. Butterfly curves of the type shown in Fig. 3.29 for each SRAM cell type are obtained from voltage sweeps at the appropriate terminals.

C-V characteristics are measured with a capacitance meter typically operating at frequencies <50 MHz or with various charge-based capacitance techniques. Gate, overlap, and diffusion capacitances (C_g, C_{ov}, and C_j) are measured at fixed voltage biases on test structures specifically configured for capacitance measurements.

Ring oscillators with inverter stages are designed with a single metal wiring layer for early readout in the manufacturing flow. An inverter RO with MOSFET gate load (FO = 1 to 4) gives the inverter delay and IDDQ. The RO data are correlated with MOSFET *I-V* and *C-V* characteristics to extract *R* and *C* parasitics.

As wafers move forward in the manufacturing line and more interconnect layers become available, complex circuits requiring several metal interconnect layers can be tested. Passive test structures are characterized to obtain *R* and *C* of each metal layer for different line-widths with serpentine and comb structures. Via resistances are measured in long series connected chains of vias. An expanded suite of ROs including stacked gates and stages with wire loads are measured for model-to-hardware correlation of more complex logic gate configurations. Critical dimension (CD) measurements are done, at least for minimum line-widths, on each layer to ensure design dimensions are faithfully printed on the wafer.

A yield model for a product chip is generated using the PLY data. The model is extended to include future projected yield based on defect learning in the manufacturing line and fixing design-related weaknesses. The PLY model is correlated with the defect-limited yield of the product as production continues.

Electrical test yield on large area test structures such as SRAM and DRAM arrays and banks of series connected scannable CSEs may also be measured. In order to get comprehensive yield learning, the area covered should be a significant fraction of the product chip area. As the area available in the scribe-line is limited, electrical testing for such defect monitoring test structures is typically carried out on dedicated test chips. These test chips may be periodically processed for monitoring the health-of-line (HOL) in manufacturing.

7.4.2 Silicon Process-Split Hardware

For high performance and high-cost chips fabricated on dedicated wafers, process optimization is carried out on special wafer lots early in the manufacturing cycle. By varying the lithographic exposure dose in selected columns on a wafer, chips with different nominal L_p values are fabricated. This scheme is illustrated in Fig. 7.28a with a wafer map showing stripes of nominal L_p and others with L_p values having $\pm 1\sigma L_p$ and $\pm 2\sigma L_p$ deviation from nominal. All other processing steps being identical on all chips, the power/performance/yield shifts in different columns or stripes are expected to be primarily from L_p differences.

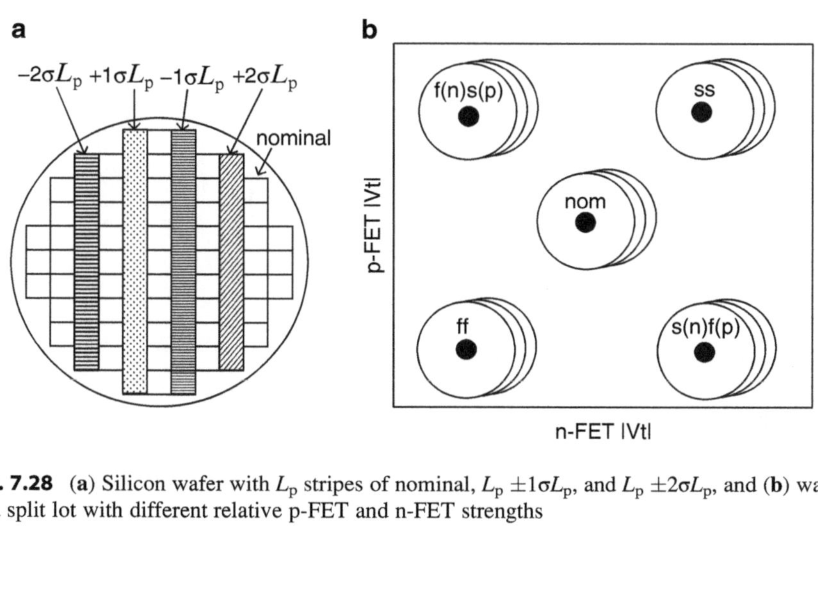

Fig. 7.28 (**a**) Silicon wafer with L_p stripes of nominal, $L_p \pm 1\sigma L_p$, and $L_p \pm 2\sigma L_p$, and (**b**) wafers in a split lot with different relative p-FET and n-FET strengths

For optimization of V_{tn} and V_{tp}, and p/n ratio, wafers from a single lot are split into groups, each group with nominally identical V_{tn} and V_{tp}. In Fig. 7.28b five groups of wafers cover nominal and four extreme corners for V_{tn} and V_{tp}, similar to the simulation corners in Fig. 6.17. Here f(n) represents a fast n-FET or lower V_{tn} and s(p) denotes a slow p-FET or higher $|V_{tp}|$.

The values of L_p for chips from different L_p stripes on a single wafer are offset with respect to each other as specified. However, systematic variations from lot-to-lot (L2L), and wafer-to-wafer (W2W) must be taken into account when analyzing data on chips from many wafers. As an example, if the nominal L_p in a lot is shifted by $+2\sigma$, the $+2\sigma$ stripe will be at $(L_p + 4\sigma L_p)$, and out of range in a well-tuned manufacturing process. Across-wafer (AcW) variations may result in L_p variations on chips within the same stripe on a wafer.

An example stacked bar chart indicating % failing chips in different L_p stripes is shown in Fig. 7.29. The bars indicate three types of fail, ATPG, IDDQ, and T_{cmin}. The ATPG fails may be attributed to defects. The IDDQ fails clearly increase at shorter values of L_p, and T_{cmin} fails increase at longer L_p as expected. If the trends cannot be explained by process changes, for example, an increase in IDDQ fails in longer L_p stripes, there is likely to be an error in test, data analysis, graphics, or some other issue. This type of graphical illustration helps in visualizing yield trends. For detailed characterization, all significant fail categories should be included.

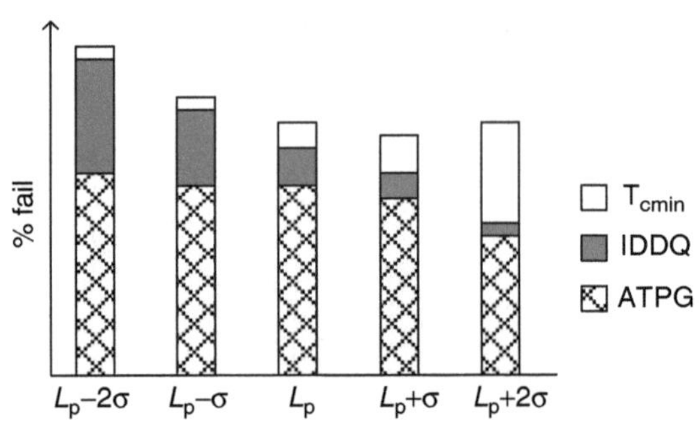

Fig. 7.29 Stacked bar chart of % of failing chips in each L_p stripe. Failing tests are ATPG, IDDQ, and T_{cmin}

7.4.3 Embedded Process Monitors

Signal propagation delays of series connected logic gates or circuit blocks are measured using embedded process monitors described in Chap. 5. Average delays of logic gates are obtained from process monitors comprising nominally identical gates in a chain or in a ring oscillator configuration. The data collected from these monitors are correlated with model predictions, with silicon technology test structures in the scribe-line, and with chip power and f_{max}. Selection of logic gate designs and locations of these monitors vary among different product chips. We will use the monitor locations shown in Fig. 7.30 to illustrate the use of these monitors and refer to them as ROs for convenience.

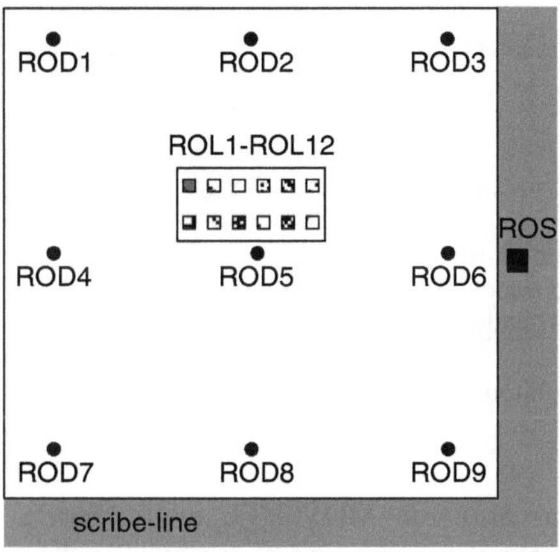

RO #	description
ROL1	inverter (FO =3)
ROL2	NAND3B (FO =3)
ROL3	NOR2T (FO =3)
ROL4	low V_t inverter (FO =3)
ROL5	high V_t inverter (FO =3)
ROL6	narrow inverter (FO =3)
ROL7	wide inverter (FO =3)
ROL8	inverter $+C_wR_w$
ROL9	Inverter +Cw
ROL10	NPG_1
ROL11	PPG_1
ROL12	critical path
ROD	inverter (FO =3)
ROS	inverter (FO =3)

Fig. 7.30 Placement and description of ckt_stgs of silicon process monitors (ring oscillators)

A circuit block with 12 ROs (ROLs 1–12) of different logic gate designs is located in the center of each chip. In addition there are nine ROs of a single design (RODs 1–9) distributed across the chip in a regular grid. Silicon technology test structures are located in the scribe-line where there is one RO (ROS) comprising a ckt_stg design similar to that in ROD. The ckt_stg design in ROD is a regular V_t inverter (FO = 3), a good representation of all regular V_t static CMOS logic gates. Ckt_stgs in ROL1 through ROL11 include NAND and NOR gates, inverters of different V_t values and finger widths, inverters with wire loads, passgates, and a representative critical data path. A hybridized combinational logic path in RO12 imitates anticipated critical path composition.

Applications of such on-chip process monitors are shown in Fig. 7.31. In this eight-sided view, different aspects of test and evaluation from silicon technology to models, product design, and power-performance trade-offs from the data collected are displayed. We will illustrate the benefits in each of these areas with graphical techniques. For this purpose it is necessary to obtain model based τ_p target values from circuit simulations with full parasitic extracted netlists at all test corners. It is highly desirable to generate the targets at different values of L_p (nominal, $\pm\sigma$, $\pm 2\sigma L_p$ and $\pm 3\sigma L_p$) for each test corner.

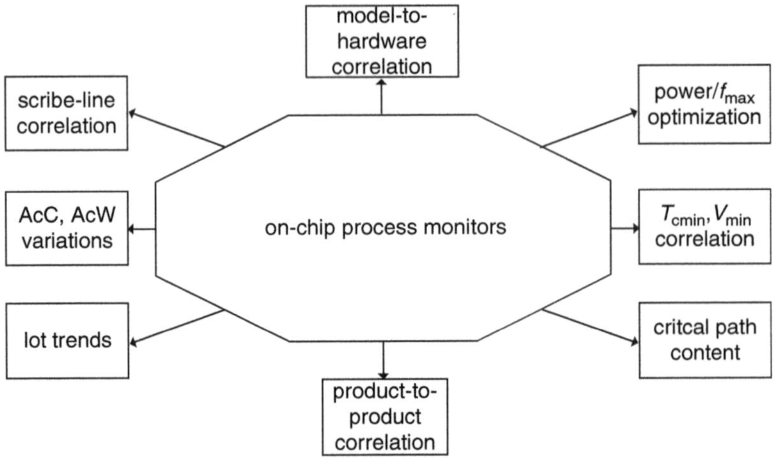

Fig. 7.31 Applications of silicon process monitors

The silicon foundry provides information on MOSFETs, interconnects, and small circuit blocks collected during manufacturing and compares the data with model expectations. It is important to establish that these data correlate with on-chip monitors. Generally even if ROS has the same circuit schematic as ROD (inverter FO = 3), the physical layouts, local pattern densities, and power distribution grids of ROS and ROD are likely to be different. The power supply voltage and silicon temperature for the scribe-line tests are also typically different than for on-chip tests. These differences are accounted for by normalizing the measured data to the targets obtained from circuit simulations.

The scatter plots of normalized τ_p values for the nine RODs are plotted vs. normalized τ_p values of ROS in Fig. 7.32. Scatter plots of this type are described in more detail in Sect. 9.5. Multiple plots of this type provide a quick view of model-to-hardware correlation, process centering, across-chip variation, and scribe-line to product chip offsets, all on a single page!

In Fig. 7.32, the placement of the plots corresponds to the physical location of RODs on the chip with reference to the ROS. The dashed line indicates 1:1 correspondence between RODs and ROS, and the solid line is the best fit to the measured data on many chips. For the plots in Fig. 7.32, all RODs have a smaller τ_p value than ROS, with the difference increasing away from scribe-line. On an average, the on-chip inverters have a smaller delay than the inverters in the scribe-line. Fig. 7.32 also illustrates across-chip (AcC) variation in inverter delays.

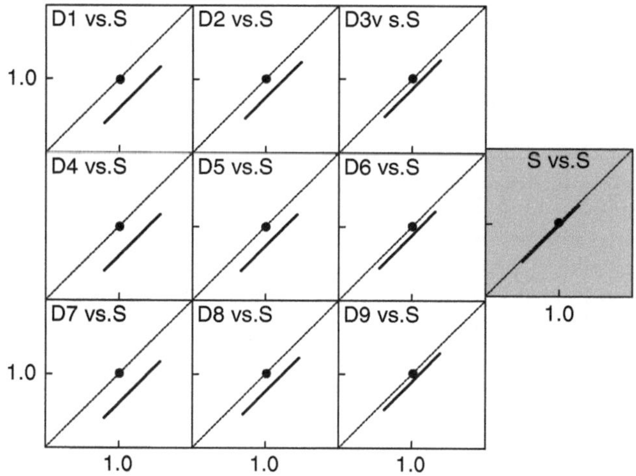

Fig. 7.32 Normalized values of τ_p for on-chip RODs (D1 through D9) vs. scribe-line ROS(S)

Across-chip (AcC) and across-wafer (AcW) variations are illustrated with wafer maps of the type shown in Fig. 7.33. In Fig. 7.33a, b the gray-scale shading in each chip location represents average inverter τ_p of all nine RODs on each chip. The wafer map in Fig. 7.33a shows no AcW variation whereas the wafer map in Fig. 7.33b shows radial and top-to-bottom AcW variations. In Fig. 7.33c, d, each ROD is represented by a square within each chip. There is only AcC variation in Fig. 7.33c, and both AcC and AcW variations in Fig. 7.33d which result in changes in the AcC variation pattern with the chip location on the wafer. Note that the appearance of such wafer maps changes with the range in each bin and the number of bins. Stacked wafer maps showing average τ_p at each location indicate systematic trends in AcC and AcW, but may hide maverick wafers.

Normalized τ_p values of logic gates collected from on-chip ROLs for wafer lots processed over several months are shown in Fig. 7.34. This box and whiskers plot (Sect. 9.2) shows that ROL8 is systematically slower than model predictions and that ROL11 is faster and has a wider spread than other ROs. This may be due to model-to-hardware offsets or process variations.

MOSFET and other parameters may be tracked on a chip-by-chip basis using τ_p data from a suite of delay chains or ROs and the methodology described in Sect. 5.6.3. In Fig. 7.35, regular, high, and low V_t inverter τ_p values normalized to nominal regular inverter delay τ_{p0} are plotted. In Fig. 7.35a, target τ_p values are obtained from circuit simulations by varying L_p. In Fig. 7.35b data collected from silicon are fit to a trend line. It becomes immediately apparent that high V_t inverters are slower than model predictions in the hardware and a possible cause of lower f_{max}. In such a case, the failing paths are likely to be those comprising high V_t logic gates with sufficient timing slack in design, but not in hardware.

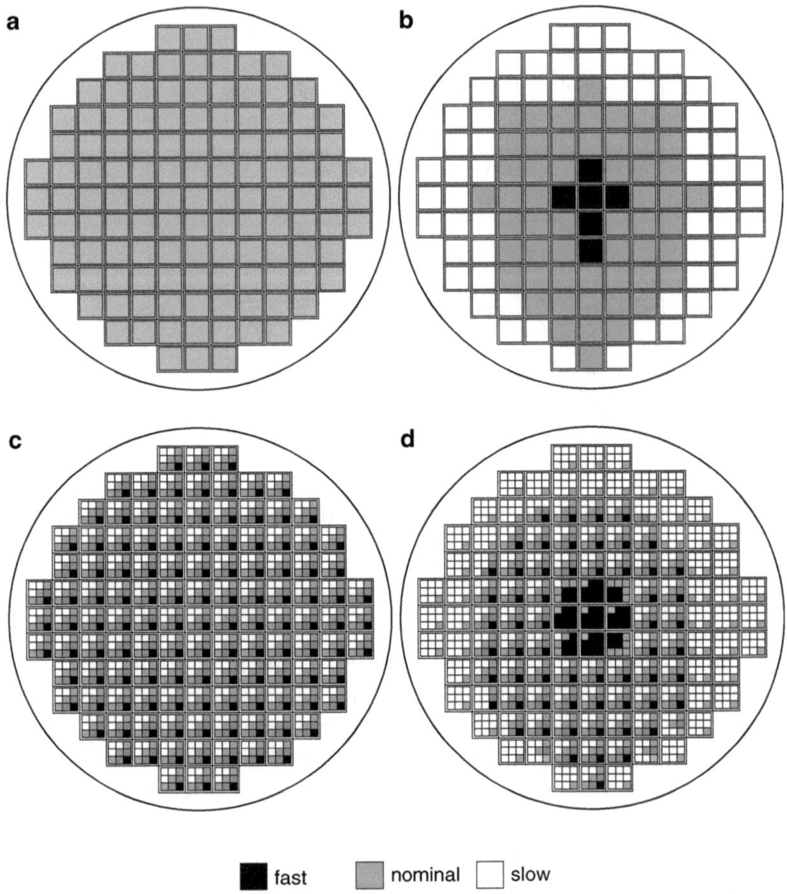

Fig. 7.33 AcW and AcC variations in inverter τ_p: (**a**) average of nine RODs per chip showing no AcW variation, (**b**) average of nine RODs per chip showing AcW variation, (**c**) nine RODs across chip showing AcC variation but no AcW variation, and (**d**) nine RODs across chip showing both AcC variation and AcW variation

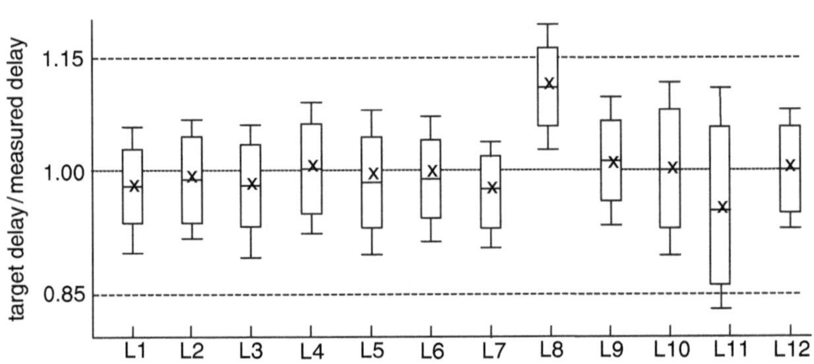

Fig. 7.34 Box and whiskers chart of normalized τ_p values of ROLs (L1 to L12) for several lots

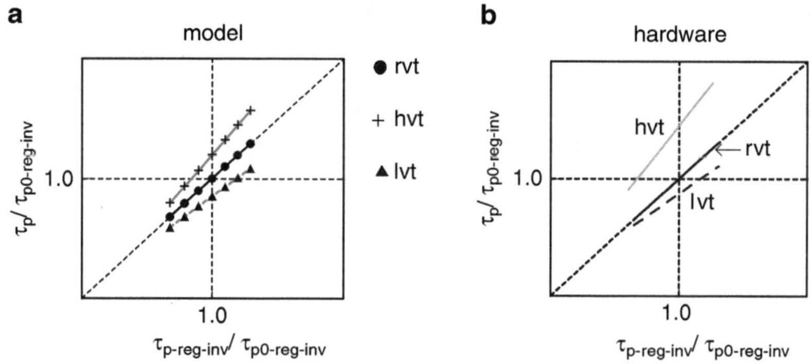

Fig. 7.35 Inverter τ_p values of high V_t (hvt), regular V_t (rvt), and low V_t (lvt) normalized to nominal regular V_t inverter are displayed for the full range of L_p variations: (**a**) model predictions, and (**b**) hardware data

Initial indications of chip power and performance are obtained by plotting chip IDDQ as a function of average inverter τ_p of the nine distributed RODs for all chips passing low frequency tests as shown in Fig. 7.36. Nominal model targets are indicated by dashed lines in the plot. As f_{max} and power data becomes available, acceptance ranges of IDDQ and τ_p are set as shown. Chips falling outside the box are rejected for potential P_{off} and f_{max} fails early in the test sequence.

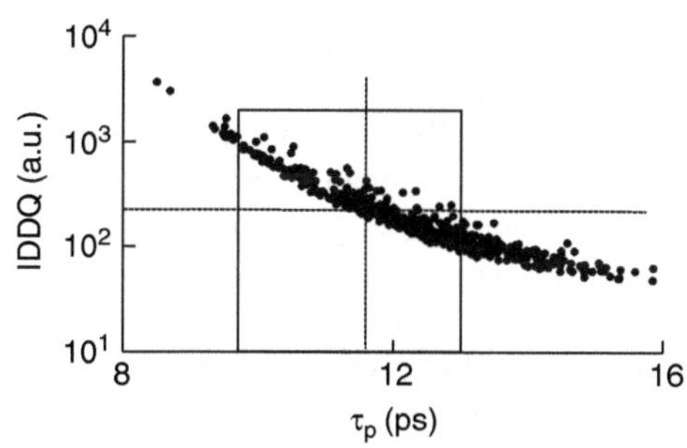

Fig. 7.36 Chip IDDQ vs. average inverter τ_p. Dashed lines show nominal model targets, and the box defined by solid lines indicates acceptance range in hardware

In Fig. 7.37 simulated values of T_{cmin} and corresponding f_{max} are plotted as a function of data path delay for the 20 inverter (FO = 4) path described in Sect. 3.4.2. The range of variation is $\pm 1\sigma L_p$ and $\pm 1\sigma Vt$. A cycle time increment of two ps in simulations matches the typical cycle time step setting of the ATE for T_{cmin} search during hardware test. A linear fit of the simulated data gives R^2 of 0.96. In hardware test, T_{cmin} or f_{max} may be plotted in a similar fashion vs. average τ_p of RODs, or τ_p of ROL12 which corresponds to the anticipated critical path. The spread in the data is generally higher in the hardware as a result of AcC variations.

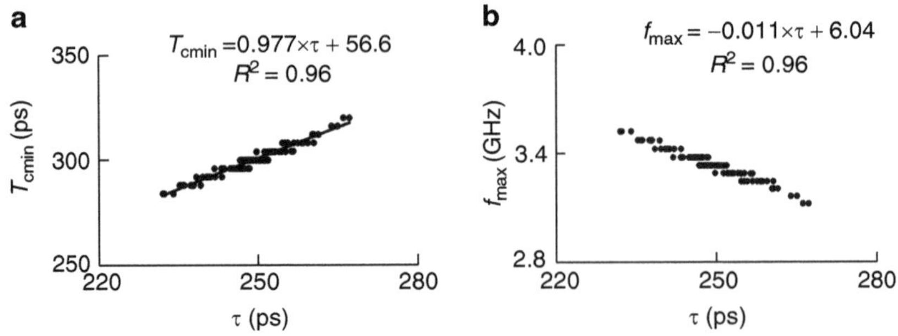

Fig. 7.37 Scatter plot of (**a**) T_{cmin} and (**b**) f_{max} as a function of path delay. 45 nm PTM HP models @1.0 *V*, 25 °C

Although ROL12 may have been intended to represent logic circuit compositions of typical cycle limiting paths, the actual cycle limiting paths on a chip may be different. Such differences are more likely to occur when there are mismatches between model and hardware arising from biases in EDA timing tools or from silicon process drifts. The dominating logic circuit configurations in cycle limiting paths can be inferred by comparing the V_{DD} and temperature sensitivities of T_{cmin} with the V_{DD} and temperature sensitivity of different ROLs [9] as illustrated below.

Circuit simulations of ckt_stgs in ROLs are carried out at V_{DD} values of 1.0 V and 0.9 V. The L_p values are varied within the $\pm 3\sigma L_p$ range to represent variations observed in the hardware. In Fig. 7.38a, ratio of % change in τ_p ($\delta\tau_p = \Delta\tau_p/\tau_p$), to % change in V_{DD} ($\delta V_{DD} = \Delta V_{DD}/V_{DD}$) is plotted as a function of τ_p for regular, high, and low V_t inverters (FO = 3). Ideally, in the vicinity of nominal V_{DD}, a 1 % change in V_{DD} is expected to result in a 1 % change in τ_p, i.e., $\delta\tau_p/\delta V_{DD} \approx 1.0$, increasing with increase in V_t/V_{DD}. By lowering V_{DD} or raising V_t, the sensitivity to V_{DD} and hence $\delta\tau_p/\delta V_{DD}$, is increased as seen in Fig. 7.38a for different V_t inverters. In Fig. 7.38b, $\delta\tau_p/\delta V_{DD}$ for a regular V_t (FO = 3) inverter is compared with an inverter with 25 % wire *RC* load, adjusted to give nearly the same nominal delay. The wire *RC* loaded inverter delay is less sensitive to V_{DD} because of the fixed *RC* load. The wire loaded inverter, however, has a larger sensitivity to temperature than the gate loaded inverter as described in Sect. 5.5.

Similar plots to those shown in Fig. 7.38 can be generated from hardware test data of τ_p for different ROLs at two values of V_{DD}. Similarly, the ratios of % change in T_{cmin} (δT_{cmin}) to % change in V_{DD} (δV_{DD}), $\delta T_{cmin}/\delta V_{DD}$ are plotted as a function of T_{cmin}. A comparison of the range of $\delta\tau_p/\delta V_{DD}$ with $\delta T_{cmin}/\delta V_{DD}$ provides insight into the circuit contents of cycle limiting paths. A major advantage of this technique is that it is based on data collected in routine characterization tests and no prior knowledge of chip design or silicon process centering is required.

The methodology described here can be extended to temperature sensitivities by comparing $\delta\tau_p/\delta T$ of ROLs comprised of different ckt_stgs designs with $\delta T_{cmin}/\delta T$. The V_{DD} and temperature values for this type of analysis are selected to get adequate resolution for making comparisons without deviating too far from the operating conditions of interest.

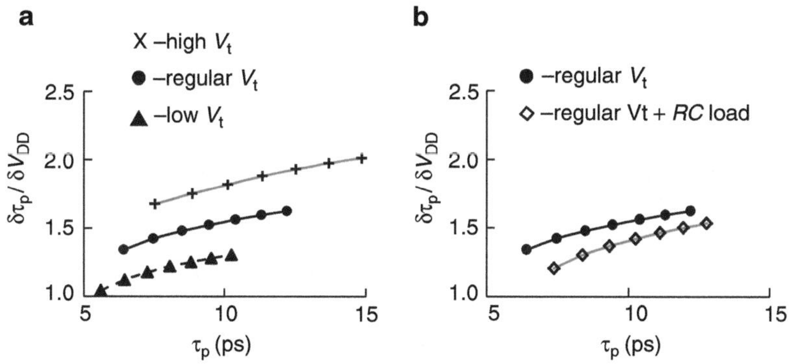

Fig. 7.38 Ratio of % change in τ_p, ($\delta\tau_p$) and % change in V_{DD} (δV_{DD}) as a function of τ_p for: (**a**) regular, high, and low V_t inverters (FO = 3), and (**b**) regular V_t inverter (FO = 3) and inverter with 25 % wire *RC* load. 45 nm PTM HP models @ 25 °C

Placement of identical RO designs on different product chips facilitates product-to-product comparisons at the same technology node. The periods of a representative set of ROs, measured at the same V_{DD} and temperature on chips of different designs, serve as references for comparing key parameters such as power, f_{max}, and V_{min}. These types of comparisons help separate differences in silicon process centering from differences in design margins and operating conditions. If embedded RO designs are faithfully migrated from one generation to the next, such comparisons can be extended to chips at different technology nodes as well.

7.4.4 Aggregate Behavior

Correlation of circuit delays obtained from ROs with chip T_{cmin} or f_{max} illustrates how the aggregate behavior of a complex chip can be described by a single circuit block. Characterization of chip power is carried out in a fashion similar to that described in Sect. 4.4. Voltage and temperature dependence of P_{off} and total power of a chip reflect the behavior of a single logic gate or a small circuit block. For a steady-state switching activity, with only the clock tree running, the switching capacitance and background leakage power can be determined from the measurement of total power as a function of frequency. This facilitates calibration of EDA power tools used for estimating chip power.

Based on the electrical test data on a sample of representative chips from the manufacturing line, empirical models for T_{cmin} and power as a function of V_{DD}, temperature, and silicon process parameters can be generated. The physical behavior of the chip must be consistent with that of the dominant circuit topologies, which can be easily validated. These models are then used to predict T_{cmin} and power under different operating conditions, providing valuable inputs for determining test corners and for debugging test codes and procedures.

7.4.5 Silicon Manufacturing Process Window

Silicon manufacturing process is tuned to maximize chip yield while meeting functional specifications. One example of determining silicon manufacturing process window is shown in Fig. 7.39. Average L_p for chips on a wafer or a column stripe are obtained from metrology data collected during manufacturing. In Fig. 7.39 chip yield, f_{max} and P_{off} are plotted as a function of average L_p value. As L_p is increased, functional yield improves, P_{off} remains low and f_{max} decreases. In this hypothetical plot, the range of L_p over which yield, P_{off} and f_{max} targets are all met is highlighted along the x-axis. This range is continuously monitored during production and the process tuned to keep the hardware within the targeted window.

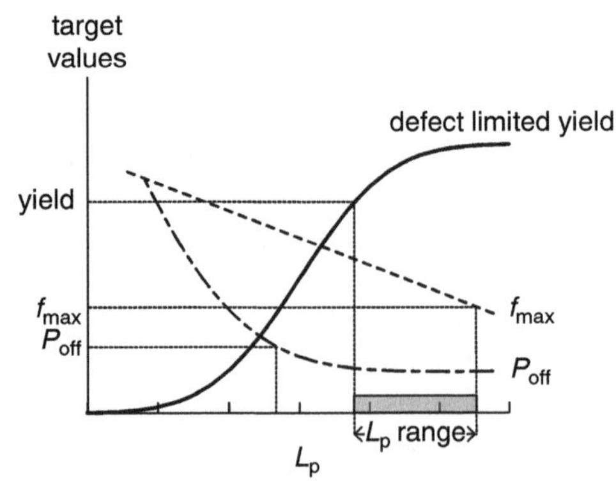

Fig. 7.39 Illustration of manufacturing process window (L_p range) to meet yield, f_{max} and P_{off} specifications

7.5 Adaptive Testing and Binning

The term adaptive testing covers a broad range of test techniques for improving product yield and performance and for improving test efficiency, based on test data and statistical data analysis. Making use of increasing compute capacity, data storage, and software tool capabilities, decisions on test definitions, test flow and acceptance and rejection criteria for individual chips or wafers can be made on the fly [10].

The idea is to generate a set of algorithms or equations based on characterization data collected on the product, historical learning, and engineering acumen to make decisions concerning each chip or wafer during test.

Some of the features of adaptive testing are:

- Real time monitoring of test results to dynamically change test conditions for each chip, or to eliminate or add tests

- Tuning V_{DD} or T_c of each chip or V_{DD} of each voltage island within a chip, based on early test results for optimum performance
- Comparing with test results from previous test steps to modify tests in subsequent test steps and to add additional screening for maverick lots
- Using statistical data analysis to identify marginal chips, trends in chips across wafer and across lots and to correlate test data from scribe-line to packaged chip
- Pre-dispositioning of chips (or lots) based on early test data (e.g., slow lot, n/p ratio offset)
- Retesting chips with questionable or unexpected test results
- Correlating among testers for integrity of interposers, thermal resistance, and measurement repeatability

Adaptive testing benefits are illustrated with three examples:

Example 1: The V_{DD} of individual chips may be tuned to achieve the f_{max} target while meeting P_{off} specifications. This is discussed in the context of a twenty inverter (FO = 4) path assuming that the f_{max} and P_{off} of the path are representative of the product chip f_{max} and P_{off}.

Consider two chips, one with nominal L_p of 0.045 μm and a second chip in which L_p of all the MOSFETs is shifted by $(+2\sigma L_p)$ due to silicon process variations. Using the circuit simulation methodology described in Sect. 3.4.2, f_{max} values for the inverter path are determined for discrete V_{DD} values ranging from 0.7 to 1.2 V. The data are shown in Fig. 7.40a where equations describing a linear relationship between f_{max} and V_{DD} are displayed on the plot. A desired f_{max} of 3.0 GHz is met for the nominal circuit at $V_{DD} = 0.96$ V. For the slow circuit, V_{DD} is raised to 1.08 V to meet the 3.0 GHz specification. In Fig. 7.40b, P_{off} values for the two circuits are plotted as a function of V_{DD} and the data fit to a power law. The P_{off} for the nominal circuit at $V_{DD} = 0.96$ V is 131 nW. For the second circuit, P_{off} is 113 nW at $V_{DD} = 1.08$ V, slightly lower than the P_{off} of the first circuit.

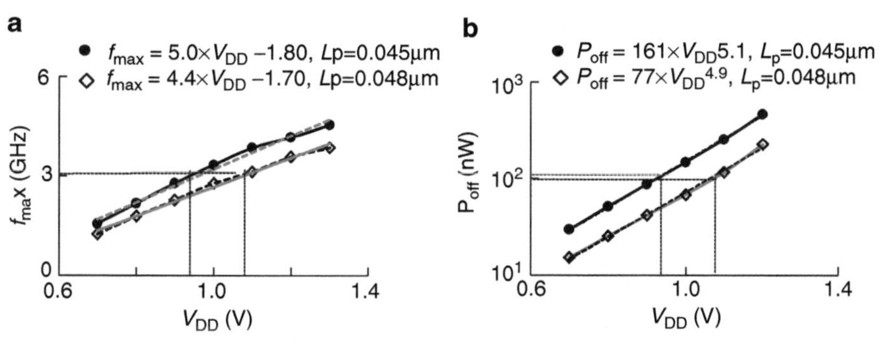

Fig. 7.40 For a 20 inverter (FO = 4) path with $L_p = 0.045$ μm (nominal) and $L_p = 0.048$ μm $(+2\sigma\ L_p)$: (**a**) f_{max} vs. V_{DD}, and (**b**) P_{off} vs. V_{DD}

Example 2: Statistical yield analysis and early test data on wafers or lots are used in determining the number of tests to be conducted. If a wafer/lot arrives with good PLY data from the silicon manufacturing line, with key device parameters well within specifications and test results of a selected few locations on the wafer give good yields, than a reduced set of test vectors and test corners may be applied to all the other chips on the wafer/lot, thereby reducing test time and cost. Conversely, if the manufacturing line and initial test sampling indicate yield below a predetermined threshold, then more rigorous testing is conducted on all other chips. This prevents partially defective or marginally functional chips from being shipped to the customer.

Example 3: Higher SRAM cell failure rates are observed in memory banks. These results are used to turn-on full bitmapping of SRAM arrays. In post-processing of the data, the bit maps are compared to the PLY and parametric data to facilitate identification of root cause (defects or process centering).

A common practice to mitigate the impact of chip-to-chip variations in performance is binning. CMOS products are often specified by frequency, power, or a combination of the two. The price paid by the customer for the highest performance chips or products is higher than that for lower performing chips. For this purpose, chips are placed in different speed bins after testing. From a cost perspective, it is good economics to sell all defect-free chips.

Assignment of speed bins again requires good engineering judgment. Too many bins adds to the complexity of handling tests and supply chain procedures. Too few bins means that a large fraction of the chips in a bin have a potential for better performance.

All chips must be ultimately tested against their specifications. It is therefore important to speed bin chips as early in the test process as possible to avoid repeated testing. On-chip ring oscillators and critical path monitors are good indicators of the chip performance. A correlation plot of $f_{\max}$ as a function of RO/CPM frequency obtained from early characterization data is useful in separating fast and slow parts.

A histogram of frequencies is on many chips measured at a constant V_{DD} shown in Fig. 7.41. The bins can also be defined with two or more parameter ranges, such as $f_1 < f < f_2$, $V_1 < V_{\mathrm{DD}} < V_2$, and $P_1 < P < P_2$. Some selections may be excluded on the basis of low frequency/high power.

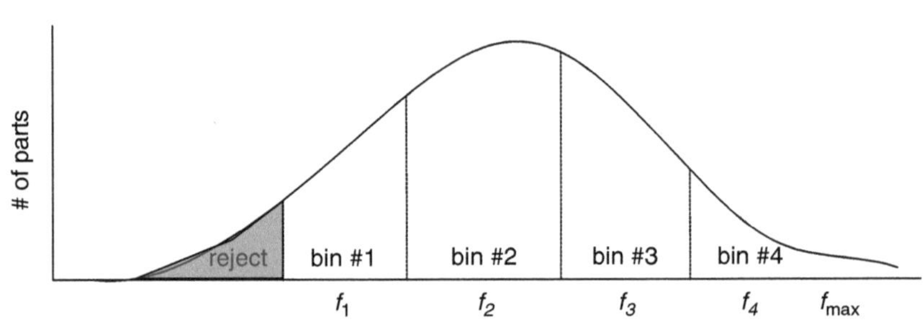

Fig. 7.41 Frequency distribution showing four speed bins

Test complexity increases with the number of binning parameters. Binning for power is useful if low power parts can be shipped with lower cost cooling arrangements (fins, blowers). Higher power parts can be typically operated at a higher frequency. These parts may be provided with more extensive cooling (multiple blowers, water or refrigerant cooling) and sold at a higher price.

Parts in each bin need to be tested at the assigned frequency and voltage for the bin. Guard-bands for each bin are also determined separately. This increases the burden on test. Hence, the net economic benefit of binning varies from product-to-product.

7.6 Summary and Exercises

Essential elements of digital CMOS chip testing are described. These include DC parametric, structural fault detection and ATPG, scan, BIST, and boundary scan. Yield modeling for defects, and for cycle time and power are discussed along with methods of failure analysis. An approach to CMOS chip characterization leveraging the embedded monitors covered in Chap. 5 is described, and modeling the aggregate behavior of the chip is introduced.

Working through the exercises in this chapter will help develop an appreciation of how the methodologies and techniques developed in Chap. 2 through Chap. 6 can be applied to digital CMOS chip test and characterization. Exercises 7.1–7.7 deal with measurements, faults, and defect-limited yield. Exercises 7.8–7.13 relate to test and characterization, and exercises 7.14 and 7.15 to adaptive tests and binning.

7.1. A CMOS chip is tested on two different test stations at identical V_{DD} and temperature set-points. The data are shown in the Table below:

Parameter	Tester #1	Tester #2
V_{DD} set-point	1.00 V	1.00 V
Temperature set-point	25 °C	25 °C
f_{max}	2.12 GHz	2.15 GHz
IDDQ	1.55 A	1.75 A
IDDA at $f = 2.0$ GHz	32.1 A	32.3 A

(a) What are the likely offsets in the two testers?
(b) What data would you need to collect to calibrate the testers?

7.2. A NAND2 gate has an open connection from p-FET P1 to Z. For what inputs would output Z be erroneous? How does the outcome depend on the previous states of A and B?

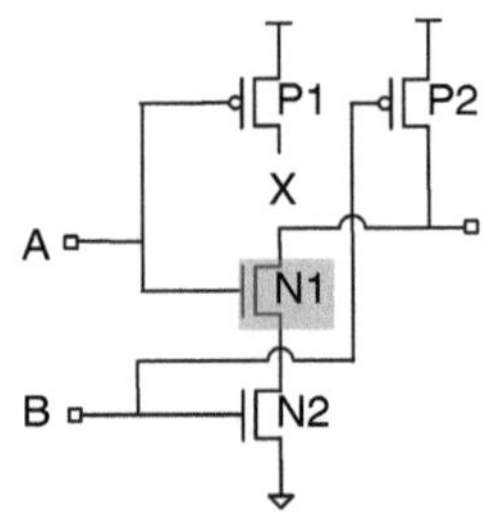

7.3. Draw a truth table of the circuit shown below.
List possible stuck-at-faults and determine fault equivalence.

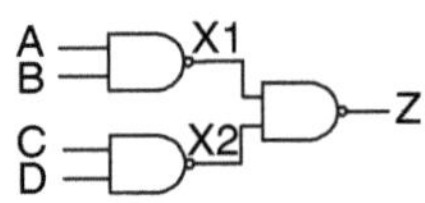

7.4. There is a high resistance in series with the output of one of the three identical NAND2 gates as shown below.
(a) Select the NAND2 widths to get $\tau_{pd} \approx \tau_{pu}$ with $(W_n + W_p) = 1$ μm. Determine R_s to get a 20 % increase in signal delay from A to Z.
(b) Under what conditions will this fault be detectable?

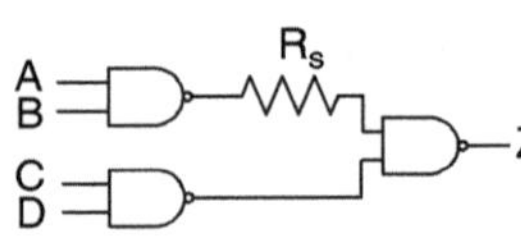

7.5. Replace the 20 inverter (FO = 4) path in Fig. 3.32 with four series connected inverters (FO = 2) to represent the circuit configuration for a scan chain.
(a) Determine f_{max} at the nominal corner. How does it compare with the 20 FO4 path?
(b) You are asked to recommend a frequency for scan test. What factors do you need to consider?

7.6. The defect density DD for a CMOS technology ready for manufacturing qualification is running at 0.005 per mm^2.
(a) Calculate the expected yield of a chip 20 mm^2 in area with cluster parameter α_c values of 0.5, 1.0, and 2.0.
(b) Repeat above for a 200 mm^2 chip. What is the desired value of the DD to get 90 % yield for this chip for α_c = 1?

7.7. A microprocessor chip in 45 nm technology that is 5 cm^2 in area has 2×10^9 MOSFETs of average width 0.2 μm (lots of SRAM!). There are ten levels of metal on this chip. In volume manufacturing the defect-limited yield has leveled off at 80 %
(a) Estimate the total length of wire on the chip.
(b) Assuming that wire opens account for 15 % of the defect-limited yield loss what is the linear opens defect density (in opens/km) for the chip wiring?
(c) What is the total MOSFET gate width on the chip?

(d) Assuming no redundancy what is a reasonable upper limit on the number of gate-dielectric shorts per unit width (in shorts/km) for this chip?

7.8. Two different product chips are designed in the same technology using the same models, EDA tools, and similar design margins. Product #1 is designed to operate at twice the frequency of product #2. Product #1 has a larger fraction of CLY fails than product #2.
 (a) How would you compare the performance of the two products in hardware?
 (b) List possible causes and electrical data needed to check for obvious differences between the two products.
 (c) If the main difference is AcC variation, prepare one chart with relevant data to illustrate and explain the differences.
 (d) If the main difference is chip operating temperature, prepare one chart with relevant data to illustrate and explain the differences.

7.9. A wafer lot having five process splits was planned with V_{tn} and V_{tp} offsets of ± 0.04 V (see Fig. 7.28b). Silicon foundry data indicate that this lot has V_t centering issues. The V_{tn} is centered 0.02 V higher and $|V_{tp}|$ centered 0.01 V lower than proposed.
 (a) List the V_{tn} and V_{tp} offsets from nominal values for the split-lots.
 (b) Which splits are outside the normal process variation range of ± 0.04 V?
 (c) This lot presents an opportunity to see the effect of a larger V_t process window. How would you filter the data to determine if a larger process window has any impact on functional yield?

7.10. RO monitors are configured to be read out by the ATE. In a new chip design, data collected by the ATE give zero frequency for most of the ROs except in some chips ROL10 and ROL11 read correctly, validating the output circuit and wiring. The design team has validated the design is correct.
 (a) What are the possible reasons for the ROs readings to be zero? How do you propose to debug these chip monitors?
 (b) How would you proceed to fix this problem?

7.11. Three different functional workloads are used for determining f_{max} of a microprocessor chip. The f_{max} values are found to be different by as much as 5 %.
 Which embedded monitors are most useful for identifying major differences in the workloads. What can you infer from the data collected from the monitors?

7.12. (a) Set up a simulation using the circuit scheme shown below to inject a noise pulse in the middle of the 20 inverter (FO = 4) data path. Set $C_L = 4 \times C_{in}$.

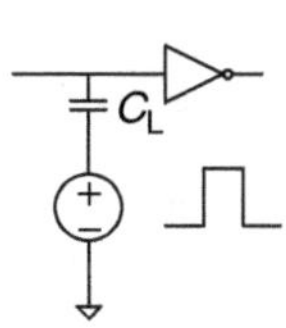

 (b) Measure T_{cmin} as a function the rise time and the arrival time of the pulse to imitate the impact of coupled noise from a neighboring signal wire.

7.13. The presentation of chip IDDQ versus average τ_p of distributed ROD data as shown in Fig. 7.36 has proven to be an extremely useful early indicator of chip performance.
   (a) For a 51 stage RO comprised of inverters (FO = 4) and a NAND2 enabling gate, simulate and plot IDDQ versus τ_p in response to independently varying V_{tn}, V_{tp}, and L_p, one at a time over their $\pm 3\sigma$ ranges.
   (b) Run Monte Carlo simulations allowing all three parameters to vary systematically but independently over their $\pm 3\sigma$ ranges and again plot IDDQ vs. τ_p.
   (c) What differences might you expect between your simulated data and data collected on CMOS chips in manufacturing test?

7.14. A chip is designed to manage power during operation by dynamically adjusting the frequency of operation.
   (a) What is the saving in power from lowering the frequency by 20 % for 30 % of the time during operation?
   (b) By what fraction can V_{DD} be lowered during the low-frequency operation? What is the net saving in power in this case?
   (c) What additional tests are required to qualify for operation at lower frequency and V_{DD}?

7.15. Run Monte Carlo simulations for 1,000 cases to determine values of τ_p for an inverter (FO = 4) using a delay chain configuration. Generate a histogram for τ_p. Assume that the dominate source of variation is L_p in its $\pm 3\sigma$ range.
   (a) Divide the frequency range into four equal intervals. How many samples are present in each bin?
   (b) What would be the frequency range in the bins if equal samples are required for each bin?

References

1. Bushnell ML, Agrawal VD (2000) Essentials of electronic testing for digital, memory and mixed-signal VLSI circuits. Springer, Boston
2. Abramovici M, Breuer MA, Friedman AD (1994) Digital systems testing & testable design. Wiley, New York
3. Weste NH, Harris D (2010) CMOS VLSI design: a circuit and systems perspective, 4th edn. Addison-Wesley
4. Bhushan M, Ketchen MB (2011) Microelectronic test structures for CMOS technology. Springer, New York
5. Schroder DK (2006) Semiconductor material and device characterization, 3rd edn. Wiley, Hoboken
6. Nigh P, Gattiker A (2004) Random and systematic defect analysis using IDDQ signature analysis for understanding fails and guiding test decisions. Proceedings of the international test conference, ITC'04, pp 309–318
7. Xanthopoulos T (ed) (2009) Clocking in modern VLSI systems. Springer, Boston
8. Tsang JC, Kash JA, Vallett DP (2000) Time resolved optical characterization of electrical activity in integrated circuits. Proc IEEE 88:1440–1459
9. Gattiker A (2011) Ying and Yang of embedded sensors in post-scaling era. IEEE 29th VLSI test symposium, pp 324–327
10. Maxwell P (2010) Adaptive test directions. IEEE international test conference ITC'10, pp 12–16

Reliability

8

Contents

8.1 Reliability and End-of-Life ... 286
    8.1.1 Accelerated Stress Tests and Failure Rates ... 288
8.2 CMOS Circuit Performance Degradation Mechanisms ... 292
    8.2.1 Bias Temperature Instability ... 292
    8.2.2 Hot Carrier Injection ... 300
    8.2.3 Time-Dependent Dielectric Breakdown ... 301
    8.2.4 Electromigration ... 302
    8.2.5 Soft Errors ... 303
8.3 Managing Reliability ... 303
    8.3.1 Voltage Screening ... 304
    8.3.2 Burn-In ... 305
    8.3.3 Guard-Banding ... 306
8.4 Summary and Exercises ... 309
References ... 310

Long-term reliability and operating margins to guarantee continued performance within specifications over the lifetime of CMOS chips are established during test. Models describing various degradation mechanisms in MOSFET and wire interconnect properties over time are provided by the silicon manufacturer and generally included in EDA tools to ensure adequate design margins. Models for failure rates of silicon process-induced defects are generated using data from accelerated stress testing of a representative sample of chips. Accelerated voltage and temperature stress testing known as "burn-in" may be carried out to eliminate chips with potential defects from the manufacturing test flow. Guard-bands between test and field operating conditions are put in place to ensure chip functionality over lifetime.

In CMOS circuits, MOSFET current drive strength degrades over time once the power supply is turned on and switching activity begins. There are several aging mechanisms, each with its own characteristic rate and dependencies on voltages, currents, and temperature.

M. Bhushan, M.B. Ketchen, *CMOS Test and Evaluation*,
DOI 10.1007/978-1-4939-1349-7_8

Aging properties of CMOS product chips are partially accounted for by the margins between manufacturing test conditions and chip specifications. Typically a chip must pass all tests at a reduced voltage and higher frequency and temperature than it will experience after shipment to customers. These "guard-bands" must be determined prior to full-scale production. Voltage and temperature stress tests cause marginal chips to fail; these chips can then be removed from the test flow. Such tests, if too severe, may inadvertently degrade good chips, reducing their useful life span. From an engineering point of view, production test specifications can be defined based on reliability modeling and early test data. However, the cost of test, field repairs, and replacements, as well as profit margins and acceptable customer satisfaction levels are important considerations in defining manufacturing test specifications.

An introduction to reliability models and accelerated stress testing is given in Sect. 8.1. Performance degradation mechanisms in CMOS circuits are described in Sect. 8.2. Managing reliability including voltage screening, burn-in, and guard-banding are discussed Sect. 8.3. The references cited for an introduction to reliability and degradation mechanisms in CMOS circuit components provide a comprehensive overview of these topics [1–3].

8.1 Reliability and End-of-Life

Reliability is technically defined as the probability that a product will perform all of its functions under a range of specified operating conditions over its stated lifetime. Poor screening in manufacturing tests can cause large fallouts in the early life of a product. Degradation in the properties of components and random environmental changes can slowly alter the failure rate of the initial survivors over the course of product life. Near and beyond the specified end-of-life (EOL), multiple wear-out mechanisms come into play, and the product failure rate begins to increase rapidly. Reliability modeling and proper screening in manufacturing tests ensure that the product failure rate will remain at an acceptable level over stated lifetime.

Failure rate of parts under use conditions is typically described by a "bathtub" curve shown in Fig. 8.1. A high failure rate in early life is observed in many products and is known as "infant mortality." The failure rate drops significantly

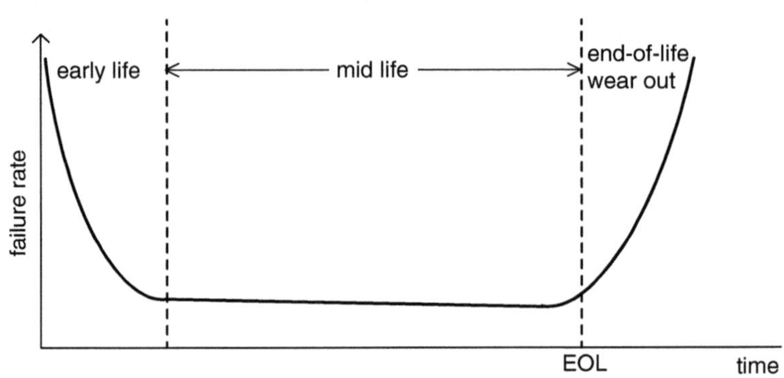

Fig. 8.1 Bathtub curve showing failure rate of a product over time

once the defective parts are identified. Over most of mid-life, the failure rate slowly decreases. Towards the EOL, the failure rate rises again as parts begin to wear out.

Different regions of the bathtub curve in Fig. 8.1 are modeled with a Weibull distribution [1] giving the probability density for failure at time t as

$$F(t, \eta_w, \beta_w) = \left(\frac{\beta_w}{\eta_w}\right)\left(\frac{t}{\eta_w}\right)^{(\beta_w - 1)} \exp\left\{-\left(\frac{t}{\eta_w}\right)^{\beta_w}\right\}, \tag{8.1}$$

where β_w is a shape factor and η_w is a scaling factor.

The shape of the failure probability density plot changes with β_w and η_w. The values of β_w and η_w are empirically determined to fit a wide range of probability density histogram shapes obtained from reliability testing. For $\eta_w = 1$,

- $\beta_w < 1$ indicates the failure probability density rapidly decreases in the early stages, with most of the parts falling in the "infant mortality" category
- $\beta_w = 1$ indicates failure probability density is low and continues to decrease exponentially with time, and that failures are caused by random events
- $\beta_w > 1$ indicates the onset of wear-out, with failure probability density rapidly increasing over time

Weibull distributions for different η_w and β_w values are shown in Fig. 8.2. In Fig. 8.2a the failure probability density is plotted for $\eta_w = 1$ and β_w values of 0.5, 1.0, and 10. In an ideal case, with $\beta_w = 1$, failures in the field are governed by random events internal or external to the parts in use. It is assumed that any defective parts are identified prior to shipping and removed from the supply chain. With $\beta_w = 0.5$, there is significant fallout in the early stages of use. This is not a favorable situation considering the cost of returns, recalls, and replacement within the product guarantee period, as well as the reputation of the product in the market place. With $\beta_w = 10$, failures begin to occur after some time has elapsed. It is desirable to have the onset of high failure rates after the specified EOL of a product. The sum of these three plots gives a "bathtub" like distribution.

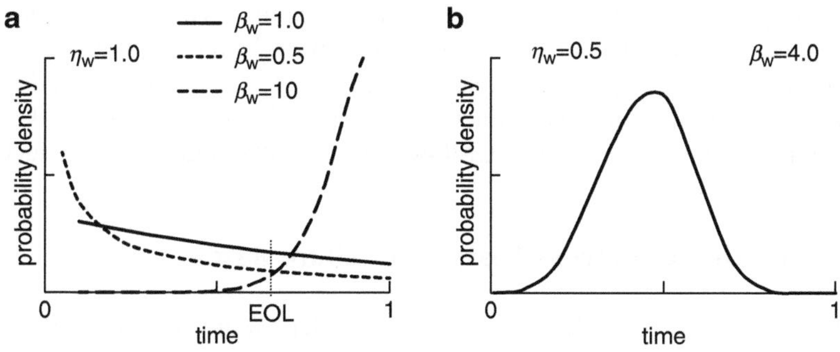

Fig. 8.2 Weibull distribution shapes for different values of η_w and β_w: (**a**) $\eta_w = 1$, $\beta_w = 0.5$, 1.0, 10, and (**b**) $\eta_w = 0.5$, $\beta_w = 4$

In Fig. 8.2b, with $\beta_w = 4$ and $\eta_w = 0.5$, the failure probability has a normal distribution. The advantage of a Weibull distribution model is that from the fitting parameters η_w and β_w failure predictions can be made with a small sample of parts. For this purpose, a special Weibull graphical plotting paper is available for analyzing failure data [1].

CMOS chip failures in the field may be attributed to:

- Degradation of silicon manufacturing induced weak points in circuits
- Degradation in the properties of circuit components
- Test escapes

The occurrence of silicon manufacturing weak points is dependent on the quality control in the foundry and the maturity of CMOS technology at a given node. Marginally passing defects may be systematic or random. Systematic defects originate from photomask and process induced line-width variations or aggressive physical layouts pushing the limits of silicon processing for some features, shapes, or device dimensions. Random defects originate from particulates, lack of line-width control, poor alignment, and other variations in the manufacturing line.

Degradation mechanisms for the properties of circuit components are well characterized during technology development and qualification and accounted for in circuit design margins. However, systematic and random variations in MOSFET and interconnect properties may push some circuit blocks to failure with use.

Test escapes occur when the manufacturing tests do not cover all possible environmental conditions and switching combinations that may lead to failures in the field. This includes incomplete testing or partial fault coverage, with some circuits on the chip not being tested at all. Occasionally the field conditions may be more severe than the conditions under which chips were tested and qualified.

For failure rate estimations of packaged chips, the degradation mechanisms in both chip and package have to be considered. Package corrosion, increase in package-to-chip resistances, and mechanical failures in flip-chip connections or wire bonds may take place with use or even on the shelf, depending on the temperature, humidity, and other environmental conditions. At printed circuit board and system level, failures in I/O interfaces, external wiring, and communication to other components come into play. Here we will only examine the failure mechanisms induced by silicon manufacturing defects and by degradation in CMOS circuit components. The general principles of product reliability described here can be extended to packaged CMOS chips and other electronic products.

8.1.1 Accelerated Stress Tests and Failure Rates

Manufacturers must be able to model and predict failure rates over the life of product chips prior to shipment. Failure data from previous generations of products in the field are collected and analyzed for learning and for making improvements. However, chip designs and specifications change with each release, and CMOS technology continuously evolves. Some failure mechanisms may no longer apply and new mechanisms may emerge over time. With typical CMOS chip lifetime of 5–11 years, reliability models based on data collected from field failures are generally not adequate to predict the reliability of future products.

In CMOS chips, failures may occur because of weaknesses in the silicon process such as narrowing of line-widths of interconnect wires and vias resulting in electrical opens due to electromigration of material over time. Current drive strengths of MOSFETs also degrade over time and may cause timing failures in critical paths with inadequate timing margins. These potential failure mechanisms, described in more detail in Sect. 8.2, are accelerated with temperature and voltage.

Accelerated stress testing is a common practice used in manufacturing of electronic products. Product functionality is evaluated under elevated operating conditions to accelerate the dominant failure mechanisms. By increasing the probability of failure, the total test time is a relatively small fraction of the expected lifetime of the product in the field. The failure data under these elevated conditions is analyzed and used for predicting the expected failure rate of the product under use conditions.

An acceleration factor, A_F, is defined as ratio of failure rate under accelerated test condition to the failure rate under nominal operating conditions. The acceleration factors for each of the failure mechanisms are functions of temperature and voltage.

All temperature activated mechanisms follow an Arrhenius equation with characteristic activation energy E_a. The temperature dependence of the failure rate λ is expressed as

$$\lambda = c1 \times \exp\left(-\frac{E_a}{kT}\right), \tag{8.2}$$

where $c1$ is a prefactor, and temperature is in °K. The acceleration factor at an elevated temperature T_2 over a nominal use temperature of T_1 is given by

$$A_{FT} = \exp\left(\frac{E_a}{kT_1} - \frac{E_a}{kT_2}\right). \tag{8.3}$$

From the above expression, with $T_2 = 393$ °K (120 °C), $T_1 = 323$ °K (50 °C), and $E_a = 0.2$ eV, we get $A_{FT} = 3.6$.

Similarly, if the voltage dependence of a failure mechanism is exponential, the voltage acceleration factor for a stress voltage V_{stress} over a nominal operating voltage V_{DD} is expressed as

$$A_{FV} = \exp\{c2 \times (V_{stress} - V_{DD})\}, \tag{8.4}$$

where $c2$ is an empirically determined constant.

If there is only a single dominant failure mechanism and both temperature and voltage are elevated, the combined acceleration factor is the product of voltage acceleration factor A_{FV} and temperature acceleration factor A_{FT}.

$$A_F = A_{FV} \times A_{FT}. \tag{8.5}$$

When there are two or more competing failure mechanisms, each accelerated by both temperature and voltage, their combined A_F is dependent on their relative failure rates under nominal and accelerated stress conditions. In practice, it is generally difficult to separate the acceleration factors of such competing mechanisms, and an empirically estimated combined acceleration factor may be used instead.

Reliability is expressed in terms of a failure unit, failure in time (FIT), defined as the number of failures per 10^9 (billion) hours of use [1]. FIT as determined under accelerated stress conditions is expressed as

$$\text{FIT} = \frac{\#\ \text{of failures}}{(\#\ \text{tested} \times \text{hours} \times A_\text{F})} \times 10^9. \tag{8.6}$$

When there are n different mechanisms contributing to failure, and FIT for each mechanism can be independently determined, the aggregate FIT is given as a sum

$$\text{FIT} = \sum_{i=1}^{n} \text{FIT}_i. \tag{8.7}$$

Hence the failure mechanism with the largest FIT dominates.

Another measure of failure rate is mean time between failures (MTBF) which is the inverse of the number of failures in a million hours

$$\text{MTBF} = \frac{10^6}{\#\ \text{of failures}}. \tag{8.8}$$

MTBF is used for field repairable products and generally not applicable to CMOS chips. Instead the mean time to failure (MTTF) is used. MTTF is defined as the average length of time for a chip to fail. For an exponential failure probability distribution ($\beta_\text{w} = 1$ in Eq. 8.1), MTTF is given by η_w.

During silicon technology development, accelerated stress tests are conducted on discrete circuit components (interconnects, MOSFETs), ring oscillators, and small circuit blocks. Feedback to the technology development team helps them build a more robust technology. Reliability models of different aging mechanisms based on these tests are ported to circuit simulators used in chip design. Circuit design and physical layout guidelines include the effects of these aging mechanisms to ensure full functionality of the chip throughout designated lifetime. The guidelines are based on an acceptable statistical failure probability.

Prior to full-scale production, a statistically representative sample of CMOS chips is subjected to accelerated stress tests. These chips are stressed for a long time (up to 1,000 h or 1 % of product lifetime) and tested at suitable intervals. In these stress tests, all degradation mechanisms act simultaneously and the dominant mechanism prevails. Silicon process variations may also influence the dominant mechanism on different chips.

The desirable FIT under normal use conditions is 100 or less. With FIT = 100, if 1,000 chips are tested for 1,000 h (~42 days), A_F must be ≥ 100 to observe at least ten failures. If $A_\text{F} \approx 10$, on average only one chip will fail in test, and the results of this stress experiment will not be statistically significant.

Designing stress test experiments requires a good understanding of the dominant failure mechanisms and their acceleration factors at a given technology node from individual component reliability models. Such models are generally not available for chip design-specific failures. The sample size and number of stress hours for the experiments are strongly influenced by the cost of chips, test equipment, test time and characterization, all in the context of the reliability requirements.

Figure 8.3 shows a bimodal distribution of failure rates where two distinct mechanisms are contributing to failures. A Weibull model of such a distribution would require different fitting parameters η_w and β_w for different regions. The sample set may be divided into subsets, and each subset stressed at different voltage and temperature to accelerate these widely different failure mechanisms. The disadvantage in this approach is that the sample size in each subset is reduced and may not be sufficient to build statistically accurate reliability models.

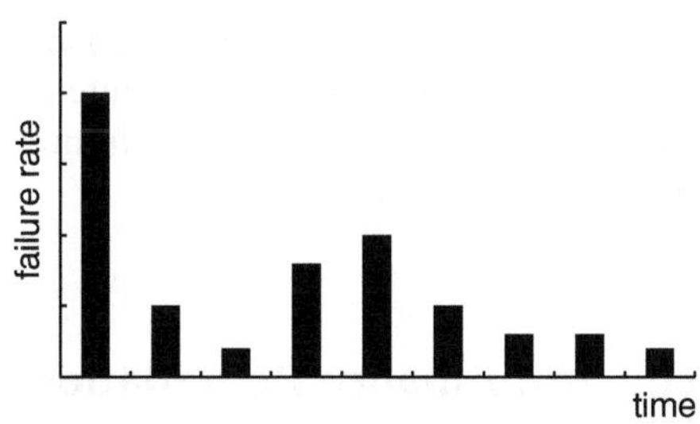

Fig. 8.3 A bimodal distribution of failure rate as a function of time, indicating two different failure modes

Selection of stress conditions needs careful consideration as chip power increases at higher V_{DD} and temperature. Nominal test conditions, and two different accelerated stress conditions are shown in Table 8.1. Increase in P_{off} during stress test relative to the nominal use conditions is illustrated with simulated data from our standard inverter using 45 nm PTM HP models. At high voltage and temperature, there is more than 15× increase in P_{off}. Note that PTM HP models may not be accurate for V_{DD} >1.2 V.

Table 8.1 Normalized P_{off} under accelerated stress conditions for the standard inverter. 45 nm PTM HP models

Test	V_{DD} (V)	Temperature (°C)	P_{off} ratio to nominal
Nominal	1.0	40	1
Accelerated stress 1	1.3	80	9
Accelerated stress 2	1.3	120	15

During accelerated stress testing functional tests are run to toggle node voltages. The test patterns also validate functionality, and a failing chip is identified immediately and removed from the test flow for diagnostics. Clock frequency is lowered to reduce AC power. Alternatively, AC power consumption is reduced by sequentially exercising smaller units on the chip. Chips are then periodically cooled to nominal test temperature and tested under nominal conditions for full characterization. As aging processes occur at a logarithmic rate in time, testing intervals are designed to appear linear on a logarithmic scale, e.g. test after 1, 10, 100, and 1,000 h of stress time.

Voltage and temperature stress tests may require new test boards and test equipment to handle higher power required for maintaining a constant chip temperature. This is particularly important in high performance chips with high leakage currents. The chip temperature is monitored using data collected from embedded temperature monitors. Special cooling equipment may be needed to prevent overheating and thermal runaway at chip hot spots.

Accelerated tests expose both defect-induced fails and fails due to known aging mechanisms described in Sect. 8.2. Failed chips are removed from the sample and analyzed for root cause. Any systematic failure mechanisms are identified and actions taken to prevent such failures in present or future designs. As an example, a physical layout style may be more susceptible to electromigration failures because of bends and corners in a high current density wires.

8.2 CMOS Circuit Performance Degradation Mechanisms

Impurities in the gate-dielectric material cause an increase in threshold voltage in MOSFETs in the on-state. This effect is known as negative bias temperature instability (NBTI) in p-FETs and positive bias temperature instability (PBTI) in n-FETs. Hot-carrier injection (HCI) effects reduce mobility over time degrading the current drive of both p-FETs and n-FETs. These effects increase circuit delay and reduce timing margins. Time-dependent dielectric breakdown (TDDB) of the MOSFET gate-dielectric is another reliability concern.

Electromigration in metal wires is a well-known phenomenon. Agglomeration of metal in the direction of current flow over time can create opens in the conducting paths and shorts between neighboring wires, thereby causing circuits to fail. This effect is enhanced at higher temperatures. Electromigration-induced failures are more likely to occur in DC paths, in interconnects with high current densities and in hot spot areas of the chip. Mismatch in mechanical properties and thermal expansion coefficients of dielectric and metal layers in an interconnect stack can create opens with thermal cycling from stress-induced voiding (SIV).

8.2.1 Bias Temperature Instability

Charge carriers, holes in p-FETs and electrons in n-FETs, damage the gate-dielectric to silicon interface in the presence of an electric field between the gate and silicon. Silicon–hydrogen bonds at the Si–dielectric interface are broken generating traps whenever a negative voltage is applied to the gate of a p-FET or a positive voltage is applied to the gate of an n-FET. These traps at the interface capture minority carriers in the channel. The V_t of MOSFETs increases (more positive for n-FET and more negative for p-FET) over time, and current drive and transconductance g_m are lowered. This phenomenon is known as bias temperature instability (BTI).

BTI occurs only when a MOSFET is in the on-state, corresponding to a negative V_{gs} for p-FETs and a positive V_{gs} for n-FETs. All CMOS technologies exhibit degradation in p-FET characteristics over time from NBTI. With the introduction of HK gate-dielectric at the 45 nm technology node and beyond, degradation from PBTI has been observed in n-FETs. BTI is therefore an important consideration in setting product specifications and guard-banding as discussed in Sect. 8.3.3.

Several different models for estimating the shift in threshold voltage $\Delta|V_t|$ have been proposed. In general, $\Delta|V_t|$ for a MOSFET in the on-state for time t is expressed as

$$\Delta|V_t| = c3 \times \exp\left(\frac{-E_{abti}}{kT_j}\right) \times \left(\frac{V_{DD}}{t_{ox}}\right) \times t^{n1}, \tag{8.9}$$

where T_j is the temperature of the MOSFET, E_{abti} is the activation energy for BTI, and $c3$ and n_1 are constants. The degradation proceeds at a higher rate in early life of the product. This is illustrated in Fig. 8.4 assuming $n_1 = 0.2$ in Eq. 8.9. In this linear-log plot, product EOL is assumed to be 100,000 h (11.4 years). The first 10 % degradation in V_t occurs in the first 1 h and 50 % of the EOL degradation occurs in the first 3,000 h (~4 months). As the V_t increases, IDDQ and P_{off} are lowered.

Fig. 8.4 $\Delta|V_t|$ in arbitrary units (a.u.) due to BTI as a function of time

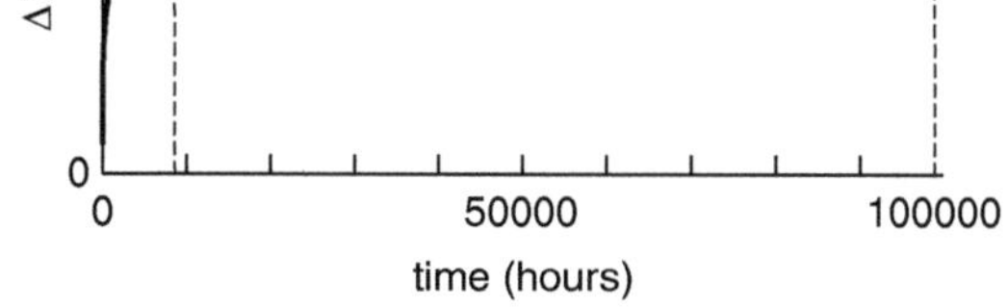

When a MOSFET is turned off and $V_{gs} = 0$, hydrogen atoms begin to diffuse back to the Si–SiO_2 interface, annealing the broken Si–H bonds. This self-healing or relaxation reduces $\Delta|V_t|$. The change in $\Delta|V_t|$ during relaxation after stressing for a time t_{stress} is proportional to the number of traps generated and is modeled as

$$\Delta|V_t| = \frac{-c4}{\left\{1 + c5\left(\frac{t_{relax}}{t_{stress}}\right)^{n2}\right\}}, \tag{8.10}$$

where $c4$, $c5$, and n_2 are constants and t_{relax} is the time elapsed after removing the stress with the MOSFET in the off-state. The shift of $|V_t|$ with periodic stress and recovery is illustrated in Fig. 8.5. Here the recovery is not completed within one cycle; hence $\Delta|V_t|$ steadily increases with time.

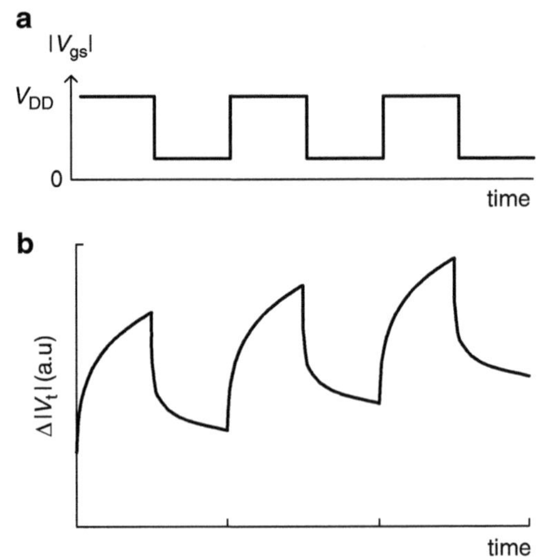

Fig. 8.5 (**a**) $|V_{gs}|$ vs. time and (**b**) corresponding $\Delta|V_t|$ vs. time showing reduction due to relaxation when $|V_{gs}| = 0$

BTI degradation is measured after the stress voltage is lowered to the operating V_{DD}. The initial recovery of $|V_t|$ is rapid, with time constants in the μs range or even lower. Typical measurement times in dedicated test structures are of the order of milliseconds, during which time some healing may already have occurred. Test equipment for measuring $\Delta|V_t|$ within a few hundred nanoseconds after stress in now available for MOSFETs and ring oscillators. Measurements of embedded RO frequencies and f_{max} and V_{min} of CMOS chips for estimating the degradation due to BTI have a longer wait time and include significant relaxation. Partial recovery from BTI degradation also occurs by annealing at a few hundred °C. For preserving simplicity in the following discussions for comparing AC and DC operating conditions, the effect of relaxation is not considered.

In logic gates, BTI degradation in MOSFETs and, in turn, increase in V_t and delay over time are functions of the duty cycles of the input voltages. In a chain of two inverters shown in Fig. 8.6a, with input node A at "0", P1 and N2 are in the on-state and subjected to NBTI and PBTI degradation, respectively. After a long time if A transitions to a "1", the PD transition in the first inverter and PU transition in the second inverter are mainly driven by P2 and N1 which are not affected by BTI. The net degradation in delay will be very small with the transition of A to a "1", although there will be a delay increase during a subsequent transition of A back to "0". Similarly, in Fig. 8.6b, input A remains at "1" for a long time and N1 and P2 are degraded. When input A switches to a "0", PU and PD delays are only marginally affected. In Fig. 8.6c, with a periodic input at A having a 50 % duty cycle, all four MOSFETs are subjected to BTI degradation for half the time. In this case, delays for all the transitions are affected. If the frequency of the input signal is, for example, in the kHz to MHz range, there will be some recovery after each transition due to BTI relaxation, and the delay degradation is reduced from what would otherwise be expected.

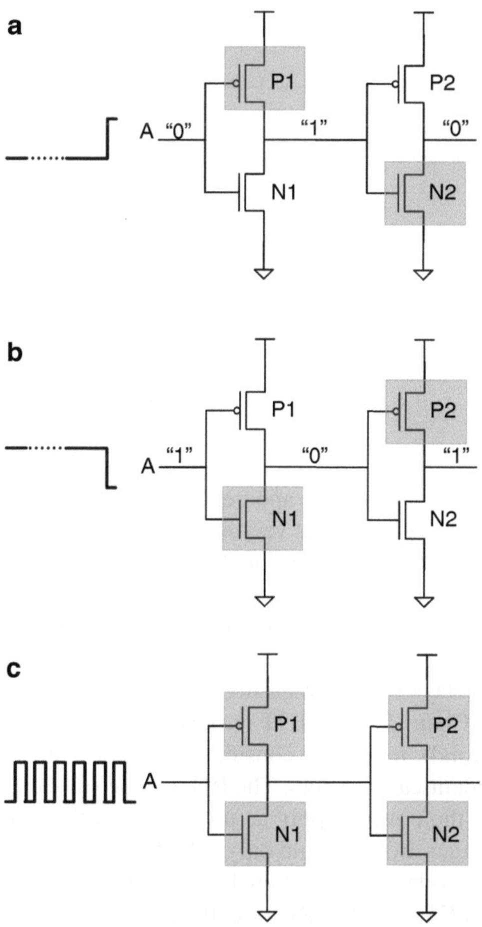

a

MOSFET	NBTI	PBTI
P1	X	
N1		
P2		
N2		X

b

MOSFET	NBTI	PBTI
P1		
N1	X	
P2		X
N2		

c

MOSFET	NBTI	PBTI
P1	X	
N1		X
P2	X	
N2		X

Fig. 8.6 Two series connected inverter stages with MOSFETs undergoing BTI degradation shaded in *gray* and indicated in the table: (**a**) input at "1", (**b**) input at "0", and (**c**) periodic input with 50 % duty cycle

A 6T SRAM cell, with "0" at its left node is shown in Fig. 8.7. If the cell remains in this state for a long time or if a "0" is rewritten in the cell repeatedly, V_t of n-FET NL will increase due to PBTI, and that of p-FET PR due to NBTI. The access pass-transistors NPL and NPR are turned on only during read or write operation and will experience a much smaller V_t shift than NL and PR. With a weaker NL, the read current will decrease, increasing the read time. On the other hand, write time to change the state of the cell to a "1" on the left-node ("0" on the right node) will be reduced slightly. However, the write time to flip the cell to again write a "0" on the left-node without a long wait time will be longer. The SRAM cell will remain symmetric if alternating "0" s and "1" s are written in the cell at regular intervals. In practice the state of a cell will be somewhere between this and a static case.

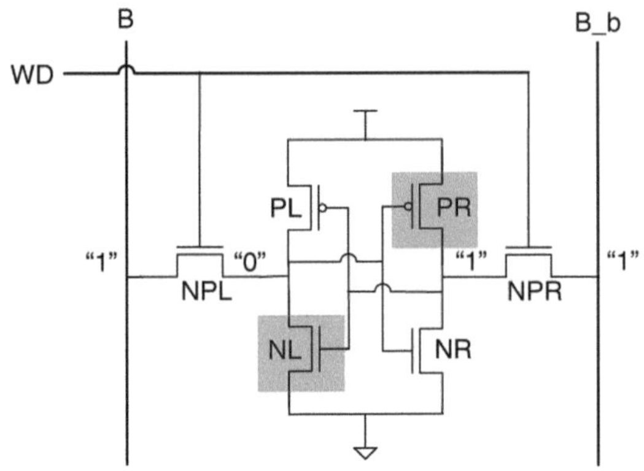

Fig. 8.7 Asymmetric BTI degradation in a 6T SRAM cell at pre-charge with "0" in its left node

From the two examples discussed above it is apparent that the impact of BTI degradation varies with circuit topology and input vectors. These effects should be taken into account when modeling BTI degradation for chip timing. The magnitude of the effect is different for n-FETs and p-FETs at different CMOS technology nodes. Generally circuits with higher sensitivity to *p*/*n* ratio are more susceptible to failure when only NBTI degradation is present.

The impact of BTI on circuit delays may be measured with delay chain or ring oscillator monitors described in Chap. 5. As PBTI is only present in advanced technology nodes (45 nm and beyond), let us first consider the effect of only NBTI on an RO comprising nominally identical inverters. The RO may be stressed in two different modes. In a static mode, the power supply is turned on, but the RO is not enabled. In this case, all node voltages are static and only p-FETs in alternate stages degrade over time. In a dynamic mode, the RO is enabled and the node voltages switch every half cycle. All p-FETs in the RO are then subjected to NBTI degradation 50 % of the time. From Eq. 8.9 with $n_1 = 0.2$ and a duty cycle of 50 %, and assuming no relaxation, $\Delta|V_t|$ is 87 % of the value for a p-FET under static stress.

The effect of stress modes on an RO is illustrated in Table 8.2 with circuit simulations using 45 nm PTM HP models. Inverter (FO = 4) IDDQ and PD and PU delays τ_{pd} and τ_{pu} are determined for the nominal case. In static stress mode, an increase in $\Delta|V_{tp}|$ of 0.02 V is assumed, and this shift in V_t applies to p-FETs in alternate stages. The τ_{pu} of alternate stages is increased in a full cycle while τ_{pd} of these stages is nearly unchanged. Note that there is no change in IDDQ prior to switching because the p-FETs in the off-state contributing to IDDQ are not degraded.

Table 8.2 Simulated effect of NBTI degradation in a standard inverter (FO = 4) and in RO period. 45 nm PTM HP models @1.0 V, 25 °C

Stress mode	$\Delta\|V_t\|$ (V)	IDDQ (nA)	RO period (ps)
None	0.000	6.7E-9	593
Static	0.020	6.7E-9	603
Dynamic	0.017	6.2E-9	609

In the dynamic mode, with 50 % duty cycle, all the p-FETs undergo a $|V_t|$ increase of 0.017 V, 87 % of the DC stress value. The IDDQ is lowered and there is a larger increase in RO period. The net change in RO period is ~1.7 % for static stress and 2.7 % for dynamic stress. The shift in $|V_t|$ under dynamic stress conditions would be smaller when a recovery model is incorporated.

In order to improve measurement accuracy of RO frequency degradation, ROs are stressed at a higher V_{DD}, but V_{DD} is lowered for RO frequency measurements. Higher V_{DD} accelerates the stress while the % shift in frequency with $\Delta|V_t|$ is larger at lower V_{DD}. The measurements must be performed immediately after lowering the V_{DD} to minimize BTI recovery after stress.

The sensitivity to V_t shifts in passgate circuits is higher than in static CMOS logic gates as discussed in Sect. 5.6.3 [4]. A scheme for directly estimating $\Delta|V_t|$ with an RO comprising a p-passgate is shown in Fig. 8.8. The schematic of the RO ckt_stg in stress mode is shown in Fig. 8.8a. With $V_{DD} = 0$, the gate terminal of the p-passgate is set at a negative voltage V_{bti} ($-1.5\times$ nominal operating V_{DD}) during stress. In this method of stressing, with source and drain at GND and gate at a negative potential, a p-FET sees the full NBTI stress. In the RO frequency measurement mode shown in Fig. 8.8b, V_{DD} is turned on and the p-FET gate is connected to GND. Measurement of RO frequencies are made before and after stress. The relative increase in average delay after stress gives a measure of NBTI degradation.

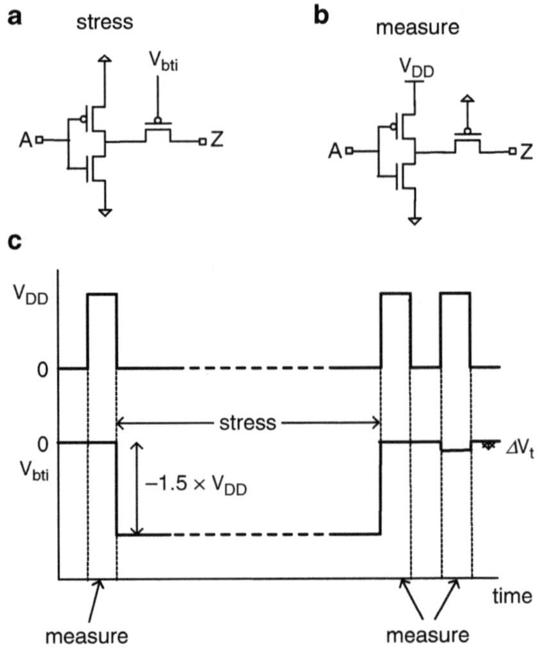

Fig. 8.8 Circuit schematics of a ckt_stg for determining $\Delta|V_t|$ from BTI: (**a**) stress mode, (**b**) measurement mode, and (**c**) voltage waveforms in stress and measurement modes

A direct measure of $\Delta|V_t|$ is obtained by applying a negative bias to the p-passgate of the RO during the post-stress measurement such that its delay matches the pre-stress delay. Timing diagrams for V_{DD} and V_{bti} are shown in Fig. 8.8c. The value of $|V_{bti}|$ to achieve matched delays is $\Delta|V_{tp}|$. A similar scheme with an n-passgate may be used for measuring PBTI degradation. The gate terminal of the n-passgate is tied to V_{DD} during RO frequency measurements and raised to a higher positive value V_{bti} during stress.

BTI degradation in large logic circuit and memory blocks or a CMOS chip is observed by measuring shifts in V_{min} and f_{max} which track with V_t. V_{min} increases and f_{max} decreases after stress. Stressing with a typical functional workload changes node voltages and stresses different MOSFETs at different rates. This method of applying stress may give a better representation of field usage than static stress. In reality, the chip usage may vary substantially in the field, with different exposures to BTI.

Model validation of NBTI effects is carried out with accelerated stress tests. With $E_{abti} = 0.05$ eV, $t_{ox} = 1.2$ nm, $n_1 = 0.2$, and $c3 = 20$ in Eq. 8.9, $\Delta|V_t|$ estimations at different values of V_{DD} and temperature for a stress time of 1 h are listed in Table 8.3. In this example, in order to measure a $|V_t|$ shift of ~0.02 V, the stress conditions would be set at 1.5 V, 120 °C.

Table 8.3 Estimated ΔV_t in 1 h from Eq. 8.9 at different values of V_{DD} and temperature

V_{DD} (V)	Temperature (°C)	$\Delta\lvert V_t\rvert$ (V)
1.0	25	0.007
1.0	120	0.012
1.5	120	0.023
1.8	120	0.097

BTI degradation may be monitored in the field using BTI-sensitive RO monitors [5]. A circuit schematic for this scheme is shown in Fig. 8.9. There are two nominally identical ROs. The reference RO1 is power gated and V_{DD} is applied only during a short measurement interval. RO2 is directly tied to the chip power grid and undergoes BTI degradation. The RO frequencies are measured prior to shipment and these values are used as reference. In the field, the RO frequencies are periodically sampled. The shift in their frequency difference Δf gives a measure of circuit delay degradation. RO2 may remain in a quiescent or oscillating state to measure delay shift corresponding to DC or AC stress. Such embedded monitors may be placed in areas of high power densities to get the worst-case degradation in the field.

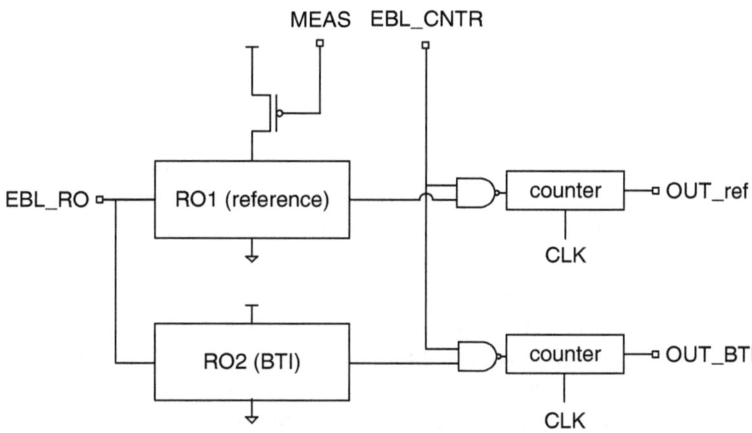

Fig. 8.9 Circuit schematic for monitoring on-chip BTI degradation in circuit delays during its lifetime

There are several different ways to utilize BTI monitor data. The monitor may be read only prior to shipping and the data stored in a database. When a chip fails in the field due to a timing error and is returned to the supplier, the monitors may be measured again to assist in diagnostics. It should be noted that because of BTI recovery, the measurements made during diagnostics may not reflect the situation in the field at the time of failure.

In another scenario, the initial frequencies of the ROs are stored on the chip in an EPROM. The ROs are sampled periodically and the measured frequencies compared with the stored values. If the frequency degradation over time exceeds a specified threshold, an alert warning may be sent to the supplier.

In a more sophisticated approach, the chip V_{DD} may be increased on the fly to overcome degradation due to BTI. Since the IDDQ is lowered over time, increase in total power with V_{DD} is somewhat mitigated by the decrease in P_{off}. Alternatively, the operating frequency may be lowered to compensate for the degradation. BTI monitors must be characterized to calibrate the frequency shift with shift in f_{max}. Because of different circuit topologies coming into play, the RO frequency and f_{max} degradation may not have a 1:1 correlation. If multiple BTI monitors are embedded on the chip, statistical data on many chips can be used to find which monitor design has the best correlation with f_{max}. In general, this approach requires considerable resources in hardware, software, test, and characterization. It is practical only when there is an economic advantage in having a modest performance enhancement in the early life of the chips.

At the time of shipping, the chip has a higher potential performance than at EOL. The anticipated BTI degradation in chip performance is included in guard-banding. The guard-band may be applied to chip voltage or frequency or distributed among both as discussed in Sect. 8.3.3.

8.2.2 Hot Carrier Injection

Hot carriers are produced when charge carriers (electrons in n-FETs and holes in p-FETs) are accelerated in the electric field across the channel and may attain an energy higher than the bandgap of the gate-dielectric. These hot carriers may get injected into the gate-dielectric on the drain side and cause damage by generating traps in the dielectric or at the silicon–dielectric interface. Hot carriers may originate in the channel (CHC) or in the substrate (SHC). Degradation from CHC is modeled when the channel is conducting ($V_{gs} > V_t$), and also under burn-in conditions in a nonconducting ($V_{gs} = 0$, $V_{ds} \sim V_{DD}$) state. The result is a decrease in the drain current and transconductance of MOSFETs.

Under nominal operating conditions, HCI degradation occurs only during switching. The equivalent time over a cycle during which this degradation takes place is illustrated in Fig. 8.10. During this time, $V_{gs} > V_t$ and $V_{ds} \geq V_{DD}/2$. HCI degradation is more severe in highly loaded gates, where the signal rise and fall times are a significant fraction of the cycle time. HCI is stronger in n-FETs than in p-FETs.

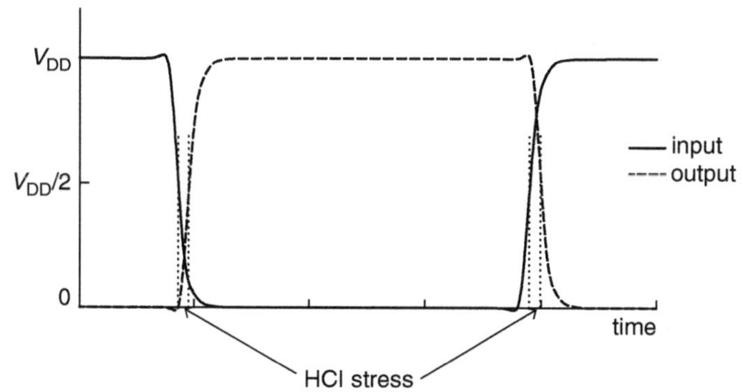

Fig. 8.10 Input and output voltage waveforms of an inverter showing when conducting HCI degradation occurs in time

The degradation in drain current in the saturation region is expressed as

$$\Delta\ |I_{ds}| = -c6 \times L_p^{-n3} \times \exp\left(\frac{V_{ds}}{V_o}\right) \times t_{eq}^{n4}, \tag{8.11}$$

where $c6$, $n3$, V_o, and $n4$ are constants. The degradation is higher in narrow channel MOSFETs and increases with V_{ds}. The equivalent time, t_{eq}, is computed as the running total of the time a MOSFET is subjected to HCI during switching transients.

Conducting HCI is very sensitive to voltage but is nearly independent of temperature, although in some cases the degradation may decrease at higher temperatures. Nonconducting HCI occurs when a MOSFET is in the off-state with $V_{ds} = V_{DD}$. Nonconducting HCI is significant during stress or burn-in at elevated voltages and temperatures and with high I_{off}.

Reliability models for HCI are based on measurements made on individual devices. The CMOS technology target is to engineer MOSFETs such that HCI remains below some predetermined limit at normal operating voltages.

8.2.3 Time-Dependent Dielectric Breakdown

When a high electric field is applied across the gate-dielectric layer, damage to the dielectric occurs by electron and hole trapping. If the density of the traps becomes high enough, a conducting path through the dielectric is created. A leakage current then flows through the dielectric in addition to the tunneling current. If the magnitude of this additional current is small, it is termed a soft breakdown. In this case the circuit may continue to function normally. A hard breakdown occurs when the leakage is very high and results in thermal runaway and local melting of the dielectric material. TDDB is a consequence of a high electric field across the gate-dielectric.

It occurs across the channel when the MOSFET is on and at the gate-drain overlap region when the MOSFET is off.

The MTTF for TDDB is given by

$$\mathrm{MTTF} = c7 \times \exp\left(\frac{c8 \times t_{\mathrm{ox}}}{V_{\mathrm{DD}}}\right) \times \exp\left(\frac{E_{\mathrm{td}}}{k\mathrm{T}}\right) \tag{8.12}$$

where E_{td} is the activation energy, and $c7$ and $c8$ are constants. A maximum V_{DD} value, V_{max}, for a chip is based on the TDDB models provided by the foundry. The value of V_{max} is obtained for the maximum T_j, power-on-hours (POH) and gate-dielectric area, and includes any voltage overshoot during switching. It is one of the considerations in setting an upper limit to adaptive V_{DD} for slow circuits (Sect. 7.5).

8.2.4 Electromigration

A major source of reliability concern in interconnects is the phenomenon of electromigration (EM). At high current densities ($>10^5\mathrm{A/cm}^2$), electrons flowing through metal transfer some of their momenta to atoms, resulting in accumulation of material on the anode side. The resistance of metal wire increases on the cathode side from thinning or development of micro-cracks. Ultimately, a metal void may be created with the wire no longer providing a conducting path. If the metal movement is in the direction of a neighboring wire, such as at a corner, the two wires may become electrically shorted. In other situations, increase in wire resistance may cause local heating leading to thermal runaway. Increase in wire resistance may also cause failures in timing of critical paths, although such paths may still be functional at lower frequencies.

Electromigration is most severe in DC current carrying wires such as those in the power grid and in any circuit drawing DC current. CMOS circuits driving capacitive loads may also have an average DC current component. Based on the models for electromigration for a technology, design guidelines for maximum current carrying ability of wires are provided by the silicon foundry. As electromigration is strongly temperature dependent, current limits for wires are set at the maximum operating temperature of the product. MTTF is estimated from Black's equation

$$\mathrm{MTTF} = c9 \times J^{-n5} \exp\left(\frac{E_{\mathrm{ae}}}{kT_j}\right), \tag{8.13}$$

where $c9$ is a constant, J is the current density, E_{ae} is the activation energy for electromigration, and the exponent $n5$ is dependent on the film composition. For Cu interconnects, $E_{\mathrm{ae}} \approx 0.95$ eV and $n5 \approx 2$ [1].

Assuming that the electromigration models provided by the foundry faithfully represent the hardware, wire widths should be adequate by design to prevent failures due to electromigration. Silicon process induced variations in wire widths and narrowing of wires at corners and bends may cause wires to be more susceptible to electromigration failures on some chips. Another source of electromigration failures is from landing of particulates on wires during fabrication, creating weak links. Extended stress tests may highlight such weaknesses in silicon manufacturing.

8.2.5 Soft Errors

Soft errors originate with the impact of high energy radiation. These errors are randomly occurring recoverable events that affect memory cells, clocked storage elements, and combinational logic by upsetting a node voltage. When a high-energy charged particle strikes the silicon surface, it generates electron–hole pairs along its trajectory through silicon. These electrons and holes can be collected at the diffusion nodes of MOSFETs causing a current spike. If the transient current integrated over time or the equivalent total charge deposited at a node exceeds a critical charge, Q_{crit}, the node voltage can change causing a single event upset (SEU). A memory cell may flip causing the stored datum to change. However, error detection and correction codes in memory array operations minimize the impact of soft errors. Flip-flops and latches are becoming increasingly prone to soft errors with scaling. Soft errors in combinational logic are rare as a glitch in a node voltage has to travel through to a clocked storage element (register file or latch) to be detected.

Soft error rate (SER) is a function of radiation dose, circuit cross section, node voltages, and node capacitances. The primary types of radiations of concern are alpha particles and cosmic rays. Removal of radioactive isotopes from chip packaging materials has reduced alpha particle emission. Secondary emission of neutrons from cosmic ray bombardment remains as a major source of soft errors in CMOS circuits. The radiation dose is low at sea-level but increases with altitude and is much higher in outer space. Circuit cross sections have been decreasing with scaling, reducing the probability of a radiation strike. At the same time, node voltages and capacitances are also getting smaller, reducing Q_{crit}, and there is a net increase in SER with scaling.

Accelerated testing to estimate FIT of soft errors requires a source of high energy neutrons. Such tests can only be conducted in special facilities on a representative sample set. Failure data may also be collected at high altitudes from a large number of samples for a long enough time to be statistically significant.

8.3 Managing Reliability

Product reliability is addressed in several different steps as outlined in Fig. 8.11. A high degree of fault coverage in test minimizes the probability of test escapes. As fault coverage may not reach the ideal goal of 100 %, it is preferable to give minimum coverage to all areas of the chip rather than having 99 % coverage with some areas left completely untested. Voltage screening is a fast and effective method of eliminating defects, preferably early in the test flow, with only voltage acceleration. Burn-in is carried out if both voltage and temperature accelerations are required to remove weak parts. Guard-bands are margins applied to the field operating conditions with respect to manufacturing test conditions. These margins account for variability in test equipment and field operating conditions, and degradation in circuits over time.

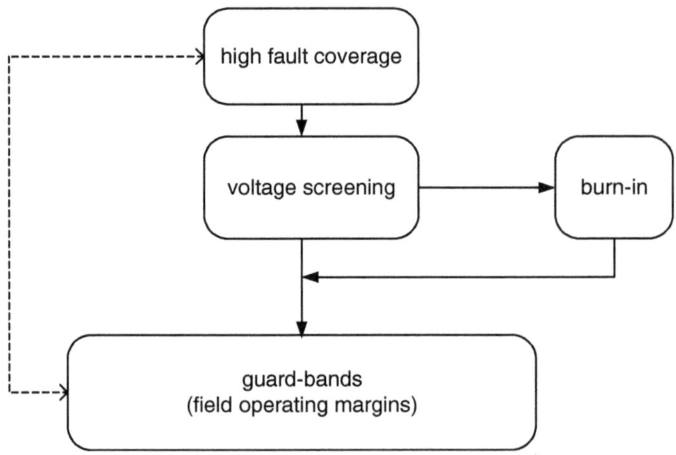

Fig. 8.11 Reliability improvement through test coverage, voltage screening, burn-in (optional), and guard-bands

8.3.1 Voltage Screening

Voltage screening at nominal test temperatures can accelerate failures due to poor gate-dielectric integrity, resistive vias, and weak links in wire interconnects. Variations in gate-dielectric thickness reduce the time to breakdown in thin regions as described in Sect. 8.2.3. Insufficient fill material in inter-level vias, and very narrow regions in wires increase wire path resistance. Such defects are more susceptible to electromigration-induced failures, beginning early in the life cycle, because of increased current density through narrow regions (Sect. 8.2.4).

Accelerated voltage stress is carried out at the nominal test temperature. In dynamic voltage stress (DVS), test vectors are applied at elevated power supply voltages, typically 1.5× to 1.7 × V_{DD} to stress all node voltages for a short time, typically for a few hundred milliseconds. The outputs are validated for each test pattern. In enhanced voltage stress (EVS), V_{DD} is bumped up to 2× for a static voltage stress, again for a short duration, after performing DVS and test vectors re-applied.

Voltage screening can sometimes "heal" defective parts during test. Metal weak links shorting neighboring wires at different potentials may evaporate as the current density is increased at elevated voltages. An increase or a decrease in yield after voltage screening gives a clear indication of process-induced defects.

8.3.2 Burn-In

Accelerated stress testing described in Sect. 8.1.1 is also employed for weeding out weak parts from the supply chain. The early failures shown in Fig. 8.1 are forced to occur during stress tests. This saves the replacement cost of field returns and protects product reputation at the time of introduction in the market. Chips may be stressed at high voltage only or at elevated voltage and temperature, known as burn-in (BI).

BI is a short accelerated stress applied to good parts identified after initial test results. BI is most effective when there is a distinct population of chips failing early in the life cycle. The failure rate with and without BI for an ideal situation is illustrated in Fig. 8.12. In the absence of BI, the "infant mortality" is high. With BI, the initial failure rate is low and continues to decrease with time. The failure rate increases again once circuit degradation and other wear-out mechanisms come into play. Ideally, specified EOL of the product chip is less than the time at which failure rate begins to increase.

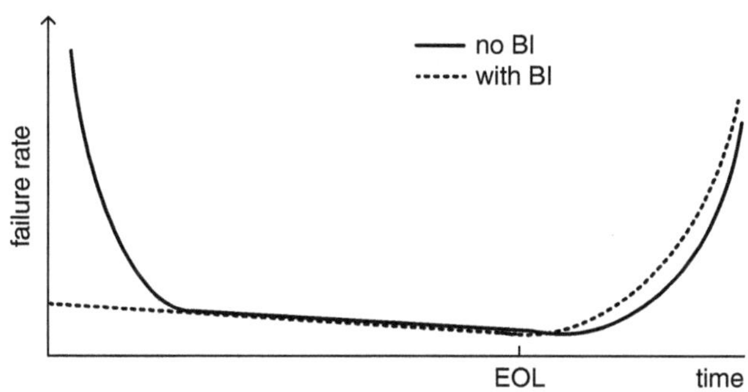

Fig. 8.12 Failure rate over time without BI and with BI

BI stress conditions and time duration are determined from accelerated stress tests on a representative sample population. Stressing the chip at high voltages and temperatures increases the power consumption in both standby and operational mode. Special boards for BI, cooling equipment and power supplies are required to maintain reasonable throughput in production. Hence the cost of BI must be weighed against the anticipated field failure rates.

If excessive power consumption limits the clock frequency, a pulsed voltage technique may be used. The power supply is pulsed from a low value (nominal V_{DD}) to a higher stress voltage V_{stress}, while functional testing is done at V_{DD}. Pulse voltage range and duty cycle are computed to obtain the ratio of desired stress time to actual clock time. Pulsed test reduces the effective stress time.

While electromigration-induced failure rate increases towards EOL, BTI degradation takes place primarily early in the life of a product (Fig. 8.4). Chips going through BI may see a large fraction of total expected BTI degradation before being deployed in the field. Although the performance is degraded after BI, the probability of failure in the field from BTI is reduced. The differential RO pair concept described in Sect. 8.2.1 may be used to monitor BTI degradation on the chip after BI, using pre-BI measurement as a reference. As the cycle limiting paths may be different after BI, the shift in chip V_{min} and f_{max} typically have larger spreads than the shifts in RO frequencies. The RO measurement accuracy being higher, it is easier to observe and model BTI degradation after BI from embedded ROs than from the chip V_{min} and f_{max}.

BTI degradation during burn-in may be healed by a bake-out at a temperature in the 200 to 400 °C range for several minutes. The product chip is then subject to degradation in the field, but the initial operating margin is higher. Implementation of the bake-out procedure is dependent on the type of packaging. Low cost organic packages cannot be subjected to such high temperatures. In this case, most of the BTI degradation occurs before shipping the product, and guard-bands may be reduced.

As the number of possible scenarios is large, a conservative method to determine the failure rate near EOL is to examine the worst-case for each degradation mechanism for the total number of power-on-hours (POH). As an example, BTI degradation may be higher when the chip is in a DC state (no recovery) and at a higher temperature whereas CHC degradation occurs only when the circuits are switching. Worst-case analysis may require building different stress models for different mechanisms. Such careful analysis adds to the cost of testing. There is a reasonable amount of engineering judgment applied to BI. Previous history of similar products is an important consideration.

The lifetime of a product under field use conditions is reduced with BI. There is some risk that BI may partially damage or generate weak spots on the chip and cause failures in the field. BI strategy for each product should therefore be determined in the early characterization phase. The decision to include BI in the test flow and the stress conditions are included in guard-bands. Defect density improves as silicon technology matures, and loss from systematic defects is eliminated with fixes in photomasks, fabrication processes, or in circuit designs. Hence, the need for BI may change over time, or the BI conditions may be optimized in mid-production if necessary.

8.3.3 Guard-Banding

The probability of field fails can be greatly reduced by testing and qualifying chips under conditions worse than those encountered in the field, a strategy known as guard-banding. Some field environmental specifications may be set by customer expectations while others are set by the product manufacturer. As an example, ambient temperature and altitude range at which a product is guaranteed to function may be based on market research, while the manufacturer may have the flexibility to set the operating power supply voltage internal to the chip. These scenarios vary from product to product.

Guard-bands are applied to cover variations in test equipment and environmental conditions, resolution of test equipment, uncertainty related to step size in measurement, variations in field usage, applications not covered in manufacturing tests, and chip and package degradation over time. The test engineering team needs to determine "worst-case" test conditions from known field specifications and sources of statistical variations in operating conditions such as voltage, temperature, and external noise, along with their impact on chip functionality. Defining "worst-case" in the field is more involved than defining the "worst-case" corner in circuit simulations described in Sect. 6.4.1.

As an example, guard-band determination for external power supply voltage is illustrated in Fig. 8.13. In this example, variations of V_{DD} in production tests, field operating conditions, and degradation in circuit performance over time are considered. With multiple sources of variations in power supply voltage on the test floor, and in the field, the V_{DD} distributions are assumed to be Gaussian. The widths of the distributions shown in Fig. 8.13 are arbitrary and used for illustration purposes only.

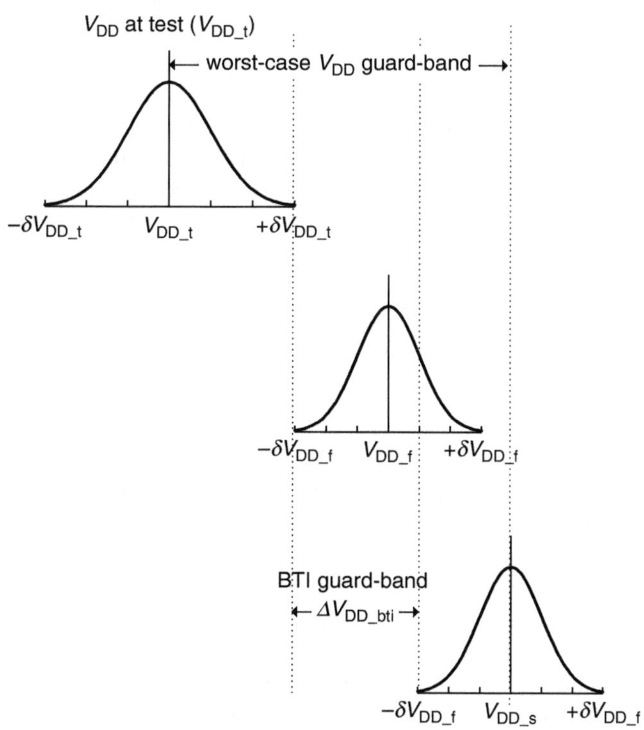

Fig. 8.13 Example of worse-case guard-band for V_{DD}

In manufacturing test, the nominal power supply voltage for all chips is set at $V_{DD\text{-}t}$. Variations in power supplies among different ATEs on the test floor and temporal fluctuations in $V_{DD\text{-}t}$ from noise result in a spread of $\pm\delta V_{DD\text{-}t}$. Similar variations in the field result in a spread of $\pm\delta V_{DD\text{-}f}$ in the assigned field power supply voltage $V_{DD\text{-}f}$. Since a chip passing all frequency requirements at $(V_{DD\text{-}t}+\delta V_{DD\text{-}t})$ may be assigned a voltage $(V_{DD\text{-}f}-\delta V_{DD\text{-}t})$ in the field, $V_{DD\text{-}f}$ is set at

$$V_{DD\text{-}f} = V_{DD\text{-}t} + \delta V_{DD\text{-}t} + \delta V_{DD\text{-}f}. \tag{8.14}$$

The degradation in MOSFET drive strengths from BTI and HCI degradation is modeled as an equivalent V_t shift over time, requiring a corresponding "worst-case" increase in power supply voltage $\Delta V_{DD\text{-}bti}$, to ensure that chips will continue to function correctly near EOL. Using the worst-case for each of the situations, chip specifications for field power supply voltage $V_{DD\text{-}s}$ should be set at

$$V_{DD\text{-}s} = V_{DD\text{-}t} + \delta V_{DD\text{-}t} + \delta V_{DD\text{-}f} + \Delta V_{DD\text{-}bti}. \tag{8.15}$$

There will be an increase in chip power in the field at $V_{DD\text{-}s}$ from the power at the test voltage of $V_{DD\text{-}t}$. In order to meet power specification in the field, manufacturing tests for power are conducted at $V_{DD\text{-}s}+\delta V_{DD\text{-}f}$ while performance is specified at $V_{DD\text{-}f}-\delta V_{DD\text{-}f}$.

In practice, the worst-case scenario is overly pessimistic if $\pm 3\sigma$ ranges are considered for each of the distributions. It penalizes >99.98 % of good parts which have to operate at a lower frequency than their f_{max} at the operating V_{DD} and at a higher power. Increase in power comes with additional penalty of operation at a increased T_j, resulting in higher degradation in circuit components.

If the sources of variation are independent, the net σ is obtained as a root square sum (RSS). When applied to test floor and field operating conditions,

$$\sigma V_{DD} = \sqrt{\left(\sigma V_{DD\text{-}t}^2 + \sigma V_{DD\text{-}f}^2\right)}\ . \tag{8.16}$$

If $\sigma V_{DD\text{-}t} = \sigma V_{DD\text{-}f}$, then $\sigma V_{DD} = \surd 2 \times \sigma V_{DD\text{-}t}$, a value reduced by $1/\surd 2\times$ from $2 \times \sigma V_{DD\text{-}t}$.

Alternatively, manufacturing tests may be conducted at a higher frequency than in the field, while keeping the power supply voltage the same in test and in the field. The frequency guard-band is a good substitute for the voltage guard-band described above. It has the advantage that P_{off} is not increased. A product may be tested with a combination of frequency and voltage guard-bands with the manufacturing test conditions optimized for screening at wafer, package, and system tests. The situation is further complicated when applied to binning of parts.

Excessive guard-bands result in shipping products with specifications set much below their actual performance level. Lower guard-bands increase the probability of failure in the field. Guaranteed reliability of a product, customer expectations, product reputation and marketing, and cost of repairs and replacements are all important criteria in determining product specifications. After collecting all the necessary data and examining it, good engineering judgment is applied in setting guard-bands. It is important to have physical insight into the items being accounted for in the guard-bands, but attempts at exhaustive high precision modeling are generally an over kill!!

8.4 Summary and Exercises

Reliability models for failure rate predictions using accelerated stress tests are key to reducing field failures. MOSFET performance degradation due to BTI and embedded monitor designs to track circuit performance with aging are described. Other degradation mechanism include HCI, TDDB, electromigration, and SIV, each with a different rate of acceleration with voltage and temperature. Screening of weak parts during test is carried out with high voltage tests and burn-in. Variations in chip qualification tests and field operating conditions are accommodated through guard-banding.

Exercises 8.1, 8.2, and 8.3 deal with degradation and accelerated stress tests. Exercises 8.4 through 8.7 are related to BTI and HCI degradation mechanisms. Exercises 8.8, 8.9, and 8.10 cover BI and guard-banding.

8.1. Failure rate is likely to increase rapidly near and beyond EOL as shown in Fig. 8.1. If the failure rate increases exponentially with time after EOL, with 100 parts in the field graphically illustrate the % of remaining parts as a function of time beyond EOL.

8.2. Activation energies for several different degradation mechanisms are shown in the table below:

Mechanism	E_a (eV)
BTI	0.05
HCI	−0.2
TDDB	0.8
Electromigration	0.6

(a) What are the temperature acceleration factors for these mechanisms between 50 and 120 °C?

(b) What should the relative failure rates from these mechanisms be at 50 °C be so that the rates become equal at 120 °C?

8.3. A FIT of 100 is desired for a product after 500 h of accelerated testing. How many samples are needed to have 95 % confidence in the measured FIT?

8.4. A p-passgate circuit is used as an NBTI monitor. Simulate an RO with 50 stages of the p-passgate circuit and a NAND2 enable gate. Show that an increase of 0.03 *V* in $|V_t|$ shift of 0.03 V is equivalent to a −0.03 V bias on the gate of the p-passgate.

8.5. The $|V_t|$ shift for an n-FET and a p-FET occurs over time as shown in Fig. 8.4. Plot the corresponding change in IDDQ of a standard inverter.

8.6. A microprocessor chip has a power management scheme to change V_{DD} based on a BTI monitor reading.
   (a) Using the inverter model of the chip, estimate the fractional change in P_{off} and total power with V_t shift.
   (b) At what point in the lifetime of the product does the power management scheme provide no measurable benefit (i.e., 1 %)?

8.7. HCI degradation takes place only when both V_{gs} and V_{ds} are high as indicated in Fig. 8.10. Assume that this situation exists with V_{gs} between $0.4 \times V_{DD}$ and $0.6 \times V_{DD}$. Simulate a 51 stage inverter (FO = 4) RO and compute % time during which HCI degradation occurs.

8.8. Draw qualitative plots of failure rate observed in accelerated testing to recommend (1) no BI, (2) short BI, (3) long BI. Also show a failure rate plot where BI will impact EOL and is not recommended.

8.9. In calculating the guard-band for BTI degradation, it is assumed that all MOSFETs degrade at the same rate over lifetime. Field data show that over half of the chips used in a specific product are turned on 25 % of the time. By how much can the BTI guard-band be reduced based on this information, without taking BTI relaxation into account. What information would you need on the usage to include relaxation? What will be the impact of your recommendations?

8.10. List factors to be included in guard-bands and categorize them as equipment, environment, test coverage, and device degradation. Only a 5 % guard-band has been factored into the economic model. How can the 5 % guard-band limit be maintained through test coverage at different test stops?

References

1. NIST/SEMATECH e-Handbook of statistical methods. http://www.itl.nist.gov/div898/handbook/. Accessed 21 July 2014
2. Bernstein JB, Gurfinkel M, Li X, Walters J, Shapira Y, Talmor M (2006) Electronic circuit reliability modeling. Microelectron Reliab 46:1957–1979
3. Weste NH, Harris D (2010) CMOS VLSI design: a circuit and systems perspective, 4th edn. Addison-Wesley, Boston
4. Ketchen MB, Bhushan M, Bolam R (2007) Ring oscillator based test structure for NBTI analysis. In: Proceedings of the 2007 I.E. international conference on microelectronic test structures, Tokyo, pp 43–47
5. Stawiasz K, Jenkins KA, Lu PF (2008) On-chip circuit for monitoring frequency degradation due NBTI. In: 46th Annual international reliability symposium, pp 532–535

Basic Statistics and Data Visualization

9

Contents

9.1 Basic Statistics 312
 9.1.1 Probability 314
 9.1.2 Statistical Distributions 315
 9.1.3 Sample Size Effects 318
 9.1.4 Non-normal Distributions 321
9.2 Data Filtering, Correlation, and Regression 323
9.3 Statistical Variations 326
 9.3.1 Range of Systematic and Random Variations 327
 9.3.2 Sensitivity Analysis of a Function 330
9.4 Bayesian Statistics 333
9.5 Data Visualization 334
9.6 Summary and Exercises 343
References 344

Relational databases for storing design and electrical test data coupled with software tools for statistical analysis and graphics have provided a high degree of automation for post-processing of data for rapid feedback and debug. However, domain expertize is a valuable asset and in many cases an essential ingredient for determining the root cause of the underlying behavior. A brief overview of statistical methods including probability, distributions, correlation, and regression analysis is given. The use of prior knowledge obtained from circuit simulations and from test results on other product chips is emphasized. Examples of visualizing and summarizing techniques take advantage of building expectations from models and circuit simulations, and exploit visual pattern recognition capabilities of humans.

Mathematical statistical methods treat data in an objective manner, restricting the analysis to available relevant data. Descriptive statistics provides a simple summary of the values of a variable, including sample size, central tendency or mean value, and the range and shape of its distribution. Inferential statistics draws conclusions based on the data. Statistical data mining together with predictive

M. Bhushan, M.B. Ketchen, *CMOS Test and Evaluation*,
DOI 10.1007/978-1-4939-1349-7_9

analytics help find patterns and correlations in large volumes of data and apply this information to the probability of related events in the future.

The growth in compute power and in memory storage systems has expanded the ability to collect, analyze and interrelate large volumes of data with relative ease. This has led to development of new terms such as "big data," "smarter planet," and "smart objects" to mention a few. Internet of things (IOT) is another area exploiting the data collected from a large number of devices, sensors, monitors, and other objects wirelessly connected to the internet. There is, in general, an increasing emphasis on coupling domain expertise, an in-depth understanding of the subject area, to statistical data analysis. This deviation from traditional mathematical statistical methods becomes necessary when handling such large and diverse data volumes.

CMOS foundries allocate sizable resources to characterization, data mining and statistical programming. CMOS product development and manufacturing efficiency is improved when engineers and scientists are cognizant of statistical data analysis methods, and the statisticians are aware of basics of CMOS circuits and technology. Interweaving of these different areas of expertise can help reduce the time to debug as well as the cost of expensive failure analysis procedures.

This chapter covers the rudiments of statistics as applied to CMOS circuits along with graphical visualization to assist in debug, characterization, and model-to-hardware correlation. Some aspects of statistics are used in earlier chapters, and a basic knowledge of statistics has been assumed. For completeness, a formal treatment of basic statistics is included in Sect. 9.1. Data collection and reduction techniques are discussed in Sect. 9.2. Statistical variations in CMOS circuits and correlations are covered in Sect. 9.3. Bayesian statistics with a priori knowledge is introduced in Sect. 9.4. Ten examples illustrating data visualization methods are presented in Sect. 9.5.

There are many excellent textbooks on statistics for experiments [1, 2], statistical process control [3–5], and the six sigma approach to quality control in manufacturing [6, 7]. Data visualization from a historical perspective, from the use of carefully prepared engraved wooden blocks for printing to the ease of today's PowerPoint charts and the related pitfalls are wonderfully illustrated in books authored by Edward Tufte [8–10]. A highly condensed version of statistical data analysis and characterization methodology for CMOS applications is useful for a quick overview [11].

9.1 Basic Statistics

Statistical data analysis methods vary with the type of data, available data volume, number of variables, precision with which conclusions are to be drawn, and many other factors. Data generated with circuit simulations is more straightforward to analyze when the number of parameter variables is limited, and their values predefined in the simulation setup. Extensive data are collected in CMOS manufacturing as every chip undergoes electrical tests at several stages. A large sample size is one good representation of device or circuits characteristics; however, the sources and ranges of parameter variations over time may also be considerable. The sample size is often smaller for detailed characterization and for chips from process split lots. In such cases, it is important to ensure that the conclusions drawn are statistically significant.

The first step in data analysis is to ensure the validity of the data. As an example, τ_p (average circuit delay) values determined from ring oscillators on 25 functional chips from a single lot are shown in Fig. 9.1a. From similar data collected on many chips during normal production testing, the nominal expected values are $\tau_p = 11.5$ ps, $\sigma\tau_p = 0.7$ ps, and chips with $\tau_p > 13$ ps have a very high probability of failing the lower acceptance limit of f_{max}. In view of this prior knowledge, it becomes apparent that the full range of 12.67 ps for τ_p in this small sample size shown in Fig. 9.1b is unacceptably large. One may note that chip # 25 has a τ_p value of 4.67 ps but is failing the f_{max} test. This suggests the presence of third harmonic in the ring oscillator from which the value of τ_p was obtained. The correct value of τ_p would then be 14.01 ps (=3 × 4.67), consistent with the chip not meeting the f_{max} criteria.

a

chip #	τ_p (ps)	f_{max} pass/fail
1	11.59	pass
2	10.24	pass
3	10.20	pass
4	12.36	pass
5	8.78	pass
6	13.86	fail
7	8.16	pass
8	9.76	pass
9	16.66	fail
10	8.94	pass
11	14.12	fail
12	8.53	pass
13	9.88	pass
14	11.88	pass
15	8.09	pass
16	14.50	fail
17	9.97	pass
18	12.69	pass
19	10.93	pass
20	10.02	pass
21	8.94	pass
22	17.14	fail
23	12.58	pass
24	8.22	pass
25	4.67	fail

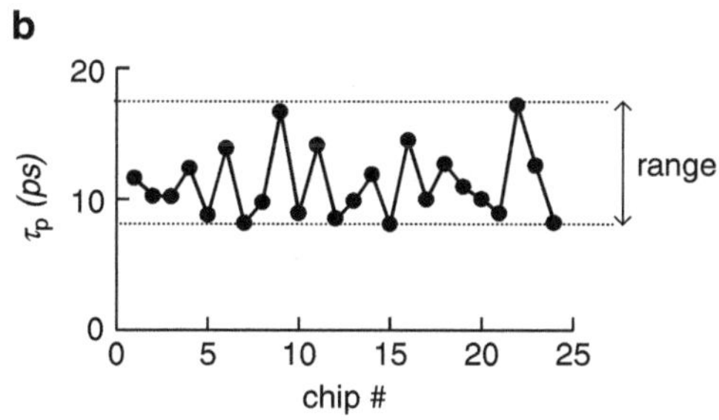

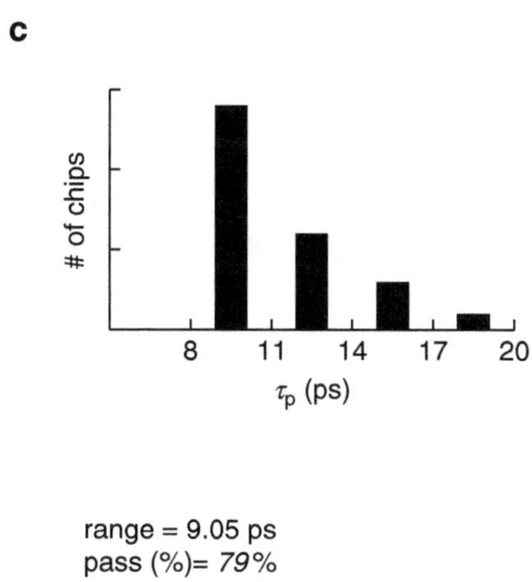

Fig. 9.1 (**a**) τ_p Values measured on a process monitor and f_{max} pass/fail data from 25 chips, (**b**) τ_p vs. chip # showing the range of variation, and (**c**) histogram of τ_p values

However, the τ_p range of 9.05 ps, after excluding chip # 25 from the analysis, is still larger than $6\sigma\tau_p$ of 4.2 ps. In addition, the histogram shown in Fig. 9.1c is far from a bell-shaped curve that is characteristic of random statistical variations (Sect. 9.1.2).

One may speculate that the chips are from a silicon process split lot with intentional L_p and V_t offsets. Hence, the probability of chips passing the f_{max} test computed from this data set is not representative of production hardware, and may lead to erroneous conclusions if presented without any process information.

The example presented above shows the importance of using prior knowledge obtained from circuit simulations and hardware data as an integral part of data analysis. This approach is very useful when the tendency in the data is deterministic and not solely governed by contributions from many independent sources of variations.

In the following subsections, the basic principles of statistical analysis are described and applied to data samples of different sizes. Examples of deviations from normal statistical distributions are demonstrated with circuit simulations, along with cases where knowledge of physical behavior of circuits is used for presorting the data.

9.1.1 Probability

Probability defines the occurrence of a chance event from a number of possibilities. Mathematically, if there are N observations with possible discrete outcomes of A or B, then the probability of finding a value A is given by

$$p(A) = \frac{n(A)}{N}, \tag{9.1}$$

where $n(A)$ is the number of observations with value A, and

$$p(A) + p(B) = 1. \tag{9.2}$$

The probability of not getting A is

$$\overline{p}(A) = 1 - \frac{n(A)}{N}. \tag{9.3}$$

Example: A value of "1" is assigned to each CMOS chip passing a test and a value of "0" to each chip failing the test. If the probability of a chip passing the test, $p(1) = 0.8$, then $p(0) = 0.2$. In a sample of 1,000 chips, 800 chips are expected to pass the test.

When two events A and B are independent, the joint probability of A and B is given by

$$p(A \text{ and } B) = p(A) \times p(B) = p(A \cap B), \tag{9.4}$$

and is equal to or smaller than the probability $p(A)$ or $p(B)$.

If the events A and B are independent and mutually exclusive i.e. $p(A \cap B) = 0$, the probability of either A or B is given by

$$p(A \text{ or } B) = p(A) + p(B) = p(A \cup B). \tag{9.5}$$

The symbols ∪ (=or) and ∩ (=and) are mathematical representations of logical functions. If events A and B are independent but not mutually exclusive, the probability of either A or B is given by

$$p(A \text{ or } B) = p(A) + p(B) - p(A \cap B). \tag{9.6}$$

Example: A microprocessor chip has four cores and the probability of a core failing a test due to a particulate defect is 0.1. From the expression in Eq. 9.4, the probability of all four cores failing simultaneously on a single chip is 0.0001. If it is known that there is exactly one fault-causing defect per chip, the occurrences of a particulate landing on each core are mutually exclusive and the probability of at least one core failing is 0.4.

Conditional probability $p(A|B)$ is defined as the probability of A given the occurrence of event B, and expressed as

$$p(A|B) = \frac{p(A \cap B)}{p(B)}. \tag{9.7}$$

Example: The probability of τ_p of an inverter measured on a process monitor on fault-free chips will be below target is 0.8. The probability that a chip with τ_p below target fails f_{max} test is 0.9. The probability of a chip failing to meet both τ_p and f_{max} targets is 0.72 (=0.9 × 0.8).

Probability estimations may be treated as purely chance events or combined with deterministic data.

9.1.2 Statistical Distributions

Consider repeated measurements of the period T_p of a ring oscillator on a selected chip at several different test stations. The measured values may be distributed over a small range because of random tester-to-tester variations in power supply voltage and chip temperature. When data for T_p are collected for the same RO design on many chips from many wafers, the range of T_p values becomes larger because of silicon process induced variations in properties of MOSFETs and parasitic components. Statistical variations of this type typically follow a bell-shaped curve as described below.

When there are many different independent sources of variations contributing to the value of a parameter and no single source dominates, the normalized probability density $p(x)$ of a continuous variable x is given by a Gaussian function

$$p(x) = \frac{1}{\sqrt{2\pi\sigma^2}} e^{-\left[\frac{(x-\mu)^2}{2\sigma^2}\right]}, \tag{9.8}$$

where μ is the mean, and σ^2 is the variance given by

$$\sigma^2 = \int_{-\infty}^{\infty} p(x)(x-\mu)^2 dx. \tag{9.9}$$

Equation 9.8 holds even when the distributions of the independent sources of variations are not normally distributed. The probability distribution of x from Eq. 9.8, known as a Gaussian or normal distribution, is shown in Fig. 9.2. The distribution is symmetrical about μ, with probabilities of x having a value $<\mu$ or $>\mu$ of 0.5 each. Although the variance σ^2 is useful in mathematical computations, the standard deviation σ, with the same units as μ, is more commonly used in describing the range of variations.

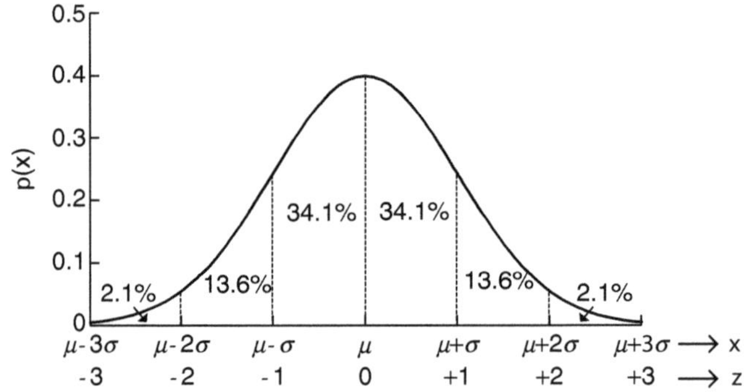

Fig. 9.2 Probability density function for $p(x)$ with mean value μ and standard deviation σ. The indicated % areas under the curve give the % probability of finding x within each σ interval

It is convenient to define a normalized parameter

$$z = \frac{(x-\mu)}{\sigma}, \tag{9.10}$$

such that the distribution of z has a mean value $\mu = 0$ and a standard deviation $\sigma = 1$. A z distribution is known as the unit normal distribution, and as with $p(x)$

$$\int_{-\infty}^{+\infty} p(z)dz = 1. \tag{9.11}$$

The probability of finding a value $<z$ is expressed as a cumulative distribution function

$$F(z) = \int_{-\infty}^{z} p(z)dz \tag{9.12}$$

The cumulative distribution function (CDF) $F(z)$ is plotted in Fig. 9.3. The probability of finding a value of $z < 0$ is 50 % and that of $z < +1$ is 84.1 %.

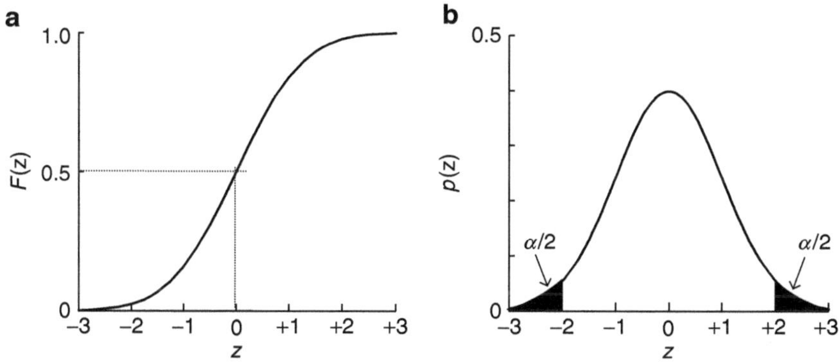

Fig. 9.3 (**a**) Cumulative distribution function (CDF) $F(z)$ as a function z, and (**b**) unit normal distribution with tail areas of $\alpha/2$ each for $|z|>2$

Statistical tables give the tail area $\alpha/2$ under the unit normal distribution for a value $>z$. As the distribution is symmetric about $z = 0$, the tail area for a value $>z$ is equal to the tail area for a value $<-z$. The tail areas for $|z| >2$ are shown as dark regions in Fig. 9.3b. Expressed in %, the probability of having a value $>|z|$ is $100 \times \alpha\%$ and the probability of having a value $<|z|$ is $100 \times (1 - \alpha)$ %. As an example, for $z = 1.96$, $\alpha/2 = 0.025$ (2.5 %) and hence there is a 5 % probability of having $|z| >1.96$, or a 95 % probability of having $|z| <1.96$.

It is often of interest to know the % of samples within a range expressed in $>|z|$. A few selected values are listed in Table 9.1. Only 0.27 % of the samples lie outside $|z| > 3$, justifying the $\pm 3\sigma$ range used in many of the simulation examples. The table also lists sample sizes to find (on average) one value $>|z|$. As an example, the number of samples needed to find one instance of $|z| > 6$ is, on average, 500 million, a very rare event. The statistical tables for tail area are also useful for calculating the probability of finding a value between any two values of z, z_1 and z_2, or between $\pm z$. The probability of having a value of z between +1 and +2 is 13.6 % [(95.47 % − 68.3 %)/2] as indicated in Fig. 9.2.

Table 9.1 Probability of observations within and outside |z|

|z|	% of samples within <|z|	Probability of occurrence for >|z|, one in
0.50	38.3	1.6
0.67	50.0	2.0
1.00	68.3	3.1
1.50	86.6	7.5
2.00	95.47	22
3.00	99.73	370
4.00	99.997	33,000
5.00	99.99997	3,300,000
6.00	99.9999998	500,000,000

9.1.3 Sample Size Effects

In almost all practical cases, the sample size n is finite. The sample mean $\overline{x}$, and standard deviation s for a finite sample size n are given by

$$\overline{x} = \frac{\sum x_i}{n} \tag{9.13}$$

$$s = \sqrt{\frac{\sum (x_i - \overline{x})^2}{(n-1)}} \tag{9.14}$$

where x_i represent individual datum with $i = 1$ to n. In statistical analysis, it is of interest to know how close $\overline{x}$, and s are to μ and σ of an ideal normal distribution.

With a finite sample size, the data are plotted as a histogram, with the value of the parameter on the X-axis and the frequency of occurrence or number of samples within a selected interval referred to as a bin. The number of bins k is typically selected such that $n = 2^k$. Example histograms for sample sizes $n = 500$, 100, 25, and 10 are shown in Fig. 9.4a–d. The data in each plot are selected at random from a parent population which is normally distributed. The ideal normal distribution for parameter x is also shown as a solid line curve in each plot. The shape of the histograms increasingly deviates from that of a normal distribution as n is reduced.

The sample mean $\overline{x}$, and standard deviation s for $n = 500$, 100, 25, and 10 corresponding to the histograms shown in Fig. 9.4 are listed in Table 9.2. For the parent population $\mu = 0$ and $\sigma = 0.024$ are known. The values of $(\overline{x} - \mu)$ and $(s - \sigma)$ are computed in units of σ. As an example, for $n = 100$, the mean $\overline{x}$ is 0.042σ larger than μ and $s = \sigma$. The parameters *skewness* and *kurtosis* in Table 9.2 are measures of deviation from normality as explained in more detail in Sect. 9.1.4.

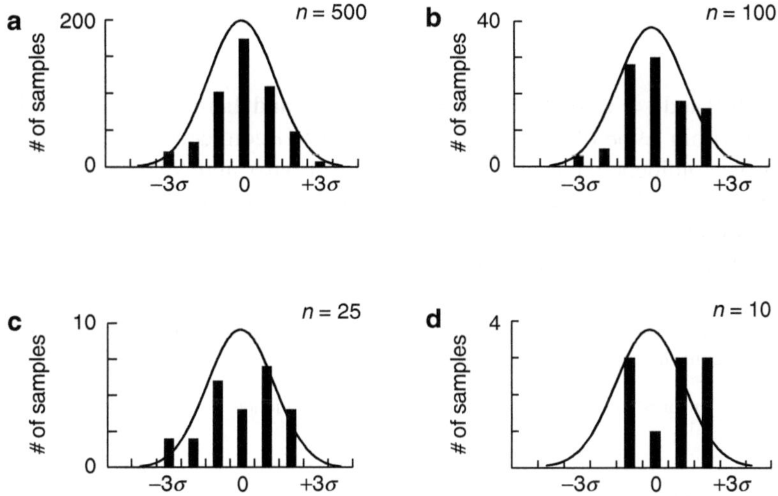

Fig. 9.4 Histograms for randomly selected data from a normal distribution: (**a**) $n = 500$, (**b**) $n = 100$, (**c**) $n = 25$ and (**d**) $n = 10$. A normal distribution curve is superposed in each case

Table 9.2 Mean $\overline{x}$ and s and deviation from μ and σ in units of σ (=0.024) for $n = 500$, 100, 25 and 10 in the example distributions in Fig. 9.4

n	$\overline{x}$	s	$\frac{(\overline{x}-\mu)}{\sigma}$	$\frac{(s-\sigma)}{\sigma}$	Skewness	Kurtosis
500	0.000	0.025	0.000	0.042	−0.26	0.38
100	0.001	0.024	0.042	0.000	−0.27	−0.05
25	0.002	0.030	0.083	0.250	−0.62	−0.37
10	0.009	0.027	0.375	0.125	−0.48	−1.44

It is apparent from Table 9.2 that as n decreases, the deviations of $\overline{x}$ from true population mean μ, and of s from true population standard deviation σ tend to increase.

The accuracy with which μ and σ from a finite sample size can be estimated is dependent on the number of samples n and whether the parent population σ is known. Provided that this population is normally distributed, $\overline{x}$ for a sample size of n has a normal distribution with a mean value of μ and a standard deviation $\sigma\overline{x}$ given by

$$\sigma\overline{x} = \frac{\sigma}{\sqrt{n}}. \tag{9.15}$$

As n increases, $\sigma\overline{x}$ decreases and $\overline{x}$ approaches the true mean μ. Following Eq. 9.10, z for the $\overline{x}$ distribution is defined as

$$z = \frac{(\bar{x} - \mu)}{\sigma/\sqrt{n}}. \tag{9.16}$$

For $n = 100$ and $z = 1.96$, the tail area $\alpha/2$ is 0.025, and there is a 95 % probability of finding $\bar{x}$ within 1.96σ of μ. This is expressed as 95 % confidence interval for $(\bar{x} - \mu)$. Similarly, confidence intervals of 90 % ($z = 1.646$) and 99.7 % ($z = 3.00$) are commonly used. The values of n for knowing $(\bar{x} - \mu)/\sigma$ with 90 %, 95 %, and 99.7 % confidence are listed in Table 9.3. The values of $(\bar{x} - \mu)/\sigma$ in the example in Table 9.2 are well within the limits in Table 9.3.

Table 9.3 Sample size n, to determine $(\bar{x} - \mu)/\sigma$ within a given interval in units of σ, with 90, 95 and 99.7 % confidence [6]

$\frac{(\bar{x}-\mu)}{\sigma}$	n for 90 % confidence	n for 95 % confidence	n for 99.7 % confidence
±1.00	3	4	9
±0.75	5	7	16
±0.50	11	16	36
±0.44	14	20	46
±0.33	25	36	83
±0.25	44	62	144
±0.20	68	97	225
±0.10	271	385	900

If the population σ is not known, then $\bar{x}$ follows a t distribution with $(n - 1)$ degrees of freedom. A t distribution is a symmetric bell-shaped distribution with thicker tails than a normal distribution. Statistical tables for calculating the tail area in a t distribution are available.

The distribution of $(n - 1) \times s^2/\sigma^2$ follows a χ^2 (chi-square) function which has a nonsymmetric shape when n is small and becomes essentially a normal distribution for $n > 200$. Example $\chi^2/(n - 1)$ plots for $n = 5$ and $n = 200$ are shown in Fig. 9.5.

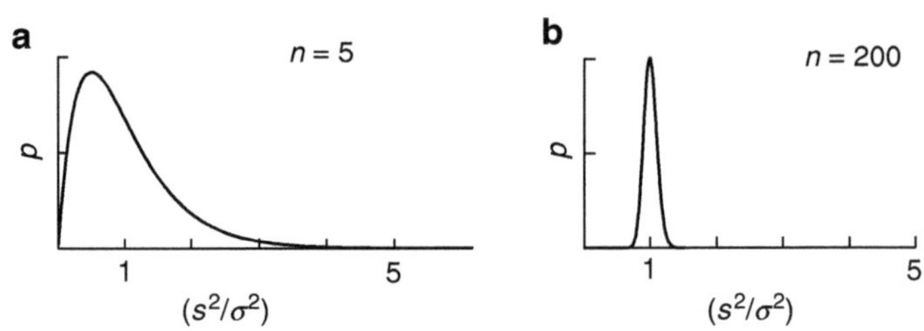

Fig. 9.5 $\chi^2/(n - 1)$ distribution for (**a**) $n = 5$ and (**b**) $n = 200$

With a long tail for $n = 5$, there is a 10 % probability of $s > 1.4\sigma$. Sample sizes for knowing $(s - \sigma)$ with 95 % confidence are listed in Table 9.4.

Table 9.4 Interval limits for $\frac{(s-\sigma)}{\sigma}$ for 95 % confidence for different values of sample size n

n	$\frac{(s-\sigma)}{\sigma}$ lower limit	$\frac{(s-\sigma)}{\sigma}$ upper limit
4	−0.43	2.73
7	−0.36	1.20
16	−0.26	0.55
20	−0.24	0.46
36	−0.19	0.30
62	−0.15	0.22
97	−0.12	0.16
200	−0.10	0.10
385	−0.07	0.07
900	−0.05	0.05
5,000	−0.02	0.02

When comparing data from different test or process splits, one would like to determine if the differences in the mean parameter values are statistically significant. If σ is known from data collected on a large population of similar tests, then a statistical table for a z distribution for the difference in normalized mean values is used. If σ is not known and the sample size is <200, a statistical table for a t distribution are used instead.

A paired t-test is used when comparing mean values of two samples in which each datum from one sample can be paired with a datum from the second sample. As an example, for comparing degradation in T_{cmin} before and after burn-in, the data for each chip in the pre-burn-in sample is paired with data for the same chip after burn-in. The test statistics are applied to the difference value for each pair.

9.1.4 Non-normal Distributions

The underlying assumption for a normal distribution is the existence of many independent sources of variations. A check for normality may be made by making a comparison with an ideal normal distribution as shown in Fig. 9.4a–d. In descriptive statistics two parameters *skewness* and *kurtosis* are computed as follows:

$$\text{Skewness} = \frac{\sum (x_i - \overline{x})^3}{ns^3} \tag{9.17}$$

$$\text{Kurtosis} = \frac{\sum (x_i - \overline{x})^4}{ns^4} - 3 \tag{9.18}$$

Ideal values of *skewness* and *kurtosis* are zero. In a nonideal distribution, the value of *skewness* gives a measure of deviation from symmetry. A positive value indicates a longer tail towards $+\infty$ and a negative value indicates a longer tail towards $-\infty$. Another deviation from normality occurs when the shape of the distribution is not Gaussian. In this case a positive value of *kurtosis* indicates the distribution to be more strongly peaked and a negative value indicates a flatter shape. With a limited number of samples, the *skewness* and *kurtosis* values in Table 9.1 give less information than that obtained from graphical views.

If one or two of the sources of variations dominate or have a strong nonlinear effect, the distributions may no longer be represented by a normal distribution. In such cases, often σ is a function of μ. As an example consider the I_{off} distribution from Monte Carlo simulations of an n-FET with random V_t and L_p variations shown in Fig. 9.6a. The distribution is highly positively skewed because of the logarithmic relationship between V_t and I_{off} (Eq. 2.4). The distribution of $\log(I_{off})$ shown in Fig. 9.6b is close to a normal distribution. Transformations of this type may be used for converting a non-normal distribution to a normal distribution [6]. In some cases with a slightly skewed distribution, standard deviations for $z > 0$ and $z < 0$ (σ^+ and σ^-) are individually determined.

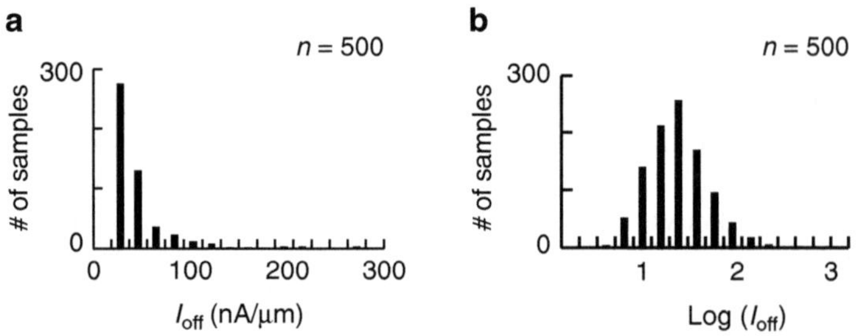

Fig. 9.6 Histogram for n-FET with random V_t and L_p variations: (**a**) I_{off} and (**b**) $\mathrm{Log}(I_{off})$

Two examples of non-normal distributions for n-FET I_{on} within a range of 1,100 to 1,800 µA/µm are shown in Fig. 9.7a, b. The n-FET I_{on} distribution in Fig. 9.7a is bimodal with two distinct peaks. The data are obtained from Monte Carlo simulations with V_t variations for ($L_p + \sigma L_p$) and ($L_p - \sigma L_p$) where $L_p = 0.045$ µm. In Fig. 9.7b, the I_{on} data are obtained from Monte Carlo simulations with V_t variations for ($L_p + 3\sigma L_p$) and ($L_p - 3\sigma L_p$). The extreme high and low values are outside the specified range and rejected. The shape of the distributions resembles a bathtub, with more samples at the extremes. Such distributions are characteristic of process split lots or of situations when there is strong bias in the data. These types of distributions cannot be represented by a single pair of $\bar{x}$ and s values.

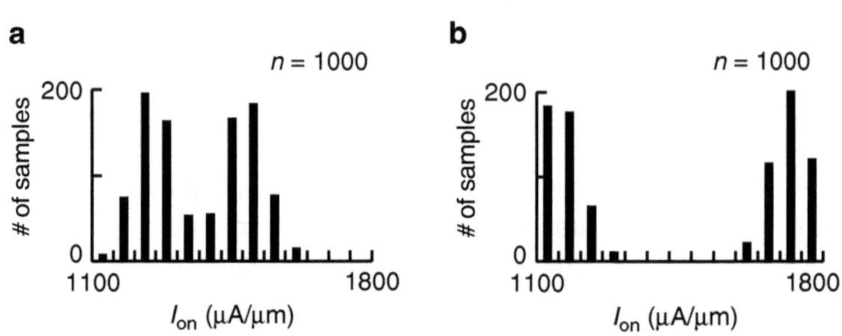

Fig. 9.7 n-FET I_{on} distribution with V_t variations for two different values of L_p: (**a**) bimodal and (**b**) bathtub

9.2 Data Filtering, Correlation, and Regression

It is important to examine the data sample carefully and to ensure the validity of all the data collected before commencing detailed analysis. Data outside an expected range, referred to as outliers or fliers, may lead to incorrect conclusions such as a larger value of $\bar{x}$. Electrical opens and shorts or data exceeding tester compliance limits are useful for yield analysis but are eliminated from the data sample used for characterization.

A commonly used methodology for filtering outliers is to reject data outside specified limits. This is exemplified with a normal distribution and a box and whiskers plot in Fig. 9.8. The data are divided into four equal parts by quartiles Q1, Q2, and Q3 such that 50 % of the data lie between Q1 and Q3. The first or lower quartile, Q1, cuts off the lower 25 % of the values, and the third quartile, Q3, cuts off the upper 25 % of the values. The range containing the middle 50 % of the data is called the interquartile range (IQR = Q3 − Q1). The whiskers are placed so that 5 % of the data are above and 5 % below the whiskers. Outliers are identified by setting filter limits defined by the lower and upper IQR multipliers, μ_l and μ_u where,

$$\text{Lower filter limit} = Q1 - \mu_l \times IQR, \text{ and}$$
$$\text{Upper filter limit} = Q3 + \mu_u \times IQR.$$

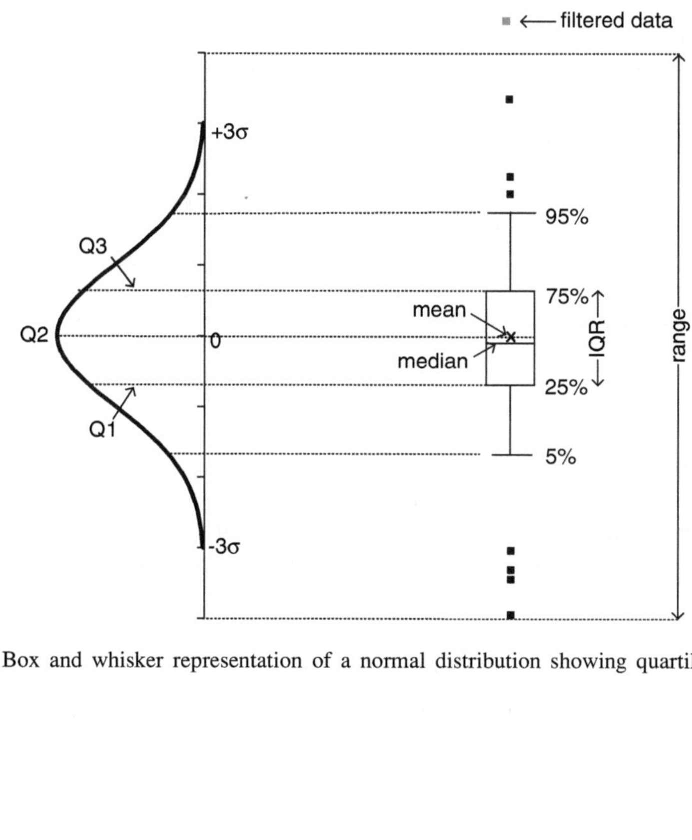

Fig. 9.8 Box and whisker representation of a normal distribution showing quartiles Q1, Q2, and Q3

In the box plot in Fig. 9.8, the vertical boundaries of the box are defined by Q1 and Q3, and Q2 is coincident with the median value of the data sample. The crossbars show the data range containing 90 % of the data (from 5 to 95 %) and values outside this range are individually shown as dark squares. If the median is not in the center of the box, the data sample is skewed. The mean of the data sample is shown as a cross in the box and may not be coincident with the median. The width of the box can be programmed to be proportional to n or $\log(n)$ to represent the total number of observations in the data sample.

If the data are normally distributed, the IQR range is within $\pm 0.69\sigma$ of the mean. The whiskers are placed at $\pm 1.64\sigma$ from the mean. The filter limits for the outliers in Fig. 9.8 are often placed at $\pm 4.0\sigma$, corresponding to $\mu_l = \mu_u = 2.4$.

The method described above works well in general. However, other situations may arise in data samples generated with circuit simulation and hardware tests. In the example shown in Fig. 9.9a, inverter delay τ_p is plotted as a function of C_L. The data are expected to lie on a straight line, and the point shown in the shaded gray region is clearly incorrect. If the data are obtained from circuit simulations, the flier is human error. If the data are obtained from electrical tests, it may be a circuit design, test or human error. The data identified by a gray box in the IDDQ histogram in Fig. 9.9b may appear to be a flier. However, it is actually within the expected range with L_p and V_t variations.

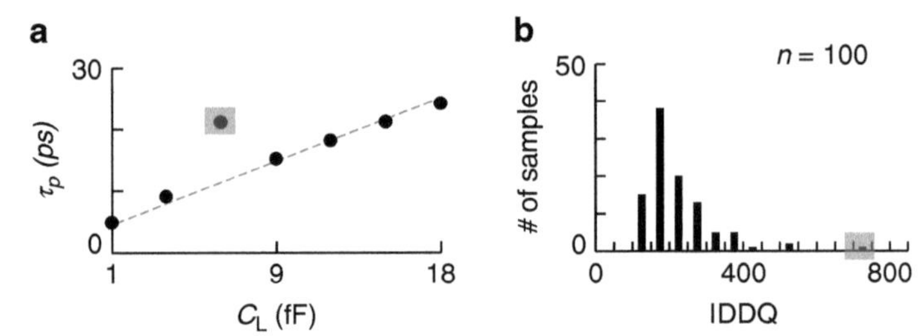

Fig. 9.9 Graphical illustration of presence of possible erroneous data shown in *gray box*: (**a**) τ_p vs. C_L and (**b**) IDDQ histogram

The dependencies of one variable on one or more variables can be represented with X–Y scatter plots. A test may be conducted to quantify the relationship of the independent variable on the X-axis to the dependent variable on the Y-axis. The correlation between two variables is expressed as the correlation coefficient ρ_{cr}

$$\rho_{cr} = \frac{ss_{xy}}{s_x s_y}, \tag{9.19}$$

where s_x and s_y are the standard deviations of X and Y and ss_{xy} is covariance defined as

$$\mathrm{ss}_{xy} = \frac{1}{(n-1)} \sum (x_i - \bar{x})(y_i - \bar{y})\,. \tag{9.20}$$

The value of ρ_{cr} lies between -1 and $+1$. The correlation is strong if $|\rho_{cr}|$ is ~1 and becomes weaker as $|\rho_{cr}|$ approaches 0. Example X and Y correlation plots are shown in Fig. 9.10a–d. In Fig. 9.10a, b T_{cmin} and f_{max} of a data path obtained from circuit simulations (Sect. 3.4.2) are plotted as a function of the simulated path delay τ. The correlation coefficient of 0.98 for T_{cmin} and -0.98 for f_{max} indicate strong positive and negative correlations, respectively.

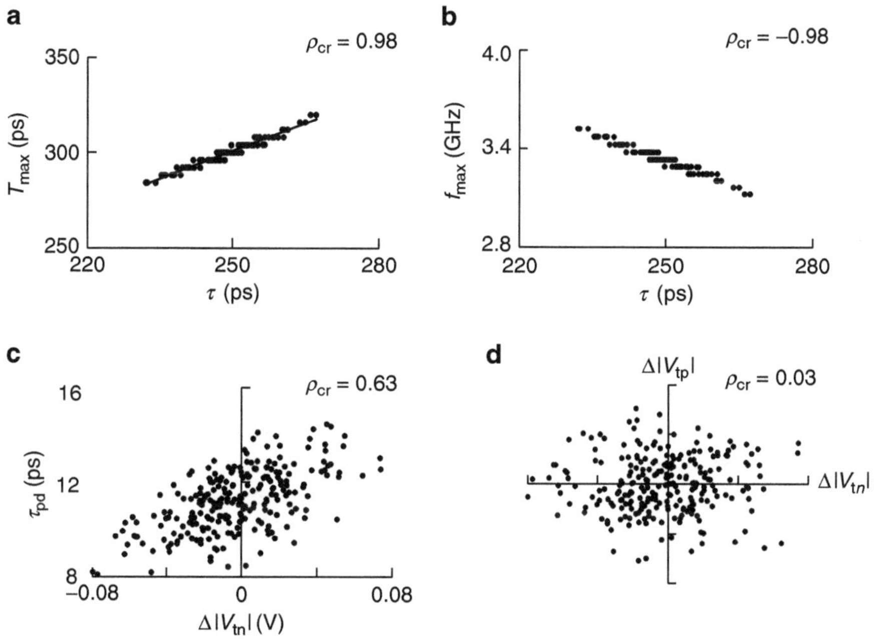

Fig. 9.10 Example correlation types: (**a**) strong positive correlation, (**b**) strong negative correlation, (**c**) weak correlation and (**d**) no correlation

In Fig. 9.10c, delays τ_{pd} and τ_{pu} of an inverter are obtained from Monte Carlo simulations with L_p, $\Delta|V_{tn}|$, and $\Delta|V_{tp}|$ variations and τ_{pd} plotted as a function of $\Delta|V_{tn}|$. The correlation of τ_{pd} with $\Delta|V_{tn}|$ is rather weak ($\rho_{cr} = 0.63$) as the value of L_p dominates the outcome. The scatter plot of $\Delta|V_{tn}|$ vs. $\Delta|V_{tp}|$ from Monte Carlo simulations for random V_t variations shown in Fig. 9.10d has $\rho_{cr} = -0.09$. The $\Delta|V_{tn}|$ and $\Delta|V_{tp}|$ in each run are randomly selected from a Gaussian distribution and as expected, $\Delta|V_{tn}|$ and $\Delta|V_{tp}|$ are uncorrelated.

When X and Y are correlated, their relationship is often described by a linear equation of the form

$$y = mx + c + \text{er}, \tag{9.21}$$

where m is the slope of the line, c is the y-intercept and er is the error or residual from the fitted line to a particular data point. The regression coefficient R^2 gives a measure of the goodness of fit, a value of 1.0 indicating a perfect fit. For a linear fit, $R^2 = \rho_{cr}^{\ 2}$. Regression analysis is also applicable when the relationship between X and Y is logarithmic, or described by some other function. The two examples in Fig. 9.11a, b show modest fits of the data to the equations.

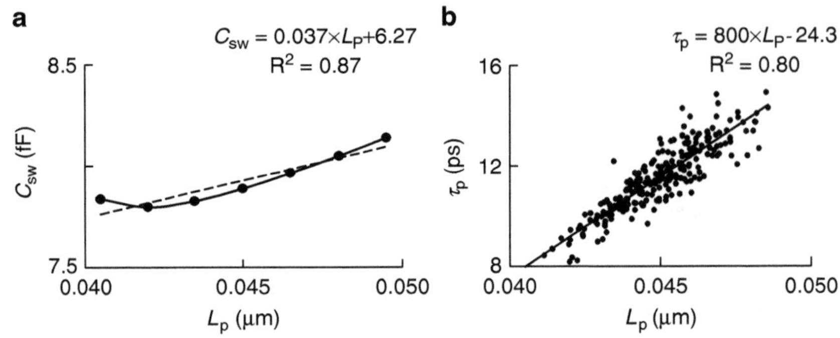

Fig. 9.11 Regression analysis for standard inverter (**a**) C_{sw} vs. L_p and (**b**) τ_p vs. L_p in the presence of random V_t variations

Software analysis packages have built-in capabilities to compute correlation coefficients and to carry out regression analysis. An understanding of the model behavior of circuits is useful in selecting X and Y parameters for correlation and the type of fit to be executed. In data mining, one may look for correlation between all pairs of measured parameters. However, a high value of ρ_{cr} for an X–Y pair of parameters must be carefully examined and validated. As an example, metal decoupling capacitor values and I_{on} values of n-FETs on the same chip may show good correlation, but this finding in all likelihood is purely accidental and not causal.

9.3 Statistical Variations

The statistical variations in a dependent variable as a consequence of variations in one or more independent variables are examined. In the first example, both systematic and random variations of independent variables V_{tn}, V_{tp}, and L_p are considered. In a second example, mathematical formulations for comparing relative sensitivities of a circuit parameter to several independent variables are described. Applications of these methods were treated in Sects. 6.4.2 and 5.6.3, respectively. A more general discussion of variations and sensitivity analysis follows.

9.3.1 Range of Systematic and Random Variations

In silicon manufacturing, quality control is exercised by centering the process such that the distribution of each of the key variables stays within specified $\pm 3\sigma$ limits. These limits match the $\pm 3\sigma$ range in the circuit models (MOSFETs and wires). Quality control thereby assures that the silicon hardware will function as predicted by the circuit models.

In circuit design, design margins are considered keeping in mind expected performance and yield. A chip design with $\pm 3\sigma$ tolerance for each key parameter is robust and ensures that all defect-free chips will function correctly. Such designs typically sacrifice performance and power targets as chips at $+3\sigma$ will have a lower frequency of operation and those at -3σ will have higher leakage currents.

Let us consider two key MOSFET properties, L_p and V_t, and examine the distribution of MOSFET parameters arising from variations in these properties. The variation in V_t has systematic and random components as described in Sect. 6.4.2. Both the systematic and random variations are assumed to be normally distributed. Systematic variation in V_t, described by σV_{ts}, is a process induced variation, and a V_t offset is ascribed to all the MOSFETs. There is an additional offset in V_t arising from random variations described by σV_{tr}. The net variation in V_t is taken as the sum of the two offsets shown in Fig. 9.12 for a MOSFET with a $\Delta V_{ts} = 2\sigma V_{ts}$ and $\sigma V_{tr} = 0.5\sigma V_{ts}$.

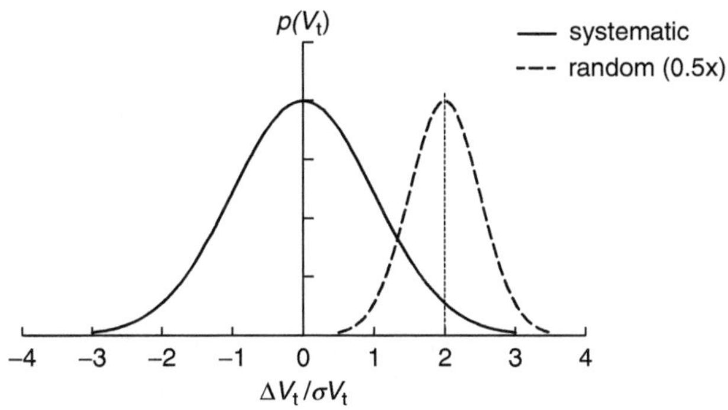

Fig. 9.12 Probability density functions for ΔV_{ts} and ΔV_{tr} with mean values of 0 and $2\sigma V_{ts}$, and with $\sigma V_{tr} = 0.5\sigma V_{ts}$

Its ΔV_t will be in the $\pm 3\sigma$ range of $0.5\sigma V_{ts}$ to $3.5\sigma V_{ts}$.

In Monte Carlo simulations, the systematic and random variations in V_t are treated as two independent distributions with the net ΔV_t defined by

$$\Delta V_t = \Delta V_{ts} + \Delta V_{tr} \tag{9.22}$$

SPICE commands for ΔV_t (*ndelvto*) and ΔL_p (*ndl*) for an n-FET are listed below:

.param ndelvto = '(ndelvtos + ndelvtor)'
.param ndelvtonom = 0.0001

.param ndelvtossigma = 0.02
.param ndelvtorsigma = 0.03
.param ndelvtos = {normal(ndelvtonom, ndelvtossigma)}
.param ndelvtor = {normal(ndelvtonom, ndelvtorsigma)}
.param pdlnom = 0.00 ndlnom = 0.00
.param pdlsigma = 0.0015u ndlsigma = 0.0015u
.param pdl = {normal(pdlnom, pdlsigma)} ndl = {normal(ndlnom, ndlsigma)}
.func normal(nom, sigma) if (run==1, nom, (gauss(sigma)))

Here *ndelvtossigma*, σV_{ts}, and *ndelvtorsigma*, σV_{tr}, are the systematic and random components of the V_{t} variations.

For an n-FET, the distribution of I_{on} is examined with variations in L_{p} and V_{t}. We have chosen I_{on} as the parameter of interest as it has a near normal distribution. For $W_n = 0.4$ μm and nominal L_{p} of 0.045 μm, $\sigma V_{\mathrm{tr}} = 0.030$ V assuming

$$\sigma V_{\mathrm{tr}} = \frac{0.004}{\sqrt{W_n L_{\mathrm{p}}}} V, \tag{9.23}$$

with $A_{\mathrm{vt}} = 0.004$ V-μm in Eq. 2.38.

In Table 9.5, mean I_{on} and σI_{on} for an n-FET are listed for different combinations of variations in L_{p}, V_{tr}, and V_{ts}. These I_{on} distributions for the different combinations follow

$$\sigma I_{\mathrm{on}}(V_{\mathrm{t}}) = \sqrt{\{\sigma I_{\mathrm{on}}(V_{\mathrm{ts}})\}^2 + \{\sigma I_{\mathrm{on}}(V_{\mathrm{tr}})\}^2}, \tag{9.24}$$

and

$$\sigma I_{\mathrm{on}}(V_{\mathrm{t}}, L_{\mathrm{p}}) = \sqrt{\{\sigma I_{\mathrm{on}}(V_{\mathrm{t}})\}^2 + \{\sigma I_{\mathrm{on}}(L_{\mathrm{p}})\}^2}. \tag{9.25}$$

The last column in Table 9.5 gives the calculated σI_{on} from Eqs. 9.24 and 9.25. The values of σI_{on} obtained from Monte Carlo simulations of 500 cases fairly closely match the σI_{on} calculated by considering each independent variable alone.

Table 9.5 Mean and σ values of I_{on} for an n-FET ($W_n = 0.4$ μm) for systematic and random variations obtained from Monte Carlo analysis (500 cases) and from Eqs. 9.24 and 9.25. 45 nm PTM HP models @ 1.0 V. 25 °C

σL_{p} (μm)	σV_{ts} (V)	σV_{tr} (V)	Mean I_{on} (μA)	σI_{on} (μA)	σI_{on} symbol	Calc σI_{on} (μA)
0.0000	0.02	0.00	529	19	$\sigma I_{\mathrm{on}}(V_{\mathrm{ts}})$	19
0.0000	0.00	0.03	531	28	$\sigma I_{\mathrm{on}}(V_{\mathrm{tr}})$	28
0.0000	0.02	0.03	532	34	$\sigma I_{\mathrm{on}}(V_{\mathrm{t}})$	34
0.0015	0.00	0.00	532	31	$\sigma I_{\mathrm{on}}(L_{\mathrm{p}})$	31
0.0015	0.02	0.00	532	37	$\sigma I_{\mathrm{on}}(V_{\mathrm{ts}}, L_{\mathrm{p}})$	36
0.0015	0.00	0.03	532	42	$\sigma I_{\mathrm{on}}(V_{\mathrm{tr}}, L_{\mathrm{p}})$	42
0.0015	0.02	0.03	532	47	$\sigma I_{\mathrm{on}}(V_{\mathrm{t}}, L_{\mathrm{p}})$	46

Next, variations in our standard inverter delay τ_p arising from independent systematic distributions in L_p and V_{ts} for n-FETs and p-FETs are examined. The L_p values of n-FET and p-FET (L_{pp} and L_{pn}) are varied independently and as well as together ($L_{pp}=L_{pn}$). Monte Carlo simulations are carried out for an inverter (FO = 4) delay chain for 500 cases at 1.0 V and 25 °C. The results are summarized in Table 9.6 along with calculated values of $\sigma\tau_p$ from equations analogous to Eqs. 9.24 and 9.25. In practice, L_p values for n-FET and p-FET nearly track together as the gate electrode is defined in the same photolithographic step. In this situation, the range of variation is wider than when L_{pp} and L_{pn} are allowed to vary independently.

Table 9.6 Mean τ_p and $\sigma\tau_p$ values for standard inverter (FO = 4) with systematic variations in L_p and V_t from Monte Carlo analysis (500 cases) and from equations. 45 nm PTM HP models @ 1.0 V. 25 °C

σL_{pp} (μm)	σL_{pn} (μm)	σV_{tp} (V)	σV_{tn} (V)	Mean τ_p (ps)	$\sigma\tau_p$ (ps)	$\sigma\tau_p$ symbol	Calc. $\sigma\tau_p$ (ps)
0.0015	0.0000	0.00	0.00	11.59	0.590	$\sigma\tau_p\ (L_{pp})$	0.590
0.0000	0.0015	0.00	0.00	11.57	0.396	$\sigma\tau_p\ (L_{pn})$	0.396
0.0015	$L_{pp}=L_{pn}$	0.00	0.00	11.56	0.982	$\sigma\tau_p(L_{pp}=L_{pn})$	0.986
0.0015	0.0015	0.00	0.00	11.60	0.720	$\sigma\tau_p\ (L_{pp,}\ L_{pn,})$	0.710
0.0000	0.0000	0.02	0.00	11.59	0.331	$\sigma\tau_p\ (V_{tsp})$	0.331
0.0000	0.0000	0.00	0.02	11.62	0.301	$\sigma\tau_p\ (V_{tsn})$	0.301
0.0000	0.0000	0.02	$\Delta\lvert V_{tp}\rvert=\Delta\lvert V_{tn}\rvert$	11.59	0.611	$\sigma\tau_p\ (\Delta\lvert V_{tsp}\rvert=\Delta\lvert V_{tsn}\rvert)$	0.631
0.0000	0.0000	0.02	0.02	11.63	0.444	$\sigma\tau_p\ (V_{tsp,}\ V_{tsn})$	0.447
0.0015	$L_{pp}=L_{pn}$	0.02	0.02	11.55	1.043	$\sigma\tau_p\ (V_{ts},\ L_{pp}=L_{pn})$	1.077
0.0015	0.0015	0.02	0.02	11.59	0.829	$\sigma\tau_p\ (V_{ts,}\ L_p)$	0.840

Simulation results from for the average delay τ_p of two series connected standard inverters (FO = 4) are shown in Fig. 9.13a. There are six fixed corners with L_{pp}, L_{pn}, $|V_{tp}|$ and $|V_{tn}|$ at $\pm\sigma$, $\pm2\sigma$, and $\pm3\sigma$ values. Monte Carlo simulations are carried out by varying L_{pp}, L_{pn}, V_{tp}, and V_{tn} independently and by varying L_p ($L_{pp}=L_{pn}$), V_{tp}, and V_{tn}. The spread in τ_p is considerably smaller in MC simulations than at fixed corners except for the 1σ case. A similar trend is seen in IDDQ in Fig. 9.13b. In both cases, setting fixed simulation corners at $\pm3\sigma$ values will be overly pessimistic.

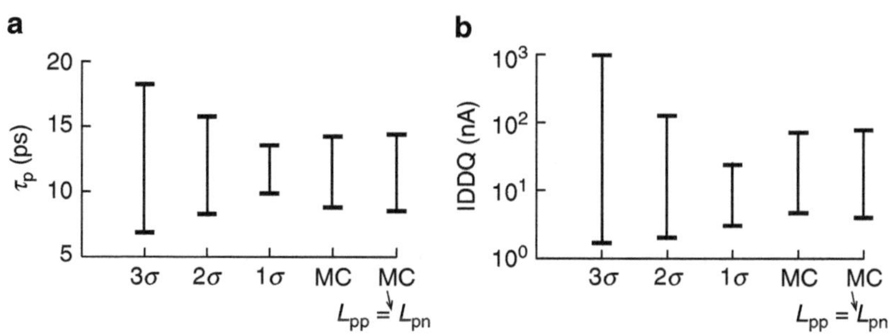

Fig. 9.13 Inverter (FO = 4) (**a**) delay range, and (**b**) IDDQ range: Monte Carlo simulations (500 cases) and fixed σ corners for L_p and V_t of n-FETs and p-FETs, 45 nm PTM HP models @ 1.0 V, 25 °C

An insight into the range of variation shown in Fig. 9.13 is obtained from the probability distribution for a multivariate sample. If L_p and V_t of n-FET and p-FET are truly independent variables, the CDF of the joint probability is given by

$$F(L_{pp}, L_{pn}, V_{tp}, V_{tn}) = F(L_{pp}) \times F(L_{pn}) \times F(V_{tp}) \times F(V_{tn}) \tag{9.26}$$

Table 9.1 shows the probability of a variable being outside a specified σ range. The probability of a single variable being outside the 3σ range is 0.0027. The probability of all four variables being simultaneously outside the 3σ range is $(0.0027)^4$ or 5.3E-11. The joint probabilities are listed in Table 9.7. It is clear that the probability of finding a wafer with L_p and V_t of n-FETs and p-FETs in a population of inverters outside of their respective $\pm 3\sigma$ ranges is extremely small. This statement holds as long as the process remains well centered at the target values of each of the parameters.

Table 9.7 Probability of four parameters being simultaneously outside σ range

σ	Probability of $>\sigma$	Probability of finding 1 in
1.0	1.0E−2	98.5
1.25	1.9E−3	514
2.0	4.1E−6	2.3E5
3.0	5.3E−11	1.9E10

9.3.2 Sensitivity Analysis of a Function

Sensitivity analysis is carried out to study how the changes in several independent input variables impact a dependent output variable. It is also referred to as "What-if-analysis" as it helps gain insight into the expected outcome due to variations in one or more critical parameters. In the case of a single input variable x, its sensitivity to the outcome y is expressed as a linear equation (Eq. 9.21). Using an inverse relationship, from an observed value of y, the corresponding values of x can be obtained, as

$$x = \frac{(y - c)}{m}, \quad \text{or } \Delta x = m^{-1} \Delta y. \tag{9.27}$$

The situation is more complex when there are n independent variables, x_i ($i = 1$ to n) influencing the outcome y such that

$$y = m_1 x_1 + m_2 x_2 + \cdots + m_i x_i + \cdots + m_n x_n + c, \tag{9.28}$$

where $m_i = dy/dx_i$.

Alternatively, Eq. 9.28 can be restructured in a differential form with normalized units as

$$\frac{\Delta y}{y_0} = m_{10} \times \frac{\Delta x_1}{x_{10}} + m_{20} \times \frac{\Delta x_2}{x_{20}} + \cdots + m_{i0} \times \frac{\Delta x_i}{x_{i0}} + m_{n0} \times \frac{\Delta x_n}{x_{n0}}. \quad (9.29)$$

Assuming x_i to be a set of independent variables, the sensitivity coefficients m_i give a measure of the sensitivity of y to x_i, without knowing the absolute value of y. In circuit simulations, with the freedom to select the values of any input variable, values of m_i may be uniquely determined. As an example, the variations of our standard inverter delay τ_p with each of the three MOSFET parameters L_p, $|V_{tn}|$, and $|V_{tp}|$ over their $\pm 3\sigma$ ranges are shown in Fig. 9.14. The dependencies of τ_p on L_p, $|V_{tn}|$ and $|V_{tp}|$ are expressed as

$$\frac{\Delta \tau_p}{\tau_{p0}} = m_{vn} \frac{\Delta |V_{tn}|}{\sigma V_{tn}} + m_{vp} \frac{\Delta |V_{tp}|}{\sigma V_{tp}} + m_l \frac{\Delta L_p}{\sigma L_p}. \quad (9.30)$$

where m_{vn}, m_{vp}, and m_l are the normalized slopes of linear fits to the data. From Fig. 9.14, one can observe that the slope $d\tau_p/dL_p$ is higher than $d\tau_p/d|V_{tn}|$ and $d\tau_p/d|V_{tp}|$.

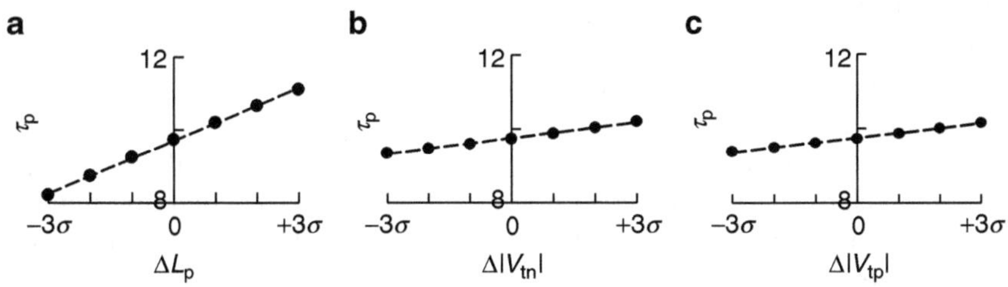

Fig. 9.14 Variation in τ_p indicating sensitivity to (**a**) L_p, (**b**) V_{tn}, and (**c**) V_{tp}

When analyzing hardware data, one would like to extract the variations of several independent parameters, which cannot be directly measured, from the measured values of a dependent parameter. This problem is addressed in Sect. 5.6.3 for detecting changes in L_p, $|V_{tn}|$, and $|V_{tp}|$ from the observed changes in τ_p values of three ROs with different sensitivities to each of these three MOSFET parameters. The methodology used in Sect. 5.6.3 is explained here with two independent variables. This allows graphic illustration of matrix elements and the condition number CN in a two-dimensional space.

Consider the measured delay of a circuit τ to be a linear function of resistances R_1 and R_2. This relationship is expressed as an equation

$$\Delta \tau = m_1 \Delta R_1 + m_2 \Delta R_2, \quad (9.31)$$

where m_1 and m_2 are constants. In order to extract R_1 and R_2 from delay measurements, we will need two circuits configured in such a way that their delays, τ_1 and τ_2 are linear functions of R_1 and R_2. Hence,

$$\Delta\tau_1 = m_{11}\Delta R_1 + m_{12}\Delta R_2, \quad (9.32)$$

$$\Delta\tau_2 = m_{21}\Delta R_1 + m_{22}\Delta R_2. \quad (9.33)$$

These relationships can be expressed as a matrix

$$\begin{vmatrix} \Delta\tau_1 \\ \Delta\tau_2 \end{vmatrix} = \begin{vmatrix} m_{11} & m_{12} \\ m_{21} & m_{22} \end{vmatrix} \begin{vmatrix} \Delta R_1 \\ \Delta R_2 \end{vmatrix}, \text{ or as a vector, } \Delta\tau = M \times \Delta R.$$

To solve for R_1 and R_2, $\Delta R = \mathbf{M}^{-1} \times \Delta\tau$. The matrix inverse $\mathbf{M}^{-1}$ is given by

$$\mathbf{M}^{-1} = \frac{1}{|\boldsymbol{M}|}\begin{pmatrix} m_{22} & -m_{21} \\ -m_{12} & m_{11} \end{pmatrix}, \quad (9.34)$$

with the determinant

$$|\boldsymbol{M}| = m_{11}m_{12} - m_{12}m_{21}. \quad (9.35)$$

For the above analysis to hold, **M** must be a non-singular matrix with its determinant $|\mathbf{M}| \neq 0$.

In an ideal and straightforward case, τ_1 is dependent only on R_1 and τ_2 only on R_2. The **M** matrix is then a diagonal matrix with off-diagonal terms $m_{12} = 0$, and $m_{21} = 0$. Variations in R_1 and R_2 are directly estimated from the variations in τ_1 and τ_2 respectively.

With non-zero off-diagonal terms, the error in estimating R_1 and R_2 from τ_1 and τ_2 can be quantified from the condition number of $|\mathbf{M}|$. The condition number is calculated from the infinity-norm of the matrix $|\mathbf{M}|$. In general, the infinity-norm of an $n \times n$ matrix with matrix elements x_{ij} is given by

$$||\boldsymbol{M}||_\infty = \max_{1 \le i \le n}\left(\sum_{j=1}^{n} |x_{ij}|\right) \quad (9.36)$$

For the 2×2 matrix, $||\mathbf{M}||_\infty$ is the maximum of the row sums $(|m_{11}| + |m_{12}|)$ and $(|m_{21}| + |m_{22}|)$. The condition number CN is the product of the infinity-norms of the matrix and that of its inverse,

$$\text{CN} = ||\boldsymbol{M}||_\infty \times ||\boldsymbol{M}^{-1}||_\infty. \quad (9.37)$$

For a diagonal and symmetric matrix with $(m_{11} = m_{22})$ and $(m_{12} = m_{21} = 0)$, $\text{CN} = 1$. This is the lowest value of CN and it gives the maximum sensitivity (minimum error) in estimating R_1 and R_2. The value of CN increases for $m_{11} \neq m_{22}$ and if $|m_{12}|$ and $|m_{21}| > 0$.

Vector representations of diagonal and non-diagonal matrices are shown in Fig. 9.15a–c. In Fig. 9.15a, for a diagonal matrix with $m_{11} = m_{22}$, vectors $\mathbf{M}_1$ and $\mathbf{M}_2$ are orthogonal and $\text{CN} = 1$. In Fig. 9.15b, again for a diagonal matrix with

$m_{11} = 4 \times m_{22}$, vectors $\mathbf{M}_1$ and $\mathbf{M}_2$ are orthogonal and CN = 4. A higher value of CN indicates lower sensitivity to one or both variables (in this case to R_2 relative to R_1). In Fig. 9.15c, $m_{11} = 2 \times m_{22}$ and $|m_{12}| = |m_{21}| = 0.25 \times m_{11}$. Vectors $\mathbf{M}_1$ and $\mathbf{M}_2$ are no longer orthogonal and CN = 2.6. This case gives a better sensitivity to

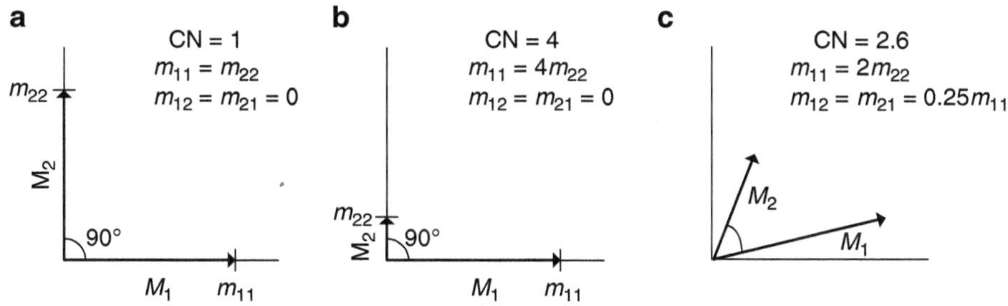

Fig. 9.15 Vectors $\mathbf{M}_1$ and $\mathbf{M}_2$ for a diagonal matrix with (**a**) $m_{11} = m_{22}$ and (**b**) $m_{11} = 4\,m_{22}$. (**c**) Vectors $\mathbf{M}_1$ and $\mathbf{M}_2$ for a non-diagonal matrix

both $\mathbf{M}_1$ and $\mathbf{M}_2$ than the case in Fig. 9.15b.

The conclusions reached by using CN in the above example of a 2×2 matrix could have been easily reached by a visual inspection of the matrix. For a $n \times n$ matrix with $n > 2$, although a visual inspection of the matrix is useful, it is best to compute CN to optimize a design for maximum sensitivity to independent variables. The matrix inverse and CN value for a matrix of arbitrary dimension can be computed using any of the available mathematical software tools [14].

9.4 Bayesian Statistics

The concept of using prior distributions to predict future outcomes was first postulated by Reverend Thomas Bayes in 18th century England. By exploiting present day computing power, Bayes formulation of probability can now be applied to a broad class of statistical problems.

Bayes formula states the conditional probability of an event A given that event B has occurred is $p(A|B)$ and can be expressed as

$$p(A|B) = \frac{p(A)p(B|A)}{p(B)}, \tag{9.38}$$

where $p(B|A)$ is the conditional probability of event B given that A occurs and $p(B)$ is the total probability of event B given all other mutually exclusive events. The probability $p(A)$ is called prior probability which is the probability of event A before event B occurs and $p(A|B)$ is posterior probability of event A, given B.

In terms of probability distributions,

$$g(\lambda|x) = \frac{f(x|\lambda)g(\lambda)}{\int_0^{\infty} f(x|\lambda)g(\lambda)d\lambda}, \qquad (9.39)$$

Where $f(x|\lambda)$ is the probability distribution for variable x given the unknown parameter λ, $g(\lambda)$ is the prior distribution model for λ and $g(x|\lambda)$ is the posterior distribution model for λ given that parameter value x has been observed.

The advantage of using Bayesian method is that it makes use of prior knowledge in statistical evaluation of current data samples. Failure and reliability predictions would specifically benefit from Bayesian methods as very little information is available in the early life of a product [12]. In another application, a Bayesian virtual probe (BVP) methodology is developed for selecting optimal locations on wafers for electrical measurements [13]. Although test structures placed within scribe-lines are uniformly distributed on the wafer by design, a reduced set of measurement sites are selected based on prior knowledge of AcW variations, thereby improving test efficiency.

9.5 Data Visualization

Examples of visualization and presentation of data collected from circuit simulations and test are provided. A strong emphasis is placed on comparing data with expected values, prior knowledge or best guess. With normalized data displayed in a standardized format, a large number of observations are summarized on a single page. Visual assimilation coupled with pattern recognition adds clarity to the conclusions drawn. Key information on the source of data and reference to models, product vintage, etc. should be included on each chart. These details are left out of the examples shown to maintain emphasis on the displayed information.

Example 1 (Data Validation) The first step in data analysis is establishing the validity of the data and the associated information. Errors may originate from circuit simulation setup, circuit design, SPICE commands, test setup or test code, as well as from measurement noise or software bugs. Erroneous data may lead to wrong conclusions as described using Fig. 9.1. It is best to carefully check the data collected on a small number of samples and compare it to known physical behavior of devices and circuits. Some other items to consider are test equipment specifications, measurement resolution, and model accuracy.

In Fig. 9.16a, AC power of a circuit is plotted as a function of power supply voltage V_{DD}. The power is expected to increase with V_{DD} as indicated by the dashed line. However, measured AC power stays constant above $V_{DD} = 1.1$ V. This is likely due to a low current clamp set-point on the power supply or an erroneous V_{DD} value.

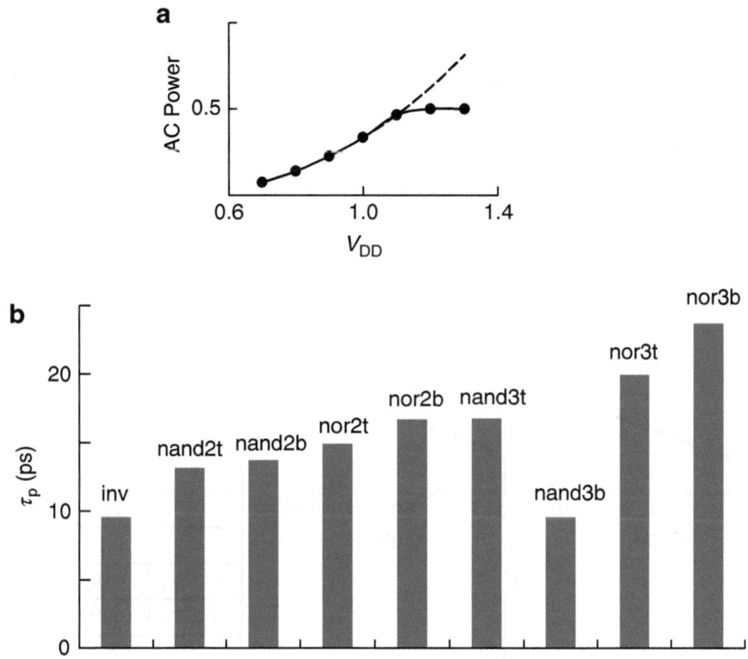

Fig. 9.16 (**a**) Measured data along with expected trend shown as a *dashed line* and (**b**) simulated τp of logic gates

In Fig. 9.16b, average delay τ_p of logic gates obtained from circuit simulations are shown in a bar chart. The logic gates are designed to have $\tau_{pd} \approx \tau_{pu}$. It becomes immediately obvious that the τ_p value corresponding to NAND3B is incorrect. It may also appear that the NAND3B value corresponds to the inverter. However, if that is not the case, circuit schematic and simulation setup should be checked to get a valid value of τ_p for the NAND3B.

Example 2 (Bivariate Data Characterization) Bivariate data, where the value of a variable is dependent on another variable, is displayed in an *X–Y* scatter plot. If there is a linear relationship between the two variables, the data are fit to an equation of the type expressed in Eq. 9.21. Given a value of the independent variable, the fitting coefficients can then be used to predict the value of the dependent variable. The regression coefficient R^2 gives the goodness of the fit—a value >0.98 indicating a strong linear dependence. Data may be fit to a power law, logarithmic or other defined relationship to get the best fit.

In Fig. 9.17a–d, the average delay τ_p obtained from circuit simulations of an inverter is plotted as a function of output load C_L, input rise time τ_{ri}, sum of n-FET and p-FET widths $(W_n + W_p)$ (with constant FO), and the width ratio W_p/W_n (with constant $W_n + W_p$). In Fig. 9.17a–c, a clear linear dependence of τ_p on C_L, τ_{ri}, and $(W_n + W_p)$ is displayed. The fitting parameters and R^2 values for all four plots are summarized in

Fig. 9.17e. The dependencies of τ_p on C_L and τ_{ri} give $R^2 > 0.95$. Even though τ_p remains nearly constant with $(W_n + W_p)$, the value of R^2 for this set is only 0.76 due a very small value of m. A linear fit for τ_p vs. W_p/W_n plot gives $R^2 = 0$, suggesting these parameters are uncorrelated, even though there is a well defined and physically meaningful dependence of τ_p on W_p/W_n. An appropriate polynomial fit would give a high degree of correlation. Hence, *X–Y* scatter plots are important for screening the relationship between two variables. A table with the fitting coefficients may be used when summarizing, for example, τ_p dependence on C_L and τ_{ri} for different circuit topologies, but are not appropriate for the data shown in Fig. 9.17c, d.

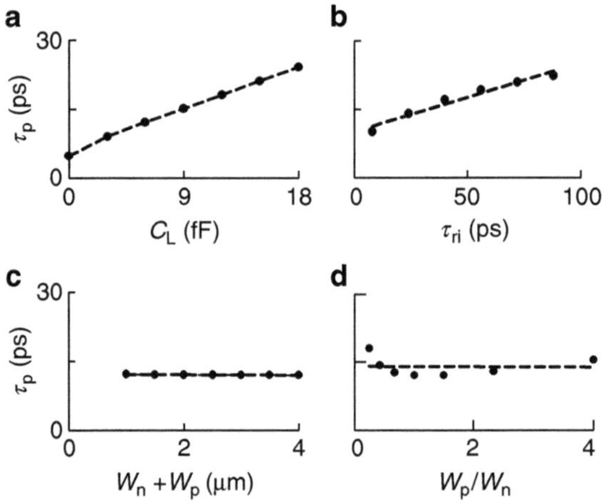

e

x	m	c	R^2
C_L	1.02	5.76	0.99
τ_i	0.15	10.15	0.96
W_n+W_p	-0.06	12.15	0.78
W_p/W_n	-0.05	14.04	0.00

Fig. 9.17 Average delay τ_p of a standard inverter as a function of (**a**) capacitance load C_L, (**b**) input rise time τ_{ri}, (**c**) $(W_n + W_p)$, and (**d**) W_p/W_n. (**e**) Linear fitting parameters for the four plots

Example 3 (Overlaying Nonlinear Characteristics) When two variables have a nonlinear relationship, *X–Y* scatter plots are extremely useful and perhaps the best way to track their behavior. MOSFET *I–V* characteristics are one example of nonlinear relationships between I_{ds} and V_{ds}. Comparing *I–V* characteristics of a large number of MOSFETs is best done by defining a few locations on the I_{ds}–V_{ds} and I_{ds}–V_{gs} plots as described in Sect. 2.2.2.

By normalizing the I_{ds}–V_{ds} plot to I_{on} and V_{DD}, and plotting I_{ds}/I_{on} vs. V_{ds}/V_{DD} MOSFET *I–V* characteristics at two different technology nodes can be compared graphically. Such a plot is shown in Fig. 9.18 for 45 and 22 nm PTM HP models. As described in Sect. 10.3.1, the change in the shape of the I_{ds}–V_{ds} plots in the two technology nodes indicates that the performance gain in going from 45 to 22 nm will be different for different circuit topologies.

A second example for comparing nonlinear characteristics is shown in Fig. 9.18b. I_{ds}–V_{gs} of an n-FET is overlaid on the C_g–V_{gs} plot to compare the onset of channel formation with the V_{tsat} value determined at a fixed I_{dsvt}. Clearly the V_{tsat} value can be shifted to the left or right. Hence, it is important to know the method of determining V_{tsat} when comparing V_{tsat} values obtained from different sources.

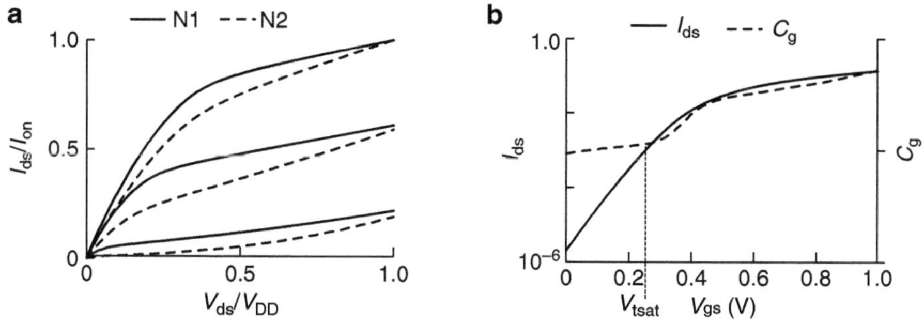

Fig. 9.18 (**a**) Overlaying normalized n-FET I_{ds}–V_{ds} characteristics (**b**) Log(I_{ds}) − V_{gs} and $C_g - V_{gs}$ showing onset of conduction relative to V_{tsat}

Example 4 (Normalized Trend Charts: C_p and C_{pk}) Silicon foundries typically use a Six Sigma methodology to center the process within specified limits. Process capability indices C_p and C_{pk} are used for monitoring parameter deviation from targets. The process mean, μ, an Upper Specification Limit, USL, and a Lower specification limit, LSL, are defined for each parameter based on model expectations or empirically determined targets. The spread around the process mean over a short period of time during which the data are collected is defined as $\pm\, 3\sigma_{short}$. The C_p index is given by

$$C_p = \frac{(\mathrm{USL} - \mathrm{LSL})}{6\sigma_{short}}. \tag{9.40}$$

The C_p index is used when a parameter distribution is perfectly symmetric about the mean. If *USL* and *LSL* are the target +3σ and −3σ limits, respectively, $C_p = 1$ indicates $\sigma_{short} = \sigma$. A $C_p > 1$ value ($\sigma_{short} < \sigma$) indicates the parameter spread is smaller than target and that the process is well controlled.

For some parameters, the distributions may not be symmetric and (USL − μ) may be different than (μ − LSL). A more convenient index is C_{pk}, defined as the smaller of the two values

$$C_{pk} = \frac{(\mathrm{USL} - \bar{x})}{3\sigma_{short}}, \tag{9.41}$$

or

$$C_{pk} = \frac{(\bar{x} - \mathrm{LSL})}{3\sigma_{short}}. \tag{9.42}$$

As with C_p, a C_{pk} value of >1 indicates the parameter is within the specified limits.

The parameter spreads for an expected symmetric distribution over a period of five quarters (Q1 to Q5) are shown in Fig. 9.19a. The whiskers show 99.7 % (6 σ short) range in each quarter. The parameter spreads for a nonsymmetric distribution are shown in Fig. 9.19b. The dark circle indicates the median value of the parameter. A histogram of C_p and C_{pk} values for a large number of parameters can be used to quickly isolate the parameters that are out of specifications.

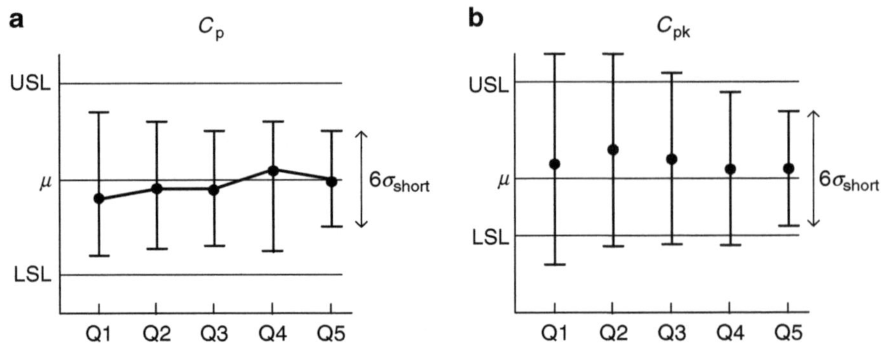

Fig. 9.19 Trend charts over five quarters (Q1 to Q5) for (**a**) C_p and (**b**) C_{pk}

Example 5 (Cumulative Distributions) Cumulative % of failing chips are plotted as a function of test sequence in Fig. 9.20 for a new chip design. A large fraction of fails occur for the first few test patterns and chips with multiple faults can be rejected early in the test sequence. A steady increase in % of fails is indicative of random faults, probably due to defects. Steps in the plot indicate systematic fails in a sizable fraction of chips for specific test patterns not exercised previously. The test sequence #s at which the steps occur may be targeted for diagnostics or the test order may be changed to eliminate more chips early in the test sequence.

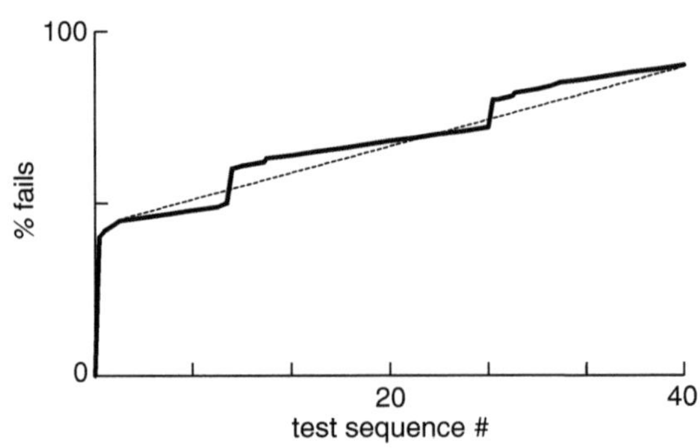

Fig. 9.20 CDF of % of fails for 40 different tests

Example 6 (Normalized X–Y Scatter Plots for Tracking) Model-to-hardware correlation, process center for different MOSFET parameters, and relative signal propagation delays of several circuit topologies can be displayed on a single page. The requirement for such a chart is that measured data on silicon process monitors (delay chains or ring oscillators) be available along with model targets for average delays τ_p at several L_p values.

In Fig. 9.21a, measured raw τ_p values of a NAND2B are plotted vs. τ_p values of an inverter for a large number of chips. Keeping the same X and Y scales, it is easy to see that, as expected, τ_p for NAND2B is larger than for the inverter. In Fig. 9.21b, the data in Fig. 9.21a are normalized to the nominal target values for the NAND2B and the inverter. The normalized delays τ_p/τ_{po} show 1:1 tracking, both speeding up and slowing down by approximately the same fraction with process variations. A linear fit of the data and the average τ_p/τ_{po} (white disc) are also shown. Dashed lines divide the square plot into four quadrants; data in the lower left quadrant are slower than nominal and in the top right quadrant are faster than nominal.

Sixteen plots of the type shown in Fig. 9.21b for different circuit configurations are combined in one chart in Fig. 9.21c. A quick view reveals the range of τ_p variations and which circuits are deviating more from the nominal model targets than others.

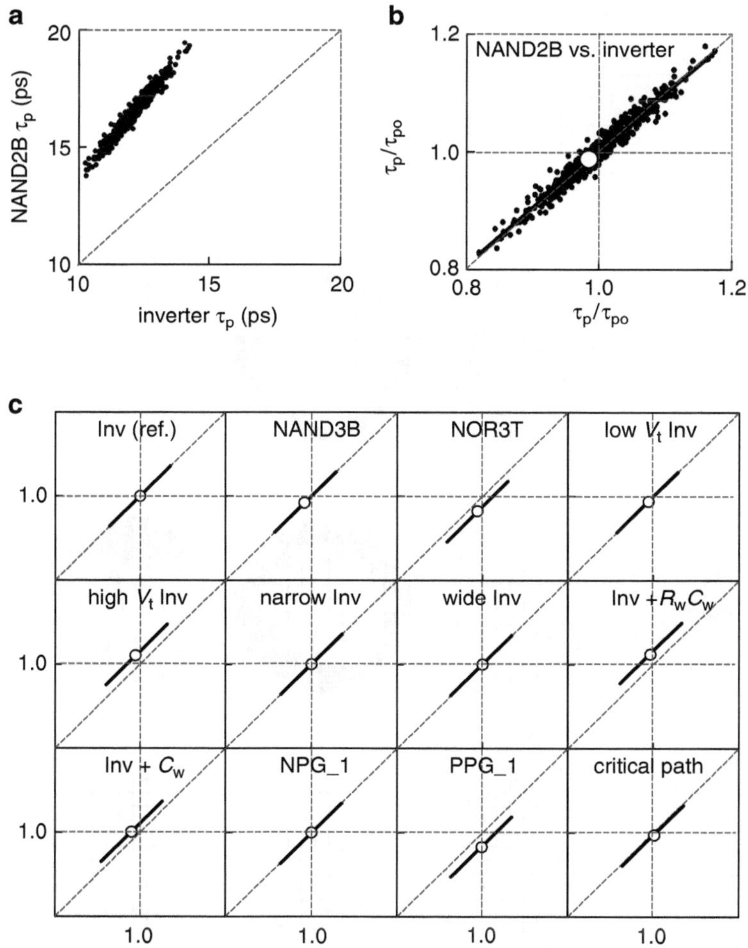

Fig. 9.21 (**a**) Measured τ_p of NAND2B vs. inverter, (**b**) normalized τ_p/τ_{po} of NANd2B vs. inverter, and (**c**) normalized τ_p/τ_{po} of 16 different circuit configurations

Example 7 (Spatial Variations) Across wafer (AcW) and across chip (AcC) mapping of measured parameters are shown in Figs. 6.12, 6.13, and 7.33. When data are available on a large number of wafers, a single stacked wafer map with the number of good chips tested in each location may be included as shown in Fig. 9.22a. It becomes apparent that the number of good chips is lower when the

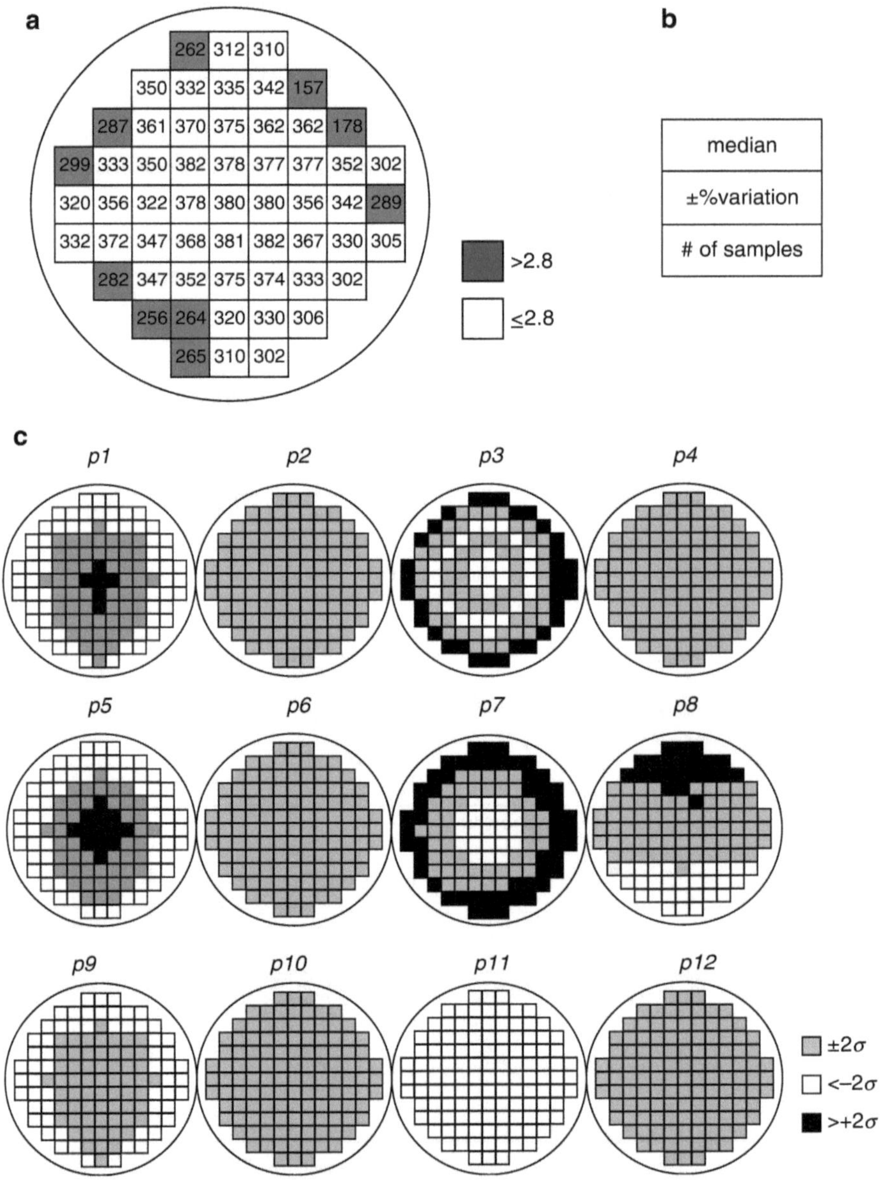

Fig. 9.22 (**a**) Stacked wafer map of 400 wafers for p/n ratios indicating the number of good chips at each location, (**b**) additional quantitative information that can be shown at each chip location, and (**c**) wafer maps of 12 parameters on a single page

parameter value (p/n ratio for example) is >2.8 and that such chips are more likely to be located near the edges of the wafer.

The AcW and AcC variations of as many as 12 to 20 parameters can be displayed on a single chart as shown in Fig. 9.22b. The bin ranges are selected to be $< -2\sigma$, $\pm 2\sigma$ and $> +2\sigma$ for each of the parameters $p1$ to $p12$. Parameters $p2, p4, p6, p10$ and $p12$ are each within $\pm 1\sigma$ of their respective target values. Parameter $p11$ is uniformly offset by more than $-2\sigma p11$. Parameters $p1$, $p2$, $p3$, $p7$ and $p9$ have radial nonuniformity with different signatures from center to edge. Parameter $p8$ shows top-to-bottom variation across wafer. For non-normal parameter distributions, $\sigma+$ and $\sigma-$ values may be used in defining the bin ranges. This type of chart highlights which parameters may require adjustments. The ordering of the wafer maps may be changed to cluster the ones that require tuning of process recipes.

Example 8 (Yield Pareto) A pareto is a type of chart which combines a bar graph with a line graph. A yield pareto shown in Fig. 9.23 displays the progression of

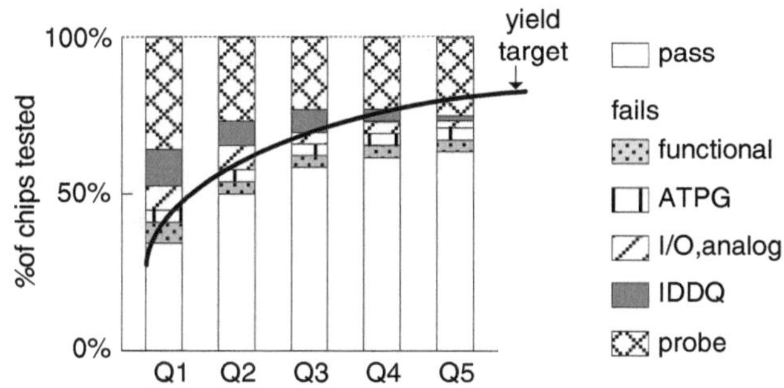

Fig. 9.23 Yield pareto showing major yield detractors as a stacked *bar* graph and yield target shown as a *line* graph

yield detractors over a period of five quarters (Q1 to Q5) in a silicon foundry. A steady improvement in defect density in the first two quarters is apparent from the reduction in probe fails (opens and shorts). IDDQ fails are practically eliminated by Q5 by re-centering the process at a longer L_p. The product yield, however, is still falling below the target. The pareto highlights probe fails continue to be the dominant yield detractor, and that gross particulate and process induced defects need to be reduced. These and other yield issues can be summarized in a single chart of this type and monitored with respect to target values.

Example 9 (Tabular Data) A tabular format is often preferable for summarizing data representing a large number of variables. A mix of tabular and graphical presentations together with color coding or shading of cells in the table can highlight the areas needing attention. An example chart is shown in Fig. 9.24.

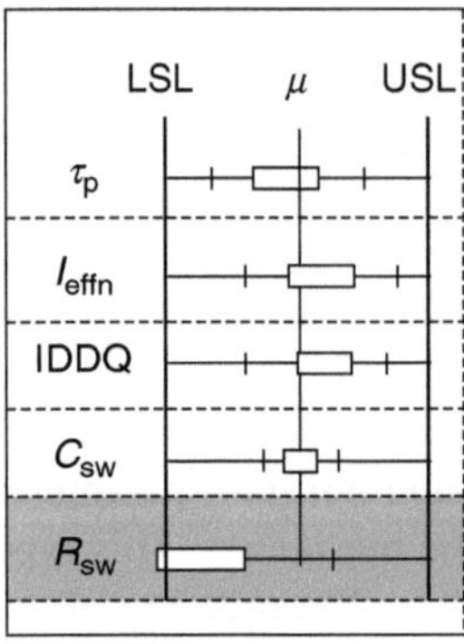

	n	$\frac{(\bar{x}-\mu)}{\sigma}$	$\frac{s}{\sigma}$	$\bar{x}$	s	μ	σ
τ_p	520	− 0.33	0.90	3.90 ps	0.27 ps	4.00 ps	0.30 ps
I_{effn}	600	+ 0.50	0.90	1.24 mA	0.07 mA	1.20 mA	0.08 mA
IDDQ	560	+ 0.67	0.83	0.24 μA	0.05 μA	0.20 μA	0.06 μA
C_{sw}	520	0.00	0.50	2.11 fF	0.03 fF	2.11 fF	0.06 fF
R_{sw}	520	− 2.30	1.60	1600 Ω	200 Ω	1887 Ω	125 Ω

Fig. 9.24 Statistical summaries of parameter measurements displayed with box and whiskers charts combined with a table showing sample sizes, means and standard deviations, as well as target μ and σ values

Statistical summaries of measured parameters are given along with a box and whiskers representation situated within USL and LSL boundaries. The table includes measured and target values of parameter means and standard deviations. The last row is shown with gray shading as this parameter spread is outside the target range.

Example 10 (Composite Summary Chart) A high-level view of a silicon manufacturing line and product performance status is shown in Fig. 9.25a–d.

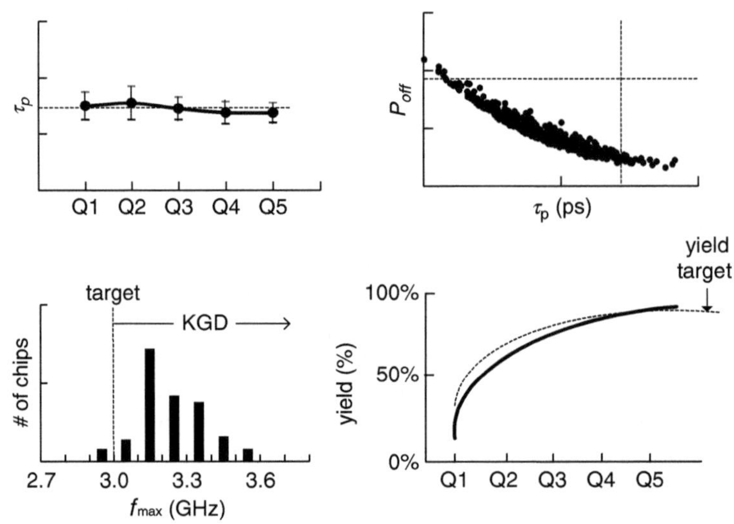

Fig. 9.25 High level view of silicon manufacturing line and product performance status over last five quarters (Q1 to Q5)

These four plots summarize the silicon line-center in terms of circuit delay, P_{off}, f_{max}, and yield over the last five quarters. Targets for each of the parameters are shown as dashed lines. The plots can easily fit on a single page of a presentation chart or report document. A quick visual scan assures the technical and management teams that the CMOS technology, product power and frequency, and the supply line are all in good health.

9.6 Summary and Exercises

The basics of probability and normal distributions are described. Non-normal distributions and correlation among measured parameters are discussed in the context of CMOS electrical test data. The impact of systematic and random variations in V_t and L_p on MOSFET and circuit delay distributions are illustrated using Monte Carlo simulations. Sensitivity analysis for measured parameters is described with a 2×2 matrix representation. Examples of data visualization and presentation exploit the use of simulated targets, normalized comparisons, and pattern recognition.

Probabilities for logic gate outputs are to be investigated in exercises 9.1 and 9.2. Exercises 9.3 through 9.9 deal with statistical parameter distributions. Data visualization is treated in exercise 9.10.

9.1. (a) Assuming the inputs of logic gates have equal probabilities of being a “1” or a “0”, what are the probabilities that the outputs of an inverter, AND2, NAND3, NOR3, and XOR2 will be a “1”.
(b) Compare logical AND, OR, and XOR functions with the expressions in Eqs. 9.4, 9.5, and 9.6.

9.2. What are the probabilities that the outputs of an inverter, AND2, NAND3, NOR3, and XOR2 would switch from a “1” to a “0”? Assume equal probabilities of the inputs being “1” or “0” for both states.

9.3. Measured f_{max} values in manufacturing test follow a normal distribution. The f_{max} specification for acceptance is set at -1.5σ.
(a) What % of chips will be rejected when using this criteria?
(b) If the target is to accept 97 % of the chips, what should the f_{max} specification for acceptance be set at in terms of σ?

9.4. Run a Monte Carlo simulation of a standard inverter chain assuming a normal distribution of p-FET V_t (V_{tp}) with a mean value of *pdelvto* = 0 V and $\sigma V_{tp} = 0.02$ V and print the *pdelvto* values for 1,000 cases.
(a) Determine the mean and σV_{tp} for the 1,000 cases and plot a histogram.
(b) Select 10 unique sets of 25 cases each. Find the mean and σV_{tp} for each of the sets and compare. Do these values fall within the error listed in Tables 9.3 and 9.4?
(c) Plot histograms of these 10 sets to make a visual comparison.

9.5. Using the simulation results of exercise 9.4, select sets with 5, 10, 20, 50, and 200 cases at random.
(a) Determine mean and σV_{tp} of these sets.
(b) Plot mean and σV_{tp} as a function of # of cases. Show the limits from statistical tables for each set on the same plot. How do they compare?

9.6. Using the setup from Exercise 9.3, run Monte Carlo simulations of 1,000 cases varying L_p, V_{tn}, and V_{tp}.
(a) Search for maximum and minimum values of τ_p. What are the corresponding values of L_p, V_{tn}, and V_{tp}?
(b) Calculate the offset from nominal in units of σ for the L_p, V_{tn}, and V_{tp} in (a).
(c) Estimate the number of cases required to obtain one instance outside of the $\pm 6\sigma$ range of τ_p.

9.7. A process split-lot has 5-way split for L_p, centered at nominal, $\pm 1\sigma L_p$ and $\pm 2\sigma L_p$. If the standard deviation of L_p within each lot is 0.5 σL_p, generate a histogram to show the overlap between splits.

9.8. Using the information in Exercises 9.6, how would you analyze the f_{max} data from the split lots? What prior knowledge is useful for this analysis?

9.9. Detailed characterization of microprocessor chips in real situations is carried out on 10 server systems.
(a) With this small sample size, what is the accuracy with which mean and σ of total power for a specific workload are known with 95 % confidence. The population standard deviation is not known.
(b) Run Monte Carlo simulations for 10 cases to measure total power of a 51 stage RO and determine mean and σ in 10 repeated simulations.
(c) The test team would like to run characterization on a larger sample. Show a chart to justify this request.

9.10. Create a chart to show qualitative and quantitative AcC variations using data obtained from at least three different methods, pointing out that consistency in the observations.

References

1. Box GEP, Hunter WG, Hunter JS (1978) Statistics for experimenters: an introduction to design, data analysis and model building. Wiley, New York
2. Montgomery DC (2001) Design and analysis of experiments, 5th edn. Wiley, New York
3. Koronacki J, Thomson JR (2001) Statistical process control: the Deming paradigm and beyond, 2nd edn. Chapman and Hall, Boca Raton
4. Burr JT (2005) Elementary statistical quality control, 2nd edn. Marcel Dekker, New York
5. Montgomery DC (2009) Introduction to statistical quality control. Wiley, New York
6. Joglekar A (2001) Statistical methods for six sigma in R and D and manufacturing. Wiley Interscience, Hoboken
7. Pande PS, Neuman RP, Cavanagh RR (2000) The six sigma way. McGraw-Hill, New York
8. Tufte E (1983) The visual display of quantitative information. Graphics, Cheshire
9. Tufte E (1997) Visual explanations. Graphics, Cheshire
10. Tufte E (1990) Envisioning information. Graphics, Cheshire

11. Bhushan M, Ketchen MB (2011) Microelectronic test structures for CMOS technology. Springer, New York
12. NIST/SEMATECH e-Handbook of Statistical Methods. http://www.itl.nist.gov/div898/handbook/. Accessed 10 May 2014
13. Zhang W, Li X (2010) Bayesian virtual probe: minimizing variation characterization cost for nanoscale IC technologies via Bayesian inference. Design automation conference DAC'10, pp 262–167
14. Online matrix calculator. http://comnuan.com. Accessed 21 Jan 2014

CMOS Metrics and Model Evaluation 10

Contents

10.1 Measurement Standards 348
10.2 Scaling Trends in CMOS Products 351
10.3 CMOS Performance Metrics 355
10.3.1 MOSFET Performance 355
10.3.2 Interconnect Performance 363
10.3.3 Logic Gate Performance 365
10.4 CMOS Power-Performance-Density Metrics 367
10.4.1 Circuit Density 368
10.4.2 Energy and Power Density 369
10.4.3 V_{DD} Dependencies of Different Metric Parameters 373
10.4.4 Summary of Performance Metrics 374
10.5 Compact Models and EDA Tool Evaluation 374
10.5.1 BSIM Models 376
10.5.2 Layout Parasitic Extraction 383
10.5.3 Timing and Power Tools 385
10.6 PD-SOI vs. Bulk Silicon Technology 386
10.7 Closing Comments on CMOS Technology Evaluation 394
10.8 Summary and Exercises 395
References 398

It is often necessary to make direct comparisons among CMOS technologies offered by different foundries at a particular technology node, among different technology nodes, or between similar technologies on different substrates, such as bulk silicon and SOI. Such comparisons are used in guiding technology development, in benchmarking and selecting the most suitable CMOS manufacturing process or foundry for a given product, and in projecting CMOS product specifications in advance of full-scale design. Quantifiable and measurable metrics for key performance tracking parameters are defined at the device and circuit level. For a correct assessment, the integrity of compact models and EDA tools needs to be validated over the full design window. The final verdict on the relative merits of

M. Bhushan, M.B. Ketchen, *CMOS Test and Evaluation*,
DOI 10.1007/978-1-4939-1349-7_10

different technologies, based on models or hardware data, can at best be obtained with some degree of uncertainty.

The international technology roadmap for semiconductors (ITRS) makes projections for key metrics almost 10 years in advance to guide and facilitate technology, process equipment and related industrial development in a timely fashion [1]. Metrics and parameter values are initially based on predictive models and later validated in the hardware as the technology progresses from early development to manufacturing.

In the first part of this chapter, performance metrics for MOSFETs, interconnects, and circuits are discussed. These metrics cover three main aspects of CMOS scaling: delay, density, and power, along with their interrelationships. Standardization of measurement techniques for clarity and accuracy is emphasized at all levels. Although the discussion here is about CMOS technologies, many of the performance metrics elements covered are suitable for other microelectronic technologies as well.

Much of the discussion on CMOS technology metrics is based on device models. It is therefore imperative that the models accurately reflect the physical behavior of the devices. Following the discussion on technology metrics, methodologies for evaluation of device models and tools used in circuit design are described. Such evaluation procedures aid in preventing errors in compact models and in the simplified models incorporated in timing and power tools from propagating to product design. It is highly desirable that model-to-hardware correlation be carried out for all EDA tools. As silicon area and test time are primary concerns in product manufacturing, an appropriate common set of test structures can serve for correlation of electrical measurements to technology models and circuit design tools.

A historical perspective on measurement standards is given in Sect. 10.1. CMOS scaling trends are described in Sect. 10.2. MOSFET and circuit performanc metrics are discussed in Sect. 10.3, and power and density metrics are covered in Sect. 10.4. Methodologies for evaluating BSIM models and EDA tools are described in Sect. 10.5. In Sect. 10.6, a brief description of MOSFET and circuit behavior in PD-SOI technology is provided to illustrate how additional complexity in technology comparisons arises as new features are incorporated. Finally, some general guidelines for model-to-hardware correlation are given in Sect. 10.7.

The ITRS roadmaps published every 2 years are excellent sources of information for tracking CMOS technology scaling [1]. CMOS scaling rules and trends are covered in textbooks [2–4]. Additional references are provided throughout the chapter.

10.1 Measurement Standards

The importance of standardized units of measure was recognized as human populations began to settle, trade, and own land. The accuracy of even the most basic units of measure such as length and time has continued to improve over many centuries. A brief historical view of some of these developments is given here to highlight the difficulties faced in reaching a consensus on measurement standards.

Initially units of length were defined in terms of human body parts, such as the arm or foot. Over a thousand years ago, the concept of a yardstick was first introduced by the Saxon King Edgar as a standard unit of length measurement. A few hundred years later, a more accurate standard, an iron rod defining a yard was placed in royal custody in England.

In 1834, the then latest version of the standard yard was destroyed in a fire. It became apparent that the length standard should be based on some fundamental quantity. The French rose to meet this challenge and replaced the yard with the meter, defined as one-tenth of a millionth ($10^{-7}\times$) of one-quarter of the circumference of the earth as measured in a survey in the 1790s. A meter rod of platinum-iridium was maintained at a fixed temperature to represent this standard. After the invention of lasers, the length standard was defined in units of wavelength of a krypton-86 laser. In 1983, the meter was defined as the length of the path travelled by light of an iodine-stabilized He-Ne laser during a time interval of 1/299792458 of a second. The accuracy of this length standard is about one part in 10^{11}.

The standard for time intervals followed a similar historical course as the length standard. The unit of time, the second, was defined as the fraction 1/86,400 of the average length of a day. Mechanical clocks based on the period of oscillation of a spring or that of a pendulum have been in use since medieval times. In the early part of the twentieth century, clocks using the natural frequency of oscillation of a quartz crystal provided better accuracy. Today, the standard for the unit of time is maintained with atomic clocks to an accuracy of a few fs (10^{-15} s).

With the discovery of electricity, standards for voltage, current, and resistance had to be defined. As technology progressed, continuous improvement and introduction of new standards became essential. Today, in most countries institutes of standards have been established by their respective governments to maintain and monitor various measurement standards. Research is conducted on standards to improve accuracy, to develop new measurement techniques, and to cover new technology areas.

The National Institute of Standards and Technology (NIST) in the USA, in close collaboration with industry, engages in defining standards and test methodologies in many areas ranging from physics, chemistry, materials science, and electronics to energy, environment, and transportation. In the area of CMOS technology, metrology tools and test structures are being developed to better characterize transistor performance and reliability. Examples of other areas covered at the time of publication of this book (2014) include linking telegraph noise to device parameters and a method to extract series resistance of transistors.

Establishing a standard methodology for measuring CMOS performance is a complex challenge. As indicated in Fig. 10.1, feature sizes and switching times are at the low end of the practical scale, adding to the difficulty of accurate measurements. The number of variables contributing to any performance metric measurement is much larger than for a single physical parameter such as length or mass. A CMOS technology performance metric may be based on component density, DC drive currents, power, and switching speed of a representative circuit at a selected operating voltage and temperature. MOSFET dimensions and parasitics change with CMOS scaling and therefore physical layouts of structures on which measurements are made cannot be fully standardized to cover all

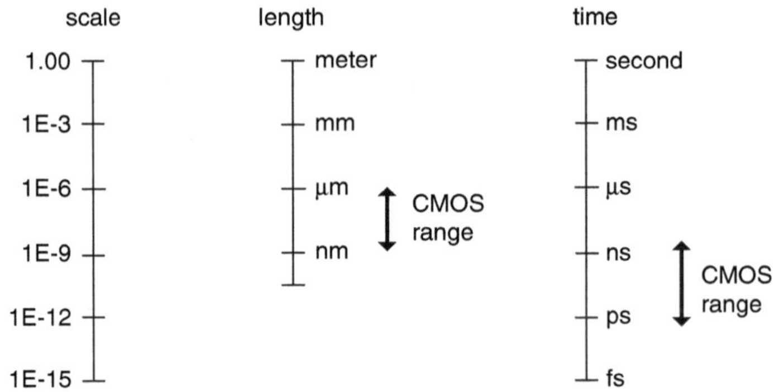

Fig. 10.1 Range of circuit dimensions and switching times in CMOS circuits

technology nodes. Some parameters such as leakage currents are very sensitive to voltage and temperature. Measurement resolution requirements for these are more stringent than for others such as drive currents.

The ITRS publishes performance, power, transistor density, and many other parameter targets for each CMOS technology node. These values are derived from models and from extrapolation of current trends based on input from industrial partners. The lengthy tables covering projections for a 10-year span mainly serve as guidelines for technology directions and development.

Measurements of CMOS performance are important for technology development, product design, and marketing. Measurements of MOSFET characteristics and CMOS circuit delays are routinely carried out by silicon technology teams. The results are typically presented at technical conferences, published in journals and documented for customer use. Test structure designs and details of measurement methodology, sample sizes for data collection and test conditions may not be readily available for others to reproduce and validate the published results.

CMOS product design teams may also evaluate technology performance using device models provided by silicon foundries. Test structures are placed on the product or scribe-line for model-to-hardware correlation and to obtain calibrated circuit performance on each chip as discussed in Chaps. 5 and 7. Reverse engineering is sometimes carried out to verify the performance of a competitor's hardware. This process is expensive and is typically attempted only by large-scale manufacturers on a very limited number of samples.

In the midst of this complexity, there is a need to establish a standard methodology for evaluating CMOS technologies. In the late 1990s, a debate emerged on the relative merits of PD-SOI and bulk CMOS technologies and continued for several years, with no industry-wide consensus. With the approaching end of the road for CMOS and a strong drive towards the development of beyond-CMOS technologies, the criticality of this issue is increasing. There are many factors to be considered in setting up benchmarking standards for a comprehensive and accurate evaluation of emerging technologies.

In the following sections, trends in scaling of CMOS products and methodologies for evaluating CMOS performance with several different metrics are described. Accuracy and standardization requirements are emphasized to remove ambiguity. The metrics may need to be modified when applied to a specific product application.

10.2 Scaling Trends in CMOS Products

Integrated circuits were first introduced in the marketplace by Fairchild Semiconductor Corp. in 1961. By integrating discrete circuit elements on a silicon substrate, the volume of electronic parts was significantly reduced. Within a few years, silicon chips were used in portable calculators and computers. In 1965, Gordon Moore projected a reduction in component dimensions to cram more function into a given space while lowering manufacturing cost. Since then silicon technology has kept up with "Moore's law" with minimum feature size reduction of ~0.7× every 2 years [5]. This is illustrated in Fig. 10.2 using published data for Intel microprocessor

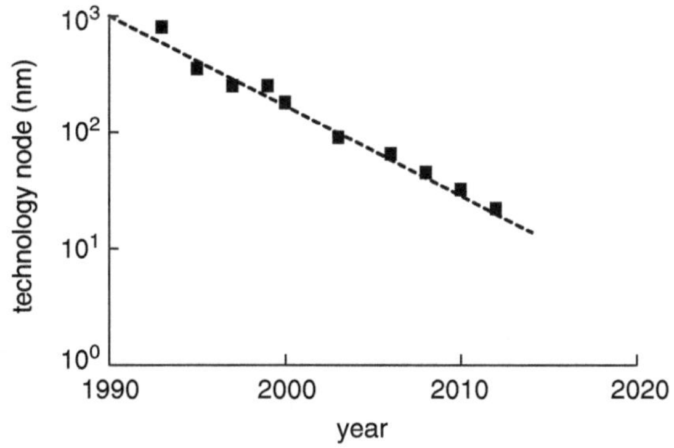

Fig. 10.2 Technology node in nm for Intel's microprocessor chips as a function of time. *Dashed line* indicates 0.7× reduction every 2 years

chips [6]. Here technology node in nm represents the minimum dimension. Taking advantage of reduction in feature sizes, DRAM cell area has continued to decrease by >30× per decade as indicated by recent data plotted in Fig. 10.3 [1].

The corresponding trend of increase in transistor density by 2× every 2 years for Intel and IBM microprocessors is shown in Fig. 10.4 along with the predictive trend line [6, 7]. Increase in circuit density has generally not resulted in smaller chip sizes. Although smaller chip size helps improve manufacturing yield, adding more functions to a chip and keeping larger chip sizes has proved to be economically more profitable. In Fig. 10.5, the chip areas for Intel's Pentium and Core processor chips are shown to illustrate the industry preference [6].

With scaling of CMOS technology, a commonly cited intrinsic MOSFET performance metric, FPG (=CV/I delay), was historically reduced at the rate of ~17 % per year prior to 2009. The rate of reduction is projected to be ~13 % per year through

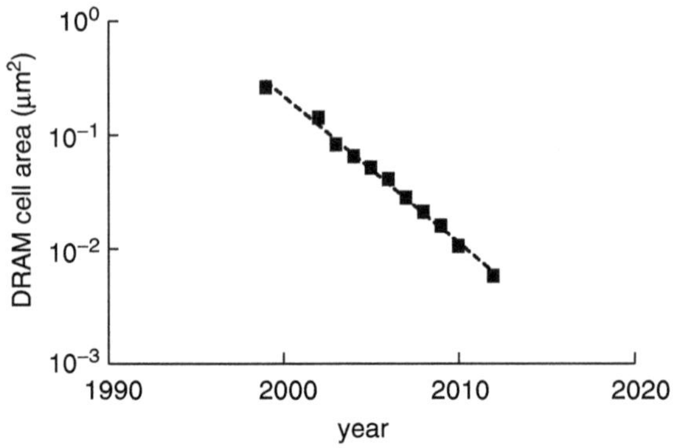

Fig. 10.3 DRAM cell area as a function of time

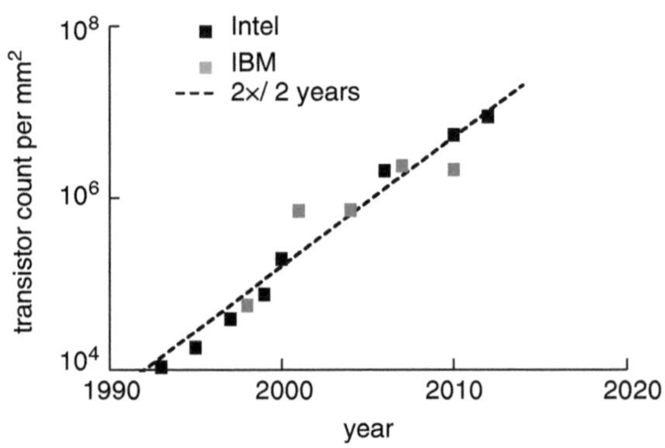

Fig. 10.4 Transistor count per mm^2 for microprocessor chips from Intel (Pentium and Core) and for IBM (Power series) [6, 7]. *Dashed line* shows $2\times$ increase every 2 years

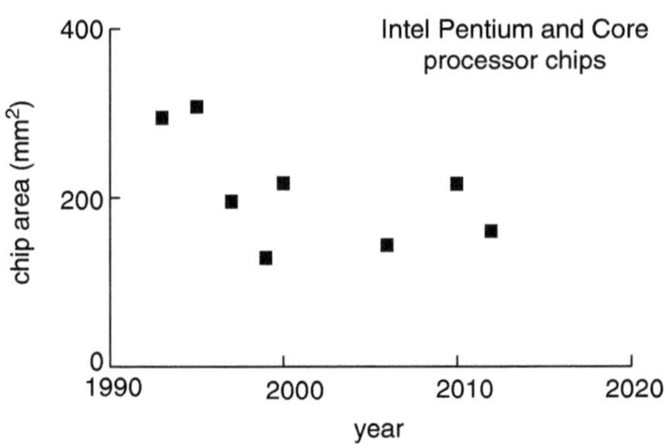

Fig. 10.5 Intel Pentium and Core processor chip area

2017, decreasing further to ~5 % per year beyond that in ITRS 2013 [1]. Reported reduction in ring oscillator delays is less than that of intrinsic MOSFET delay due to parasitic and loading effects. If a CMOS chip is faithfully migrated from one technology node to the next, it should exhibit a similar reduction in cycle time as in circuit delays. In practice, this is not always the case. Changes in chip architecture, circuit design methodology, design margins, specifications, and operating conditions may also change as a product evolves over time. The cycle time reduction has nearly leveled because of power and cooling constraints. This is suggested by the progression in Intel's microprocessor frequency as shown in Fig. 10.6.

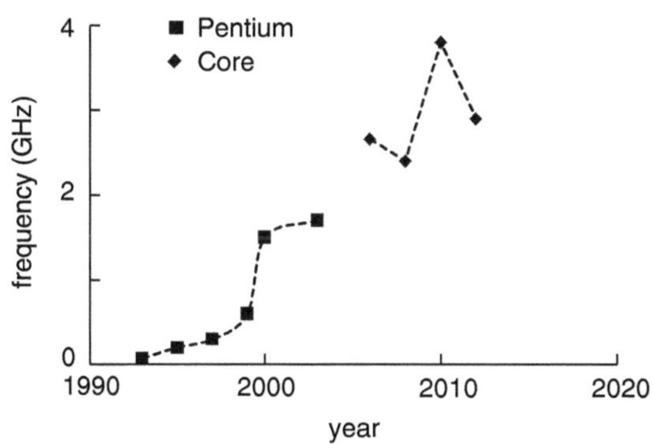

Fig. 10.6 Progression of Intel's Pentium and Core microprocessor clock frequencies [6]

A direct consequence of scaling MOSFET dimensions is an exponential increase in subthreshold and gate-oxide leakage currents. Projected trends of IDDQ for high performance devices based on ITRS roadmaps over the years are shown in Fig. 10.7.

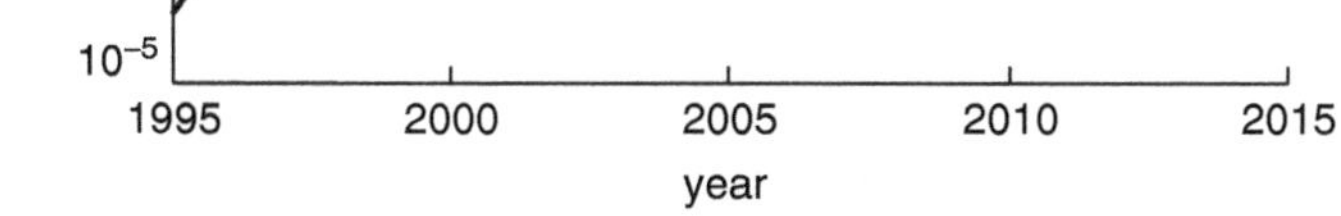

Fig. 10.7 Normalized IDDQ trend over time (ITRS roadmap projections)

The exponential rise in IDDQ could not be sustained beyond the late 1990s. In order to lower chip IDDQ while taking advantage of higher MOSFET performance with scaling, silicon manufacturers began to offer high and low V_t devices in addition to the nominal V_t offering. Managing IDDQ started to become a joint responsibility of silicon technology through MOSFET engineering, and chip design through judicious use of different MOSFET offerings. For 180 and 130 nm nodes, high V_t MOSFETs typically have $0.1 \times I_{off}$ and 10 % lower I_{on} than nominal MOSFETs, and low V_t MOSFETs typically have $10 \times I_{off}$ and 10 % higher I_{on} than nominal MOSFETs. The technology performance is gauged by nominal V_t MOSFETs which are most widely used in on-chip circuits. Circuits paths with large timing margins are then populated with high V_t MOSFETs, and those with critical timing requirements with low V_t MOSFETs. Introduction of high-K gate-dielectric and FinFETs has also helped manage the rise in gate and subthreshold leakage currents.

Beyond the 130 nm technology node, silicon foundries began to offer MOSFET pairs engineered to optimize either for low leakage with reduced performance or for high performance with relatively higher leakage. It was soon recognized that further benefit from silicon technology could be extracted by independently tuning MOSFETs used in different SRAM cell designs, and long-channel MOSFETs for analog and mixed signal I/O circuits. This has led to a proliferation of MOSFET offerings in each technology node as illustrated in Fig. 10.8.

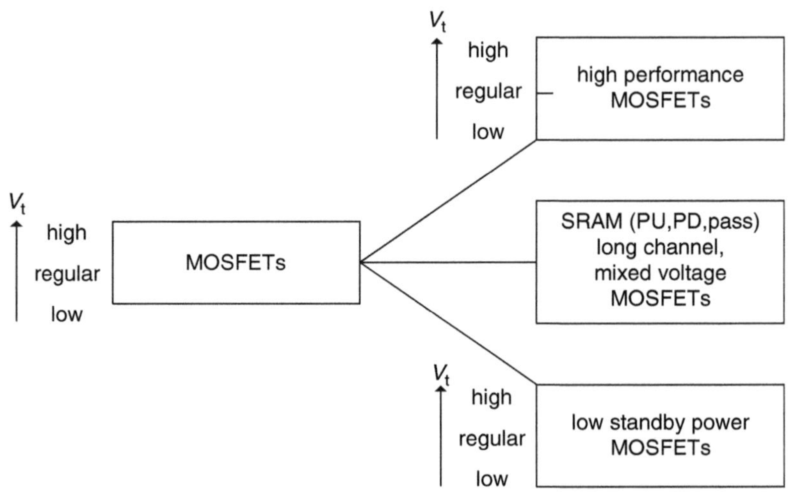

Fig. 10.8 Proliferation of MOSFET offerings for optimum performance and power with CMOS technology advancement

The increase in MOSFET offerings has a direct impact on the development of compact models and EDA tools. BSIM and parasitic extraction models are needed for each MOSFET pair type. These models are incorporated in the circuit design environment with some level of abstraction. A large fraction of digital circuit design is now fully automated and in such cases circuit designers may never manually design or tune circuit properties based on device models.

10.3 CMOS Performance Metrics

Model-based evaluation of CMOS circuit performance is important in technology development, in setting up circuit design methodology and tools, and in test debug. We use PTM HP models for 45, 32, and 22 nm technology nodes to illustrate differences among models and the limitations of relying on a single performance metric. The methodology described can be used for comparing different model releases at the same technology node as well.

10.3.1 MOSFET Performance

A measure of the current drive strength of a MOSFET can be obtained from its DC *I–V* characteristics. Leakage currents in the "on" and "off" states of a MOSFET dissipate power when a circuit is idle. MOSFET capacitances are charged or discharged with changes in V_{gs} and V_{ds}. These capacitances contribute to signal propagation delay when a CMOS circuit changes its logic state. Physical layout-dependent parasitic resistances and capacitances reduce MOSFET current drive and add wasteful capacitive load.

To lower circuit delay and enable a higher frequency of operation with minimum power consumption, the desirable properties of a MOSFET are:

- High current (I_{ds}) drive strength
- Low leakage currents (I_{off} and I_{gl})
- Low intrinsic capacitances (C_g, C_{ov}, C_j)
- Low parasitic resistances and capacitances

In view of the above requirements, setting a single performance metric for MOSFETs is not straightforward. Common reporting practices include individual parameters I_{on} ($=I_{dsat}$), I_{eff}, I_{off}, I_{gl}, and C_g, and a combined metric FPG for switching delay in the form of CV/I.

10.3.1.1 MOSFET Intrinsic Performance

We will first examine the current drive and capacitance components of MOSFET CV/I metrics at the nominal operating V_{DD} of a technology. BSIM PTM HP models describing intrinsic MOSFET properties of 45, 32, and 22 nm nodes are used for this exercise. The nominal V_{DD} values of these technologies are 1.0 V, 0.9 V, and 0.8 V, respectively. MOSFET dimensions are scaled as listed in Tables A.4 and A.6 of Appendix A.

Simulated DC I_{ds}–V_{ds} characteristics with $V_{gs}/V_{DD} = 1.00$, 0.75, and 0.50 for 1.0 μm wide n-FETs and p-FETs in the three technology models are overlaid in Figs. 10.9a and Fig. 10.9b, respectively (all p-FET currents and voltages are shown as positive). It is immediately apparent that I_{on} defined at nominal V_{DD} improves with scaling. However, I_{ds} values in the other regions of the I_{ds}–V_{ds} space are either comparable or lowered with scaling.

A visual comparison of how the shape of the *I–V* characteristics changes in the scaled models is obtained by plotting normalized values (I_{ds}/I_{on} vs. V_{ds}/V_{DD} for the

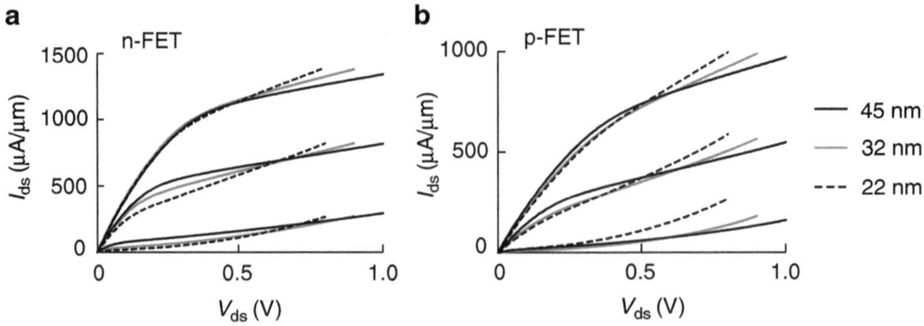

Fig. 10.9 Nominal MOSFET DC I_{ds}–V_{ds} characteristics of (**a**) n-FETs and (**b**) p-FETs in 45, 32, and 22 nm PTM HP models @ nominal V_{DD}, 25 °C

same V_{gs}/V_{DD}) as shown in Fig. 10.10a, b. As V_{ds}/V_{DD} and V_{gs}/V_{DD} are reduced, the drive current relative to I_{on} is substantially reduced in 22 nm models compared with 45 nm models.

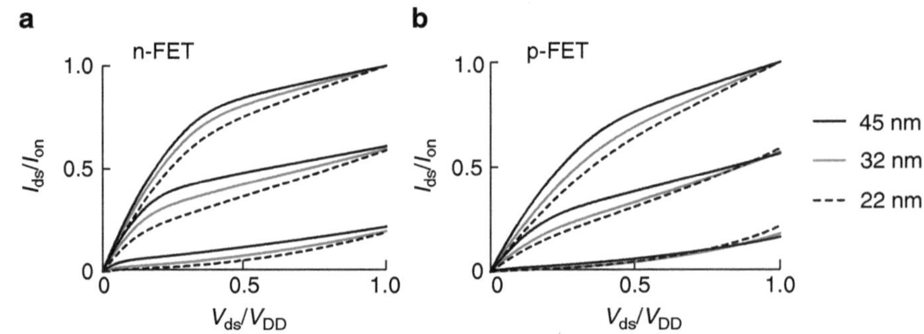

Fig. 10.10 MOSFET DC I–V characteristics scaled to I_{on} and V_{DD} of (**a**) n-FETs and (**b**) p-FETs in 45, 32 and 22 nm PTM HP models, @ nominal V_{DD}, 25 °C

The I_{ds}–V_{gs} characteristics showing the behavior in the subthreshold region in the three technology models are plotted in Fig. 10.11a, b for $V_{ds} = V_{DD}$. I_{off} for both the n-FET and p-FET increase with scaling, and V_{tsat} is reduced.

Key MOSFET parameters for the three technologies are summarized in Table 10.1 for the n-FET and in Table 10.2 for the p-FET. Both I_{eff} and I_{mid} are reduced with scaling while both I_{on} and I_{off} increase. Reviewing the trajectories of MOSFETs during switching as described in Sects. 2.2.3 and 5.6.1 it is straightforward to draw the conclusion that the current drive during switching will not be proportional to relative I_{on} values in the three technologies.

The MOSFET parameters listed in Tables 10.1 and 10.2 are for nominal values of V_t and L_p. In silicon manufacturing, fluctuations in the manufacturing processes result in variations of V_t, L_p and other key device parameters in the hardware.

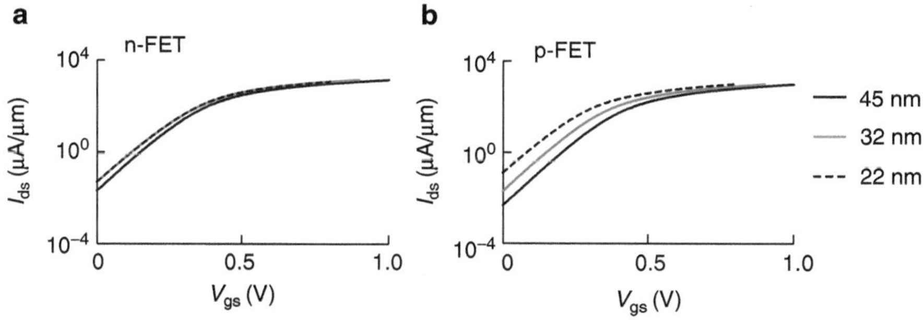

Fig. 10.11 Nominal MOSFET I_{ds}–V_{gs} characteristics of (**a**) n-FETs and (**b**) p-FETs in 45, 32, and 22 nm PTM HP models @ nominal V_{DD}, 25 °C

Table 10.1 Nominal n-FET parameters. 45, 32, and 22 nm PTM HP models @ 25 °C

Node (mm)	V_{DD} (V)	I_{on} (μA/μm)	I_{eff} (μA/μm)	I_{mid} (μA/μm)	I_{off} (nA/μm)	V_{sat} (V)
45	1.0	1,339	711	159	20	0.233
32	0.9	1,377	689	106	49	0.221
22	0.8	1,389	653	76	116	0.202

Table 10.2 Nominal p-FET parameters. 45, 32, and 22 nm PTM HP models @ 25 °C

Node (mm)	V_{DD} (V)	I_{on} (μA/μm)	I_{eff} (μA/μm)	I_{mid} (μA/μm)	I_{off} (nA/μm)	V_{sat} (V)
45	1.0	968	450	60	4.5	0.237
32	0.9	986	431	46	19	0.197
22	0.8	994	427	48	124	0.147

Consequently the average MOSFET characteristics vary from lot-to-lot and wafer-to-wafer. Random variations in V_t (Sect. 6.4), physical layout sensitivities and across-chip and across-wafer variations must also be considered. In comparing technologies we need to take the parameter distributions into account.

Let us examine three different methods for extracting representative I_{on} in different technology models at their respective nominal V_{DD} and at 25 °C. A similar analysis may be carried out for I_{eff}.

Method 1 $I_{on}(1)$ defined as I_{on} for nominal MOSFETs
This is a common practice for comparing different technologies. The value of $I_{on}(1)$ is obtained by simulating DC characteristics for technology nominal channel length and other default parameters in the model.

Method 2 $I_{on}(2)$ defined as the mean of the I_{on} distribution over $\pm 3\sigma L_p$ and a $\pm 3\sigma V_t$

This method is illustrated by simulating n-FET characteristics while varying V_t and L_p over their $\pm 3\sigma$ ranges. In Fig. 10.12, a histogram of the I_{on} distribution is shown

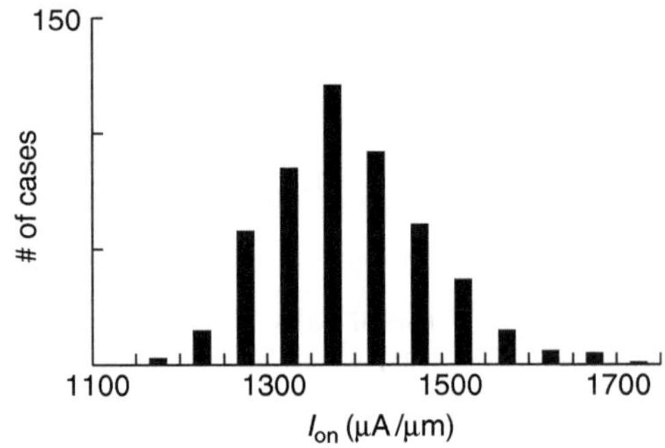

Fig. 10.12 Histogram of n-FET I_{on} values with $\pm 3\sigma$ variation in L_p and V_t. 45 nm PTM HP models @1.0 V, 25 °C

for an n-FET using the 45 nm PTM HP models and the σ values in Table A.4 of Appendix A. Variations in other parameters such as source/drain resistance and carrier mobility may be included for better representation of the hardware.

Method 3 $I_{on}(3)$ defined at fixed I_{off} (=100 nA/μm) at nominal V_{DD} and 25 °C
In order to use this method, Monte Carlo simulations are carried out over a $\pm 3\sigma$ range of L_p and V_t. The simulation results are shown in Fig. 10.13. I_{off} is plotted as a function of I_{on} for each of the technology nodes. The value of I_{on} for $I_{off} = 100$ nA/μm is obtained from the distribution of I_{on} for I_{off} in the 90 nA/μm to 110 nA/μm range. The mean of the distribution gives the $I_{on}(3)$ at a fixed I_{off}. With a limited number of samples (10–25), the shapes of the histograms do not appear to be normal.

This method of establishing I_{on} or I_{eff} at fixed I_{off} has been used by Intel for technology development with a product application focus [8]. This approach ensures that improvement in performance is not driven simply by lowering V_t and thereby increasing product chip IDDQ. As MOSFET widths are scaled, IDDQ for a truly scaled design also scales by the same factor. Setting a constant I_{off} per unit width provides more flexibility in design allowing the option of not scaling the chip area as aggressively. More circuits and functions may be added on the chip without a significant IDDQ penalty.

The n-FET I_{on} values obtained from the three methods described above are summarized in Table 10.3. There is a moderate increase in $I_{on}(1)$ and $I_{on}(2)$ with technology scaling, with these two parameters remaining within 1 % of each other. The trend in $I_{on}(3)$ is reversed. At constant I_{off}, there is a substantial decrease in I_{on} with scaling.

In an inverter (FO = 3), approximately 75 % of the switching capacitance contribution is from C_g. The gate capacitance, C_g in inversion mode is obtained

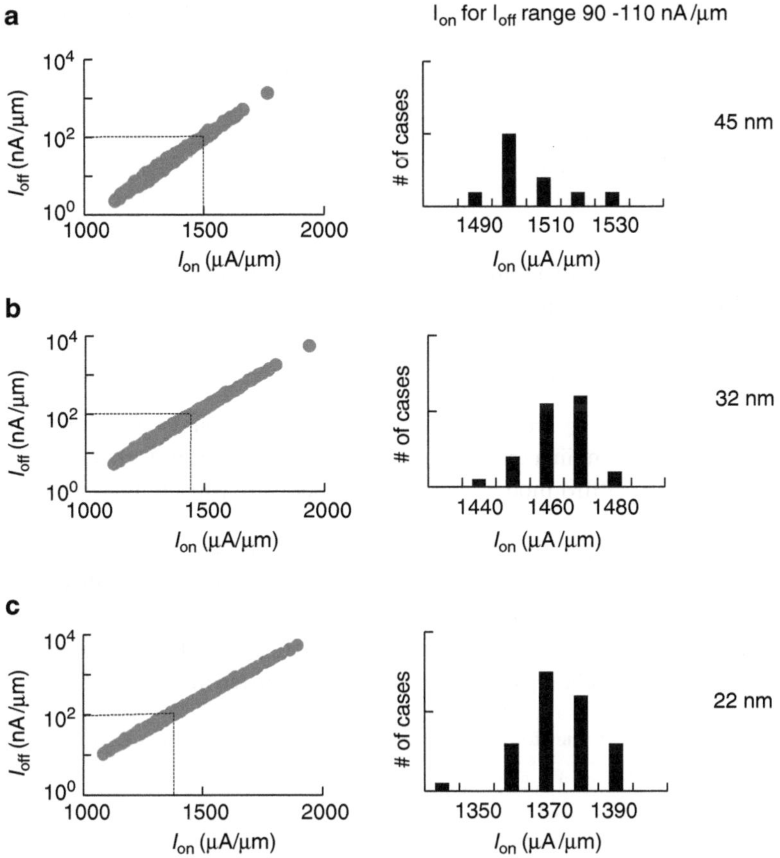

Fig. 10.13 I_{off} vs. I_{on} and I_{on} distribution for I_{off} in the range of 90–110 nA/μm in (**a**) 45 nm, (**b**) 32 nm and (**c**) 22 nm PTM models @ nominal V_{DD}, 25 °C

Table 10.3 $I_{on}(1)$, $I_{on}(2)$, and $I_{on}(3)$ for n-FETs in 45, 32, and 22 nm PTM HP models @ nominal V_{DD}, 25 °C

Node (mm)	V_{DD} (V)	$I_{on}(1)$ (μA/μm)	$I_{on}(2)$ (μA/μm)	$I_{on}(3)$ (nA/μm)
45	1.0	1,339	1,345	1,500
32	0.9	1,377	1,385	1,470
22	0.8	1,389	1,399	1,370

from simulations as described in Sect. 2.2.2. With $V_{ds}=0$ and $V_{gs}=V_{DD}$, the simulated value of C_g includes C_{ov}. Here we use this value of C_g for the *CV*/*I* metric FPG (1) with $I_{on}(1)$ and FPG(3) with $I_{on}(3)$. The results for the three technology models are summarized in Table 10.4. Both FPG(1) and FPG(3) significantly decrease with scaling as the reductions in capacitance and nominal V_{DD} values offset the small to negative increase in I_{on}.

Table 10.4 FPG (CV/I) for an n-FET corresponding to $I_{on}(1)$ and $I_{on}(3)$ in 45, 32, and 22 nm PTM models

Node (nm)	V_{DD} (V)	C_g (fF/μm)	$I_{on}(1)$ (μA/μm)	FPG(1) (ps)	$I_{on}(3)$ (μA/μm)	FPG(3) (ps)
45	1.0	1.492	1,339	1.11	1,500	0.99
32	0.9	1.346	1,377	0.88	1,470	0.82
22	0.8	1.217	1,389	0.70	1,370	0.71

It is more realistic to include all MOSFET capacitances for the CV/I delay metric. The total capacitance may be defined as

$$C_T = c_1C_g + c_2C_{ov} + c_3C_j + c_4C_p, \tag{10.1}$$

where c_1, c_2, c_3, and c_4 are constants giving the fractional contribution of each capacitance component in a circuit, for example in an inverter. Here C_g, C_{ov}, and C_j are included in the BSIM model, and C_p is the parasitic capacitance extracted from the layout.

Although the FPG ($=CV/I$) metric has been in use for many technology generations, it does not correctly represent the current drive capability and signal propagation delay for wire loaded circuits.

For analog applications, MOSFET transconductance g_m and output conductance g_{ds} are compared. The simulation results for an n-FET are listed in Table 10.5. The value of

Table 10.5 g_m and g_{ds} values for an n-FET at specified V_{ds} and V_{gs}, PTM HP models @25 °C

Node (nm)	V_{ds} (V)	V_{gs} (V)	g_m (μS/μm)	V_{ds} (V)	V_{gs} (V)	g_{ds} (μS/μm)
45	0.50	0.43	985	0.50	1.0	560
32	0.45	0.42	1,045	0.45	0.9	756
22	0.40	0.40	1,151	0.40	0.8	1,034

g_m is determined at a set of representative voltage bias values: $V_{ds} = 0.5\times\ V_{DD}$ and $V_{gs} = V_t + 0.2$ V. For g_{ds}, $V_{ds} = 0.5\times\ V_{DD}$ and $V_{gs} = V_{DD}$. Other voltage bias values may be selected for specific applications. A smaller g_{ds} and larger g_m are preferred, maximizing g_m/g_{ds}. Note that the PTM models are not optimized for analog circuit applications and the simulation results in Table 10.5 are shown only for illustration.

10.3.1.2 MOSFET Performance with Parasitics

For a realistic evaluation of MOSFET performance, parasitic resistance and capacitances must be included in the netlist. These parasitic components are extracted from the physical layout. Some examples of how MOSFET parasitics may depend on the physical layout are described below.

In Fig. 10.14a, physical layout of an n-FET is shown. This layout is derived from the physical layout of an inverter described in Chap. 2 (Fig. 2.35) and is representative of an n-FET in a logic gate. This layout is then used as a reference for comparing relative values of parasitic capacitances and series resistances in other layouts. It has maximum number of H0 contacts for its width that are placed on a

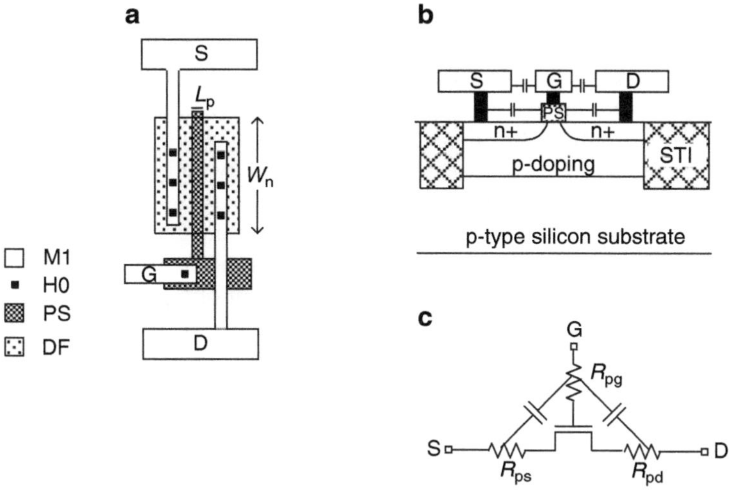

Fig. 10.14 n-FET (**a**) Physical layout, (**b**) physical cross-section indicating major parasitic capacitances, and (**c**) schematic showing parasitic resistance and capacitances

fixed grid, minimum M1 width, minimum H0 to PS (gate) spacing, minimum PS extension beyond diffusion and minimum source and drain diffusion areas. A schematic of the n-FET cross-section is shown in Fig. 10.14b. It includes the first-order parasitic capacitances. These capacitances are between H0 and PS, and between M1 metal contacts to source/drain and gate. The gate-to-drain capacitance contribution from H0 and M1 is more significant than the corresponding capacitance on the source side as it may experience Miller gain during switching.

The circuit schematic of an n-FET with dominant parasitic resistances and capacitances is shown in Fig. 10.14c. The spreading resistances of the diffusion areas and source and drain contact resistances are lumped into R_{ps} and R_{pd}. A distributed RC network for the PS layer (not shown) impacting switching delays is included in parasitic extraction models.

In Fig. 10.15, simulated values of I_{on} for an n-FET ($W_n = 0.4$ μm) are plotted as a function of R_{ps} and R_{pd}. For $R_{ps} = 60$ Ω and $R_{pd} = 0$ Ω, I_{on} decreases by ~5 %

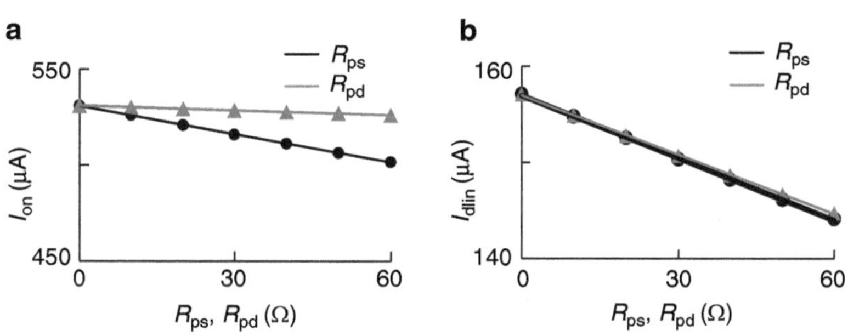

Fig. 10.15 n-FET ($W_n = 0.4$ μm) currents vs. R_{ps} and R_{pd}: (**a**) I_{on} and (**b**) I_{dlin}. 45 nm PTM HP models @ 1.0 V, 25 °C

whereas with $R_{ps} = 0\ \Omega$ and $R_{pd} = 60\ \Omega$, the decrease in I_{on} is only ~1 %. The *IR* drop across R_{ps} reduces the gate overdrive (V_{gs} at the internal node) and consequently I_{on}. As I_{on} is more weakly dependent on V_{ds}, the *IR* drop across R_{pd} has a smaller influence on I_{on}.

In Fig. 10.15b, simulated values of I_{dlin} are plotted as a function of R_{ps} and R_{pd}. In the linear region of I_{ds}–V_{ds} space, I_{ds} is modulated by both V_{gs} and V_{ds}. There is a reduction in I_{dlin} of ~8 % when either R_{ps} or R_{pd} is increased from 0 to 60 Ω.

MOSFET parasitic components are dependent on the geometrical arrangement of its constituent layers. This is illustrated with four different physical layouts of a single finger n-FET in Fig. 10.16. The standard layout of Fig. 10.14a is reproduced in Fig. 10.16a. In the physical layouts of the n-FET in Fig. 10.16b–d, the locations and/or dimensions of H0, M1, and DF layers have been modified. The physical layout in Fig. 10.16b, with a single H0 via in the source and drain contacts and higher diffusion area spreading resistances, has the highest R_{ps} and R_{pd} among all four. The PS parasitic capacitance from extension over the DF layer is also increased. In Fig. 10.16c, the diffusion area on the drain side is larger and gate-to-drain capacitance from H0 and M1 is reduced by moving the drain M1 contact metal and H0 vias away from PS layer. While the gate-to-drain parasitic capacitance is reduced, there is an increase in drain series resistance and junction capacitance. In Fig. 10.16d, the source and drain diffusion areas are increased and also

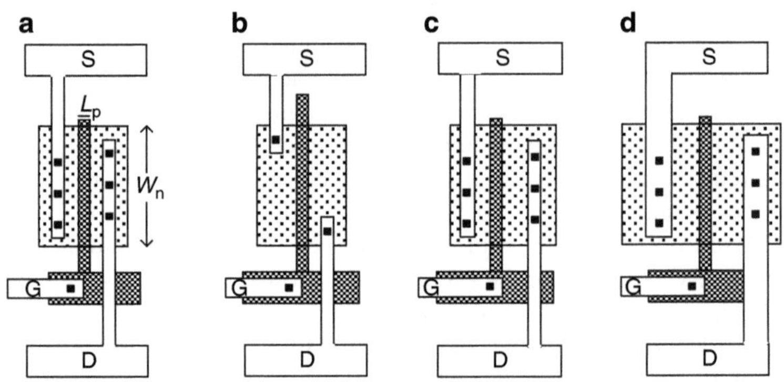

Fig. 10.16 n-FET physical layout variations relative to a standard layout: (**a**) standard layout, (**b**) higher R_{ps} and R_{pd}, lower gate-to-source/drain parasitic capacitances, (**c**) lower gate-to-drain parasitic capacitance, higher R_{pd} and C_{dd}, and (**d**) lower gate-to-drain/source capacitance, higher C_{ds} and C_{dd}, lower M1 metal source and drain series resistances and higher source/drain resistances

metal M1 layer separation is increased to reduce the gate-to-drain/source capacitance. However, source/drain series resistances and junction capacitances are higher, although M1 wire resistances in series with source and drain terminals are somewhat smaller. It is apparent that many variations in the physical layouts and in turn in the magnitude of parasitic resistances and capacitances are possible.

This sensitivity to physical layout is increased even more when stress layer for mobility enhancement are introduced in advanced technology nodes.

The physical layout styles may be optimized for best performance. Wider separation between contacts to MOSFET terminals and use of wide metal lines may reduce some parasitic resistances and capacitances at the cost of increases in logic gate areas. However, when circuit density and manufacturing cost benefits are considered, density, and yield optimization generally get priority over performance.

10.3.2 Interconnect Performance

Resistances and capacitances of metal interconnects play an important role in the overall performance of a CMOS chip [9]. Layers in the lower part of the metal stack comprise thin, narrow lines for local gate or small circuit block wiring. In the middle section of the stack, the layers are thicker and less dense for local and global interconnects. Metal layers in the top part of the stack are thicker, offering low resistance for power and clock distribution.

In each technology generation, silicon foundries offer several metal stacks varying in the number, pitch, and thickness of metal layers. The lower metal layers must scale with the transistor dimensions, typically 1× layers have the same minimum pitch as the PS layer for MOSFET gate definition. The definitions of the layers at the top of the metal stack are governed by I/O density and C4 pitch, power handling capability and packaging requirements. Generally, MOSFET scaling and packaging technology enhancements are not fully synchronized, and this has a strong influence on the metal stack definition. The number and pitch of intermediate metal layers can vary with the chip architecture, area and circuit density. For some high performance chips produced in large volumes, the cost of customization of the metal stack may be justified to achieve optimum performance.

Cross-sections of 1× and 2× metal layer stacks are shown in Fig. 10.17a, b. In this ideal case, the wire capacitance per unit length C_w of the two stacks is nearly the same, and the resistance per unit length R_w of the 1× stack is higher by a factor of four than that of the 2× stack. As interconnect wire widths and thicknesses are reduced,

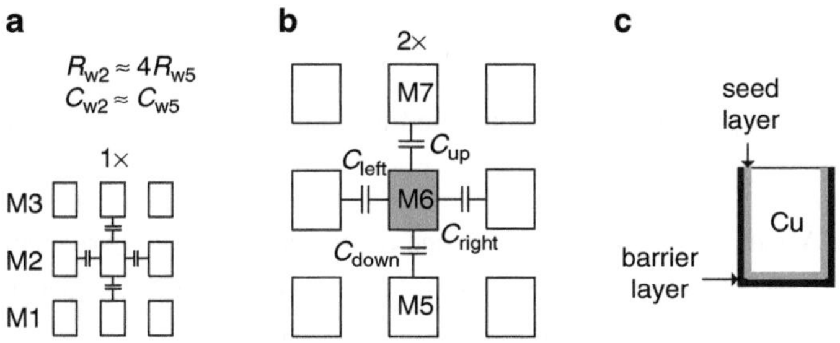

Fig. 10.17 Cross-sections of (**a**) 1×, and (**b**) 2× metal layer stacks, and (**c**) cross-section of copper metal wire with barrier and seed layers used in damascene processing

R_w and C_w do not scale in the expected geometric proportions. As an example, in damascene processing of thin electroplated Cu wires, inclusion of the barrier and seed layers (Fig. 10.17c) results in a higher resistance than that of copper metal alone.

Total geometric wire capacitance C_w is the sum of C_{up}, C_{down}, C_{left}, and C_{right} as indicated in Fig. 10.17b. The total capacitance of a signal wire in a switching transient is a function of both the geometric capacitance components and the time varying voltages on the neighboring wires. With the introduction of low dielectric constant materials and liner layers, the effective dielectric constants in the lateral and vertical directions are different, and C_{up} and C_{down} may scale differently than C_{left} and C_{right}. Coupling between neighboring wires in the same metal layer is mainly determined by lateral capacitances, C_{left} and C_{right}. Hence, the scaling of each capacitance component is examined and weighted appropriately.

The relative contributions of interconnect RC delays to the total signal propagation delay through a circuit block can vary from <5 % to as much as 30 % or more. The fractional interconnect RC delay for a wire of length l is dependent on the R_{sw} and C_{out} of the driving logic gate, C_{in} of the logic gate load, and the wire resistance $R_w l$ and its capacitance $C_w l$. The signal propagation delay τ, through a stage in an inverter chain with each inverter (FO = 3) driving a wire of length l is given by

$$\tau = R_{sw}\,(3C_{in} + C_{out} + C_w l) + R_w l \times 3C_{in} + \frac{R_w C_w l^2}{2}. \tag{10.2}$$

For short interconnecting wires with $R_{sw} \gg R_w l$, the last two terms in Eq. 10.2 containing $R_w l$ may be ignored, and only the wire capacitance contribution is considered. This is typically the case for local wire interconnects in high density thin metal layers. In medium length wires, R_{sw} is maintained greater than $R_w l$ but wire resistance contributions become appreciable. The maximum wire length is determined by signal integrity requirements, and buffers are inserted in signal wires travelling over long distances. Low resistance thick metal wires in the top layers are utilized for the global power distribution grid. When using these for clock signal distribution, wire inductance can be significant. The wire delay τ_w for $R_w l \ll 2\pi f L_w l$, where L_w is the inductance per unit length, is given by

$$\tau_w = \frac{1}{\sqrt{(L_w C_w l^2)}}. \tag{10.3}$$

Significant wire properties for short, medium, and long wires are summarized in Table 10.6.

Table 10.6 Metal wire interconnect lengths and contributing factors to signal propagation delay

Type	Length, l	Driver vs. wire	Wire density	Significant factors
Local	Short	$R_{sw} \gg R_w l$	High	C
Local	Medium	$R_{sw} > R_w l$	Medium	RC
Clock, global	Long	$R_{sw} > R_w l$	Low	IR, RC, $1/\sqrt{(LC)}$
Power				IR

With increase in circuit density and total number of transistors on a chip, the demand for wiring tracks has increased with CMOS scaling. In order to meet this demand, the number of metal layers has been increasing, from 3 to 4 in the 1990s to as many as 15 in 2014. The total number of interconnections, N can be estimated from Rent's rule

$$N = N_{\mathrm{m}}^{pr},\ 0.5 < pr < 0.85 \tag{10.4}$$

where N_{m} is the transistor count and pr is the Rent exponent. The value of pr is 0.5 for an ordered set of circuits such as a memory array and may increase up to 0.85 for random logic circuits. With 2× increase in device count per technology generation, N scales as 1.4× for memories and 1.8× for random logic. The number of wiring tracks must be increased at a rate faster than provided by technology scaling ($1/S \sim 1.4$) to accommodate the increase in the number of required interconnects.

The function of the wires is to carry a maximum number of independent signals with minimum delay. A fundamental interconnect metric can be defined as the integrated bandwidth for the minimum pitch wiring in all the metal layers. The bandwidth of a single metal layer, B_{w} in Hz/mm, is given by

$$B_{\mathrm{w}} = \frac{N_{\mathrm{w}}}{R_{\mathrm{w}} C_{\mathrm{w}}} \tag{10.5}$$

where N_{w} is the number of wiring tracks crossing per unit width. As an example, for a minimum wire pitch of 0.2 μm, $N_{\mathrm{w}} = 5{,}000$/mm. With $R_{\mathrm{w}} = 2\ \Omega/\mu\mathrm{m}$ and $C_{\mathrm{w}} = 0.2$ fF/μm, $B_{\mathrm{w}} = 12.5$/mm/ps (=12.5 THz/mm). The bandwidth of the full metal stack is the sum of bandwidths for each layer and serves as a metric for interconnect performance, IPG.

$$\mathrm{IPG} = \sum\nolimits_{i=1}^{n} B_{\mathrm{w}i}. \tag{10.6}$$

If the number of metal layers remains constant from one technology node to the next and all layer dimensions are scaled by the scaling factor S ($S \sim 0.7$), IPG as defined in Eq. 10.6 would increase as $1/S$ which is proportional to the increase in the number of wiring tracks. However, as discussed earlier, the required increase in wiring tracks is larger, and an increase in the number of metal layers becomes essential. A fraction of the wiring tracks are blocked by the placement of inter-level vias, further increasing the demand for more metal layers.

10.3.3 Logic Gate Performance

A metric based on inverter (FO = 3 or FO = 4) delay is a useful measure of CMOS technology performance. This metric is driven by MOSFET properties and associated parasitics, with minimum interconnect wire load. An inverter with a mix of logic gate load and wire capacitance or wire RC load may be used to more accurately gauge expected change in product performance.

Three circuit stages (ckt_stgs) shown in Fig. 10.18 are designed to give approximately equal delays. In Fig. 10.18a, an inverter drives three identical inverters in addition to the inverter of the following stage. In Fig. 10.18b, by reducing the MOSFETs widths in one of the load inverters by 0.5×, the inverter drives equivalent of 3.5 inverters and a fixed wire capacitance load equal to $C_{in}/2$. This inverter has ~12.5 % wire capacitance load. In Fig. 10.18c, an inverter drives three inverters through a wire *RC* load which contributes ~25 % to delay. The load inverters in each ckt_stg in turn drive fixed capacitance loads equivalent to $4 \times C_{in}$.

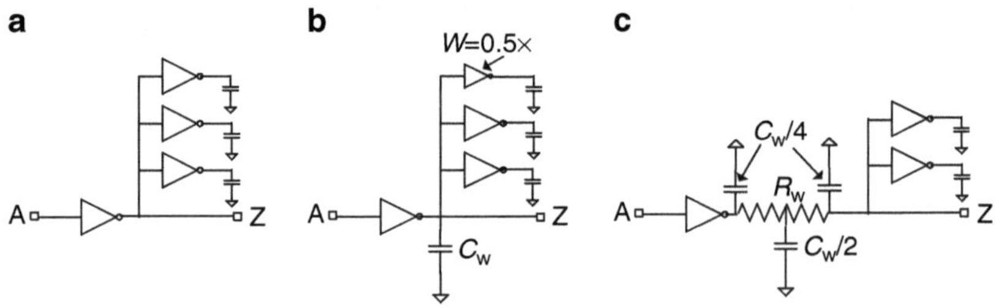

Fig. 10.18 Inverter FO = 4 ring oscillator stages for circuit performance evaluation: (**a**) inverter gate load, (**b**) inverter gate load + 12 % wire capacitance load, and (**c**) inverter gate load + *RC* wire load

One circuit performance metric CPG(1), based on MOSFET performance alone, is defined in the context of an inverter (FO = 4) ckt_stg in Fig. 10.18a.

$$\text{CPG}(1) = \tau_p \text{ (inverter FO} = 4) \tag{10.7}$$

In Table 10.7, inverter (FO = 4) design and circuit parameters for 45, 32, and 22 nm technologies are listed at nominal operating V_{DD} values. Circuit simulations

Table 10.7 Nominal design and average circuit parameters for a standard inverter (FO = 4) in 45, 32, and 22 nm PTM HP models @ 25 °C

Node (nm)	V_{DD} (V)	W_p (μm)	W_n (μm)	L_{ds} (μm)	τ_p (ps)	E_{sw} (fJ)	P_{off} (nW)	R_{sw} (Ω)	C_{sw} (fF)
45	1.0	0.60	0.40	0.120	11.68	3.94	6.7	1,481	7.88
32	0.9	0.42	0.28	0.085	10.27	2.04	10.3	2,035	5.05
22	0.8	0.30	0.20	0.060	9.14	1.07	23.4	2,728	3.35

are carried out for an RO with 50 stages and a NAND2 gate to enable the oscillations. Circuit parameters τ_p, R_{sw}, C_{sw}, P_{off}, P_{ac}, and E_{sw} are then determined as described in Sects. 2.2.5 and 4.3.2. There is a 12.1 % reduction in τ_p with scaling from 45 to 32 nm technology and a further 11.0 % reduction from 32 to 22 nm.

This improvement is typically measured as delay reduction instead of frequency gain since circuit design methodology is based on cycle time and not clock frequency. If frequency ($1/\tau_p$) is considered instead, the gain from 45 to 32 nm is 13.7 % and 12.4 % from 32 to 22 nm. This improvement in both cases is ~1.5 % higher than that for % cycle time reduction.

With reduction in τ_p, there is also a reduction in the energy per switching event, E_{sw}. If a product is operated at the same frequency in all three technologies, there is a reduction in AC power by a factor of ~3.7× in scaling from 45 to 22 nm. This reduction is offset somewhat by an increase in P_{off}.

The circuit delay components R_{sw} and C_{sw} ($\tau_p = R_{sw}C_{sw}$) show opposite trends, with R_{sw} increasing with scaling and C_{sw} decreasing corresponding to MOSFET width scaling. If MOSFET *I–V* characteristics in the three technologies at their nominal V_{DD} values were identical, R_{sw} would increase by 2× in going from 45 to 22 nm technology because of the 0.5× reduction in MOSFET widths. Here the increase in R_{sw} is only 1.84× while there is a reduction in C_{sw} by a factor of 2.35× .

For a realistic assessment of technology performance, it is recommended to include parasitic resistances and capacitances in the *CV*/*I* and CPG(1) metrics.

A logic gate delay metric may be customized for a product by comparing delays of a suite of most widely used logic gates or critical circuit blocks on a chip. A single metric for mixed combinational CMOS circuit delay MCPG(1) may be created by taking suitably weighted delays of *n* circuits:

$$\mathrm{MCPG}(1) = \sum_{i=1}^{i=n} c_i\tau_i, \tag{10.8}$$

where

$$\sum_{i=1}^{i=n} c_i = 1,$$

τ_i is the delay of circuit *i*, and c_i is a fractional weight of the *i*th circuit.

10.4 CMOS Power-Performance-Density Metrics

CMOS performance metrics based on MOSFET intrinsic performance and circuit delay (FPG and CPG) initially served well for many CMOS technology generations. With advances in technology, there has been increasing focus on power consumption and manufacturing cost reduction through circuit density enhancements. Hence, in a balanced CMOS metric, measures of power and circuit density need to be included along with circuit delay. Creating a single metric with three or more variables becomes more challenging!

In the following sections several different metrics for CMOS technology evaluation are described. A single metric or a combination of several metrics may be selected for a specific CMOS chip design and application space. These metrics serve well as guidelines, but the accuracy with which comparisons among technologies can be made has its limitations. The underlying assumptions for any metric should therefore be clearly stated and examined carefully for making a fair comparison.

10.4.1 Circuit Density

The trends in scaling of minimum contacted PS pitch and minimum SRAM cell area over many CMOS technology generations are shown in Fig. 10.19. The PS

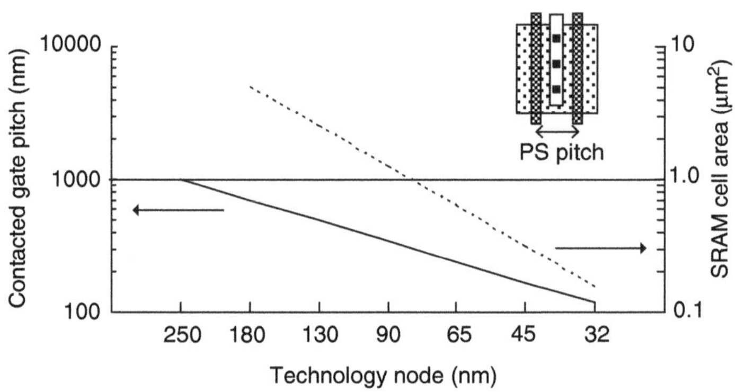

Fig. 10.19 Technology scaling trend in contacted PS pitch and SRAM cell density [1]

pitch is reduced by a factor of 0.7× and SRAM cell area by a factor of ~0.5× per technology generation every 2 years. Interconnect wire pitch, at least in the lower metal layers, tracks with the PS pitch. Physical areas of logic circuits and memory arrays are scaled to take full advantage of the feature size reduction in all layers. This is demonstrated with the physical layout of an inverter having two PS fingers in Fig. 10.20a scaled by a factor of 0.7× in Fig. 10.20b. The inverter area is faithfully scaled by 0.5× .

Although the minimum contacted PS pitch may scale by 0.7×, not all physical design rules have been scaling by the same factor. In practice, the area scaling factor of a circuit varies with physical design styles used in the standard cell library and with the power grid pitch. Redundancy requirements to improve silicon manufacturing yield take up more and more fractional area at advanced technology nodes. Placement of metal layers and inter-level vias on a fixed grid also consumes additional area. Package and I/O density has scaled at a lower rate than PS pitch, and the overall reduction in chip area is dependent on these factors as well.

Defect limited yields of SRAM and DRAM arrays are improved by eliminating defective sub-blocks during test. Hence, the effective density of good cells may be lower than the density trend shown in Fig. 10.19. There has been a steady increase in additional on-chip circuits for DFT and for circuit tuning during test to improve performance.

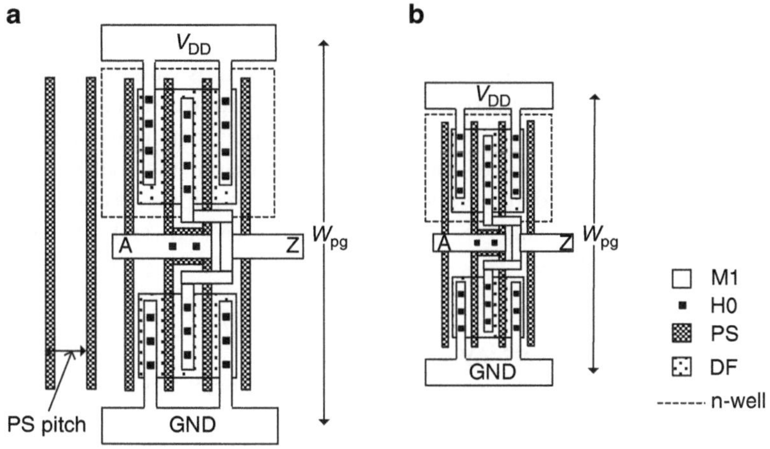

Fig. 10.20 (**a**) Physical layout of an inverter with two PS fingers, (**b**) physical layout of the inverter in (**a**) with all linear dimensions scaled by 0.7×

Inclusion of built-in redundancy, error correction schemes, and self-repairing circuits adds to the overhead, reducing the net density of circuits performing functional tasks.

In many low cost CMOS application areas, such as consumer electronics and wireless identification, the driving force for scaling is circuit density and chip area. Many of these products operate at lower frequencies, and power consumption is negligible. In such cases, technology scaling is evaluated mainly on the basis of circuit density. In larger and higher performance CMOS products, there is a definite benefit derived from increased density with scaling, but the magnitude of this gain varies among products.

10.4.2 Energy and Power Density

Leakage power and AC power density have been increasing with scaling. Beyond the 65 nm technology node, an increasing fraction of CMOS products are forced to reduce power at the expense of lower frequency of operation. Several CMOS technology metrics to assess optimum balance between power and performance have been proposed.

Recalling from Sect. 4.4, the total power is given by

$$P = (P_{sw} + P_{sc}) + P_{off} = P_{ac} + P_{off}, \tag{10.9}$$

$$P = \mathrm{IDDA} \times V_{DD} = \frac{1}{2} C_{sw} V_{DD}^{n} f_n + \mathrm{IDDQ} \times V_{DD}, \tag{10.10}$$

where f_n is the number of switching transitions per second and P_{off} is the DC power when the circuits are in the idle state. The contribution of short-circuit power

P_{sc} and the V_{DD} dependence of gate capacitance are included in the exponent n in the first term on the right-hand side of Eq. 10.12.

CMOS product applications dictate selection of the most suitable power metric for a particular product. For products where standby power is critical, a metric based on P_{off} is applied. In high performance applications, both power and delay are considered. For the purposes of technology comparison, the metrics described below are applied to a standard inverter with a current multiplier circuit to represent FO = 4. Simulated data using 45, 32, and 22 nm PTM HP and LP models for 51 stage ROs are compared.

10.4.2.1 Case 1. Circuit delay at Constant P_{off}

Low cost CMOS applications, and battery operated electronics in general require low standby or off-state power. In high performance chips, off-state power is a significant fraction of the total power and must be contained within limits set by package and system requirements. In such cases, keeping a P_{off} metric in sight is useful in assessing the application voltage for a given technology or design.

Continuing with a standard inverter (FO = 4) circuit, P_{off} variations with V_{DD} in 45, 32, and 22 nm PTM HP models are shown in Fig. 10.21a. With P_{off} in 45 nm technology as a reference, the values of V_{DD} for 32 nm and 22 nm are selected such that P_{off} is the same in all three technologies. In Fig. 10.21b, average delay τ_p is plotted as a function of V_{DD}. From this plot, τ_p values at constant P_{off} are extracted.

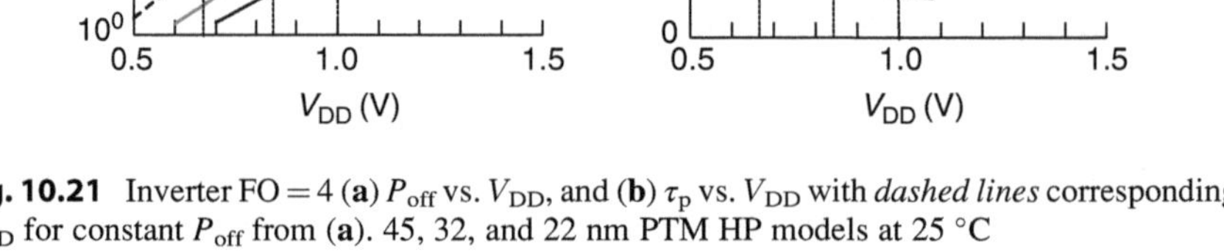

Fig. 10.21 Inverter FO = 4 (**a**) P_{off} vs. V_{DD}, and (**b**) τ_p vs. V_{DD} with *dashed lines* corresponding to V_{DD} for constant P_{off} from (**a**). 45, 32, and 22 nm PTM HP models at 25 °C

The V_{DD} and τ_p values for constant P_{off} in 45, 32, and 22 nm technologies are listed in Table 10.8. The τ_p at constant P_{off} is nearly the same in 45 and 32 nm technologies but increases by 22 % in transitioning from 32 nm node to 22 nm. A second metric for CMOS performance is defined as

$$\text{CPG}(2) = \tau_p \ \text{ at constant } P_{off}. \tag{10.11}$$

Table 10.8 Nominal design and average circuit parameters for inverter (FO = 4) at constant P_{off}. 45, 32, and 22 nm PTM HP models at 25 °C

Node (nm)	V_{DD} (V)	W_p (μm)	W_n (μm)	L_{ds} (μm)	τ_p (ps)	P_{off} (nW)	R_{sw} (Ω)	C_{sw} (fF)
45	1.00	0.60	0.40	0.120	11.68	6.7	1,481	7.89
32	0.84	0.42	0.28	0.085	11.70	6.8	2,339	5.00
22	0.67	0.30	0.20	0.060	14.33	6.8	4,351	3.29

The data shown in Table 10.8 are for the nominal MOSFET parameters at 25 °C. One may choose to evaluate the technology power/performance at a worst-case corner for P_{off} (e.g., $-2\sigma L_p$, 85 °C).

10.4.2.2 Case 2. Power × Delay Metric

A power × delay metric equivalent to the energy per switching event E_{sw} has been in use for comparing many different technologies, where

$$E_{sw} = P_{ac}\tau_p = \frac{1}{2}C_{sw}V_{DD}^n . \tag{10.12}$$

Similar to the circuit delay metric, a smaller value of E_{sw} is desirable as it indicates lower power or shorter time to propagate a signal when carrying out a logic function. One weakness of E_{sw} metric is that it does not differentiate between changes in switching power and speed. A second weakness is that it does not include P_{off}.

Circuit delay, power, and E_{sw} for a ring oscillator in 45, 32, and 22 nm HP and LP technologies are listed in Table 10.9. One can easily see from the data in Table 10.9 that although the τ_p ratio in HP and LP technologies at a particular node is >3×, the ratio of E_{sw} is ~1.5. It thus takes more switching energy to perform functions in LP technologies than in HP, but this drawback can be far outweighed by the lower P_{off}.

Table 10.9 Metric parameters for an inverter (FO = 4) ring oscillator with 51 stages. 45, 32, and 22 nm PTM HP and LP models @ 25 °C

Model	Node (nm)	V_{DD} (V)	τ_p (ps)	P_{ac} (μW)	E_{sw} (fJ)	P_{off}/stage (nW)
HP	45	1.00	11.68	337	3.94	6.7
HP	32	0.90	10.15	200	2.04	10.3
HP	22	0.80	9.15	117	1.07	23.4
LP	45	1.10	41.15	124	5.12	0.35
LP	32	1.00	35.91	74	2.68	0.49
LP	22	0.95	40.07	39	1.58	1.33

In Fig. 10.22a, E_{sw} is plotted as a function of V_{DD} for all three technologies. As expected with reduction in C_{sw} and V_{DD}, there is a significant decrease in E_{sw} with

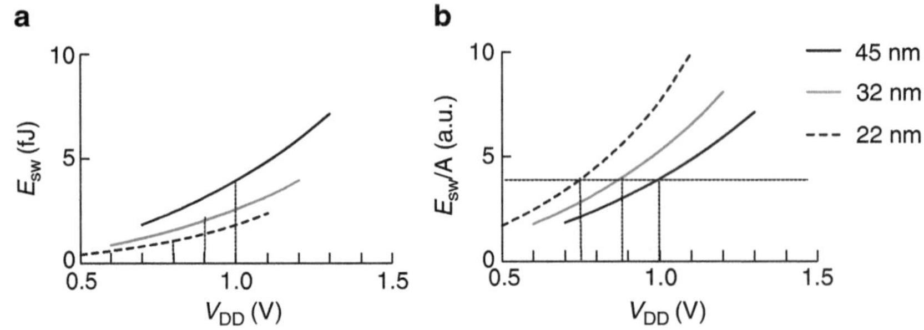

Fig. 10.22 Inverter FO = 4 (**a**) E_{sw} vs. V_{DD}, and (**b**) E_{sw}/A vs. V_{DD}. 45, 32, and 22 nm PTM HP models @25 °C

scaling. From some considerations of packaging and cooling, a practical approach is to scale at a constant power density. However total power, chip area and design content also typically change with scaling. Different chip designs at the same technology node may have very different power densities due to different circuit densities, duty factors, and operating frequencies. One metric that remains well defined at the circuit level, is independent of the overall chip design, and captures the impact of performance and area scaling, is τ_p at constant E_{sw} per unit area (E_{sw}/A).

A third performance metric CPG(3) is thus defined as

$$\text{CPG}(3) = \tau_p \text{ at constant } \frac{E_{sw}}{A}. \qquad (10.13)$$

Assuming that circuit density increases by ~2× in each technology node (scaling factor ~0.7×) E_{sw}/A, with $A = 1$, for the 45 nm technology node, is plotted as a function of V_{DD} in Fig. 10.22b. V_{DD} and circuit parameters for an inverter (FO = 4) at constant E_{sw}/A are listed in Table 10.10. Only small adjustments from the nominal technology values of V_{DD} are required to maintain a constant E_{sw}/A. With CPG(3), the delay reduction is 8.5 % from 45 to 32 nm and 3 % from 32 to

Table 10.10 Nominal design and average circuit parameters for our standard inverter (FO = 4) stage with nearly constant E_{sw}/area in 45, 32, and 22 nm PTM HP models at 25 °C

Node (nm)	V_{DD} (V)	W_p (μm)	W_n (μm)	l_{ds} (μm)	τ_p (ps)	E_{sw}/A[a] (fJ/area)	P_{off}/stage (nW)
45	1.000	0.60	0.40	0.120	11.68	3.95	6.7
32	0.880	0.42	0.28	0.085	10.69	3.97	10.3
22	0.755	0.30	0.20	0.060	10.40	3.95	17.5

[a]Inverter (FO = 4) area in 45 nm = 1

22 nm. These gains are clearly very different from those obtained from CPG(1) and CPG(2).

Note that keeping E_{sw}/A constant with scaling also implies constant power density at fixed frequency for any chip design that is uniformly scaled with no other changes. The improvement in CPG(3) with scaling provides an opportunity for higher frequency of operation at constant power density (with further V_{DD} reduction).

10.4.3 V_{DD} Dependencies of Different Metric Parameters

The key driving parameters in the metrics discussed previously are listed in Table 10.11. Two additional parameters, energy × delay and energy × power, that have been proposed to overcome the weaknesses in the E_{sw} metric are included as well. Each metric parameter is also expressed in terms of R_{sw}, C_{sw}, and V_{DD}. The V_{DD} dependencies of $1/R_{sw}$ and C_{sw} for an inverter in 45 nm PTM HP models are shown in Fig. 10.23. C_{sw} has a weak dependence on V_{DD} and $1/R_{sw}$ varies approxi-

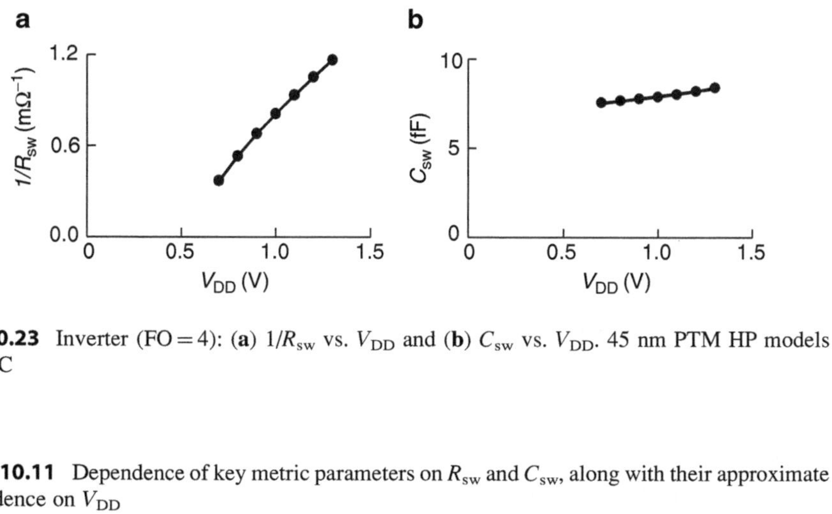

Fig. 10.23 Inverter (FO = 4): (**a**) $1/R_{sw}$ vs. V_{DD} and (**b**) C_{sw} vs. V_{DD}. 45 nm PTM HP models @25 °C

Table 10.11 Dependence of key metric parameters on R_{sw} and C_{sw}, along with their approximate dependence on V_{DD}

Metric parameter	R_{sw}, C_{sw}, V_{DD} dependence	Approximate V_{DD} dependence
Delay, τ_p	$R_{sw}C_{sw}$	${V_{DD}}^{-1}$
Power, P_{ac}	${V_{DD}}^2/R_{sw}$	${V_{DD}}^3$
Power × delay, E_{sw}	${V_{DD}}^2 \times C_{sw}$	${V_{DD}}^2$
Energy × delay, $E_{sw}\tau_p$	${V_{DD}}^2 \times R_{sw}{C_{sw}}^2$	V_{DD}
Energy × power, $E_{sw}P_{ac}$	${V_{DD}}^4 \times C_{sw}/R_{sw}$	${V_{DD}}^5$
Off-state power, P_{off}		${V_{DD}}^5$

mately linearly with V_{DD}. From these known relationships, V_{DD} exponents for different metric parameters are derived and indicated in the last column in Table 10.11. The V_{DD} exponent for P_{off} is derived from simulations in Sect. 4.3.

The V_{DD} exponents for the metric parameters in Table 10.11 vary from −1 to +5. A small change in V_{DD} affects some parameters a lot more than others. Adaptive V_{DD} methods to get best power-performance benefits in the hardware (Sect. 7.5) further impact the technology potential and the benefits derived by different CMOS products.

10.4.4 Summary of Performance Metrics

The different performance metrics described in previous sections are summarized in Table 10.12. Each n-FET and p-FET type is evaluated using the *CV*/*I* delay

Table 10.12 Summary of CMOS performance metrics based on delay under different constraints

Component	Performance metric	Parameter	Comments
MOSFET	FPG(1)	$CV/I_{on}(1)$	Nominal V_{DD}
MOSFET	FPG(2)	$CV/I_{on}(3)$	Constant I_{off}
Inverter (FO = 4)	CPG(1)	τ_p	Nominal V_{DD}
Inverter (FO = 4)	CPG(2)	τ_p	Constant P_{off}
Inverter (FO = 4)	CPG(3)	τ_p	Constant E_{sw}/A
Logic gates	MCPG(1)	$\Sigma c_i \times \tau_{pi}$	Nominal V_{DD}
Logic gates	MCPG(2)	$\Sigma c_i \times \tau_{pi}$	Constant P_{off}
Logic gates	MCPG(3)	$\Sigma c_i \times \tau_{pi}$	Constant E_{sw}/A

metric (FPG). Average delay of an inverter (FO = 4) serves well as a metric for basic static CMOS circuit performance evaluation. This may be determined at nominal V_{DD} for CPG(1), and also at V_{DD} values selected for constant P_{off} or constant E_{sw}/A for CPG(2) and CPG(3), respectively. A combination of circuit blocks may be used to represent a product chip design in delay metrics MCPG(1), MCPG(2), and MCPG(3), similar to the CPG metrics.

The final verdict on the relative benefits of a technology can only come from data collected from embedded performance monitors (delay chains and ring oscillators) and from f_{max}, V_{min}, power and circuit density of the product itself.

10.5 Compact Models and EDA Tool Evaluation

In the initial phase of a CMOS product chip architecture and design, an assessment of power and performance at the technology node of interest is made from the compact models provided by the silicon foundry. In the design implementation phase, circuits and physical layouts are optimized by incorporating these models in the EDA tools. In migrating a design from one technology node to the next, or when substituting a different model for the one already in place, it is important to compare circuit behaviors from the two sets of models. Differences in device properties, parameter distributions, physical layout ground rules, and reliability models beyond those expected from pure scaling provide an early assessment on what aspects of the design will be affected the most.

Essential to the success of this approach is that the compact models do accurately capture the physical behavior of devices and circuits over the range of application conditions. It is therefore prudent to evaluate the device models after incorporating them in the chip design environment and in EDA tools. This evaluation should be conducted over the expected range of operation for the specific chip and product design.

Models and design tools used for timing, noise, and power analysis are shown in Fig. 10.24. The models for MOSFETs (BSIM), parasitic extraction, IDDQ, and

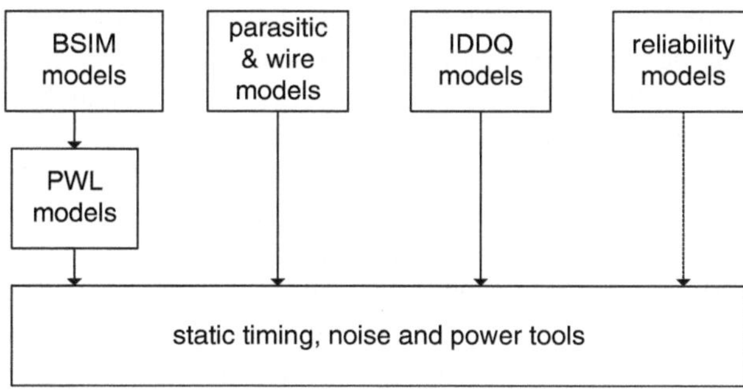

Fig. 10.24 Models and tools for timing and power

reliability mechanisms are provided by the silicon foundry. BSIM and parasitic models are used in transistor level circuit simulations of custom circuits and standard cell library books. Piecewise linear (PWL) models are generated by approximating the nonlinear characteristics of the MOSFETs as linear segments. This simplification reduces the simulation time for large circuit blocks. IDDQ models have higher accuracy in the subthreshold region than BSIM models and are useful in designs where P_{off} management is essential. Reliability models include degradation mechanisms in MOSFETs such as BTI and HCI (Chap. 8), and electromigration and self-heating limits in wires.

Trade-offs may be made between simulation time and model accuracy. Monte Carlo simulations to explore the entire process and application space are limited to a few critical circuits. Logic and memory circuit blocks are simulated only in specific corners. Static timing, noise, and power tools may incorporate PWL models to reduce simulation time. Reliability models may be used for custom circuits or for global design. In some chip designs, aging mechanisms are absorbed in the design margins and other design restrictions such as current limits for interconnects.

It is highly recommended to evaluate the models and tools using a small suite of devices and circuits [10] to:

- Verify that the device models are consistent with known physical behavior.
- Ensure that the EDA timing and power tools are consistent with the foundry compact models.
- Generate guidelines for circuit designs based on device properties in view of silicon technology changes from previous model releases.
- Determine simulation corners and design margins for variability and reliability
- Evaluate any changes recommended for optimizing CMOS product architecture, and for setting test conditions and customer specifications.

10.5.1 BSIM Models

BSIM compact models for MOSFETs are at the lowest level of the model hierarchy in a CMOS chip design. BSIM models are based on equations describing various physical effects in MOSFETs and use fitting parameters to match these effects to representative hardware. The model versions have progressed from BSIM1 to BSIM4 and to BSIM-CMG for multi-gate devices (FinFETs). There are over 200 different fitting parameters in BSIM4, and this number is growing to accommodate new effects in each successive technology generation.

BSIM models are built by making detailed measurements over a range of voltages and temperatures on a set of representative devices of different geometries and physical layout styles, and then fitting the equations describing the device behavior to the data. Statistical variation ranges (σ values) of 10–15 key MOSFET parameters are included for Monte Carlo and process corner simulations. In a mature silicon technology, representative hardware from the manufacturing line is selected to obtain the measured σ values. In early stages of technology development, the variability is derived from a previous generation with adjustments to account for any additional known sources of variations, or elimination or reduction of other sources.

The model cards (files) are incorporated in the circuit design tool environment. The first step in evaluating MOSFET (and diode) models is to generate a set of target values and X–Y plots for comparison. The target values and plots may originate from a previous generation model, or a previous version of the model for the same technology. We will denote the model which is used for comparison as the reference model, model_ref, and the one being evaluated as model_eval.

MOSFET DC characteristics are very useful for obtaining qualitative assessments of differences in current drives ($1/R_{sw}$) and in channel leakage currents of different circuit topologies. As an example, in Fig. 10.25a, I_{ds}–V_{ds} characteristics of an n-FET are plotted for model_ref (nominal 45 nm PTM HP model) and model_eval which is a 45 nm PTM HP model with longer L_p by $+1\sigma L_p$ ($L_p = 0.0465$ μm) and V_t offset by +0.02 V. Model_eval shows a decrease in I_{ds} in response to its longer L_p and higher V_t. In Fig. 10.25b, I_{ds}–V_{ds} characteristics for

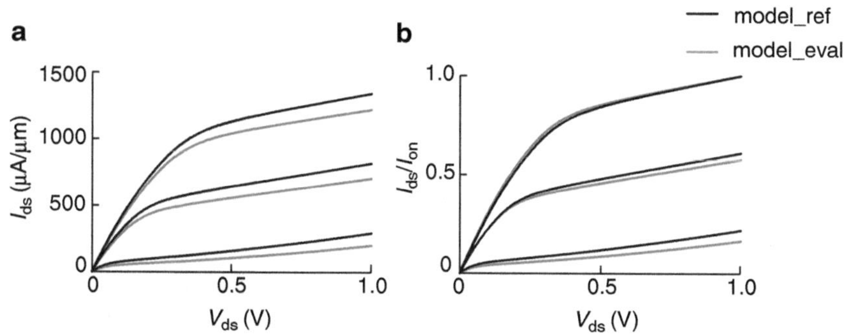

Fig. 10.25 n-FET characteristics comparing model_ref with model_eval: (**a**) I_{ds}–V_{ds} and (**b**) normalized I_{ds}/I_{on}. Model_ref = 45 nm PTM HP, model_eval = 45 nm PTM HP with $L_p = 0.0465$ μm and V_t increased by 0.02 V, @1.0 V, 25 °C

the same two cases are normalized to their respective I_{on} values. It becomes immediately apparent that I_{ds}/I_{on} in model_eval is relatively lower in the low V_{gs} region because of higher V_t. This also corresponds to an increase in I_{on}/I_{off} in model_eval. From the n-FET trajectories during switching, a larger reduction in circuit delays of n-passgates operating in the low V_{gs}–V_{ds} region compared with delay reduction in inverters is expected (Fig. 5.26).

MOSFET parameters are published for nominal and minimum allowed dimensions at nominal V_{DD} and 25 °C in a design guide. Parameter values with higher sensitivity to V_{DD} and temperature may also be listed at one or two more V_{DD} and temperature corners. These published tables are useful for validating the simulation setup as a first step towards model evaluation.

The dependencies of MOSFET characteristics on physical dimensions, layout configuration and variation in material properties, and V_{DD} and temperature are governed by the model equations and the fitting parameters. It is instructive to observe these dependencies through circuit simulations and to compare them with expected behavior. This is especially useful when large excursions from nominal occur in hardware or during test. In such cases, the model accuracy may be poor or out of the range over which the models equations are fit to the hardware measurements.

The procedure described below for evaluating the 45 nm PTM HP model of an n-FET is repeated for each MOSFET type. Graphical representations of the simulation results are easy to assimilate and the fitting parameters can be tabulated for quick inspection. L_p values are selected to cover the nominal $\pm 3\sigma$ range in the BSIM model for logic MOSFETs. MOSFET width range is selected from the physical layout styles of the library books. Typically, wide MOSFETs are drawn with multiple-fingers to reduce parasitics.

Key MOSFET parameters are plotted as a function of physical dimensions, L_p and W, and application conditions, V_{DD} and temperature T, as listed in Table 10.13. These plots provide a quick view of MOSFET physical behavior. Additional information is obtained from subthreshold slope (SS), change in V_t with

Table 10.13 MOSFET parameters to be plotted as a function of physical dimensions and application conditions

MOSFET parameters	Design parameters	Application conditions	Comments
I_{eff} (I_{on})	L_p, W	V_{DD}, T	
I_{off}	L_p, W	V_{DD}, T	I_{off} vs. V_{DD}, T
V_t	L_p, W	V_{DD}, T	V_t roll-off, $\delta V_t/\delta T$

temperature, and body coefficient (γ) and comparing these with expected values as listed in Table 10.14. If any discrepancy from the expected behavior of bulk silicon devices is observed, it should be confirmed with the silicon technology and model developers.

Table 10.14 Expected range of MOSFET properties extracted from simulations

MOSFET Parameters	Expected range	Comment
SS	70–120 mV/decade	Linear increase with temperature
$\delta V_t/\delta T$	$\leq$1 mV/°C	Reduces with higher doping
γ	1.1–1.2	Increases with doping

The variation of n-FET I_{on} as a function of $1/L_p$ is shown in Fig. 10.26a. A linear fit gives a regression coefficient of 0.986, and one can observe deviation from linearity near the $\pm 3\sigma L_p$ values. In Fig. 10.26b, V_{tlin} and V_{tsat} are plotted as a function of L_p. For smaller values of L_p, DIBL = (V_{tlin}–V_{tsat}) increases due to short-channel effects. This V_{tsat} roll-off at short-channels indicates higher I_{on} at $-3\sigma L_p$ than expected from a strictly linear relationship between I_{on} and $1/L_p$, as observed in Fig. 10.26a.

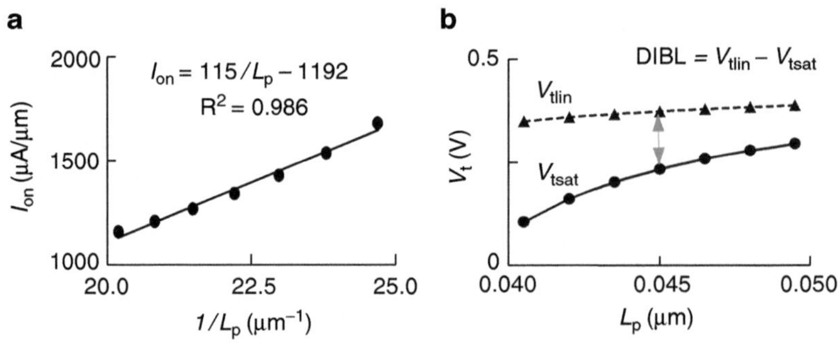

Fig. 10.26 n-FET (**a**) I_{on} vs. $1/L_p$ and (**b**) V_t vs. L_p. 45 nm PTM HP model @1.0 V, 25 °C

In Fig. 10.27a, n-FET I_{off} is plotted as a function of L_p. There is >100× increase in I_{off} as L_p decreases from 0.0495 μm (+3σ) to 0.0405 μm (−3σ). The gate capacitance C_g of a MOSFET is determined by the method described in Sect. 2.2.1. In Fig. 10.27b, inversion mode gate capacitance C_g values in fF/μm for an

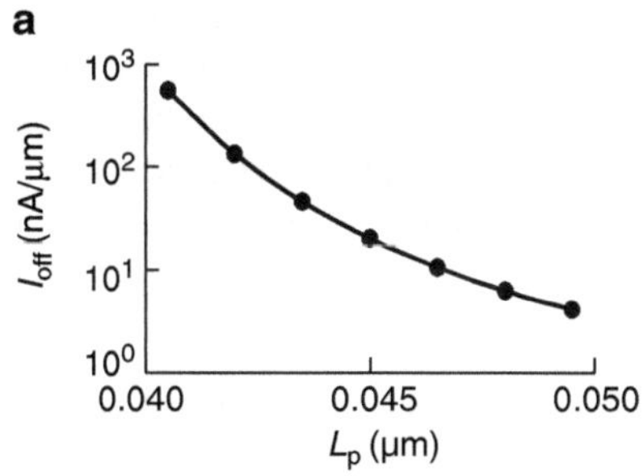

b

L_p (μm)	C_g (fF)
0.0405	1.369
0.0450	1.490
0.0495	1.630

Fig. 10.27 n-FET (**a**) I_{off} vs. L_p and (**b**) table with C_g (inversion mode) for nominal L_p and $L_p \pm 3\sigma L_p$. 45 nm PTM HP model @1.0 V, 25 °C

n-FET are listed for nominal and $L_p \pm 3\sigma\, L_p$. As expected, C_g increases with L_p in proportion to the increase in gate area.

In Fig. 10.28a, b, I_{on} and I_{off} per unit width are plotted as a function of n-FET width W_n. There is a slight decrease (~2.7 %) in I_{on} normalized to W_n for

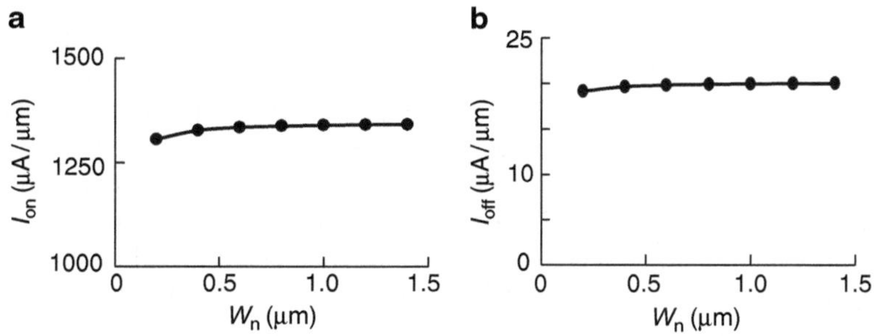

Fig. 10.28 n-FET (**a**) I_{on} per μm vs. W_n and (**b**) I_{off} per μm vs. W_n. 45 nm PTM HP model @1.0 V, 25 °C

$W_n = 0.2$ μm whereas I_{off} is essentially independent of W_n. The three n-FET physical layouts shown in Fig. 10.29a–c, having one, two, and four PS fingers all

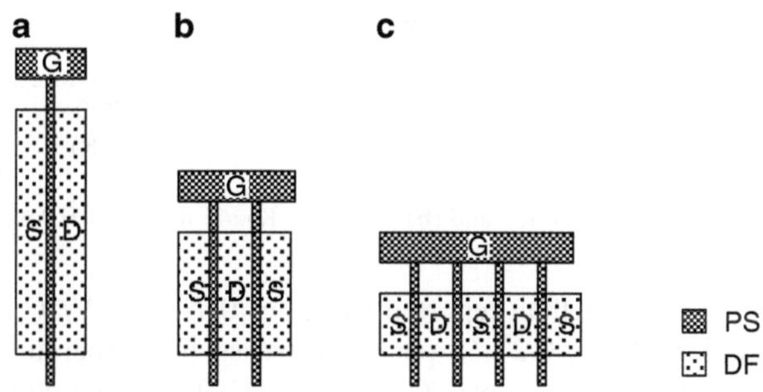

Fig. 10.29 MOSFET physical layout with DF and PS layers for equal W: (**a**) one PS finger, (**b**) two PS fingers and (**c**) four PS fingers

with the same total W, give the same I_{off}. As long as each finger width is >0.3 μm, any observed differences in their I_{on} values are due to parasitic resistances in the contacts and wiring (not shown) and not originating from the basic device model.

The temperature dependence of key n-FET properties (I_{off}, SS, V_{tsat}, and I_{on}) are shown in Figs. 10.30 and 10.31. For convenience, all parameters are plotted as a

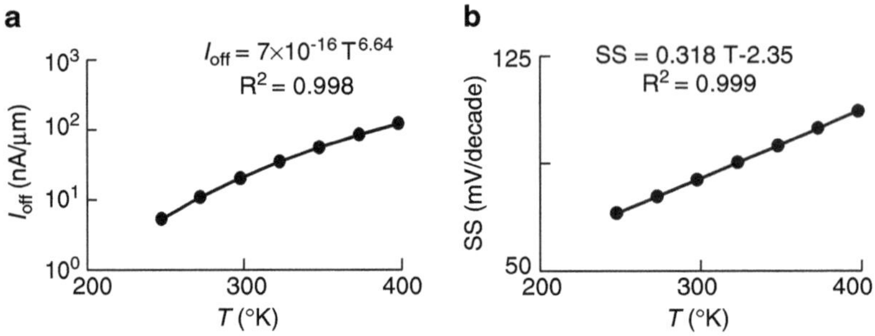

Fig. 10.30 n-FET (**a**) I_{off} vs. T (°K), and (**b**) SS vs. T (°K). 45 nm PTM HP model @1.0 V

function of temperature in °K (=temperature in °C + 273). The I_{off} vs. temperature data shown in Fig. 10.30a are fit to a power law. The temperature exponent of 6.64 indicates a rapid increase in I_{off} (P_{off}) with temperature as expected. The subthreshold slope, SS in Fig. 10.30b increases at the rate of 3.18 mV/10 °C. This parameter is proportional to temperature in °K. There is a 1.6× increase in SS as the temperature increases from −25 to 125 °C, corresponding to the temperature ratio in °K (=398/248), and consistent with the theoretical MOSFET model.

V_{tsat} and I_{on} values for the n-FET are plotted as a function of temperature in Fig. 10.31a and in Fig. 10.31b, respectively. V_{tsat} is decreasing at the rate of

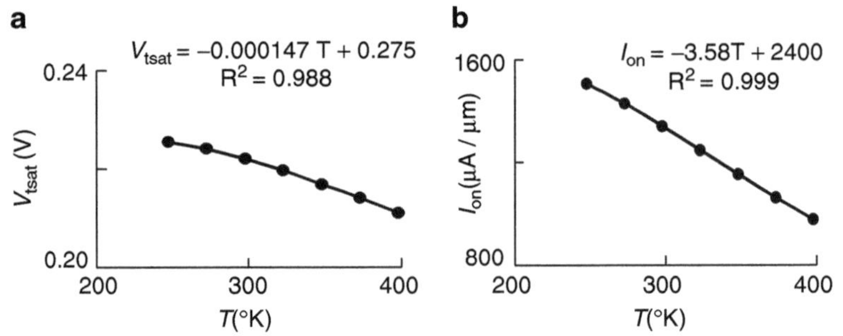

Fig. 10.31 n-FET (**a**) V_{tsat} vs. T(°K), and (**b**) I_{on} vs. T(°K). 45 nm PTM HP model @1.0 V

0.15 mV/°C. This is lower than reported values of 0.7–1.0 mV/°C in bulk CMOS [2]. The decrease in I_{on} is 0.27 % per °C increase in temperature near 25 °C, increasing to a 0.34 % decrease per °C near 100 °C. This is much higher than the expected value of ≤ 0.1 % per °C. The mobility of the MOSFET decreases with rise in temperature reducing its current drive strength. On the other hand, decrease in V_{tsat} increases the drive strength. Generally, MOSFETs are engineered so that the combined result of mobility degradation and lowering of V_t with temperature is a reduction of current drive in the on-state by ≤ 0.1 % per °C. This observed higher rate of decrease in I_{on} with temperature in the 45 nm PTM model is consistent with the lower rate of decrease in V_{tsat}. Hence, this model for the n-FET does not give the expected I_{on} behavior with temperature.

Next, key circuit parameters are evaluated for the 45 nm PTM HP models. In logic gate (and SRAM) designs, the relative strengths of p-FETs and n-FETs are important considerations in balancing gate drive strengths for PU and PD delays. If the widths of all p-FETs in a logic gate are identical as are those of all n-FETs, the ratio of their respective widths W_p/W_n is called the p/n ratio, beta ratio, or β_r:

$$\beta_r = \frac{W_p}{W_n}. \tag{10.15}$$

The design value of β_r is typically set to give minimum average delay. In addition to minimizing τ_p, it is highly desirable to have equal PU and PD delays, $\tau_{pu} = \tau_{pd}$, and equal rise and fall times. Simulated results for an inverter (FO = 3) τ_{pu}/τ_{pd} and τ_r/τ_f ratios, obtained from a delay chain configuration, are shown as a function of W_p/W_n in Fig. 10.32a, b. The ratios are equal to 1.0 near $W_p/W_n \approx 1.5$–1.6.

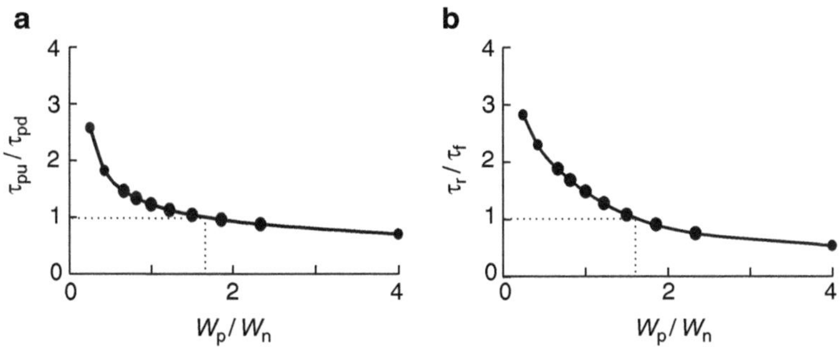

Fig. 10.32 Inverter (FO = 3) (**a**) τ_{pu}/τ_{pd} vs. W_p/W_n and (**b**) τ_r/τ_f vs. W_n/W_p. $(W_p + W_n) = 1.0$ μm, nominal 45 nm PTM HP models @1.0 V, 25 °C

Based on the simulation results for τ_{pu}/τ_{pd} and τ_r/τ_f, the standard inverter in 45 nm PTM technology was designed with $W_p/W_n = 1.5$. For $(W_n + W_p) = 1.0$ μm, this sets $W_p = 0.6$ μm and $W_n = 0.4$ μm (Sect. 2.2.3).

In Fig. 10.33a, inverter delay τ_p is plotted as a function of temperature in the range of −25 °C to 125 °C. From a linear fit of the data, τ_p increases at the rate of 0.58 % per °C change in temperature. This rate is much higher than the expected change of ≤ 0.1 % per °C in typical bulk silicon technologies and is consistent with the I_{on} observation in Fig. 10.31b. Further investigation is carried out by plotting PD and PU delays, τ_{pu} and τ_{pd}, as functions of temperature in Fig. 10.33b. The increase

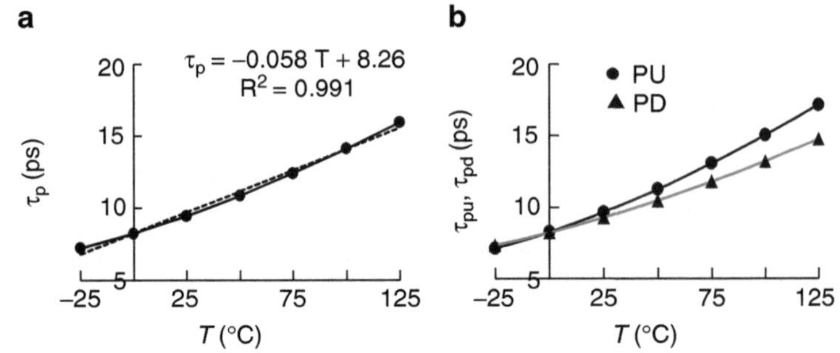

Fig. 10.33 Standard inverter (FO = 3) (**a**) τ_p vs. temperature and (**b**) τ_{pu} and τ_{pd} vs. temperature. 45 nm PTM HP model @1.0 V

in τ_{pu} with temperature is larger than in τ_{pd}, indicating a larger deterioration with temperature of the p-FET strength than that of the n-FET. This observation can be validated by examining I_{on} or I_{eff} of the p-FET as a function of temperature.

Variations in τ_p and IDDQ values of an inverter with V_{DD} are shown in Fig. 10.34a, b. In Fig. 10.34a, $\delta\tau_p/\delta V_{DD}$ which is the ratio of % change in τ_p,

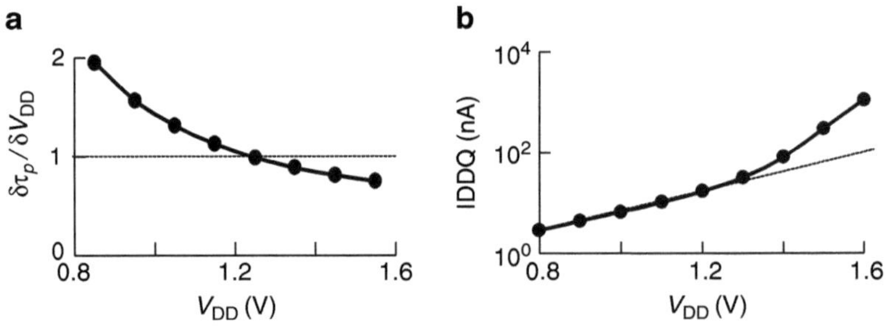

Fig. 10.34 Standard inverter (FO = 3) (**a**) $\delta\tau_p/\delta V_{DD}$ vs. temperature V_{DD}, and (**b**) average IDDQ vs. V_{DD}. 45 nm PTM HP model @25 °C

($\delta\tau_p = \Delta\tau_p/\tau_p$) to % change in V_{DD} ($\delta V_{DD} = \Delta V_{DD}/V_{DD}$) is plotted as a function of V_{DD}. At V_{DD} values of ≥1.0 V, $\delta\tau_p/\delta V_{DD}$ is expected stay at ~1.0. However, in these models $\delta\tau_p/\delta V_{DD}$ continues to decrease with V_{DD}. In Fig. 10.34b, IDDQ increases steeply for $V_{DD} > 1.3$ V, which is a deviation from the expected behavior in silicon.

The V_{DD} and temperature dependencies in the PTM models clearly deviate from expected MOSFET behavior beyond the nominal corner. These models are based on simplified BSIM equations and are intended to be used for early evaluation of a technology. They are not meant to be used for real circuit designs. However, in our experience in working with evaluating models used in the industry for CMOS chip designs, some weaknesses in the models are often revealed with this evaluation procedure. Such weaknesses become more apparent moving away from the nominal corner but still within the operating region of the product. Since BSIM models rely on fitting parameters to capture MOSFET characteristics over wide ranges of the variables, an incorrect choice of a fitting parameter may cause larger than anticipated excursions in a device property.

The BSIM model evaluation procedure described here, with no parasitic extraction, is a clean way of understanding the basic underlying model for circuit simulations. With this procedure, any discrepancies can be resolved before investing in the design infrastructure and EDA tools.

10.5.2 Layout Parasitic Extraction

A layout parasitic extraction (LPE) tool generates circuit netlists extracted from the physical layouts of circuits, which include parasitic resistances and capacitances associated with transistors and diodes as well as interconnect resistances, capacitances, and inductances. MOSFET properties modulated with layer dimensions and proximity to other physical layers, such as stress enhanced mobility, are included in the LPE models.

Interconnect resistances and capacitances of simple geometric shapes with uniform material compositions are described in Sect. 2.1.2. The resistance of a long rectangular wire can be calculated from the known sheet resistance and the number of squares ($l = w$) along its length. Calculating the capacitance of a parallel plate capacitor with a uniform dielectric material and plate dimensions much larger than the separation between the plates is also straightforward.

For the layer shapes and dimensions encountered in real CMOS physical layouts, many other effects must be taken into account. Some of these geometric effects, as illustrated in Fig. 10.35a, b, are bends in wires, inter-level vias and tapered wire

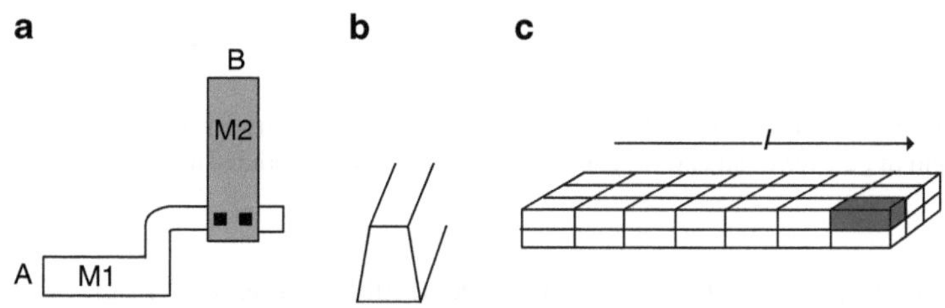

Fig. 10.35 (**a**) Metal interconnect in layers M1 and M2 with two inter-level vias. (**b**) Tapered metal layer cross-section and (**c**) layer filaments with uniform properties

cross-sections. Current crowding in corners and nonuniform resistivity of composite metal layers need to be taken into account when computing resistances.

Capacitance extraction has similar issues. Examples include fringing fields near wire edges, nonuniform dielectric materials, different dielectric material compositions in lateral and vertical directions, proximity to other metal shapes and floating metal areas generated in fill routines to obtain nearly uniform pattern densities. Parasitic *R* and *C* extraction is based on numerical methods to solve 2-dimensional (2D) or 3-dimensional (3D) electromagnetic field equations. A 3D extraction tool can account for complex variations in shapes, side-wall profiles and layered material properties in metals and dielectrics. A section of a wire is divided into filaments, as shown in Fig. 10.35c. Each filament is represented by an *R*, *L*, and *C* network. The computed values of *R*, *L*, and *C* for each filament are included in the netlist. Improvement in modeling accuracy of parasitic components comes at the cost of the having hundreds and thousands of circuit elements in a netlist. Unlike the short netlists used in simulation examples throughout this book, debugging a fully extracted netlist, with even just a few components, becomes a tedious task.

Attributes in the physical layouts may affect the extracted netlist in some cases. One example is placement of a terminal node location indicated by a cross (pin) in the physical layout in Fig. 10.36a, b. In Fig. 10.36a, metal M1 resistance in series with the gate terminal is included in the netlist. In Fig. 10.36b, the pin overlays the M1 layer and its resistance is excluded. Power grids may be eliminated from the netlist by placing large area pins to reduce the number of components in the netlist. In a robust power grid, the resultant error is small.

In another example in Fig. 10.36c, M1 to M1 capacitance can be modeled as gate-to-drain or split between gate to GND and drain to GND by the extraction tool.

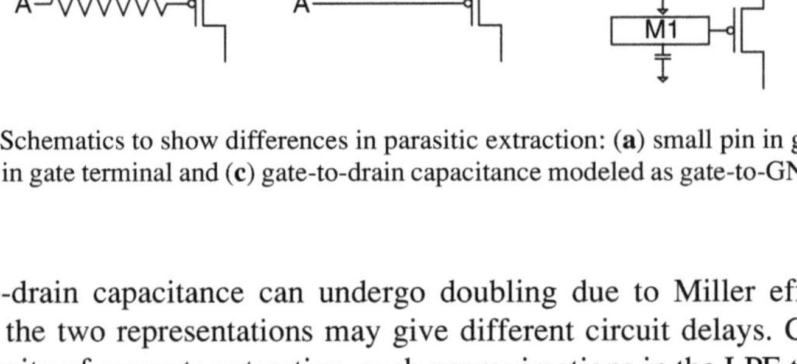

Fig. 10.36 Schematics to show differences in parasitic extraction: (**a**) small pin in gate terminal, (**b**) large pin in gate terminal and (**c**) gate-to-drain capacitance modeled as gate-to-GND and drain-to-GND

As gate-to-drain capacitance can undergo doubling due to Miller effect during switching, the two representations may give different circuit delays. Considering the complexity of accurate extraction, such approximations in the LPE tool may be justified as long as all major parasitic components are correctly represented.

Circuit design guidelines, standard cell libraries, and auto-routing tools ensure a uniform and correct methodology for physical layouts. Most designs rely on synthesizing the physical layouts directly from chip VHDL and EDA tools

For model-to-hardware correlation and extraction of device parameters from delay chains and ROs (Sect. 5.6), custom layouts are essential. In scribe-line test structures designed to be measured with a single metal layer, wire resistances in the power distribution network and in signal wires must be extracted correctly. For such test structures and embedded monitors, it is also important to understand the assumptions made in the LPE tool and models. It is worth verifying LPE tool accuracy independently by comparing the simulation results from different layouts with known expected differences.

10.5.3 Timing and Power Tools

Simulation time of large circuit blocks with BSIM models and a SPICE or other simulator becomes very long. To reduce simulation time, piecewise linear models (PWL) are generated from simplified MOSFET *I–V* and *C–V* characteristics. Examples of PWL approximations are shown in Fig. 10.37a, b. In Fig. 10.37a, an I_{ds}–V_{gs} plot is approximated by two lines of constant slope. In Fig. 10.37b, a fixed average value of C_g during a transition is estimated from a C_g–V_{gs} plot (Sect. 2.2.2).

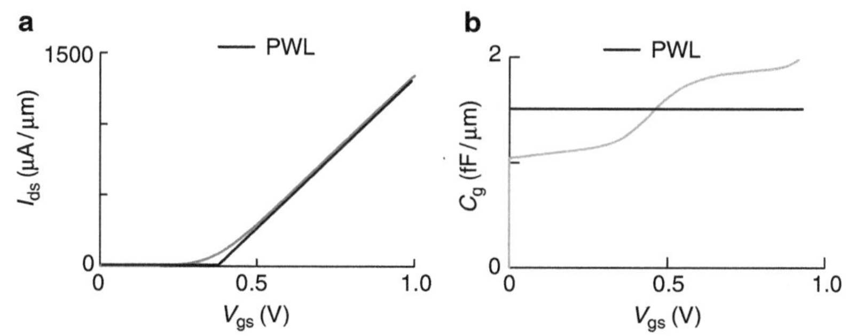

Fig. 10.37 PWL approximation of (**a**) an I_{ds}–V_{gs} characteristic and (**b**) a C_g–V_{gs} characteristic

These approximations lead to small differences ($\delta\tau$) between circuit delays obtained from SPICE simulations using BSIM models vs. the simplified models used in timing tools. Over a large number of circuit topologies, the $\delta\tau$ distribution should have a mean value of zero and an acceptable standard deviation. The accuracy of simplified models is improved if in addition to $\delta\tau$, δR_{sw}, and δC_{sw} values are also examined to ensure that the approximations in I_{ds} and C_{sw} are individually matched. Further rigor is instituted by matching simulated values of major C_{sw} components using the methodology outlined in Sect. 5.6.

At the next level in the EDA tool hierarchy, timing slack may be compared with the slack obtained from SPICE simulations (Sect. 3.4.2). Similarly, P_{off} and total power values from SPICE simulations can be compared with those obtained from EDA power tools. Thorough regression analysis of EDA tools with a small suite of carefully selected logic circuit blocks is extremely helpful in validating the assumptions made in complex timing and power tools.

10.6 PD-SOI vs. Bulk Silicon Technology

In the early 1990s, in advance of the race to higher microprocessor operating frequencies, the semiconductor industry began looking for alternative CMOS technologies. SOI technology had been developed for military applications because of its superior radiation hardness and immunity to soft-errors. It clearly had potential performance benefits over bulk silicon technology. Over time as commercial products began to be manufactured in SOI technology [11], accurate quantification of those benefits led to many debates with no clear resolution. We will use a comparison of partially depleted silicon-on-insulator technology (PD-SOI) to conventional bulk silicon to illustrate some of the complexities involved in comparing technologies.

In SOI technology, MOSFETs and other active devices are delineated in a thin silicon film electrically isolated from the bulk silicon substrate by a buried oxide layer (BOX). In PD-SOI, the silicon layer thickness is greater than the depletion layer thickness in the channel region, and there is quasi-neutral region in the body of the MOSFET. In fully depleted silicon-on-insulator technology (FD-SOI), the thin silicon layer is fully depleted. Because of a strong short-channel effect and silicon thickness control in manufacturing of FD-SOI, PD-SOI was adopted for high performance digital CMOS chips.

Schematic cross-sections of an n-FET delineated in bulk and PD-SOI are shown in Fig. 10.38. The thickness of the buried-oxide (BOX) layer in PD-SOI is typically $>$100 nm. As a result, the junction area component of the diffusion region capacitance is significantly reduced in PD-SOI. The body of the MOSFET is electrically isolated from the substrate and the floating-body potential (V_{bs}) modulates V_{t} during switching. Thermal isolation from the substrate tends to raise the body temperature from self-heating effects.

Capacitance reduction and dynamic V_{t} shifts associated with floating-body effects are two important features of PD-SOI that bring promise of significant performance improvement. Inverters and NAND2s are the predominate logic gates in most high performance bulk silicon microprocessor designs. With the benefit of hindsight over the last 20 years, it is clear that the introduction of PD-SOI has not significantly changed this design practice. Consequently, an inverter is both the most basic starting point and a representative circuit component in most cycle limiting paths. We will describe differentiating behaviors of inverters in PD-SOI and also broaden the discussion to include NAND2s, along with some comments on more complex gates.

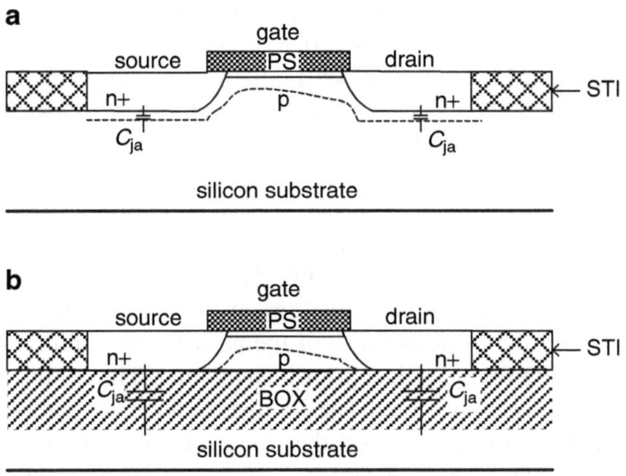

Fig. 10.38 Schematic cross-section of an n-FET in (**a**) bulk silicon and (**b**) PD-SOI

In a static state, the floating-body (FB) potential of a MOSFET is determined by a balance of charge generated by impact ionization and source-to-body, drain-to-body, and gate-to-body leakage currents. The time constants associated with such processes are relatively long, ~ 1–1,000 μs. In a switching transient, the body potential is further modulated by capacitive coupling to the gate, source and drain, a process that occurs on a very short time scale. When the body becomes more forward biased with respect to the source, V_t is lowered as shown in Fig. 10.39a. The dynamic V_t modulation during switching transients cannot be measured directly; however, it impacts measurable quantities such as circuit delay and MOSFET leakage current I_{off}.

Dynamic V_t shifts can potentially reduce switching delay. Fig. 10.39b shows voltages associated with an initially "off" n-FET of a PD-SOI inverter that has been inactive for a long period of time. Its body voltage is initially set by the static back-

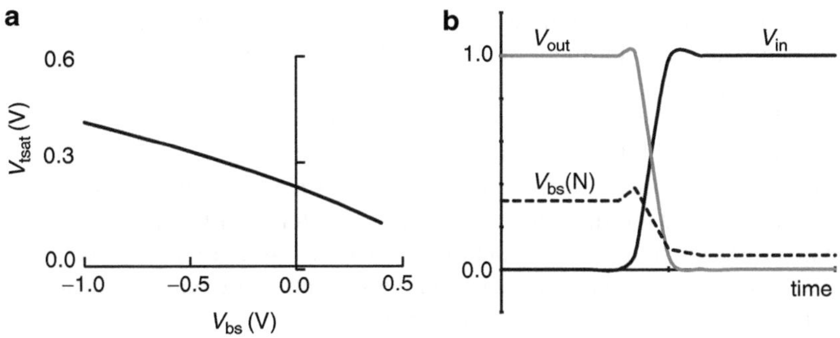

Fig. 10.39 (**a**) MOSFET V_t as a function of V_{bs} for an n-FET in PD-SOI technology, and (**b**) V_{in}, V_{out}, and V_{bs} of a PD-SOI n-FET in an inverter PD transition

to-back drain-body-source diode configuration (as well as gate-to-body leakage, if significant). This is defined as the inverter pre-switch (ip) potential. Prior to the switching event, the n-FET behaves like a bulk n-FET with a forward biased body and I_{off} consistent with the value of V_t and the subthreshold slope. As the transient progresses the body voltage changes. Initially it is pulled up by the gate-body capacitive coupling. However, once the channel forms, it is pulled down by the drain-body capacitive coupling. The net impact of these dynamic body voltage shifts and corresponding V_t shifts on the value of R_{sw} can be positive or negative. A similar picture holds for the p-FET in a PU transition. The n-FETs and p-FETs of inverters in an "equivalent" bulk technology at the ip reference condition would have static V_t values the same as those of the actively switching n-FET (PD) and p-FET (PU) in the PD-SOI case just prior to switching. The resultant difference in R_{sw} is a unique floating-body effect. The changes in the body potentials during this initial (first switch) transition also set the stage for a potentially very significant change in R_{sw} of a subsequent (second switch) transition, along with a temporary modulation of I_{off}.

A more complete picture of the floating-body potential V_{bs} and of V_t in the p-FET and n-FET of a PD-SOI inverter undergoing PD and PU transitions is presented in Fig. 10.40, where V_{bs} and V_t values are all shown as positive [12]. Prior to a PD transition, the voltage at the output node V_{out} is a "1" and the leakage current is determined by I_{offn} of n-FET N1. For the n-FET the PD transition proceeds as previously described. Immediately after the PD transition, the V_{bs} of n-FET N1* is lowered and its V_t is raised, as shown in Fig. 10.40c. The V_{bs} of p-FET P1* after the PD transition is higher and its V_t is lowered. The leakage current of the inverter, I_{offp}, is now determined by P1* in Fig. 10.40a*.

The V_{bs} and V_t values of the inverter p-FET and n-FET, for a PU transition, are shown in Fig. 10.40d. Prior to a PU transition, V_{out} is a "0" and the inverter leakage current is determined by I_{offp} of p-FET P2. Immediately after the PU transition, the V_{bs} of p-FET P2* is lowered, raising its V_t. The V_{bs} of n-FET N2* is raised and its V_t is lowered. The leakage current of the inverter, I_{offn}, is now determined by N2* in Fig. 10.40b*. For an interconnected chain of such inverters experiencing passage of a single isolated edge, the indicated δV_t(N) and δV_t(P) values will invoke a change in IDDQ of the chain that will persist until the MOSFET bodies settle back to their pre-switch potentials. Furthermore a second edge passing through the chain soon after the first will experience different switching times as the pre-second transition threshold voltages of the dominating MOSFETs will be shifted by δV_t(N) and δV_t(P) compared with those experienced by the first edge.

The V_{bs} and V_t values after an initial switching transition vary with time after the event transition and return to the pre-switch state after ~1–1,000 μs. The MOSFETs in an inverter switching at a constant rate as in a ring oscillator reach an equilibrium state within ~1 millisecond and their average V_{bs} values are typically different than those before or after an isolated switching event.

From the above discussion, it is apparent that the leakage current in the quiescent state and signal propagation delay of a circuit in PD-SOI technology vary with switching history. When a circuit switches after being in a static idle state for a few

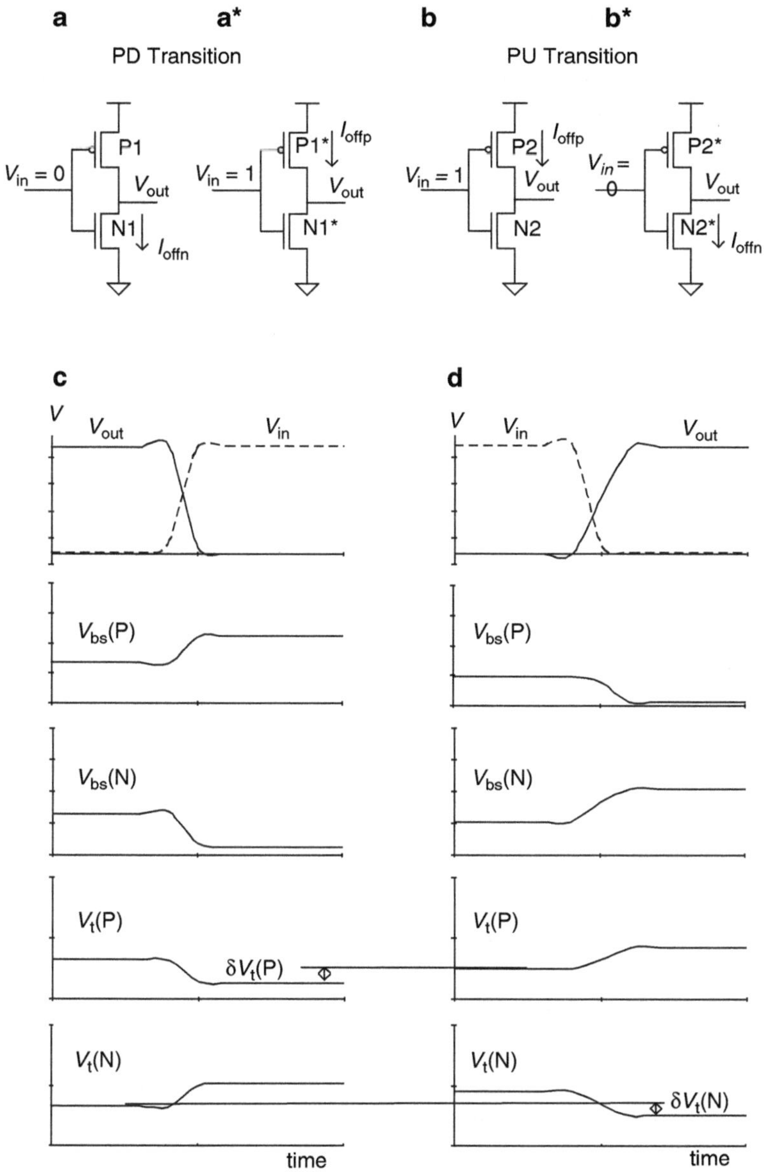

Fig. 10.40 Circuit schematic indicating I_{off} contributions for an inverter before and after (**a, a***) PD and (**b, b***) PU transitions. (**c, d**) Inverter input and output waveforms, absolute values of V_{bs} and V_t as functions of time for the transitions corresponding to (**a, a***), and (**b, b***), respectively. Reproduced from [12]

milliseconds or longer, the event is called a "first switch" or 1SW transition. When the circuit switches again within a few nanoseconds of the 1SW transition, it is called a "second switch" or 2SW transition. A periodic switching situation, as in the case of clock buffers, is a steady-state (SS). The V_{bs} and V_t values of MOSFETs in steady-state are typically between their values in the pre-1SW and pre-2SW states.

Output PD and PU signal waveforms for 1SW, 2SW, and SS transitions of an inverting logic gate are shown in Fig. 10.41 . The (1SW–2SW) switching history effects for PD and PU transitions, H_{tpd}, and H_{tpu}, can be defined as

$$H_{tpd}(\%) = 2\left(\frac{\tau_{1pd} - \tau_{2pd}}{\tau_{1pd} + \tau_{2pd}}\right) \times 100 \tag{10.14}$$

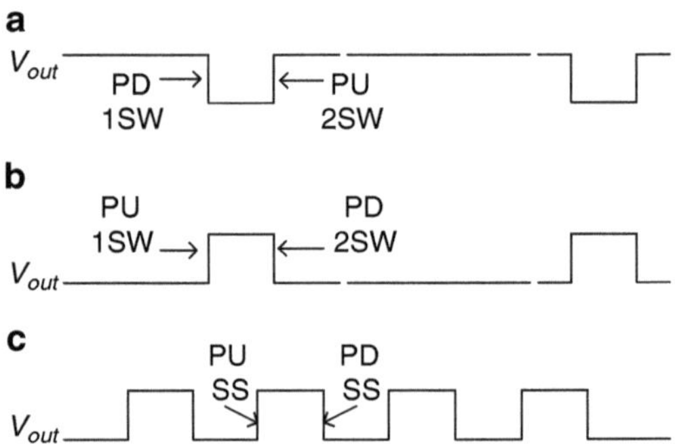

d

Transition	Delay
PD 1SW	τ_{1pd}
PD 2SW	τ_{2pd}
PU 1SW	τ_{1pu}
PU 2SW	τ_{2pu}
PD SS	τ_{sspd}
PU SS	τ_{sspu}

Fig. 10.41 Inverter output signal waveforms for (**a**) 1SW PD and 2SW PU, (**b**) 1SW PU and 2SW PD, and (**c**) SS transitions, and (**d**) corresponding circuit delay symbols. Reproduced from [12]

and

$$H_{tpu}(\%) = 2\left(\frac{\tau_{1pu} - \tau_{2pu}}{\tau_{1pu} + \tau_{2pu}}\right) \times 100 \tag{10.15}$$

where τ_{1pd} and τ_{2pd} are the 1SW and 2SW PD delays, and τ_{1pu} and τ_{2pu} are the 1SW and 2SW PU delays. The MOSFETs with the dominating influence on pre-switch IDDQ and subsequent delays, in different switching transitions of an inverter, are listed in Table 10.15.

Table 10.15 MOSFETs dominating the pre-switch IDDQ as well as the drive strength of an inverter for different switching transitions. MOSFET designations correspond to those in Fig. 10.40

Transition	MOSFET
1SW PD	N1
2SW PD	N2*
1SW PU	P2
2SW PU	P1*

The measured (1SW–2SW) history effect H_t in a circuit path comprising a chain of gates is the average of H_{tpd} and H_{tpu}. The values of H_t for PD or PU transitions are positive for $\tau_1 > \tau_2$ (1SW slower than 2SW) and negative for $\tau_1 < \tau_2$ (1SW faster than 2SW). The history effect varies with V_{DD} and, to a lesser degree, temperature. The MOSFETs in PD-SOI technology can be engineered to exhibit positive, negative, or negligibly small values of H_t in standard logic gates at a desired value of V_{DD} and temperature, although H_t values will vary with departure from these conditions. Generally, H_t for standard logic gates is designed to be ~5 % at the nominal operating conditions of a product. H_t is, however, dependent on circuit topology and some circuits may experience H_t values of >20 %.

Stacked gates, particularly those that are three or more high, can have history effects much larger than an inverter. The largest history effects are observed in single-ended passgate circuits. The reverse body effect penalty experienced by some of the MOSFETs in such gates for bulk technology is also mitigated. Under special circumstances such PD-SOI circuits can show very significant delay reductions compared to similar circuits in bulk. In a 1SW bottom-switching PD transition in a NAND4, for example, the bodies of the upper MOSFETs in the stack begin the transition close in voltage to their respective sources and are capacitively pulled towards ground as the transition proceeds. In the corresponding bulk case the bodies are all tied to ground throughout the transition, effectively rising the V_t of the upper MOSFETs and slowing the transition. In general, large history effects correlate with reduced delay under special circumstances.

There are of course a wide variety of switching history scenarios and associated history effect values. For static CMOS gates, 1SW and 2SW delays establish upper and lower limits. This is not necessarily true for all circuits, such as single-ended passgates.

Reduction in capacitance is a primary contributor to PD-SOI performance improvement. Looking at the *RC* model for an inverter we have $\tau_p = R_{sw}C_{sw} = R_{sw} \times (\text{FO} \times C_{in} + C_{out})$. In bulk silicon technology, the inverter C_{out} is approximately equal to C_{in}, so if we very optimistically assume $C_{out} = 0$ for PD-SOI, then a 20 % decrease in τ_p of a PD-SOI inverter (FO = 4) in comparison with an inverter in an "equivalent" bulk technology is expected. As a somewhat more realistic approach, the PD-SOI performance improvement can be estimated by setting all of the zero-bias diffusion area and perimeter capacitance components in the BSIM model card equal to zero in a representative bulk model. These parameters *cjs*, *cjd*, *cjsws*, *cjswd*, *cjswgs*, and *cjswgd* are indicated in bold in the model cards in Appendix B. The results of such an exercise with a delay chain are shown in Table 10.16 using the 45 nm PTM HP models at 1.0 V, 25 °C. For the inverter the impact is somewhat lower than the $C_{out} = 0$ estimate suggests. It is also important to keep in mind that the improvement is less for FO > 4 and with the addition of wire load, and more for FO < 4.

In reality, with PD-SOI there remains diminished but non-zero diffusion capacitance to ground through the BOX. In addition, there are significant internal capacitance components that come into play. For example, the internal source-to-body and body-to-drain capacitances form a series combination that can be substantial.

The improvements indicated in Table 10.16 should thus be viewed as optimistic upper limits.

Table 10.16 Reduction in τ_p and C_{sw} in response to setting diffusion area and perimeter capacitance components (C_{jx}) equal to zero for a standard inverter (FO = 4), NAND2T (FO = 4), and NAND2B (FO = 4). 45 nm PTM HP models@ 1.0 V, 25 °C

Logic gate	τ_p (ps) Reference	τ_p (ps) $C_{jx}=0$	C_{sw} (fF) Reference	C_{sw} (fF) $C_{jx}=0$	$\Delta\tau_p$ (%)	ΔC_{sw} (%)
Inverter	11.68	9.80	7.89	6.72	16.1	14.8
NAND2T	16.89	13.20	8.79	6.97	21.8	20.7
NAND2B	17.23	13.66	8.99	7.27	20.7	19.1

We should also point out that confinement in the vertical direction imposed by the BOX mitigates the challenges of design and control of device features and materials composition in both vertical and horizontal directions. This may imply that PD-SOI can be successfully scaled to smaller dimensions than bulk. This has not proved to be an insurmountable problem for bulk down to at least the 32 nm node, although PD-SOI may enjoy some advantage from, for example, potentially improved short-channel behavior.

One additional practical consideration involving PD-SOI is that *I–V* characteristics of MOSFETs in PD-SOI technology are intrinsically different under quasi-static conditions than during transient switching events because of floating-body and heating effects. The difference between MOSFET currents measured under DC and pulse excitation is larger at higher bias voltages and higher current densities where significant V_t modulation and self-heating occur. Self-heating has not been a serious detriment to PD-SOI under actual operating conditions, however, it has greatly complicated the interpretation of standard DC *I–V* characteristics and necessitated the introduction of sophisticated pulse measurement techniques to obtain information that is available from basic DC measurements in bulk technology.

In the preceding discussion we have reviewed a number of the distinguishing features of PD-SOI technology. Now we will mention some of the issues relating to bulk and PD-SOI technology comparisons. Ideally one would like to do an apples-to-apples comparison of an optimized PD-SOI technology with an "equivalent" optimized bulk technology at the same CMOS technology node. Such a comparison is not an easy task. It requires a direct comparison of representative samples of hardware (hardware based) or comparison of models whose accuracy is based on measurements of representative hardware (model based). If the performance of one technology is 2× that of another the difference is not hard to discern. If it is instead only 5–15 %, it may never be possible to unambiguously demonstrate the advantages of one over the other. In addition to the standard challenges encountered in comparing one bulk technology with another, here a different set of issues must be dealt with. DC MOSFET characteristics in PD-SOI are affected by self-heating at higher values of I_{ds}, namely I_{on} and I_{eff}. Modulation of V_t due to history effect affects I_{off}.

Circuit delays measured using ring oscillators give steady-state values which are generally not the worst-case delays in PD-SOI. Measurement of worst-case delay, typically for 1SW, requires high speed test structures [12] during development. The delay variation due to history effect in different circuit configurations may be $<5\ \%$ to $>20\ \%$ and the performance comparison is heavily influenced by the mix of circuit topologies used. Furthermore, it has in general not been possible to derive significant performance improvement leveraging a faster 2SW transition, since the same circuit has to meet timing constraints under 1SW conditions as well. In fact there are cases where a faster 2SW can lead to race conditions.

Although definitive PD-SOI/bulk technology comparisons have been elusive, it has been possible to develop methodologies using a single PD-SOI model to investigate key advantages associated with floating-body effects and reduced capacitance. These methodologies do require a physically realistic PD-SOI model [13], but are not plagued by model-to-model inconsistencies. As an example consider PD-SOI inverter switching situations shown in Fig. 10.41 that were previously described. From a switching resistance perspective one could argue that an "equivalent" bulk technology would be one in which the bodies of the p-FETs and n-FETs are tied with zero resistance to fixed ip potentials, as can be implemented in standard PD-SOI models [14]. Delay chains with and without the body ties can then be simulated and the switching resistances extracted as described in Sect. 2.2. The change in R_{sw} is then a direct measure of the change in performance due to the dynamic threshold shifts in PD-SOI. This same approach can be used to evaluate any PD-SOI logic gate with bodies tied to ip potential values. Such exercises suggest that the improvement associated with PD-SOI dynamic threshold shifts for inverters and NAND2s is in the range of zero to a few %.

Similarly, in the area of capacitance reduction with PD-SOI, a methodology has been developed that relies predominately on just the PD-SOI model, again removing the ambiguity of multiple models [14]. With this approach it was found that for earlier PD-SOI generations the actual τ_p reduction was about half that determined by setting the diffusion capacitance components in the bulk model to zero, with a very strong dependence on the source and drain-to-body capacitance.

At this point it is the authors consensus that PD-SOI can deliver improved performance at a given technology node at the level of 10 % or less, mainly from capacitance reduction. This improvement can be taken as an increase in frequency at constant power or a decrease in power at constant frequency. In selecting whether or not to use PD-SOI technology for any particular application, the added cost of manufacturing process, design complexity, and modeling complexity must be weighed against the potential performance gain.

10.7 Closing Comments on CMOS Technology Evaluation

As CMOS technologies are scaled beyond 32 nm, the focus has been shifting from circuit performance gains to density and yield. Many CMOS products continue to be manufactured in older technologies. Selecting the right technology node and foundry for a product requires a robust method of technology evaluation.

Circuit delay is an important factor in selecting, developing, and optimizing a technology for digital and analog circuit applications. However, establishing a criterion for circuit delay as a technology metric has been a challenge and there is no clear resolution to date. The ITRS sponsored by a consortium of semiconductor companies publishes a technology roadmap every year. The technology performances of different CMOS technology nodes are listed using intrinsic n-FET delay from a CV/I estimation. This metric does not address p-FET performance and parasitic delays associated with the physical layout of a logic gate. In 2003 and 2004, a ring oscillator circuit delay metric was defined for a 2-input NAND gate with FO = 3. This metric was dropped in 2005. In 2010, again a ring oscillator circuit delay for FO = 1 was included. In 2011, ring oscillator delays for both FO = 1 and FO = 4 are included. These delays are calculated from MASTAR tool supplied by ITRS but the circuit topology is not described in the reports. It is apparent from ITRS reports that there has been no general consensus in the semiconductor industry on a circuit delay metric that can be sustained over 10–20 years.

Attempts to compare circuit delays from published conference proceedings and journal articles are futile. Publications from the semiconductor industry do cite measured circuit delays in hardware obtained from ring oscillators. However, details of circuit topology, physical layouts, and parasitic components are not provided. In a race to achieve the highest technology performance, ring oscillators can be designed with minimum parasitics, not necessarily representing circuit design styles used in products. Circuit delays in a given technology vary from wafer to wafer because of variability over time in MOSFET parameters, such as threshold voltage and channel length. A large volume of hardware data must be gathered over a long time in the manufacturing cycle to obtain a nominal value for the technology. Intel has published ring oscillator delays as a function of the sum of n-FET and p-FET currents in the off-state to compare different technology nodes instead of the ring oscillator leakage in the quiescent state (IDDQ) [15]. A drawback of this methodology is that MOSFET data are measured on different test structures that may differ from corresponding MOSFETs used in the ring oscillators in the context of device widths and physical layouts, both of which affect performance at a specific technology node.

Circuit designers prefer to compare the performance, power, and area requirements of a large circuit block, such as a 32-bit adder, for assessing the benefits of a technology [16]. Such comparisons are based on models and subject to different sensitivities to circuit implementation and physical layout styles used.

In order to ship a product at the committed specifications, manufacturing test teams have often exercised the option of tuning V_{DD}. The operating V_{DD} of high performance CMOS chips is often higher than the technology nominal. In low power applications, the V_{DD} knob is even more sensitive. In addition, chip architecture, circuit design, guard-bands, and operating software all come into play in deriving maximum benefits.

The decision to develop and use a technology must be carefully weighed against the increased complexity and cost. A 50 % reduction in τ_p may be clearly demonstrated in hardware. However, to convincingly demonstrate a 10 % reduction, as in the case of a bulk to PD-SOI comparison, in the face of other rapidly evolving technology factors, is far more difficult to verify. This situation is further exasperated by the fact that generation-to-generation τ_p improvement within a single foundry is itself only 15–30 %.

As the "end of CMOS scaling" approaches, a number of new and emerging technologies are being explored and evaluated as candidates for continuing the exponential advance of digital microelectronics into the "beyond CMOS" arena [16]. However, independent of how this scenario plays out, currently available and yet more advanced CMOS technologies will be well suited to fulfill a wide variety product requirements for a very long time. Along with this, the need to select the most suitable CMOS technology for a particular application will remain as an essential and ongoing challenge. In the end, the choice of which CMOS technology to use for a given application comes down to solid informed engineering judgment based on physical insight and the available relevant data. The data, although typically incomplete, are obtained from the literature, and from developers and manufacturers of technology. In this book, through both text and exercises, it has been one of our key goals to develop and present a perspective that will help to provide the insight.

10.8 Summary and Exercises

The challenges in standardizing a CMOS metric for performance, power, and density are discussed in view of technology development trends. Methods of model-based evaluation of MOSFET and circuit performance in CMOS technologies are compared. Performance metrics are further expanded to include power, energy, and circuit densities. A methodology for evaluating compact device models and EDA tools for parasitic extraction, timing, and power, based on physical understanding of device and circuit behavior, is described. This helps in minimizing model related errors in design. The issues confronted in making accurate technology performance comparisons are exemplified in the context of PD-SOI and bulk silicon technologies.

Exercise 10.1 relates to the scaling trends in CMOS technologies. Exercises 10.2–10.4 deal with different methods used for comparing MOSFET and circuit performance. Exercises 10.5 and 10.6 involve compact models and EDA tool evaluations. Exercise 10.7 guides an investigation of performance improvement through introduction of low-k inter-level dielectric materials. The last three exercises, 10.8–10.10, address various aspects of PD-SOI and bulk silicon technology model comparisons.

10.1. (a) A chip has an area of 100 mm^2 in 180 nm technology. If the design is faithfully scaled with a scaling factor of 0.7× in each technology generation (0.5× area scaling), what will be the area of the chip after five generations at the 32 nm technology node?

(b) Estimate the chip area if limited by the I/O pitch scaling factor of 0.85× per technology node instead

(c) By what factor can the transistor count be increased if chip area remains constant at 100 mm^2

10.2. Intel has published data on key MOSFET parameters for comparing technology performance. Comparison of 65 and 45 nm technologies is found in [8]. Both MOSFET and ring oscillator data are compared. It is common practice in the industry not to release design and measurement details. What information is needed to validate the data comparison?

10.3. (a) At the 45 nm technology node, V_{tsat} is measured at $300 \times W/L_p$ nA for n-FETs and $100 \times W/L_p$ nA for p-FETs. Assuming SS = 100 mV/decade, what will be the change in V_{tsat} if measured at $250 \times W/L_p$ nA for n-FETs and $125 \times W/L_p$ nA for p-FETs?

(b) Determine model n-FET and p-FETs V_{tsat} values using the I_{dsvt} method and by $\sqrt{I_{gs}}$–V_{gs} extrapolation. How different are V_{tsat} values obtained from the two methods?

(c) What is the significance of V_{tsat}? When comparing two technologies, can you rely on published values of V_{tsat} for comparison?

10.4. Inverter (FO = 3) delays on silicon wafers from two different foundries match to within 1 %. A more detailed comparison should include R and C components of the inverter delay.

(a) How many ROs are needed to compare R_{sw}, C_{in}, C_{out}, and C_p (parasitic capacitances)?

(b) Create circuit schematics of the ckt_stgs for the proposed ROs and extract the delay parameters from simulations.

10.5. A new compact BSIM model is released for the same technology node to replace a previously released model. The impact on logic gates is to be evaluated. Develop a suite of circuits (library books and delay chains) to determine major differences in the models that have measureable effect of circuit delays and power. Summarize data in a one or two charts (you may use PTM and your own models for the comparison exercise or use the same model in two different V_{DD} and temperature corners).

10.6. (a) Generate three different physical layouts for a NAND2 logic gate with different parasitic resistances and capacitances as illustrated in Fig. 10.16.

(b) Estimate the relative differences in parasitic resistances and capacitances from the physical layouts.

(c) Run circuit simulations to determine τ_p, C_{sw}, and R_{sw} using extracted netlists from the physical layouts. If parasitic extraction models are not available, add estimated series resistance and capacitances in the schematic.

(d) Do the difference in C_{sw} and R_{sw} match expectations? If not, explain the differences.

10.7. Enhancements in dielectric and interconnect processing have led to the use of low-k dielectric materials. The benefits are easily demonstrated in circuit simulations. However, management wants to see the improvement in chip performance and net power reduction for the following cases: (1) the effective dielectric constant of a composite is reduced by 30 % in layers M1 through M4, and (2) air-gap is introduced in the top metal layers to reduce capacitance by 50 %. Split-lots with control wafers and wafers with the new dielectrics are processed to independently evaluate case (1) and case (2).

(a) For case (1) assuming 15 % wire capacitance load on a circuit stage, what are the improvements in τ_p and power? What is the fractional power reduction in case (2) if the capacitance of top layers is 8 % of total switching capacitance.

(b) With 400 chips/wafer and 65 % yield, how many wafers should be processed in each split to get the mean and σf_{max} with 2 % accuracy at 95 % confidence level.

(c) What analysis needs to be done to ensure that the performance and power delta is truly from M1–M4 capacitance reduction and not by other process variations.

(d) Generate graphs to display all of the relevant findings and summarize in two charts.

10.8. Copy the bulk silicon BSIM models and set all zero-bias diffusion capacitance parameters in the model card equal to zero to emulate PD-SOI technology.

(a) Compare the delays of logic gates (inverter, NAND4T, NAND4B, NPG_1, and TG) in the original and new models. Which circuit configurations benefit the most?

(b) Create an inverter schematic with nmos4 and pmos4 with independent power supplies for body-bias V_{bs}. Determine τ_p and P_{off} with $V_{bs} = -0.2$ V, 0.0 V, and +0.2 V and compare.

(c) Optional—If PD-SOI models are available, using the methodology in reference [14], compare performance of bulk and PD-SOI logic gates

10.9. In PD-SOI technology individual device temperature can be well above the substrate temperature due to the insulating BOX layer. Assume that each MOSFET in our standard inverter can be treated as an independent entity with a thermal resistance to the underlying substrate of $R_{th} = 4 \times 10^5$ W/°K and a heat capacity $C_{th} = 2.5 \times 10^{-14}$ J/°K.

(a) Calculate the thermal time constant of the n-FET.

(b) Assuming device characteristics, aside from self-heating, are well described with the bulk model, use that model to self consistently determine n-FET temperature, I_{off}, I_{on}, and I_{eff} under assumed conditions of (1) no self-heating, (2) DC bias, and (3) periodic switching in a 51 stage ring oscillator.

10.10. Dynamic threshold shift during switching transients can affect the switching resistance of an inverter. Set up a simulation for a single inverter with a pure capacitive load equivalent to FO4 with pmos4 and nmos4. Initially set $V_{bs} = 0$ for the n-FET and V_{DD} for the p-FET and drive the input with a rising input with $\tau_r = 16$ ps (input transition time of 20 ps).
(a) Determine the inverter switching resistance R_{sw}.
(b) Construct a piecewise linear (6 points) voltage source to drive the n-FET body during the transient that is 0.0 V at the beginning of the input transition, rises to 0.2 V after 4 ps, stays at 0.2 V for 2 ps, falls to −0.3 V after 12 more ps and remains there.
(c) Re-simulate and determine the new value for R_{sw}. By what % has R_{sw} increased or decreased?

References

1. International technology roadmap for semiconductors: ITRS. http://www.itrs.net. Accessed 21 July 2014
2. Taur Y, Ning TH (2009) Fundamentals of modern VLSI devices, 2nd edn. Cambridge University Press, New York
3. Weste NH, Harris D (2010) CMOS VLSI design: a circuit and systems perspective, 4th edn. Addison-Wesley, Reading
4. Rabaey JM, Chandrakasan A, Nikolic B (2003) Digital integrated circuits, 2nd edn. Prentice Hall, Upper Saddle River
5. Moore GE (1965) Cramming more components onto integrated circuits. Electronics 38:114–117
6. Intel chips timeline. http://www.intel.com/content/www/us/en/history/history-intel-chips-timeline-poster.html. Accessed 21 July 2014
7. IBM System P. http://en.wikipedia.org/wiki/IBM_System_p. Accessed 21 July 2014
8. Ranade P, Ghani T, Kuhn K, Mistry K, Pae S (2005) High performance 35 nm L_{GATE} CMOS transistors featuring NiSi metal gate (FUSI), uniaxial strained silicon channels and 1.2 nm gate oxide. In: Proceedings of the International electron device meeting IEDM 2005
9. Ho R, Mai KW, Horowitz MA (2001) The future of wires. Proc IEEE 89:490
10. Das KK, Walker SG, Bhushan M (2007) An integrated methodology for evaluating MOSFET and parasitic extraction models and variability. Proc IEEE 85:670–687
11. Shahidi GG (2002) SOI technology for the GHz era. IBM J Res Dev 46:121–131
12. Bhushan M, Ketchen MB (2011) Microelectronic test structures for CMOS technology. Springer, New York
13. Goo J-S, William RQ, Workman GO, Chen Q, Lee S, Nowak EJ (2008) Compact modeling and simulation of PD-SOI MOSFETs: current status and challenges. In: IEEE 2008 custom integrated circuits conference, CCIC, pp 265–272
14. Ketchen MB (2003) Competitive advantage of SOI from dynamic threshold shifts and reduced capacitance. In: Proceedings 2003 international symposium on VLSI technology, systems and applications, pp 129–132
15. Auth C, Cappelani A, Chun J-S, Dalis A, Davis A, et al (2008) 45 nm high-k + strained enhanced transistors. 2008 Symposium on VLSI technology, pp 128–129
16. Nikonov DE, Young IA (2013) Overview of beyond-CMOS devices and a uniform methodology for their benchmarking. Proc IEEE 101:2498–2533

Appendix A: MOSFET and Logic Gate Parameters (PTM HP Models)

A brief description of a CMOS physical cross section with key MOSFET and interconnect layers is presented. Example physical layouts of MOSFETs and a standard inverter provide planar views of the layers. MOSFET properties and circuit parameters for static CMOS logic gates in 45, 32, and 22 nm PTM models are summarized in Tables A.1–A.6. These summary tables are very useful for cross-checking simulated data, and for estimating static logic gate circuit parameters for a variety of load conditions. It is recommended to generate such summary tables from circuit simulations to serve as a handy reference when evaluating CMOS technologies and analyzing electrical test data.

A.1 CMOS Physical Cross Section

A typical CMOS process includes MOSFETs, diodes, resistors, and capacitors delineated in silicon along with 10 or more levels of metal. A schematic cross section of key CMOS circuit layers is shown in Fig. A.1a. MOSFETs (n-FET and p-FET) are defined in a p-type silicon substrate. A four metal layer stack with metal layers M*X* ($X = 1, 2, 3, T$) and via layers H*X* ($X = 0, 1, 2, 3$) is shown. I/O connections are made to the top metal layer (MT) through controlled collapse chip connections (C4s) or by wire bonding.

M. Bhushan, M.B. Ketchen, *CMOS Test and Evaluation*,
DOI 10.1007/978-1-4939-1349-7

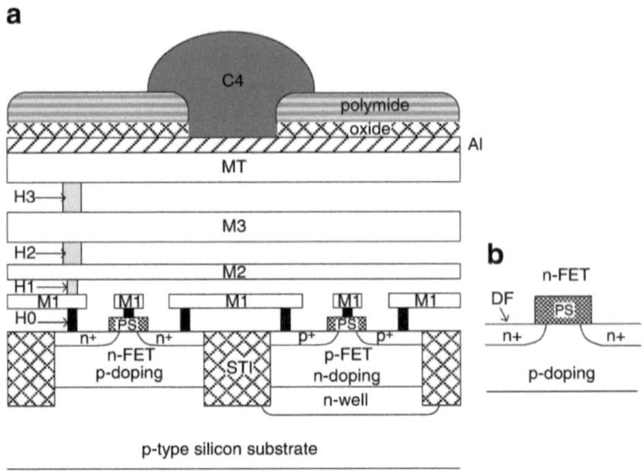

Fig. A.1 (**a**) Schematic of the physical cross section of a CMOS circuit with four metal layers and a C4 I/O, and (**b**) n-FET cross section with PS (gate) and DF (diffusion) layers

The cross section of an n-FET is shown in Fig. A.1b. Gate area is defined by the PS layer, and active n^+ silicon diffusion areas for source and drain are defined by the DF layer. Properties of the n-FET and p-FET are engineered with dopant profiles, and material properties and dimensions of constituent layers. The body of an n-FET is p-doped silicon. The body of a p-FET is formed in an n-well on a p-type silicon substrate. In a twin-well (twin-tub) process, both n-FET and p-FET bodies are isolated from the silicon substrate and can be biased independently. Contact to the p-type body of the n-FET is made through a p^+ region and to the n-type body of the p-FET through an n^+ region (Fig. 2.5).

Minimum allowed metal wire pitch for the lower layers (typically M1 and M2) is typically matched with the MOSFET gate pitch. The minimum pitch is increased in the upper layers in the metal stack with the top layers matching the I/O contact pitch. Via dimensions for interconnecting vertically adjacent layers also increase with the wire widths.

Minimum allowed widths, spacings, and layer thicknesses may be equal in a pair of orthogonal metal layers and integral multiples of the lowest (M1 and M2) layers. These are denoted by $n\times$ where n is the multiplier. An example layout of metal wires for two $1\times$, two $2\times$, and two $4\times$ layers is shown in Fig. A.2. The preferred directions of wiring for vertical neighboring layers are orthogonal. Silicon foundries have multiple offerings of metal layer stacks with different numbers of layers and pitches in each technology generation. In high performance microprocessor chips and other special applications, metal layer stack definition may be customized for optimum performance.

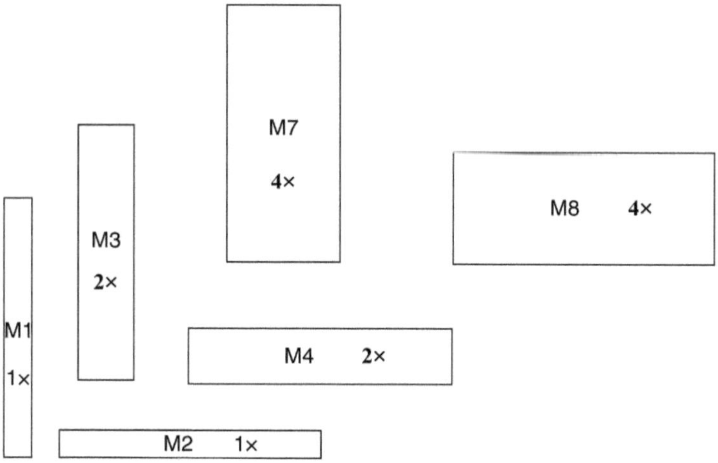

Fig. A.2 Metal layers M1 through M8 and preferred orientation for minimum width wires

A.2 MOSFET and Circuit Parameters for 45 nm PTM HP Models

MOSFET and circuit parameters listed in this section are obtained from the 45 nm PTM HP models. These models include parasitic capacitances of the diffusion (DF layer). Parasitic resistances and capacitances associated with interconnect wires and vias are not included. The tables are intended to serve as examples of key parameters for model evaluation.

Physical layouts of an isolated n-FET and p-FET, each with a single PS finger are shown in Fig. A.3. DC parameters for an n-FET and p-FET in the 45 nm HP models at 1.0 V, 25 °C with nominal values of L_p and V_t are listed in Table A.1.

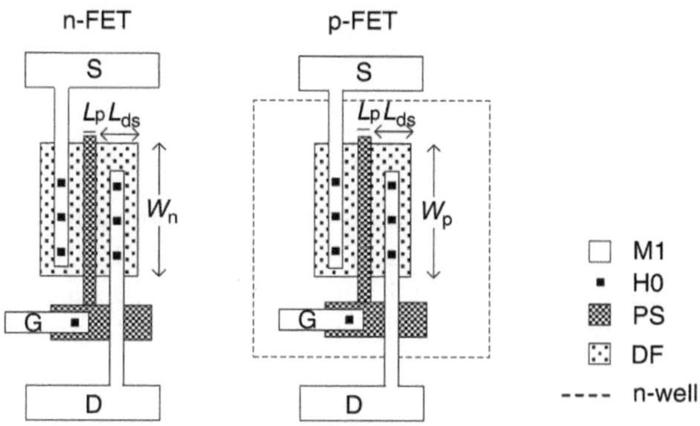

Fig. A.3 Physical layouts of an n-FET and p-FET

Table A.1 Parameters for 1.0 μm wide single finger n-FETs and p-FETs. 45 nm PTM HP models @ 1.0 V, 25 °C

Parameter	n-FET	p-FET
Gate length, L_p (μm)	0.045	0.045
Diffusion length, L_{ds} (μm)	0.120	0.120
I_{on} (μA/μm)	1,339	968
I_{eff} (μA/μm)	711	450
I_{off} (nA/μm)	20	4.5
I_{gl} (nA/μm)	0.62	2.52
V_{tsat}* (V)	0.233	0.237
V_{tlin}* (V)	0.374	0.411
C_g (inversion) (fF/μm)	1.49	1.53

*V_t measured at $I_{dsvt} = 300W_n/L_p$ nA for n-FET and $I_{dsvt} = 100W_p/L_p$ nA for p-FET

A standard inverter layout, with one PS finger, having $W_p = 0.6$ μm and $W_n = 0.4$ μm is shown in Fig. A.4. An extracted netlist would include resistances and capacitances associated with PS and DF layers, H0 vias, and M1 wires.

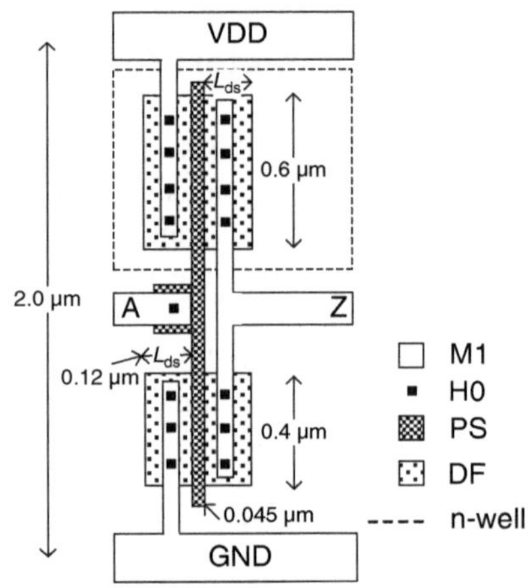

Fig. A.4 Layout of a standard inverter with one PS finger. $W_n = 0.4$ μm and $W_p = 0.6$ μm

The design and circuit parameters of this standard inverter are listed in Table A.2. Circuit simulations are carried out using a ring oscillator with 50 inverter stages and a NAND2 as described in Sect. 2.2.5. The MOSFET widths are selected to give $\tau_{pu} \approx \tau_{pu} \approx \tau_p$. The W_p/W_n ratio of 1.5 nearly matches I_{effn}/I_{effp} (=1.58) from Table A.1.

Table A.2 serves as a useful reference for estimating inverter delays under different loads without running many circuit simulations. The values of τ_p (FO = 3) is nearly 2× that of τ_p (FO = 1). The increase in τ_p with each additional FO can be estimated as

Table A.2 Design and nominal circuit parameters for the standard inverter (FO = 3). 45 nm HP PTM models @ 1.0 V, 25 °C

Parameter	Description	Value
W_p	p-FET width	0.60 μm
W_n	n-FET width	0.40 μm
L_{ds}	Diffusion region length	0.12 μm
IDDQ	Average off-state leakage current	6.57 nA
τ_p (FO = 1)	Average PU and PD delay	5.27 ps
τ_p (FO = 3)	Average PU and PD delay	9.55 ps
C_{in}	Input capacitance	1.50 fF
C_{out}	Output capacitance	1.83 fF
R_{sw} (FO = 3)	Average switching resistance	1,500 Ω
C_{sw} (FO = 3)	Average switching capacitance	6.37 fF
τ_f, τ_r (FO = 3)	Signal rise, fall time	~16 ps
E_{sw} (FO = 3)	Average energy per switching event	3.18 fJ

$$\Delta\tau_p/\text{FO} = \frac{1}{2}\left[\tau_p(\text{FO} = 3) - \tau_p(\text{FO} = 1)\right]. \tag{A.1}$$

The $\Delta\tau_p$/FO for the inverter at $V_{DD} = 1.0$ V, 25 °C is 2.14 ps which compares well with the value of 2.13 ps/FO obtained from delay chain simulations in Sect. 2.2.4. The R_{sw} and C_{sw} values in Table A.2 can be used to estimate delays with C and RC loads in an inverter chain for a wire length l using the following equations:

$$\tau_p = R_{sw}(C_{in} + C_{out} + C_w l) + (R_{sw}C_w + R_w C_{in})l + \frac{R_w C_w l^2}{2}. \tag{A.2}$$

A similar approach may be used for other logic gates.

In Table A.3, n-FET and p-FET widths for equal PD and PU delays ($\tau_{pd} \approx \tau_{pu}$) for an inverter and stacked logic gates (NANDs and NORs) are listed. The sum of the widths of an n-FET and p-FET ($W_n + W_p$) is 1.0 μm. The average delay τ_p (FO = 3) increases with stack height and is higher with the bottom input switching. When all logic gates are equally weighted, the average value of W_p/W_n is 2.14 and the average FO = 3 delay is 16.31 ps. Such tables are useful for comparing W_p/W_n and τ_p values in different models issued for the same technology node or for different technology nodes.

Table A.3 MOSFET widths for $\tau_{pu} \approx \tau_{pd}$ and average delay τ_p for logic gates (FO = 3) with $(W_n + W_p) = 1.0$ μm. 45 nm HP PTM models @ 1.0 V, 25 °C

Logic gate	W_p (μm)	W_n (μm)	W_p/W_n	Average delay, τ_p (ps)
Inverter	0.60	0.40	1.50	9.55
NAND2T	0.50	0.50	1.00	13.13
NAND2B	0.50	0.50	1.00	13.71
NOR2T	0.75	0.25	3.00	14.90
NOR2B	0.76	0.24	3.17	16.68
NAND3T	0.43	0.57	0.75	16.75
NAND3B	0.46	0.54	0.85	18.43
NOR3T	0.80	0.20	4.00	19.93
NOR3B	0.80	0.20	4.00	23.69

A.3 Scaled Parameters for 45, 32, and 22 nm PTM Models

Nominal power supply voltages, L_p and V_t for MOSFETs in 45, 32, and 22 nm PTM HP models are listed in Table A.4. The power supply voltage is reduced with scaling. Nominal values of L_p correspond to the technology node name. The $\pm 3\sigma$ range in L_p is assumed to be ± 10 % of the nominal L_p value. Systematic process variations in V_t are considered to be the same at all three technology nodes. As MOSFET dimensions are scaled by 0.7×, the spread in V_t due to random variations increases by 1.4× per technology generation.

Table A.4 Nominal V_{DD}, L_p, and values of σL_p and σV_t used in circuit simulations

Node	V_{DD} (V)	L_p (μm)	σL_p (μm)	Systematic σV_{ts} (V)	Random σV_{tr} (V)
45 nm	1.0	0.045	0.00150	0.02	$0.004/\sqrt{(WL_p)}$
32 nm	0.9	0.032	0.00105	0.02	$0.004/\sqrt{(WL_p)}$
22 nm	0.8	0.022	0.00071	0.02	$0.004/\sqrt{(WL_p)}$

In Table A.5, I_{eff}, I_{off}, and V_{tsat} for n-FETs and p-FETs at the three technology nodes are listed. The MOSFET widths in 45 nm technology are $W_n = W_p = 1.0$ μm. The scaled widths in 32 and 22 nm technology nodes are 0.7 and 0.5 μm, respectively. Note that the current drive of the MOSFETs (I_{eff}) decreases as the width is scaled, but I_{off} increases.

Table A.5 n-FET and p-FET parameters for scaled device widths. 45, 32, and 22 nm PTM HP models @ 25 °C

Node	V_{DD} (V)	Width (μm)	n-FET, I_{eff} (μA)	n-FET, I_{off} (nA)	n-FET, V_{sat} (V)	p-FET, I_{eff} (μA)	p-FET, I_{off} (nA)	p-FET, V_{sat} (V)
45 nm	1.0	1.00	711	20	0.233	450	4.5	0.237
32 nm	0.9	0.70	482	34	0.221	302	13	0.197
22 nm	0.8	0.49	320	57	0.202	209	61	0.147

In Table A.6, the physical dimensions of an inverter and its delay parameters for the three technology nodes are given. The delay and P_{off} for scaled designs show opposite trends, with higher P_{off} at the 22 nm node than at the 45 nm node. The energy per switching event E_{sw} decreases with scaling.

Table A.6 Design and average circuit parameters for standard inverters (FO = 3) with scaled device widths. 45, 32, and 22 nm PTM HP models @ 25 °C

Node	V_{DD} (V)	W_p (μm)	W_n (μm)	L_{ds} (μm)	τ_p (ps)	R_{sw} (Ω)	C_{sw} (fF)	P_{off} (nW)	E_{sw} (fJ)
45 nm	1.0	0.60	0.40	0.120	9.55	1,499	6.37	6.57	3.18
32 nm	0.9	0.42	0.28	0.085	8.67	2,106	4.12	9.99	1.72
22 nm	0.8	0.30	0.20	0.060	7.83	2,825	2.77	22.3	0.88

Appendix B: BSIM4 PTM Models

The Berkeley short channel IGFET model (BSIM) is an analytical model comprising a set of transistor physics-based equations for MOSFET circuit simulations [1, 2]. The equation set has been modified over time to incorporate new effects and to obtain the best empirical fit to measured transistor characteristics. Values of the fitting parameters in the equations are generated for each technology process definition. Some of these parameters can be easily modified during circuit simulations allowing the user to observe changes in circuit behavior over a range of MOSFET geometries and process variations.

The Berkeley group has released several versions of BSIM models since 1987. The compact models used in this book for the 45, 32, and 22 nm technology nodes are based on BSIM4 and take the physical effects in the sub-100 nm regime into account. Other model formulations are BSIMSOI for silicon-on-insulator, BSIM-CMG for common multi-gate and BSIM-IMG for independent multi-gate transistors.

A simplified set of predictive technology models (PTM) is released by Arizona State University for circuit simulations in advance of full-scale technology development [3, 4]. At the time of publication of this book, model cards for technology nodes from 7 nm through 180 nm had been made available [3]. BSIM4 model cards for 45 nm PTM high performance (HP) and low power (LP) NMOS and PMOS devices in metal gate/HK/strained silicon technology are listed below. These version V2.1 models were released on 15 November 2008.

The following modifications in the model cards have been made for compatibility with LTspice:

- Model level changed from 54 to 14
- Model names changed from nmos to mnmos and from pmos to mpmos (LTspice requires MOSFET component model names to begin with the letter "m")

M. Bhushan, M.B. Ketchen, *CMOS Test and Evaluation*,
DOI 10.1007/978-1-4939-1349-7

References

1. Sheu BJ, Scharfetter DL, Ko P-K, Jeng M-C (1987) BSIM: Berkeley short-channel IGFET model for MOS transistors. IEEE J Solid State Circuits SC-22:558–566
2. BSIM group. http://www-device.eecs.berkeley.edu/bsim/. Accessed 21 Jul 2014
3. Predictive technology models website. http://ptm.asu.edu/latest.html. Accessed 21 Jul 2014
4. Cao Y, Sato T, Orshansky M, Sylvester D, Hu C (2000) New paradigm of predictive MOSFET and interconnect modeling for early circuit simulation. Proceedings of the custom integrated circuit conference, pp 201–204

B.1 NMOS (HP)

```
* PTM High Performance 45nm Metal Gate / High-K / Strained-Si
* nominal Vdd = 1.0V

   .model  mnmos  nmos  level = 14

+version = 4.0        binunit = 1          paramchk= 1          mobmod  = 0
+capmod  = 2          igcmod  = 1          igbmod  = 1          geomod  = 1
+diomod  = 1          rdsmod  = 0          rbodymod= 1          rgatemod= 1
+permod  = 1          acnqsmod= 0          trnqsmod= 0
+tnom    = 27         toxe    = 1.25e-009  toxp    = 1e-009     toxm    = 1.25e-009
+dtox    = 2.5e-010   epsrox  = 3.9        wint    = 5e-009     lint    = 3.75e-009
+ll      = 0          wl      = 0          lln     = 1          wln     = 1
+lw      = 0          ww      = 0          lwn     = 1          wwn     = 1
+lwl     = 0          wwl     = 0          xpart   = 0          toxref  = 1.25e-009
+xl       = -20e-9
+vth0    = 0.46893    k1      = 0.4        k2      = 0          k3      = 0
+k3b     = 0          w0      = 2.5e-006   dvt0    = 1          dvt1    = 2
+dvt2    = 0          dvt0w   = 0          dvt1w   = 0          dvt2w   = 0
+dsub    = 0.1        minv    = 0.05       voffl   = 0          dvtp0   = 1e-010
+dvtp1   = 0.1        lpe0    = 0          lpeb    = 0          xj      = 1.4e-008
+ngate   = 1e+023     ndep    = 3.24e+018  nsd     = 2e+020     phin    = 0
+cdsc    = 0          cdscb   = 0          cdscd   = 0          cit     = 0
+voff    = -0.13      nfactor = 2.22       eta0    = 0.0055     etab    = 0
+vfb     = -0.55      u0      = 0.054      ua      = 6e-010     ub      = 1.2e-018
+uc      = 0          vsat    = 170000     a0      = 1          ags     = 0
+a1      = 0          a2      = 1          b0      = 0          b1      = 0
+keta    = 0.04       dwg     = 0          dwb     = 0          pclm    = 0.02
+pdiblc1 = 0.001      pdiblc2 = 0.001      pdiblcb = -0.005     drout   = 0.5
+pvag    = 1e-020     delta   = 0.01       pscbe1  = 8.14e+008  pscbe2  = 1e-007
+fprout  = 0.2        pdits   = 0.08       pditsd  = 0.23       pditsl  = 2300000
+rsh     = 5          rdsw    = 155        rsw     = 80         rdw     = 80
+rdswmin = 0          rdwmin  = 0          rswmin  = 0          prwg    = 0
+prwb    = 0          wr      = 1          alpha0  = 0.074      alpha1  = 0.005
+beta0   = 30         agidl   = 0.0002     bgidl   = 2.1e+009   cgidl   = 0.0002
+egidl   = 0.8        aigbacc = 0.012      bigbacc = 0.0028     cigbacc = 0.002
+nigbacc = 1          aigbinv = 0.014      bigbinv = 0.004      cigbinv = 0.004
+eigbinv = 1.1        nigbinv = 3          aigc    = 0.02       bigc    = 0.0025
+cigc    = 0.002      aigsd   = 0.02       bigsd   = 0.0025     cigsd   = 0.002
+nigc    = 1          poxedge = 1          pigcd   = 1          ntox    = 1
+xrcrg1  = 12         xrcrg2  = 5
+cgso    = 1.1e-010   cgdo    = 1.1e-010   cgbo    = 2.56e-011  cgdl    = 2.653e-010
+cgsl    = 2.653e-010 ckappas = 0.03       ckappad = 0.03       acde    = 1
+moin    = 15         noff    = 0.9        voffcv  = 0.02
+kt1     = -0.11      kt1l    = 0          kt2     = 0.022      ute     = -1.5
+ua1     = 4.31e-009  ub1     = 7.61e-018  uc1     = -5.6e-011  prt     = 0
+at      = 33000
+fnoimod = 1          tnoimod = 0
+jss     = 0.0001     jsws    = 1e-011     jswgs   = 1e-010     njs     = 1
+ijthsfwd= 0.01       ijthsrev= 0.001      bvs     = 10         xjbvs   = 1
+jsd     = 0.0001     jswd    = 1e-011     jswgd   = 1e-010     njd     = 1
+ijthdfwd= 0.01       ijthdrev= 0.001      bvd     = 10         xjbvd   = 1
+pbs     = 1          cjs     = 0.0005     mjs     = 0.5        pbsws   = 1
+cjsws   = 5e-010     mjsws   = 0.33       pbswgs  = 1          cjswgs  = 3e-010
+mjswgs  = 0.33       pbd     = 1          cjd     = 0.0005     mjd     = 0.5
+pbswd   = 1          cjswd   = 5e-010     mjswd   = 0.33       pbswgd  = 1
+cjswgd  = 5e-010     mjswgd  = 0.33       tpb     = 0.005      tcj     = 0.001
+tpbsw   = 0.005      tcjsw   = 0.001      tpbswg  = 0.005      tcjswg  = 0.001
+xtis    = 3          xtid    = 3
+dmcg    = 0          dmci    = 0          dmdg    = 0          dmcgt   = 0
+dwj     = 0          xgw     = 0          xgl     = 0
+rshg    = 0.4        gbmin   = 1e-010     rbpb    = 5          rbpd    = 15
+rbps    = 15         rbdb    = 15         rbsb    = 15         ngcon   = 1
```

B.2 PMOS (HP)

```
* PTM High Performance 45nm Metal Gate / High-K / Strained-Si
* nominal Vdd = 1.0V

.model  mpmos  pmos  level = 14

+version = 4.0         binunit = 1             paramchk= 1             mobmod  = 0
+capmod  = 2           igcmod  = 1             igbmod  = 1             geomod  = 1
+diomod  = 1           rdsmod  = 0             rbodymod= 1             rgatemod= 1
+permod  = 1           acnqsmod= 0             trnqsmod= 0
+tnom    = 27          toxe    = 1.3e-009      toxp    = 1e-009        toxm    = 1.3e-009
+dtox    = 3e-010      epsrox  = 3.9           wint    = 5e-009        lint    = 3.75e-009
+ll      = 0           wl      = 0             lln     = 1             wln     = 1
+lw      = 0           ww      = 0             lwn     = 1             wwn     = 1
+lwl     = 0           wwl     = 0             xpart   = 0             toxref  = 1.3e-009
+xl          = -20e-9
+vth0    = -0.49158    k1      = 0.4           k2      = -0.01         k3      = 0
+k3b     = 0           w0      = 2.5e-006      dvt0    = 1             dvt1    = 2
+dvt2    = -0.032      dvt0w   = 0             dvt1w   = 0             dvt2w   = 0
+dsub    = 0.1         minv    = 0.05          voffl   = 0             dvtp0   = 1e-011
+dvtp1   = 0.05        lpe0    = 0             lpeb    = 0             xj      = 1.4e-008
+ngate   = 1e+023      ndep    = 2.44e+018     nsd     = 2e+020        phin    = 0
+cdsc    = 0           cdscb   = 0             cdscd   = 0             cit     = 0
+voff    = -0.126      nfactor = 2.1           eta0    = 0.0055        etab    = 0
+vfb     = 0.55        u0      = 0.02          ua      = 2e-009        ub      = 5e-019
+uc      = 0           vsat    = 150000        a0      = 1             ags     = 1e-020
+a1      = 0           a2      = 1             b0      = 0             b1      = 0
+keta    = -0.047      dwg     = 0             dwb     = 0             pclm    = 0.12
+pdiblc1 = 0.001       pdiblc2 = 0.001         pdiblcb = 3.4e-008      drout   = 0.56
+pvag    = 1e-020      delta   = 0.01          pscbe1  = 8.14e+008     pscbe2  = 9.58e-007
+fprout  = 0.2         pdits   = 0.08          pditsd  = 0.23          pditsl  = 2300000
+rsh     = 5           rdsw    = 155           rsw     = 75            rdw     = 75
+rdswmin = 0           rdwmin  = 0             rswmin  = 0             prwg    = 0
+prwb    = 0           wr      = 1             alpha0  = 0.074         alpha1  = 0.005
+beta0   = 30          agidl   = 0.0002        bgidl   = 2.1e+009      cgidl   = 0.0002
+egidl   = 0.8         aigbacc = 0.012         bigbacc = 0.0028        cigbacc = 0.002
+nigbacc = 1           aigbinv = 0.014         bigbinv = 0.004         cigbinv = 0.004
+eigbinv = 1.1         nigbinv = 3             aigc    = 0.010687      bigc    = 0.0012607
+cigc    = 0.0008      aigsd   = 0.010687      bigsd   = 0.0012607     cigsd   = 0.0008
+nigc    = 1           poxedge = 1             pigcd   = 1             ntox    = 1
+xrcrg1  = 12          xrcrg2  = 5
+cgso    = 1.1e-010    cgdo    = 1.1e-010      cgbo    = 2.56e-011     cgdl    = 2.653e-010
+cgsl    = 2.653e-010  ckappas = 0.03          ckappad = 0.03          acde    = 1
+moin    = 15          noff    = 0.9           voffcv  = 0.02
+kt1     = -0.11       kt1l    = 0             kt2     = 0.022         ute     = -1.5
+ua1     = 4.31e-009   ub1     = 7.61e-018     uc1     = -5.6e-011     prt     = 0
+at      = 33000
+fnoimod = 1           tnoimod = 0
+jss     = 0.0001      jsws    = 1e-011        jswgs   = 1e-010        njs     = 1
+ijthsfwd= 0.01        ijthsrev= 0.001         bvs     = 10            xjbvs   = 1
+jsd     = 0.0001      jswd    = 1e-011        jswgd   = 1e-010        njd     = 1
+ijthdfwd= 0.01        ijthdrev= 0.001         bvd     = 10            xjbvd   = 1
+pbs     = 1           cjs     = 0.0005        mjs     = 0.5           pbsws   = 1
+cjsws   = 5e-010      mjsws   = 0.33          pbswgs  = 1             cjswgs  = 3e-010
+mjswgs  = 0.33        pbd     = 1             cjd     = 0.0005        mjd     = 0.5
+pbswd   = 1           cjswd   = 5e-010        mjswd   = 0.33          pbswgd  = 1
+cjswgd  = 5e-010      mjswgd  = 0.33          tpb     = 0.005         tcj     = 0.001
+tpbsw   = 0.005       tcjsw   = 0.001         tpbswg  = 0.005         tcjswg  = 0.001
+xtis    = 3           xtid    = 3
+dmcg    = 0           dmci    = 0             dmdg    = 0             dmcgt   = 0
+dwj     = 0           xgw     = 0             xgl     = 0
+rshg    = 0.4         gbmin   = 1e-010        rbpb    = 5             rbpd    = 15
+rbps    = 15          rbdb    = 15            rbsb    = 15            ngcon   = 1
```

B.3 NMOS (LP)

```
* PTM Low Power 45nm Metal Gate / High-K / Strained-Si
* nominal Vdd = 1.1V

.model  mnmos  nmos  level = 14

+version = 4.0          binunit = 1            paramchk= 1            mobmod  = 0
+capmod  = 2            igcmod  = 1            igbmod  = 1            geomod  = 1
+diomod  = 1            rdsmod  = 0            rbodymod= 1            rgatemod= 1
+permod  = 1            acnqsmod= 0            trnqsmod= 0
+tnom    = 27           toxe    = 1.8e-009     toxp    = 1.5e-009     toxm    = 1.8e-009
+dtox    = 3e-010       epsrox  = 3.9          wint    = 5e-009       lint    = 0
+ll      = 0            wl      = 0            lln     = 1            wln     = 1
+lw      = 0            ww      = 0            lwn     = 1            wwn     = 1
+lwl     = 0            wwl     = 0            xpart   = 0            toxref  = 1.8e-009
+vth0    = 0.62261      k1      = 0.4          k2      = 0            k3      = 0
+k3b     = 0            w0      = 2.5e-006     dvt0    = 1            dvt1    = 2
+dvt2    = 0            dvt0w   = 0            dvt1w   = 0            dvt2w   = 0
+dsub    = 0.1          minv    = 0.05         voffl   = 0            dvtp0   = 1e-010
+dvtp1   = 0.1          lpe0    = 0            lpeb    = 0            xj      = 1.4e-008
+ngate   = 1e+023       ndep    = 3.24e+018    nsd     = 2e+020       phin    = 0
+cdsc    = 0            cdscb   = 0            cdscd   = 0            cit     = 0
+voff    = -0.13        nfactor = 1.6          eta0    = 0.0125       etab    = 0
+vfb     = -0.55        u0      = 0.049        ua      = 6e-010       ub      = 1.2e-018
+uc      = 0            vsat    = 130000       a0      = 1            ags     = 0
+a1      = 0            a2      = 1            b0      = 0            b1      = 0
+keta    = 0.04         dwg     = 0            dwb     = 0            pclm    = 0.02
+pdiblc1 = 0.001        pdiblc2 = 0.001        pdiblcb = -0.005       drout   = 0.5
+pvag    = 1e-020       delta   = 0.01         pscbe1  = 8.14e+008    pscbe2  = 1e-007
+fprout  = 0.2          pdits   = 0.08         pditsd  = 0.23         pditsl  = 2300000
+rsh     = 5            rdsw    = 210          rsw     = 80           rdw     = 80
+rdswmin = 0            rdwmin  = 0            rswmin  = 0            prwg    = 0
+prwb    = 0            wr      = 1            alpha0  = 0.074        alpha1  = 0.005
+beta0   = 30           agidl   = 0.0002       bgidl   = 2.1e+009     cgidl   = 0.0002
+egidl   = 0.8          aigbacc = 0.012        bigbacc = 0.0028       cigbacc = 0.002
+nigbacc = 1            aigbinv = 0.014        bigbinv = 0.004        cigbinv = 0.004
+eigbinv = 1.1          nigbinv = 3            aigc    = 0.015211     bigc    = 0.0027432
+cigc    = 0.002        aigsd   = 0.015211     bigsd   = 0.0027432    cigsd   = 0.002
+nigc    = 1            poxedge = 1            pigcd   = 1            ntox    = 1
+xrcrg1  = 12           xrcrg2  = 5
+cgso    = 1.1e-010     cgdo    = 1.1e-010     cgbo    = 2.56e-011    cgdl    = 2.653e-010
+cgsl    = 2.653e-010   ckappas = 0.03         ckappad = 0.03         acde    = 1
+moin    = 15           noff    = 0.9          voffcv  = 0.02
+kt1     = -0.11        kt1l    = 0            kt2     = 0.022        ute     = -1.5
+ua1     = 4.31e-009    ub1     = 7.61e-018    uc1     = -5.6e-011    prt     = 0
+at      = 33000
+fnoimod = 1            tnoimod = 0
+jss     = 0.0001       jsws    = 1e-011       jswgs   = 1e-010       njs     = 1
+ijthsfwd= 0.01         ijthsrev= 0.001        bvs     = 10           xjbvs   = 1
+jsd     = 0.0001       jswd    = 1e-011       jswgd   = 1e-010       njd     = 1
+ijthdfwd= 0.01         ijthdrev= 0.001        bvd     = 10           xjbvd   = 1
+pbs     = 1            cjs     = 0.0005       mjs     = 0.5          pbsws   = 1
+cjsws   = 5e-010       mjsws   = 0.33         pbswgs  = 1            cjswgs  = 3e-010
+mjswgs  = 0.33         pbd     = 1            cjd     = 0.0005       mjd     = 0.5
+pbswd   = 1            cjswd   = 5e-010       mjswd   = 0.33         pbswgd  = 1
+cjswgd  = 5e-010       mjswgd  = 0.33         tpb     = 0.005        tcj     = 0.001
+tpbsw   = 0.005        tcjsw   = 0.001        tpbswg  = 0.005        tcjswg  = 0.001
+xtis    = 3            xtid    = 3
+dmcg    = 0            dmci    = 0            dmdg    = 0            dmcgt   = 0
+dwj     = 0            xgw     = 0            xgl     = 0
+rshg    = 0.4          gbmin   = 1e-010       rbpb    = 5            rbpd    = 15
+rbps    = 15           rbdb    = 15           rbsb    = 15           ngcon   = 1
```

B.4 PMOS (LP)

```
* PTM Low Power 45nm Metal Gate / High-K / Strained-Si
* nominal Vdd = 1.1V

.model  mpmos  pmos  level = 14

+version = 4.0          binunit = 1             paramchk= 1             mobmod  = 0
+capmod  = 2            igcmod  = 1             igbmod  = 1             geomod  = 1
+diomod  = 1            rdsmod  = 0             rbodymod= 1             rgatemod= 1
+permod  = 1            acnqsmod= 0             trnqsmod= 0
+tnom    = 27           toxe    = 1.82e-009     toxp    = 1.5e-009      toxm    = 1.82e-009
+dtox    = 3.2e-010     epsrox  = 3.9           wint    = 5e-009        lint    = 0
+ll      = 0            wl      = 0             lln     = 1             wln     = 1
+lw      = 0            ww      = 0             lwn     = 1             wwn     = 1
+lwl     = 0            wwl     = 0             xpart   = 0             toxref  = 1.82e-009
+vth0    = -0.587       k1      = 0.4           k2      = -0.01         k3      = 0
+k3b     = 0            w0      = 2.5e-006      dvt0    = 1             dvt1    = 2
+dvt2    = -0.032       dvt0w   = 0             dvt1w   = 0             dvt2w   = 0
+dsub    = 0.1          minv    = 0.05          voffl   = 0             dvtp0   = 1e-011
+dvtp1   = 0.05         lpe0    = 0             lpeb    = 0             xj      = 1.4e-008
+ngate   = 1e+023       ndep    = 2.44e+018     nsd     = 2e+020        phin    = 0
+cdsc    = 0            cdscb   = 0             cdscd   = 0             cit     = 0
+voff    = -0.126       nfactor = 1.8           eta0    = 0.0125        etab    = 0
+vfb     = 0.55         u0      = 0.021         ua      = 2e-009        ub      = 5e-019
+uc      = 0            vsat    = 90000         a0      = 1             ags     = 1e-020
+a1      = 0            a2      = 1             b0      = 0             b1      = 0
+keta    = -0.047       dwg     = 0             dwb     = 0             pclm    = 0.12
+pdiblc1 = 0.001        pdiblc2 = 0.001         pdiblcb = 3.4e-008      drout   = 0.56
+pvag    = 1e-020       delta   = 0.01          pscbe1  = 8.14e+008     pscbe2  = 9.58e-007
+fprout  = 0.2          pdits   = 0.08          pditsd  = 0.23          pditsl  = 2300000
+rsh     = 5            rdsw    = 250           rsw     = 75            rdw     = 75
+rdswmin = 0            rdwmin  = 0             rswmin  = 0             prwg    = 0
+prwb    = 0            wr      = 1             alpha0  = 0.074         alpha1  = 0.005
+beta0   = 30           agidl   = 0.0002        bgidl   = 2.1e+009      cgidl   = 0.0002
+egidl   = 0.8          aigbacc = 0.012         bigbacc = 0.0028        cigbacc = 0.002
+nigbacc = 1            aigbinv = 0.014         bigbinv = 0.004         cigbinv = 0.004
+eigbinv = 1.1          nigbinv = 3             aigc    = 0.0097        bigc    = 0.00125
+cigc    = 0.0008       aigsd   = 0.0097        bigsd   = 0.00125       cigsd   = 0.0008
+nigc    = 1            poxedge = 1             pigcd   = 1             ntox    = 1
+xrcrg1  = 12           xrcrg2  = 5
+cgso    = 1.1e-010     cgdo    = 1.1e-010      cgbo    = 2.56e-011     cgdl    = 2.653e-010
+cgsl    = 2.653e-010   ckappas = 0.03          ckappad = 0.03          acde    = 1
+moin    = 15           noff    = 0.9           voffcv  = 0.02
+kt1     = -0.11        kt1l    = 0             kt2     = 0.022         ute     = -1.5
+ua1     = 4.31e-009    ub1     = 7.61e-018     uc1     = -5.6e-011     prt     = 0
+at      = 33000
+fnoimod = 1            tnoimod = 0
+jss     = 0.0001       jsws    = 1e-011        jswgs   = 1e-010        njs     = 1
+ijthsfwd= 0.01         ijthsrev= 0.001         bvs     = 10            xjbvs   = 1
+jsd     = 0.0001       jswd    = 1e-011        jswgd   = 1e-010        njd     = 1
+ijthdfwd= 0.01         ijthdrev= 0.001         bvd     = 10            xjbvd   = 1
+pbs     = 1            cjs     = 0.0005        mjs     = 0.5           pbsws   = 1
+cjsws   = 5e-010       mjsws   = 0.33          pbswgs  = 1             cjswgs  = 3e-010
+mjswgs  = 0.33         pbd     = 1             cjd     = 0.0005        mjd     = 0.5
+pbswd   = 1            cjswd   = 5e-010        mjswd   = 0.33          pbswgd  = 1
+cjswgd  = 5e-010       mjswgd  = 0.33          tpb     = 0.005         tcj     = 0.001
+tpbsw   = 0.005        tcjsw   = 0.001         tpbswg  = 0.005         tcjswg  = 0.001
+xtis    = 3            xtid    = 3
+dmcg    = 0            dmci    = 0             dmdg    = 0             dmcgt   = 0
+dwj     = 0            xgw     = 0             xgl     = 0
+rshg    = 0.4          gbmin   = 1e-010        rbpb    = 5             rbpd    = 15
+rbps    = 15           rbdb    = 15            rbsb    = 15            ngcon   = 1
```

Glossary

Symbols

A	Area (cm^2)
A_F	Acceleration factor (none)
A_{FV}	Voltage acceleration factor (none)
A_{FT}	Temperature acceleration factor (none)
A_{vt}	Constant for random V_t variation (V)
α	Number of stages in a circuit (none)
α	Tail area in a unit normal distribution (none)
α_c	Clustering parameter (none)
B_w	Bandwidth of a metal interconnect (Hz/mm)
β	MOSFET gain factor ($[\Omega\text{-V}]^{-1}$)
β_r	p/n ratio: W_p/W_n (none)
β_w	Weibull distribution shape parameter (none)
c_{in}	Input capacitance of a logic gate per unit width (F/cm)
c_{out}	Output capacitance of a logic gate per unit width (F/cm)
C	Capacitance (F)
C_d	Depletion layer capacitance (F)
C_{db}	MOSFET drain-to-body capacitance (F)
C_{down}	Inter-level wire capacitance to layer below (F)
$C_{ds(d)}$	Source (drain) diffusion capacitance (F)
C_g	Gate-to-substrate capacitance of a MOS capacitor (F)
C_{gs}	Gate-to-source capacitance (F)
C_{in}	Equivalent input capacitance of a logic gate (F)
$C_{js(d)}$	Source (drain)-to-body junction area capacitance (F)
$C_{jswgs(d)}$	Source (drain)-to-gate perimeter capacitance (F)
$C_{jsws(d)}$	Source (drain)-to-STI perimeter capacitance (F)
C_{left}	Wire capacitance to adjacent wire to the left (F)
C_L	Load capacitance (F)
CN	Condition number (none)
C_{out}	Equivalent output capacitance of a logic gate (F)
C_{ov}	MOSFET overlap capacitance (F)
C_{ox}	Oxide capacitance per unit area (F)

M. Bhushan, M.B. Ketchen, *CMOS Test and Evaluation*,
DOI 10.1007/978-1-4939-1349-7

C_p	Parasitic interconnect capacitance (F)
C_p	Six Sigma index for process control (none)
C_{pk}	Six Sigma index for process control (none)
C_{right}	Wire capacitance to adjacent wire to the right (F)
C_{sb}	MOSFET source-to-body capacitance (F)
C_{sc}	Equivalent short-circuit capacitance (F)
C_{sw}	Switching capacitance of a logic gate (F)
C_{up}	Inter-level wire capacitance to layer above (F)
C_w	Wire capacitance per unit length (F/cm)
d	Film thickness (cm)
delvto	V_t adder (V)
DD	Defect density (cm^{-2})
$\Delta\|V_t\|$	Change in the magnitude of V_t (V)
E_a	Activation energy (eV)
E_{sw}	Energy per logic gate switching event (J)
ε	Dielectric constant (none)
ε_o	Vacuum permittivity ($=8.85 \times 10^{-14}$ F/cm) (F/cm)
f	Frequency of oscillation (Hz)
f_{max}	Maximum frequency of oscillation (Hz)
f_n	Number of switching transitions (Hz)
FIT	Failure in time (1/h)
FO	Fan out (none)
g	Inductance correction factor (none)
g_{ds}	Output conductance (A/V)
g_m	Transconductance (A/V)
GND	Ground potential (V)
γ	Area efficiency of a defect monitor (none)
γ	MOSFET body-effect coefficient (none)
h	Dielectric thickness (cm)
H	History effect (%)
H_t	History effect for 1SW–2SW transitions (%)
H_{tpd}	PD history effect for 1SW–2SW transitions (%)
H_{tpu}	PU history effect for 1SW–2SW transitions (%)
I	Current (A)
IDDQ	Quiescent current of a circuit (A)
$IDDQ_m$	Measured quiescent current of a circuit (A)
IDDA	Active current of a circuit (A)
$IDDA_m$	Measured active current of a circuit (A)
I_{dlin}	MOSFET drain-to-source linear current (A)
I_{ds}	MOSFET drain-to-source current (A)
I_{dsat}	MOSFET drain-to-source saturation current (A)
I_{eff}	MOSFET effective current $= (I_{hi} + I_{lo})/2$ (A)
I_{effn}	n-FET effective current $= (I_{hi} + I_{lo})/2$ (A)

I_{effp}	p-FET effective current $= (I_{hi} + I_{lo})/2$ (A)
I_{gl}	MOSFET gate-oxide leakage current (A)
I_{hi}	MOSFET I_{ds} at $V_{ds} = V_{DD}/2$, $V_{gs} = V_{DD}$ (A)
I_{lo}	MOSFET I_{ds} at $V_{ds} = V_{DD}$, $V_{gs} = V_{DD}/2$ (A)
I_{off}	MOSFET drain-to-source leakage current (A)
I_{offn}	n-FET drain-to-source leakage current (A)
I_{offp}	p-FET drain-to-source leakage current (A)
I_{on}	MOSFET on current (A)
I_{onn}	n-FET drain-to-source saturation current (A)
I_{onp}	p-FET drain-to-source saturation current (A)
I_{sc}	Short-circuit current of a logic gate (A)
k	Boltzmann constant (eV/K)
κ	Cell (bin) size in a histogram (variable)
$\lambda(t)$	Failure rate function
l	Wire length (cm)
L	Inductance (H)
L_{ds}	Source/drain diffusion length (cm)
L_{eff}	MOSFET effective channel length (cm)
L_p	MOSFET gate length (cm)
L_{pn}	n-FET gate length (cm)
L_{pp}	p-FET gate length (cm)
L_w	Inductance per unit length (H/cm)
m_l	Sensitivity coefficient for L_p (none)
m_{vn}	Sensitivity coefficient for V_{tn} (none)
m_{vp}	Sensitivity coefficient for V_{tp} (none)
μ	Sample mean (variable)
μ_{eff}	Carrier mobility (cm^2/V-s)
μ_l	Lower filter multiplier (none)
μ_o	Permeability of free space (H/cm)
μ_u	Upper filter multiplier (none)
n	Number of data points or samples (none)
n	Exponent (none)
n_{sq}	Number of squares in a film (none)
N	Number of elements (none)
N_w	Number of interconnects (none)
NOM	Nominal simulation corner (none)
η_w	Weibull distribution scaling factor (none)
$p(x)$	Probability density of x (none)

pr	Rent exponent (none)
P	Power dissipation (W)
P_{ac}	Active power dissipation (W)
P_{max}	Maximum power dissipation (W)
P_{off}	Standby power dissipation (W)
P_{res}	DC power due to resistive paths across power supply (W)
P_{sc}	Short-circuit power (W)
Q1	First quartile in descriptive statistics (variable)
Q2	Second quartile in descriptive statistics (variable)
Q3	Third quartile in descriptive statistics (variable)
Q_{crit}	Critical charge to initiate a soft error
r_{sw}	Switching resistance of a logic gate × width (Ω)
R	Resistance, Ω
R_{eff}	Effective inverter resistance calculation from I_{eff} (Ω)
R_{pg}	Power grid resistance in series (Ω)
R_{pd}	Parasitic resistance in series with MOSFET drain (Ω)
R_{ps}	Parasitic resistance in series with MOSFET source (Ω)
R_s	Parasitic series resistance (Ω)
R_{ds}	Source-drain series resistance (Ω)
R_{sh}	Shunt resistance (Ω)
R_{sw}	Switching resistance of a logic gate (Ω)
R_{th}	Thermal resistance (°C/W)
R_w	Wire resistance per unit length (Ω/cm)
P	Resistivity (Ω cm)
ρ_{cr}	Correlation coefficient (none)
ρ_{sh}	Sheet resistance (Ω/□)
s	Sample standard deviation (variable)
s	Spacing between wires in the same metal layer (cm)
S	Primary CMOS technology scaling factor (none)
S_h	Stack height in logic gates (none)
S_k	Secondary CMOS technology scaling factor (none)
SS	MOSFET subthreshold slope (V/decade)
σ	Standard deviation (variable)
t	Time (s)
t_{ox}	Gate_dielectric thickness (cm)
T	Temperature (°C)
T_c	Clock period (s)
T_{cmin}	Minimum clock period (s)
TCR	Temperature coefficient of resistance (Ω/°C)
T_d	Time delay (s)
T_f	Pulse fall time (s)
T_h	Hold time of a latch (s)
T_j	Silicon temperature (°C)
T_{jitter}	Clock jitter time (s)

T_μ	DAT-to-CLK delay to metastability (s)
T_{on}	Pulse on time (s)
T_p	Period of oscillation (s)
T_r	Pulse rise time (s)
T_s	Setup time of a latch (s)
T_{skew}	Clock skew (s)
T_w	Pulse width (s)
τ	Delay of a logic gate (s)
τ_{1pd}	1SW PD delay of a logic gate (s)
τ_{1pu}	1SW PU delay of a logic gate (s)
τ_{2pd}	2SW PD delay of a logic gate (s)
τ_{2pu}	2SW PU delay of a logic gate (s)
τ_f	Signal fall time (s)
τ_{fi}	Input signal fall time (s)
τ_{fo}	Output signal fall time (s)
τ_p	Average of pull-down (PD) and pull-up (PU) delays (s)
τ_{pd}	Pull-down (PD) delay (s)
τ_{pu}	Pull-up (PU) delay (s)
τ_r	Signal rise time (s)
τ_{ri}	Input signal rise time (s)
τ_{ro}	Output signal rise time (s)
V	Voltage (V)
V_{ds}	MOSFET drain-to-source voltage (V)
V_{DD}	Power supply voltage (V)
V_F	Forward voltage of a diode (V)
V_{gs}	MOSFET gate-to-source voltage (V)
V_{in}	Voltage of an input circuit node (V)
V_{out}	Voltage of an output circuit node (V)
V_t	MOSFET threshold voltage (V)
V_{tn}	n-FET threshold voltage (V)
V_{tp}	p-FET threshold voltage (V)
V_{tlin}	MOSFET threshold voltage in linear mode (V)
V_{tsat}	MOSFET threshold voltage in saturation mode (V)
w	Wire width (cm)
W	MOSFET width (cm)
W	Logic gate width ($=W_p+W_n$) (cm)
W_n	n-FET width (cm)
W_p	p-FET width (cm)
x	Observed values in a data sample (variable)
XL	Capacitive load multiplier
y	Transformed variable (variable)
Y	Yield (none)
z	Transformed variable (none)

Acronyms

ALU	Arithmetic logic unit
ASIC	Application-specific integrated circuit
ATE	Automated test equipment
BC	Best case simulation corner
BI	Burn-in
BIST	Built-in self-test
BSIM	Berkeley short-channel IGFET models
BTI	Bias temperature instability
C4	Controlled collapse chip connection
CHC	Channel hot carriers
CMOS	Complementary metal-oxide-semiconductor
CMP	Chemical mechanical polishing
CPG	Circuit performance gauge
CSE	Clocked storage element
DD	Defect density
DECAP	Decoupling capacitor
DIBL	Drain-induced barrier lowering
DFM	Design for manufacturing
DFS	Dynamic frequency scaling
DFT	Design for testability
DRAM	Dynamic random access memory
DRC	Design rule checker
DVFS	Dynamic voltage and frequency scaling
EDA	Electronic design automation
EM	Electromigration
EOL	End-of-life
ESD	Electrostatic discharge
FD-SOI	Fully depleted silicon on insulator
FIB	Focused ion beam
FPG	MOSFET performance gauge
GCB	Global clock buffer
GIDL	Gate-induced drain leakage
HCI	Hot carrier injection
HDL	Hardware description language
HK	High-k (gate dielectric material)
HOL	Health-of-line
HP	High performance
HVM	High volume manufacturing
ILD	Inter-level dielectric
I/O	Input/output
IPG	Interconnect performance gauge
ITRS	International Technology Roadmap for semiconductors
IQR	Interquartile range

JTAG	Joint test access group
KGD	Known good die
LADA	Laser-assisted device alteration
LBIST	Logic built-in self-test
LCB	Local clock buffer
LCP	Locating critical path (buffer)
LFSR	Linear feedback shift register
LG	Logic gate
LGXL	Logic gate with fanout = XL
LP	Low power
LPE	Layout parasitic extraction
LSL	Lower specification limit
LSSD	Level-sensitive scan design
MTBF	Mean time between failures
MISR	Multiple input shift register
MOS	Metal-oxide-semiconductor
MOSFET	Metal-oxide-semiconductor field-effect transistor
MTTF	Mean time to failure
NBTI	Negative bias temperature instability
n-FET	n-type MOSFET (NMOS)
p-FET	p-type MOSFET (PMOS)
PBTI	Positive bias temperature instability
PD	Pull down
PDK	Process design kit
PD-SOI	Partially depleted silicon on insulator
PICA	Pico-second imaging circuit analysis
PLL	Phase-locked loop
PLY	Photo limited yield
POH	Power-on hours
PRPG	Pseudo random pattern generator
PVT	Process voltage temperature
PU	Pull up
RDF	Random dopant fluctuation
RSNM	Read static noise margin
RIE	Reactive ion etching
RO	Ring oscillator
ROI	Return on investment
RTL	Register transfer level
SCE	Short-channel effect
SE	Storage element
SER	Soft error rate
SEU	Single event upset
SHC	Substrate hot carriers
SMU	Source measure unit

SNM	Static noise margin
SOI	Silicon on insulator
SPICE	Simulation program with integrated circuit emphasis
SRAM	Static random access memory
STI	Shallow trench isolation
TDDB	Time-dependent dielectric breakdown
TFI	Thin film interposer
USL	Upper specification limit
VCO	Voltage-controlled oscillator
VLSI	Very-large-scale integration
WC	Worst case simulation corner
WSNM	Write static noise margin

Index

A

Accelerated stress tests, 8, 243, 285, 288–292, 298, 304, 305, 309
Acceleration factor, 289–291, 309
Adaptive testing, 9, 13, 126, 242, 278–281
Automated test equipment (ATE), 8, 168, 212, 242, 245, 275, 283, 308
Automated test pattern generation (ATPG), 7, 244, 246–251, 265, 270, 281

B

Bayesian statistics, 312, 333–334
Berkeley short-channel IGFET models (BSIM), 13, 17, 19, 27, 35, 37–39, 45, 50, 53, 76, 130, 183, 348, 354, 355, 360, 375–383, 385, 391, 407
Bias temperature instability (BTI)
  monitor, 296–297, 300, 306, 309, 310
  NBTI, 215, 292, 293, 295
  PBTI, 215, 292, 293, 295, 298
Binning, 5, 7, 13, 126, 202, 224, 225, 242, 265, 278–281, 308
Boundary scan, 254, 255, 258–259, 281
Box and whiskers, 208, 273, 274, 323, 341, 342
Built-in self-test (BIST), 7, 244, 254, 255, 257, 281
Burn-in (BI), 5, 8, 13, 135, 243, 263, 285, 286, 300, 301, 304–307, 309, 321

C

Chip-to-chip (C2C) variations, 208, 212, 216, 280
Circuit performance metric
  delay metric, 371–373
  density metric, 348, 367–374
  energy metric, 395
  power metric, 348
Clock distribution, 86, 87, 91–93, 121, 151, 153, 172, 213, 216, 221, 363
Clocked storage elements (CSE), 2, 7, 12, 85, 86, 90, 93–102, 109, 121, 170, 244, 246, 251, 255–257, 265, 269, 303
Clock gating, 93, 154, 213
Clock jitter, 10, 122, 170, 172, 227
CMOS scaling rules, 36, 127, 348
CMOS scaling trends, 126, 346, 351–354, 395
Complementary metal-oxide-semiconductor (CMOS) cross-section, 19, 383, 399–401
Condition number (CN), 188, 191, 331–333
Confidence interval, 320, 321
Correlation, 5–6, 10–13, 19, 27, 37, 38, 43, 55, 64, 74–76, 120, 140, 149, 159–162, 177, 179, 193, 194, 197, 216, 218, 220, 242, 244, 254, 264, 266–270, 272, 277, 279, 280, 300, 311–313, 323–326, 336, 343, 348, 350, 385, 391
Critical path monitor (CPM), 160, 173–174, 197, 216, 220, 255, 280
Current multiplier, 65, 66, 156, 177, 230, 370
Cycle time limited yield (CLY), 261, 263–265, 283

D

Data filters, 10, 166, 283, 323–326
Data visualization, 11, 13, 160, 208, 217, 311–344
Defects, 5, 7–9, 12, 13, 15, 86, 93, 103, 125, 126, 128, 136–140, 148, 153–156, 182, 242, 244, 246–253, 260–266, 269, 270, 280–282, 285–288, 292, 304–306, 315, 327, 338, 341, 368

M. Bhushan, M.B. Ketchen, *CMOS Test and Evaluation*,
DOI 10.1007/978-1-4939-1349-7

Delay chain, 17, 36, 37, 64–71, 75, 76, 80, 82, 132, 136–138, 143, 150, 151, 156, 160, 162–166, 170, 171, 173–175, 177–193, 195, 197, 198, 216, 222, 223, 227, 229, 230, 238, 255, 273, 284, 296, 329, 338, 374, 381, 385, 391, 393, 396, 403
Delay parameters, τ_p, C_{in}, C_{out}, R_{sw}, 183, 396, 405
Design for Testability (DFT), 3, 5, 241, 242, 254–255, 265, 368
Dynamic random access memory (DRAM), 86, 102, 103, 108, 121, 235, 269, 351, 352, 368

E
Edge detector, 164–166, 170, 173, 197
Edge exclusion, 237
Electromigration (EM), 8, 209, 211, 215, 263, 289, 292, 302–304, 306, 309, 375
Electronic design automation (EDA) tools, 4–5, 10, 14, 55, 135, 140, 149, 152, 153, 156, 174, 177, 193, 203, 204, 220, 258, 276, 277, 283, 285, 347, 348, 354, 374–386, 395

F
Failure analysis, 7, 8, 194, 242, 265–267, 281, 312
Fanout (FO), 2, 60, 112, 132, 165, 213, 253, 297, 329, 358, 400
Fault models, 1, 93, 246, 250, 251, 265
Flip-flops
  negative edge-triggered, 98, 99, 101
  positive edge-triggered, 98, 99, 168
Floorplanning, 220, 221
f_{max}, 86, 109, 114, 120, 122, 123, 157, 162, 170, 173, 174, 194, 197, 210, 218, 244, 260, 271, 273, 275–283, 295, 298, 300, 306, 308, 313–315, 325, 343, 344, 374
Focused ion beam (FIB), 267
Frequency divider, 73, 91, 168, 169, 192

G
Guard-banding, 1, 13, 202, 203, 211, 245, 259, 281, 285, 286, 293, 300, 304, 306–310, 395

H
Hold time, 100–101, 122
Hot carrier injection (HCI), 211, 292, 300–301, 308–310, 375

I
IDDQ, 7, 9, 12, 54, 64, 68–71, 73–75, 125–157, 168, 174, 178–179, 210, 213, 218, 246–248, 250–254, 260, 263, 268, 270, 275, 284, 293, 297, 300, 310, 324, 329, 341, 353–354, 358, 375, 382, 390, 394
IDDQ test, 126, 128, 140, 151, 246, 251–254
I_{gl}, 41, 42, 128–132, 355
Interconnects, 13, 14, 18–20, 28–30, 32–34, 53, 66–68, 85, 86, 92, 120, 127, 174, 179, 181, 188, 197, 207, 210, 211, 213, 215, 223, 248, 250, 251, 262, 266, 268, 272, 285, 288–290, 292–293, 302, 304, 348, 363–365, 368, 375, 383, 388, 397, 399–401
Inverter, 2, 17, 95, 130, 165, 213, 252, 291, 315, 358, 397
I/O, 7, 9, 12, 17, 19, 28, 30, 86–88, 153, 160–163, 168, 177, 212, 213, 242–245, 254, 258, 259, 288, 354, 363, 368, 396, 399, 401
I_{off}, 24, 25, 41–46, 62, 80, 129–136, 157, 163, 211, 218, 219, 237, 252, 268, 301, 322, 354–359, 374, 377–380, 387–389, 392, 397, 404

J
Jitter, 10, 93, 100, 122, 123, 170, 172, 173, 227

K
Known good die (KGD), 7, 245, 257

L
Layout parasitic extraction (LPE), 80, 383–385
Level sensitive scan designs (LSSD), 133
Lot-to-lot (L2L) variations, 77, 163, 208, 209, 216, 226, 270
LTspice, 12, 18, 36–40, 42, 45, 46, 48, 49, 51–53, 66, 69, 75–79, 110, 111, 114, 115, 117, 118, 134, 184, 232, 235, 407

M
Manufacturing window, 278
Mean, 13, 37, 77, 79, 81, 122, 150, 162, 198, 203, 208, 216, 224, 230, 280, 290, 311, 316, 318, 319, 321, 324, 327–329, 337, 342–344, 357, 358, 382, 385, 397
Metal oxide semiconductor field effect transistor (MOSFET)
  capacitances, 17, 18, 26–27, 32, 46, 50, 51, 55, 56, 59, 60, 66, 70, 76, 79–80, 85, 126–127, 141–145, 149, 152, 156, 163,

179, 181, 189, 207, 210, 212, 213, 355, 358, 360–364, 366, 378, 383
C_g-V_{gs} characteristics, 46
I_{ds}-V_{ds} characteristics, 24, 62, 376
I_{ds}-V_{gs} characteristics, 41, 356, 357
performance metric, 128, 348, 349, 351, 355, 358, 363, 366, 374, 393
Model-to-hardware correlation, 5–6, 10, 13, 19, 37, 38, 64, 74–76, 140, 149, 159–161, 177, 179, 194, 203, 216, 242, 268, 272, 312, 338, 348, 350, 385
Monte Carlo (MC) simulations, 46, 77–79, 82, 109, 112, 122, 133, 136, 230, 231, 233–235, 238, 239, 284, 322, 325, 327–329, 343, 344, 358, 375

N
Non-normal distribution, 321–322, 343
Normal distribution, 77, 133, 224, 227, 230, 288, 316–323, 328, 343

P
P_{ac}, 82, 140, 141, 148–150, 152, 156, 197, 218, 225, 226, 366, 369
Parasitic extraction, 38, 80, 195, 271, 354, 361, 375, 383, 384, 395, 396
Partially depleted silicon-on-insulator technology (PD-SOI)
floating body, 135, 166, 386–388, 392, 393
history effect, 166, 390–393
self-heating, 386, 392
Phase-locked loops (PLLs), 87, 91, 92, 153, 172
PICA imaging, 219, 220, 237
Picosecond imaging circuit analysis (PICA), 218–220, 266
P_{off}, 73–74, 140, 146–153, 156, 218, 225, 226, 236, 237, 263–264, 275, 277–279, 291, 293, 300, 308, 310, 343, 366, 367, 369–375, 380, 386, 397, 405
Power
AC, 3, 9, 12, 27, 30, 125–128, 140–146, 148, 151, 155, 213, 226, 259–260, 292, 295, 334, 367, 369
DC, 3, 12, 67, 125, 126, 140, 141, 146–148, 212, 242, 349, 369
gating, 154, 213, 299
leakage, 71, 126–128, 140, 146, 154, 215, 277, 369
management, 12, 125, 126, 128, 151–155, 214, 310
Predictive technology models (PTM), 12, 19, 35, 37–38, 41, 43–49, 51, 53–54, 57–59, 62–64, 67–69, 71, 76, 77, 79, 120, 121, 128, 130, 134–139, 141, 143–147, 149, 152, 175, 178, 183, 184, 188, 189, 191, 192, 194–196, 213, 224, 228, 230–235, 252, 253, 276, 277, 291, 297, 328, 329, 336, 355–361, 366, 370–373, 376–382, 391, 392, 396, 399–407
Probability, 13, 77, 79, 100, 224, 262, 263, 286–289, 303, 306, 308, 311, 313–318, 320, 327, 330, 333, 334, 343
Process split, 267, 269–270, 283, 312, 314, 321, 322, 344
PVT monitors, 159–199, 216–217, 254, 255

R
Random dopant fluctuation (RDF), 77, 163, 203, 211
Register files, 86, 87, 101–102, 171, 244, 255, 303
Regression, 57, 147, 311, 323–326, 335, 378, 386
Reliability, 5, 11–13, 26, 125, 126, 127, 155, 160, 211, 215, 243, 285–310, 334, 349, 375, 376
Resistive faults, 137, 261
Ring oscillator, 17, 36, 37, 71–76, 80, 91, 141, 143, 144, 148, 149, 156, 160, 162, 166, 167, 177, 178, 182, 186, 187, 192, 193, 195, 197, 216, 232, 255, 268, 270, 271, 280, 290, 295, 296, 313, 315, 338, 349, 353, 366, 371, 374, 388, 393, 394, 396, 397, 402

S
Scribe-line, 5–6, 9, 11, 75, 162, 168, 178, 195, 199, 206, 207, 210, 212, 216, 217, 222, 223, 237, 241–243, 267–273, 279, 334, 350, 385
Sensitivity analysis, 12, 182, 187, 188, 197, 326, 330–333, 343
Setup time, 95, 100, 101
Shmoo plot, 264, 265
Short-circuit current, I_{sc}, 142, 143
Silicon process monitor, 5–6, 12, 159–170, 173, 174, 179, 194, 197, 199, 255, 271, 272, 338
Simulation corners, 54, 82, 223, 225–227, 270, 329, 376
SKITTER, 170, 172, 173, 177, 214

Soft-errors, 103, 250, 303, 386
Spatial variations
 across chip (AcC), 208, 216, 217, 219, 223, 272, 274, 340, 357
 across reticle (AcR), 208, 216
 across wafer (AcW), 217, 270, 273, 340, 357
Standard deviation, 37, 77, 78, 122, 163, 208, 211, 224, 227, 228, 230, 316, 318, 319, 322, 324–325, 342, 344, 385
Static noise margin (SNM), 12, 107–112, 121–123, 235, 238, 239
Static random access memory (SRAM), 12, 86, 102–112, 121–123, 133, 152, 186–187, 199, 202, 211, 235, 238, 239, 268, 269, 280, 282, 296, 354, 368, 381
Stuck at fault
 Stuck-at-0, 138, 139, 246, 250
 Stuck-at-1, 137, 139, 246, 250
 Stuck-open, 248
 Stuck-short, 248
Switching energy, E_{sw}, 144, 366, 367, 371–374, 405
Systematic variations, 82, 133, 186, 195, 210, 211, 223, 233, 234, 238, 270, 327, 329

T

T_{cmin}, 109, 114–117, 119–123, 170, 173, 174, 188, 214, 232, 233, 259–260, 264, 265, 270, 275–277, 283, 321, 325
Temperature monitor, 159, 174–177, 196–198, 214, 237, 242, 255, 292
Test challenges, 1, 9
Test economics, 8–9, 121
Test overview, 4–5
Test structures, 5, 9–12, 36–37, 75, 76, 80, 81, 135, 149, 160, 162, 195, 206, 215, 216, 222, 223, 242, 243, 267–269, 271, 294, 334, 348–350, 385, 393, 394
Test types, 5–8, 244
Thermal imaging, 8, 216, 220
Time-dependent dielectric breakdown (TDDB), 211, 292, 301–302, 309

V

Variability, 7, 20, 86, 134, 159, 201, 242, 288, 312, 356
 characterization, 201, 215–220
 random, 77, 78, 80, 104, 109, 112, 114, 186, 202, 203, 205, 211–212, 223, 227–235, 238, 239, 288, 313, 315, 322, 325–330, 343, 357, 404
 systematic, 37, 77–80, 82, 104, 112, 114, 119, 133, 136, 153, 163, 186, 195, 202, 203, 205, 210, 211, 220, 222, 231–234, 236, 238, 239, 253, 270, 273, 288, 326–330, 343, 404
V_{min}, 86, 109, 114–123, 194, 197, 210, 218, 232–234, 238, 244, 259–260, 263, 277, 295, 298, 306, 374
Voltage monitor, 159–199, 216–217, 254
Voltage screening, 263, 286, 304–305

W

Wafer stripe, 269, 270, 278
Wafer-to-wafer variations (W2W), 182, 208, 216, 219, 222, 226, 270, 394
Weibull plot, 287, 288, 291

Y

Yield, 1, 7, 10, 11, 13, 14, 21, 104, 126, 128, 159, 161, 162, 201, 202, 205, 207–208, 218, 222–225, 227, 242, 244, 260–267, 269, 270, 278, 280–283, 305, 323, 327, 341, 343, 351, 363, 368, 394, 397

MIX
Papier aus verantwortungsvollen Quellen
Paper from responsible sources
FSC® C105338

If you have any concerns about our products,
you can contact us on
ProductSafety@springernature.com

In case Publisher is established outside the EU,
the EU authorized representative is:
Springer Nature Customer Service Center GmbH
Europaplatz 3, 69115 Heidelberg, Germany

Printed by Libri Plureos GmbH
in Hamburg, Germany